4th International Conference on Advanced Material for Better Future (ICAMBF 2019)

Insight Toward Novel Research in Field of Materials for a Better Life

IOP Conference Series: Materials Science and Engineering
Volume 858

Surakarta, Indonesia
7 - 8 October 2019

ISBN: 978-1-7138-3616-2
ISSN: 1757-8981

Printed from e-media with permission by:

Curran Associates, Inc.
57 Morehouse Lane
Red Hook, NY 12571

Some format issues inherent in the e-media version may also appear in this print version.

Institute of Physics
Dirac House, Temple Back
Bristol BS1 6BE UK

Phone: 44 1 17 929 7481
Fax: 44 1 17 920 0979

techtracking@iop.org

Additional copies of this publication are available from:

Curran Associates, Inc.
57 Morehouse Lane
Red Hook, NY 12571 USA
Phone: 845-758-0400
Fax: 845-758-2633
Email: curran@proceedings.com
Web: www.proceedings.com

TABLE OF CONTENTS

CIVIL AND CONSTRUCTION MATERIALS (CCM)

MATERIALS CHEMISTRY AND PHYSICS (MCP)

DRUG AND BIOMATERIALS TECHNOLOGY (DBT)

ELECTRICAL, OPTICAL AND MAGNETIC MATERIALS (EOM)

POLYMER & COMPOSITE MATERIALS (PCM)

FUNCTIONAL MATERIALS (FM)

METALLURGY & ALLOY MATERIALS (MAM)

MISCELLANEOUS TOPICS (MT)

IOP Conf. Series: Materials Science and Engineering **858** (2020) 011001 doi:10.1088/1757-899X/858/1/011001

IOP Conf. Series: Materials Science and Engineering **858** (2020) 011001 doi:10.1088/1757-899X/858/1/011001

PREFACE

4th International Conference on Advanced Material for Better Future 2019 (ICAMBF 2019)

The 4th International Conference on Advanced Material for Better Future 2019 (ICAMBF 2019) was held on October 7-8, 2019 in Surakarta, Central Java, Indonesia. ICAMBF 2019 is an international scientific forum that was initiated to overcome material problems as well as factors affecting them that have emerged as the main issue and challenge attracting more attention from various scientists, government, and regulatory bodies worldwide. The ICAMBF 2019 was attended by leading scientists from countries near and abroad who work in the field of material.

This event based on theme "**Insight Toward Novel Research in Field of Materials for a Better Life**" that aims to provide a forum for scientists, researchers, and engineers within the nation and from around the world to present, discuss, share, and exchange their knowledge, experiences, and research finding on all aspects of material. Thus, this conference can give an excellent overview in all scope of fields including Functional Material, Material Chemistry and Physics, Advanced Nanomaterials, Polymer and Composites, Drug and Biomaterials Technology, Electrical, Optical and Magnetic Materials, Advanced Materials for Energy, Advanced Materials for Energy, Materials In Industry, Civil and Construction Materials, Metallurgy and Alloy Materials.

We are pleased to our outstanding invited speakers Prof. David Black, Prof. Yuichi Sato, Prof. Ari Handono Ramelan, Dr. B. Muthuraaman, Prof. Nurul Taufiqu Rochman, Dr. MD. Rahim Bin Sahar FS and other participated speakers for sharing the best knowledge in Advanced Material.

We also thanks to all researchers and student who participate in conference by submitting their papers to convey the current science discussed at ICAMBF 2019, all of the organizing committee wish you get awesome conference experience, and reviewers for their valuable revision of these papers.

Prof. Ir. Ari Handono Ramelan, M.Sc (Hons), Ph.D

Chairman of ICAMBF 2019

IOP Conf. Series: Materials Science and Engineering **858** (2020) 011002 doi:10.1088/1757-899X/858/1/011002

CONFERENCE COMMITTEE/ORGANIZER

Chairman	Prof. Ir. Ari Handono Ramelan, M.Sc., (Hons), Ph.D.	Sebelas Maret University, Indonesia
Co-Chairman	Dr. Sayekti Wahyuningsih, MSi	Sebelas Maret University, Indonesia
Secretary	Ahmad Marzuki, SSi, PhD	Sebelas Maret University, Indonesia
Scientific Committee	Prof. Venty Suryanti. M.Phil. PhD	Sebelas Maret University, Indonesia
	Dr. Desi Suci Handayani, MSi	Sebelas Maret University, Indonesia
	Prof. Ir. Ari Handono Ramelan, M.Sc., (Hons), Ph.D.	Sebelas Maret University, Indonesia
	Dr. Sayekti Wahyuningsih, MSi	Sebelas Maret University, Indonesia
	Ahmad Marzuki, SSi, PhD	Sebelas Maret University, Indonesia
	Dr. Agus Supriyanto, S.Si. M.Si	Sebelas Maret University, Indonesia
	Dr. Fahru Nurosyid, M.Si	Sebelas Maret University, Indonesia
	Dr. Tetri Widiyani, M.Si.	Sebelas Maret University, Indonesia
	Dr. Artini Pangastuti, M.Si.	Sebelas Maret University, Indonesia
	Dr. Ahmad Ainurrofiq, M.Si. Apt	Sebelas Maret University, Indonesia
	Dr.rer.nat Witri Wahyu Lestari, M.Sc.	Sebelas Maret University, Indonesia
	Teguh Endah Saraswati, M.Sc, Ph.D	Sebelas Maret University, Indonesia
	Prof. David St Clair Black	University of New South Wales, Australia
	Prof. Yuichi Sato	Akita University, Japan
	Dr. B. Muthuraaman	University of Madras, India
	Prof. Nurul Taufiqu Rochman	LIPI, Indonesia
	Dr. MD. Rahim Bin Sahar FS	Universiti Teknologi Malaysia, Malaysia

IOP Conf. Series: Materials Science and Engineering **858** (2020) 011003 doi:10.1088/1757-899X/858/1/011003

No	First Name	Surname	Institution	Email
1	Ary	Setyawan	Universitas Sebelas Maret	cenase@yahoo.com
2	Rengga	Putra	Universitas Sebelas Maret	cenase@yahoo.com
3	Erhar	Augusto	Universitas Sebelas Maret	erhar.augusto@gmail.com
4	Sabdoyo	Harimurti	Universitas Sebelas Maret	sabdoyono.w@gmail.com
5	Fajar	Nugroho	Universitas Sebelas Maret	marnis_ri@yahoo.com
6	Ade	Lisantono	Universitas Atma Jaya Yogyakarta	adelisantono@mail.uajy.ac.id
7	John	Hatmoko	Universitas Atma Jaya Yogyakarta	john.trihatmoko@uajy.ac.id
8	Lina	Mahardiani	Universitas Sebelas Maret	lina80_ssa@yahoo.com
9	Rahmat	Hidayat	Universitas Sebelas Maret	sayekti@mipa.uns.ac.id
10	Patiha	Patiha	Universitas Sebelas Maret	maulidan@mipa.uns.ac.id
11	Zudah	Kubro	IPB	zudah_simaatul@apps.ipb.ac.id
12	Himawan	Himawan	Universitas Sebelas Maret	himawanhim.11@gmail.com
13	Kiagus	Dahlan	IPB	kdahlan@apps.ipb.ac.id
14	Budi	Hastuti	Universitas Sebelas Maret	Budihastuti@staff.uns.ac.id
15	Mazlan	Mazlan	Universitas Sebelas Maret	edi.pramono.uns@staff.uns.ac.id
16	Nadiyah	Diyanahesa	Universitas Sebelas Maret	agusf22@staff.uns.ac.id
17	Windy	Lestari	Universitas Sebelas Maret	edi.pramono.uns@staff.uns.ac.id
18	Agus	Supriyanto	Universitas Sebelas Maret	agusf22@staff.uns.ac.id
19	Husna	Syaima	Universitas Sebelas Maret	sentotbr@yahoo.com
20	Sayekti	Wahyuningsih	Universitas Sebelas Maret	sayekti@mipa.uns.ac.id
21	Tsalatsa	Salsabila	Institut Teknologi Sepuluh Nopember	sheru@chem-eng.its.ac.id
22	Fitri	Aini	Universitas Sebelas Maret	sayekti@mipa.uns.ac.id
23	Pranoto	Pranoto	Universitas Sebelas Maret	pakpranotomipa@staff.uns.ac.id
24	Maryam	Utami	Universitas Sebelas Maret	windyayulestari7@gmail.com
25	Shanghnesy	Nirvana	Universitas Sebelas Maret	shanghnesyjovita@gmail.com
26	Tutik	Murniasih	Research Center for Oceanography-Indonesian Institute of Sciences	tutikmurniasih71@gmail.com
27	Wisnu	Kundarto	Universitas Sebelas Maret	wisnukundarto@staff.uns.ac.id
28	Hisreidi	Funome	Universitas Sebelas Maret	Hisreidifunome@gmail.com
29	Ariyanti	Ariyanti	Stikes Kendal	riri99.cettaazzahra@gmail.com
30	Yc	Danarto	Universitas Sebelas Maret	yc.danarto@gmail.com
31	Artini	Pangastuti	Universitas Sebelas Maret	estu_rn@staff.uns.ac.id
32	Michael	Michael	Atma Jaya Catholic University of Indonesia	prasetyanto@atmajaya.ac.id
33	Jessica	Evelyn	Atma Jaya Catholic University of Indonesia	prasetyanto@atmajaya.ac.id
34	Yuliana	Suselo	Universitas Sebelas Maret	yulianaheri@staff.uns.ac.id
35	Ahmad	Marzuki	Universitas Sebelas Maret	amarzuki@mipa.uns.ac.id
36	Nurma	Sari	Universitas Lambung Mangkurat	nurmasari01@ulm.ac.id
36	Nurhayati	Nurhayati	Institut Teknologi Bandung	vsuendo@chem.itb.ac.id
37	Syahrul	Fatrozi	Universitas Sebelas Maret	msoerya@staff.uns.ac.id
38	Siti	Alaa	Universitas Mataram	siti.alaa@unram.ac.id
39	Dwi	Hatmoko	Universitas Sebelas Maret	pakpranotomipa@staff.uns.ac.id
40	Faisol	Mukarrom	Universitas Sebelas Maret	pakpranotomipa@staff.uns.ac.id
41	Pranoto	Pranoto	Universitas Sebelas Maret	pakpranotomipa@staff.uns.ac.id
42	Endang	Susilowati	Universitas Sebelas Maret	endwati@yahoo.co.id
43	Fatimah	Prasetyaningrum	Universitas Sebelas Maret	hastuti.uns@gmail.com
44	Florentinus	Firdiyono	Indonesian Institute of Sciences	ffirdiyono@yahoo.com
45	Fariza	Yunita	Indonesian Institute of Sciences	fariza.ekayunita@yahoo.co.id
46	Yulinda	Lestari	Metallurgy and Materials Engineering Department, University of Indonesia	yulinda.lestari17@gmail.com
47	Eni	Febriana	Research Center for Metallurgy and Material - Indonesian Institute of Sciences	enif001@lipi.go.id
48	Eko	Sulistiyono	Research Center for Metallurgy and Material - Indonesian Institute of Sciences	eko221068@gmail.com
49	Efendi	Mabruri	Research Center for Metallurgy and Materials, Indonesian Institute of Sciences	efendi_lipi@yahoo.com
50	Ahmad	Marzuki	Universitas Sebelas Maret	amarzuki@mipa.uns.ac.id
51	Hartatik	Hartatik	Universitas Sebelas Maret	hartatik.uns@gmail.com
52	Rahmawati	Rahmawati	Universitas Sebelas Maret	sayekti@mipa.uns.ac.id
53	Risti	Suryantari	Parahyangan Catholic University	ristisuryantari@unpar.ac.id

IOP Conf. Series: Materials Science and Engineering **858** (2020) 011004 doi:10.1088/1757-899X/858/1/011004

Peer review statement

All papers published in this volume of *IOP Conference Series: Materials Science and Engineering* have been peer reviewed through processes administered by the proceedings Editors. Reviews were conducted by expert referees to the professional and scientific standards expected of a proceedings journal published by IOP Publishing.

The Implementation of Road Evaluation and Monitoring System at The City Road Conditions

Martri Daniyati[1], Ary Setyawan[1,2], F. Pungky Pramesti[1,2] and Bayu Setyawan[1]
[1] Civil Engineering Postgraduate Department
Sebelas Maret University, Jl. Ir. Sutami 36 A Solo 57126 Indonesia
[2] Roadmate Research Group, Civil Engineering Department
Sebelas Maret University, Jl. Ir. Sutami 36 A Solo 57126 Indonesia

E-mail: arysetyawan@staff.uns.ac.id

Abstract. The rapid development of new city road infrastructure causes the ignorance of Indonesian people to think about maintaining road infrastructure that has been built. This is indicated by the weak effort in maintaining road infrastructure by the agency, for example: the absence of a good historical record of a road section when it was started to be built until it is maintained, there is no continuity and integration in recording the history of road segments. It is needed an action to solve problems by forming a road maintenance management system. This research involves two types of primary data consisting of system framework data and case study data. The first serves as a determinant of the direction of computer programming logic that can function as a blueprint for the process of storing, processing and presenting case study data, while case study data is factual field data obtained by conducting a direct survey on four city streets, namely, Joko Tingkir street, Monginsidi street, Ir. Juanda and Kyai Mojo street. All damage images, type and segment coordinates of the field are recorded using the Road maintenance and Monitoring System application (REMS). The Pavement Condition Index (PCI) value of each road section are displayed and generated by the Android REMS application will be then processed to determine the maintenance and rehabilitation great strategy.

1. Introduction

At the preparation of a road maintenance management system, data inventory is needed, the data required must contain at least damage conditions, class, geometry and location of the road. The data contained in this inventory are important variables and parameters that can be collected, compiled and processed in a database system that functions as an input form, data processing and presentation, along with recommendations in providing maintenance implementation options as one of the supporting tools for decision makers.

1.1. Pavement condition index

The PCI method is a method developed by the United States Army, this method uses statistical measurements and requires visual observation. The measurement results of this method are index numbers from 0 to 100 which indicate the damage conditions of a road section. This method is a process of evaluating the procedures contained in ASTM D 5340 rules, used throughout the world by considering the functional parameters of the road that have a significant influence on the strength of the road structure [1]. Basically, in evaluating the performance of road conditions can be seen into 2 (two) points of view, namely, functional performance and structural performance. Functional performance is more directed at driving comfort, assessment based on methods and structural performance is more directed at the structure of the road section. Ideally, an assessment by looking at functional performance and structural performance must be carried out to guarantee the longevity of pavement service life, but in Indonesia it is only assessed by looking at functional performance, even though it is carried out by various methods of assessment simultaneously [2]

1.2. The type of pavement deterioration
A road infrastructure that is completed, with increasing time will experience a decrease in quality. Determination of the average PCI value can only be obtained after identification of damage to each observed road segment, according to Shahin (2005) [3], the following are the types of road damage:

1. Alligator Cracking, Fine cracks, longitudinal hair cracks run parallel to each other with fissures that connect each other.
2. Bleeding, Asphalt overweight is seen on the surface of the wear layer, asphalt can be sticky on the vehicle's wheels or shoes.
3. Block Cracking, Cracks form square blocks and are interconnected.
4. Bumps & Sags. The pavement is a pavement which affects the quality of driving.
5. Corrugation, Crimped asphalt which affects the quality of driving.
6. Depression, Localized hardness areas experience a slight decrease in elevation.
7. Edge Cracking, Pavement cracks on the edge of the road due to an old asphalt binder mixture.
8. Joint Reflection Cracking, Crack on areas where there is no filler or cracks on areas filled with fillers on the joint.
9. Lane / Shoulder Drop Off, There is a high difference between the edge of the pavement and the shoulder of the road.
10. Longitudinal & Transverse Cracking, Crack on areas where there is no filler or cracks on areas filled with filler.
11. Patching and Utility Cut Patching, On the road, patches with reduced quality and comfort in the road are disturbed.
12. Polished Aggregate, Aggregate already looks smooth and worn.
13. Potholes, A hole occurred in the road section.
14. Railroad Crossing. There are railroad crossings that cause a decrease in driving comfort.
15. Rutting, Decreased pavement surface that occurs in the vehicle's wheel grooves.
16. Shoving, Urgent asphalt causes a decrease in driving comfort.
17. Slippage Cracking, Cracks occur that cause deformation.
18. Swell, Asphalt expands / bumps causing a slight decrease in driving comfort
19. Weathering and Raveling, The aggregate or asphalt binder starts to detach. diameter.

1.3. Road Evaluation and maintenance system
REMS is an acronym for Road Evaluation and Monitoring System, an android application system that is useful to simplify the calculation of PCI values of roads without having to do graphical calculations on PCI nomograms. This application is also equipped with a photo damage documentation feature and location recording. This study uses the help of the REMS application to make it easier for authors to calculate PCI values in each section of the road. According to Setyawan (2017) [4], REMS applications involve various types of sensors and programs that can help detect road damage and simultaneously record data in the form of geographic information systems [5,6].

2. Experimental
Research conducted is descriptive quantitative based on primary data obtained directly from the field (based on factual conditions) and secondary data obtained from interviews from relevant sources. Primary data and secondary data will be combined and processed to complement each other. Data that has been processed are expected to produce a conclusion from the comparison of the data conducted.

The places used in this study are 4 urban road sections under the auspices of the Public Works and Public Housing Agency of the Surakarta City Government. Basic data collection uses the scope of the project level and has the heaviest axle load (MST) of 8 tons, the category of strategic roads, the type of flexible pavement that supports economic activities, offices, tourism and education and is traversed by freight transportation. Following are the 5 roads that were observed:

1. Joko Tingkir Street (Length: 950m; Width: 7m)
2. Ir. Juanda Street (Length: 3,200m; Width: 10m)
3. Monginsidi Street (Length: 1820m; Width: 9.5m)
4. Kyai Mojo Street (Length: 1320m; Width: 7m)

This research involves two types of primary data consisting of system framework data and case study data. System framework data serves as a determinant of the direction of computer programming logic that can function as a blueprint for the process of storing, processing and presenting case study data, while case study data is factual field data obtained by conducting a direct survey on four city streets, namely, Joko Tingkir street, Monginsidi Street, Ir. Juanda and Kyai Mojo street. The system framework data is in the form of a basic algorithm in the form of a logic logic blueprint regarding:

1. Data of observed road segment objects.
2. Segment coordinate data for each road segment.
3. Road damage data.
4. PCI value data.

3. Results and discussions

This study uses the results of recording geographic data in the form of coordinates obtained from the reading of the Global Positoning System (GPS) on a smartphone. The use of geographic information systems is very helpful in the delivery of integrated information as conveyed by Pantha (2010) [5] which states that geographic information systems can be used to manage and describe various types of data in an integrated or separate manner which in turn can help the decision making process in planning road maintenance. One element in the geographic information system is the coordinate system, in this study the coordinates recorded are the coordinates of the damage points on the road, the coordinates of the stationing point of the road every 100 meters to divide the road segments into segments. The results of the recording of this road damage point will be linked to the station stationary segment and will eventually be displayed in a web page display along with other road segments observed. The results of this merger can facilitate the user in analyzing and identifying patterns of damage applied. The reults of using Road Evaluation and Monitoring system are presented in Project Module, Stationing Module, PCI module and Recapitulation Module.

3.1. Project level module

The module shown in Figure 1 presents four observed roads, namely Joko Tingkir Street, Monginsidi Street, Ir. Juanda and Kyai Mojo street. Each observed road section is given additional information in the form of naming the project or work package, type of maintenance, dimension data of the road section, classification data and administrative location of the road section.

IOP Conf. Series: Materials Science and Engineering **858** (2020) 012001 doi:10.1088/1757-899X/858/1/012001

Show 10 row every page Find

No	Project Name	Project Type	Pavement Name	Length	Width	Function	Province	City	District	Action
1	Joko Tingkir Project	Routine Maintenance	Jl. Joko Tingkir	0.95 km	7.00 m	Kolektor 3	Jawa Tengah	Kota Surakarta	Laweyan	
2	Monginsidi Project	Routine Maintenance	Jl. Monginsidi	1.82 km	9.50 m	Kolektor 3	Jawa Tengah	Kota Surakarta	Banjarsari, Jebres	
3	Juanda Project	Routine Maintenance	Jl. Ir. Juanda	3.20 km	10.00 m	Kolektor 3	Jawa Tengah	Kota Surakarta	Jebres	
4	Kyai Mojo Project	Routine Maintenance	Jl. Kyai Mojo	1.32 km	7.00 m	Kolektor 3	Jawa Tengah	Kota Surakarta	Pasar Kliwon	

Show page 1 of 1 Previous 1 Next

Figure 1. The Project level module.

3.2. The stationing module

The Stationing Module is presented in an interactive digital map format that can be clicked on each station to find out the station's STA information, in addition a table containing station details and coordinates are presented. The module in Figure 2 shows that the Joko Tingkir road section has 19 stations and 18 segments. The other road segments are presented at the similar pattern.

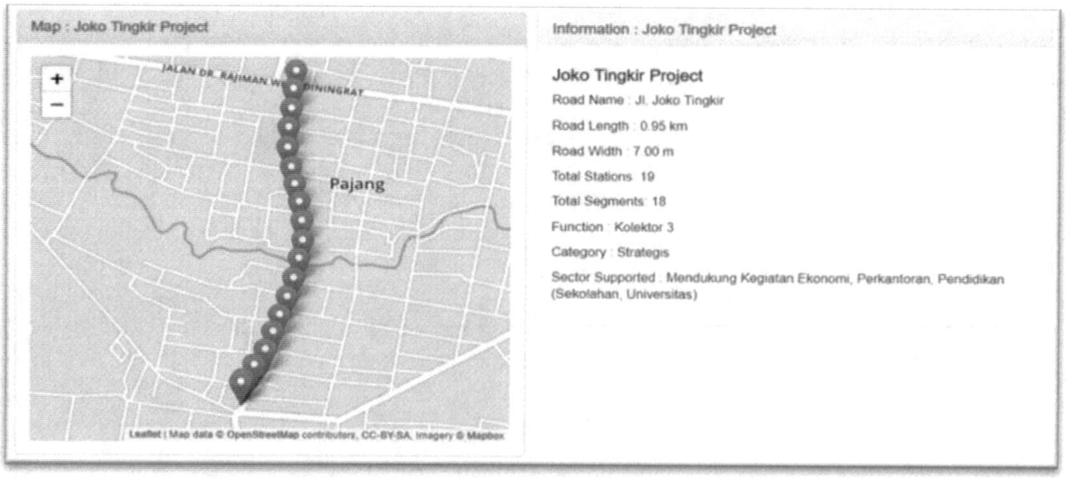

Figure 2. The typical of stationing module at Joko Tingkir street.

3.3. The road deterioration module

With this module, users can identify images, the amount and type of damage that occurs on each road segment that has been classified by segment. Modules are presented in the form of interactive digital maps, tables and pie charts. Figure 3 shows that, Joko Tingkir road suffered 10 types of damage. The most common types of damage are patching and utility cut patching and polished aggregate types.

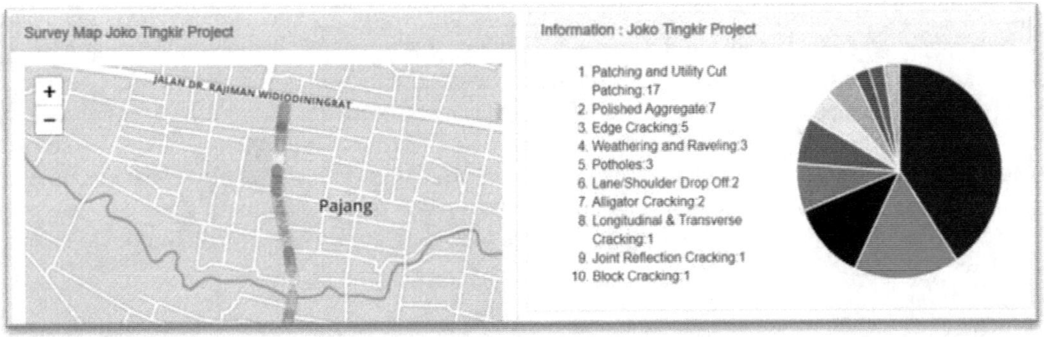

Figure 3. The road deterioration module at the specific road segment of Joko Tingkir street..

3.4. The pavement condition index module

With this module, the pavement condition index value can be observed in detail by clicking on the graph nodes. In addition to the graph also provided a table containing the segment along with the PCI value of the segment. Figure 4 shows that the Joko Tingkir road has a total PCI index of 89.16.

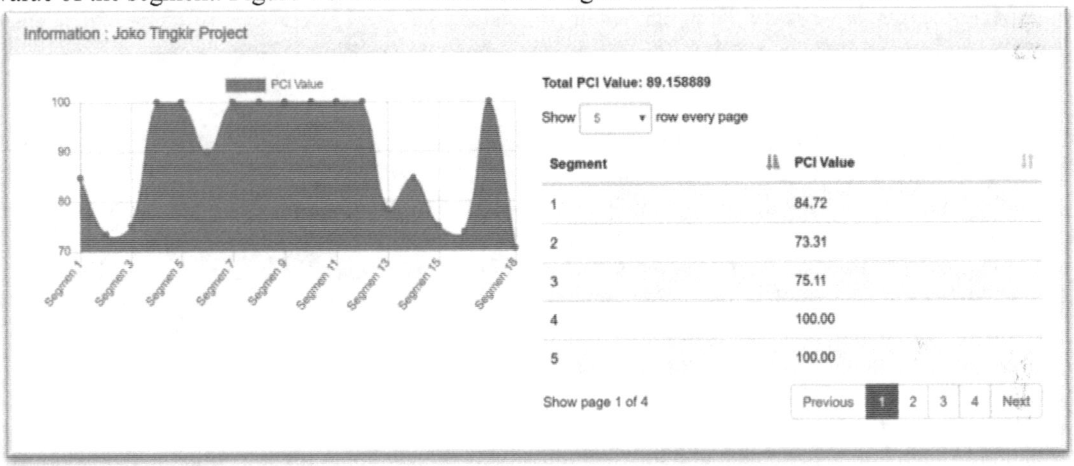

Figure 4. The pavement condition index module at the specific road segment of Joko Tingkir street. .

3.5. *Recapitulation module*

In Table 1 shows the recapitulation of maintenance costs on the observed roads. In the damage survey that has been carried out, the most amount of damage is on the Ir. Juanda street, this is because the length of the track is also the longest with a PCI value of 79.60.

Table 1. The recapitulation of road database and PCI values.

Road Segment	Length (km)	Width (m)	The number of detrioration	PCI Value
Joko Tingkir	0.95	7	42	89.16
Monginsidi	1.82	9.5	54	89.54
Ir. Juanda	3.2	10	102	79.6
Kyai Mojo	1.32	7	93	52.4

In this study also analyze that the value of PCI can not directly affect the amount of maintenance costs or can be said there is no relationship, the underlying thing is not a strong connection between the value of PCI and maintenance costs is, the calculation of maintenance costs is very dependent on the area or volume of damage and also influenced by the maintenance method and the standard price of the selected work unit. The choice of maintenance method and standard unit price of work are additional factors that can obscure the value of PCI as a representation of pavement conditions if brought into the context of calculating maintenance costs [7]. From these findings, it can be shown that road section documentation to be used as a database as a tool for road management and maintenance is very useful in conducting deeper analysis of existing variables. With the ongoing survey and input of new roads, in the end it can sharpen the user analysis in processing the available road segment database.

4. Conclusion

From the results of the discussion in this study, we obtained several important conclusions, namely documenting any road pavement conditions regularly or planned is highly recommended because this is an easy and inexpensive step for stakeholders in order to establish a good road maintenance database that can ultimately help in the process of analysis and planning for maximum maintenance. Road maintenance database system that is maintained in quality as well as by presenting data in an orderly and integrative manner can guarantee reservation of road pavement conditions in accordance with expected levels and enable the achievement of sustainable road infrastructure management and ultimately can have a large cost-saving impact and can provide services to society in prime.

References

[1] Boyapati B and Kumar RP. Prioritisation of Pavement Maintenance based on Pavement Condition Index. Indian Journal of Science and Technology. 2015 8 14.
[2] Suryoto, Siswoyo DP dan Setyawan. A The Evaluation of Functional Performance of National Roadway Using Three Types of Assessment Method. 2017 171 pp 1435-1442.
[3] Shahin MY. *Pavement Management for Airports, Roads, and Parking Lots. 2nd ed.* LLC: Springer Science and Business Media; 2005.
[4] Setyawan A, Al-Alimi MF, Handayani D dan Sumarsono. A *The Design of Road Evaluation and Monitoring System based on Geographical Information System.* MATEC Web of Conference. 2017 **138** 07010.
[5] Pantha RB, Yatabe R, Bhandary NP *GIS-based highway maintenance prioritization model: an integrated approach for highway maintenance in Nepal mountains.* Journal Transport of Geography. 2010 **18** pp 426-433.
[6] Karim F, Rubasi K dan Saleh A. The Road Pavement Condition Index (PCI) Evaluation and Maintenance: A Case Study of Yemen. Organization, Technology and Management in Construction. De Gruyter. 2016 8 pp 1446-1455.
[7] Bayu Setyawan, Ary Setyawan, Budi Yulianto, The development of road evaluation and monitoring system using database application, AIP Conference Proceedings **2114**, 060014 (2019)

IOP Conf. Series: Materials Science and Engineering **858** (2020) 012002 doi:10.1088/1757-899X/858/1/012002

The effect of vehicle speed on the stiffness modulus of conventional asphalt and bioasphalt

R A S S Putra[1], A Setyawan[2] and F P Pramesti[2]
[1] Masters Students in Infrastructure Maintenance and Rehabilitation, Faculty of Engineering, Sebelas Maret University, Surakarta, Indonesia
[2] Lecturer in The Masters of Infrastructure Maintenance and Rehabilitation, Faculty of Engineering, Sebelas Maret University, Surakarta, Indonesia

E-mail: renggaaldian@student.uns.ac.id

Abstract. Stiffness modulus is the ratio between stress and strain found in a layer. The value of the stiffness modulus is related to the performance of the layer in bearing the load. High modulus of stiffness makes the asphalt mixture stiff and resistant to deformation, while a decrease in the modulus of stiffness will make the pavement layer vulnerable to crack. The purpose of this study was to analysis the characteristic differences between damar asphalt mixture and conventional asphalt mixture in term of the stiffness of asphalt mixtures due to the variations in vehicle speed at the specified temperature. Speed variations are 10, 20, 30, 40, 50 and 60 kph and the temperature reference are 20 and 40°C. Data analysis is performed by comparing the stiffness values of damar asphalt mixtures and conventional asphalt mixture. The results of this study indicate that damar asphalt mixes are more resistant to changes in temperature and vehicle speed than conventional asphalt.

1. Introduction

Flexible pavement is pavement that uses asphalt as a binder. The layers of pavement are carrying and spreading the burden of traffic to the compacted soil [1]. Each layer of pavement that is burdened will experience a response to the vehicle load in the form of stress and strain in each layer of pavement. Stiffness modulus is the ratio between stress and strain found in a layer. Stiffness modulus is a value that indicates the stiffness of a material. The higher the stiffness modulus of a material, the more difficult it is to change shape because the strain that occurs is very small.

The theory of linear viscoelasticity states that the relaxation modulus E (t) and complex modulus are constant, thus time and frequency can be interconverted. The stiffness modulus change during the loading time is not a constant function [2]. This theory is supported by the opinion of Van der Poel who mentioned the stiffness modulus as stiffness of bitumen (Sb) which is a comparison between the stresses on asphalt, which is a function of the length of loading (frequency) applied, the temperature difference with T800 and the penetration index. T800 is the temperature when penetration reaches 800 [3]. So determination of the stiffness modulus of asphalt mix is significantly influenced by pavement temperature and time loading period [4][5].

Determination of temperature in this study is based on the results of research conducted Taherkhani et al showed that in dry condition, the resilient modulus increases with increasing loading frequency, while, in saturated condition, a slight increase of resilient modulus with loading frequency was observed only at 40°C. It is also found that, in both the dry and saturated conditions, the effect of

loading frequency on the resilient modulus decreases with decreasing temperature. Therefore, the temperature of 40°C was recommended for investigating the effect of loading frequency on the resilient modulus [6]. The temperature range in research conducted by taherkhani is -5, 5, 20 and 40°C. So the researchers decided to use temperatures of 20 and 40°C as a reference, another consideration because temperatures in Indonesia ranged between these ranges.
Similar research has been conducted by Lubis to determine the AC-WC stiffness modulus with the result that the modulus of stiffness decreased with increasing concrete asphalt temperature [7]. Sjahdanulirwan conducted a literature review on the effect of loading time on the stiffness of concrete asphalt according to direct methods and indirect methods. For the direct methods used Marshall test and for the indirect methods used Brown & Brunton method and Shell Bitumen method [8]. Riwibowo conducted research on service life which in the process alluded to the stiffness of the asphalt mixture. The results of this study concluded that the pavement temperature and loading time can affect the service life of the road, where the loading time variable is more influential than the pavement temperature [9]. Speed is related to the service life of the road because differences in vehicle speed cause variations in the modulus of elasticity of material that affects the design life [10]. The higher the speed of the vehicle, the damage caused by the road will be lower so that the ability of the road to serve traffic will be longer [11]. Therefore, researchers are interested to know the effect of speed on modulus of stiffness, especially on damar asphalt and its differences with conventional asphalt. Researchers through this study will compare the results of vehicle speed variations at temperatures of 20 and 40°C against the mix stiffness of conventional asphalt and bioasphalt using analytical approach to determine the difference.

2. Experimental

The result from the laboratory test analized using the Nottingham design method to obtain the value of asphalt mixture stiffness (S_{mix}) based on an analysis of variations in vehicle speed and temperature. Tests and methods carried out in this study are described in sections 2.1 to section 2.3:

2.1. Bitumen

This research uses two different types of asphalt, the intended asphalt is damar asphalt and conventional asphalt. For research purposes, the two asphalt physical properties are tested, such as penetration, softening points, ductility, specific gravity and flash points.

2.1.1. Damar asphalt is asphalt resin is a binding agent with a mixture of resin, fly ash, cooking oil and latex with a composition of pure resin 100 g is added resin powder 350 g, fly ash 150 g, cooking oil 205 g, and 4% latex. The main physical properties of damar asphalt can be seen in Table 1.

Table 1. Main physical properties of damar asphalt binder.

Type of Testing	Unit	Specification	Test Results
Penetration	0,1 mm	40-59	43
Softening Point	°C	51-63	57,5
Ductility	cm	min. 100	115,5
Specific Gravity	g/cm³	min. 1	0,97
Flash Point	°C	min. 200	260

2.1.2. Conventional asphalt used in this research is 60/70 penetration asphalt which is generally used in road engineering in Indonesia. The main physical properties of conventional asphalt can be seen in Table 2.

Table 2. Main physical properties of conventional asphalt binder.

Type of Testing	Unit	Specification	Test Results
Penetration	0,1 mm	60-79	61-62
Softening Point	°C	48-58	50,1-54,5
Ductility	cm	min. 100	100-140
Specific Gravity	g/cm³	min. 1	1,020-1,031
Flash Point	°C	min. 200	350,4

IOP Conf. Series: Materials Science and Engineering **858** (2020) 012002 doi:10.1088/1757-899X/858/1/012002

2.2. Asphalt Mixture

This research uses two different types of asphalt mixture, the intended asphalt is damar asphalt mixture and conventional asphalt mixture. For research purposes, the two asphalt mixture tested are physical properties test (volumetric) to get density, VIM, VMA, VFB and marshall test to get stability and flow from sample. Sample testing was carried out on three samples at room temperature conditions on damar asphalt mixture and conventional asphalt mixture.

2.2.1. Damar mixture is a mixture of road pavement consisting of coarse aggregate, fine aggregate, filler and damar asphalt binding material with a certain ratio and mixed in hot conditions. The damar asphalt mixture used in this study was AC-WC, AC-BC and AC-Base with the main physical properties of each mixture can be seen in Table 3.

Table 3. Main physical properties of damar asphalt mixture.

No.	Type of Test	AC-WC[a]	AC-BC[b]	AC-Base[c]	Spec.
1	Stability (Kg/cm^2)	1181,564	1181,564	1181,564	>800 kg
2	Flow (mm)	4,603	4,603	4,603	2-4 mm
3	Density (%)	2,410	2,410	2,410	-
4	VIM (%)	3,79	3,79	3,79	3-5%
5	VMA (%)	12,494	12,494	12,494	>15%
6	VFB (%)	81,033	81,033	81,033	>65%
7	Marshall Quotient (Kg/mm)	268,847	268,847	268,847	-

[a] Asphalt concrete wearing coarse
[b] Asphalt concrete binder coarse
[c] Asphalt concrete base coarse

2.2.2. Conventional mixture is a mixture of road pavement consisting of coarse aggregate, fine aggregate, filler and asphalt binder with a certain ratio and mixed in hot conditions. The conventional asphalt mixture used in this study is AC-WC, AC-BC and AC-Base with the main physical properties of each mixture can be seen in Table 4.

Table 4. Main physical properties of conventional asphalt mixture.

No.	Type of Test	AC-WC[a]	AC-BC[b]	AC-Base[c]	Spec.
1	Stability (Kg/cm^2)	1307,36	1307,36	2143,92	>800 kg
2	Flow (mm)	2,05	2,05	3,15	2-4 mm
3	Density (%)	2,4	2,4	2,4	-
4	VIM (%)	4,71	4,71	4,16	3-5%
5	VMA (%)	17,15	17,15	19,60	>15%
6	VFB (%)	72,75	72,75	78,85	>65%
7	Marshall Quotient (Kg/mm)	642,21	642,21	681,17	-

[a] Asphalt concrete wearing coarse
[b] Asphalt concrete binder coarse
[c] Asphalt concrete base coarse

2.3. Nottingham Design Method

2.3.1. Calculating Bitumen Stiffness (Sb). Strain depends on the temperature and the length of time the stress is applied, therefore the simple concept of Modulus Young being the ratio of stress to strain and having a fixed value for certain materials clearly does not apply. Van der Poel introduced Stiffness of Bitumen (Sb) as a substitute for young modulus which is a comparison between the stresses on asphalt, which is a function of the length of loading (frequency) applied, the temperature difference with the T800 and the Penetration Index. T800 is the temperature when penetration reaches 800.

The loading time is the length of load received by the road pavement when the wheels of the vehicle cross the road pavement [12]. Vehicles with the same load will produce different levels of damage due to speed differences. The smaller the speed value will result in a longer loading time. Loading time is closely related to the wavelength of vertical compressive stress, depth of interest, and vehicle speed

[13]. The equation between loading time, layer thickness and vehicle speed are presented in equation (1):

$$\log t = 5 \times 10^{-4}\, h - 0{,}2 - 0{,}94 \log v \quad (1)$$

where: t = loading time (s); h = layer thickness (mm) and v = vehicle speed (kph)
The following equation has been derived by Ullidtz for calculating the value of bitumen stiffness under certain conditions [14]:

$$Sb = 1{,}157 \times 10^{-7} \times t^{-0{,}368} \times 2{,}718^{-PIr}\,(SPr - T)^5 \quad (2)$$

where: Sb = bitumen stiffness (MPa); t = loading time (s); PIr = recovered penetration index (0,1 mm); SPr = recovered softening point (°C) and T = design temperature (°C). The limitation for finding the modulus of stiffness of bitumen by Ullidtz is the value of t = 0,01 to 0,10 s; PIr = -1 to +1 and (SPr – T) = 20 to 60°C

The Recovered Penetration Index can be calculated using the following formula:

$$PIr = \frac{27{,}00 \times \log P_i - 21{,}65}{76{,}35 \times \log P_i - 232{,}82} \quad (3)$$

where: PIr = recovered penetration index (0,1 mm) and P_i = initial penetration (0,1 mm)

The Recovered Softening Point (SPr) can be calculated using the following formula:

$$SPr = 98{,}4 - 26{,}35 \times \log 0{,}65 \times Pi \quad (4)$$

where: SPr = recovered softening point (°C) and P_i = initial penetration (0,1 mm)

2.3.2. Calculating Mix Stiffness (S_{me}). Modulus of stiffness of asphalt mixture (Sme) can be calculated from asphalt stiffness (Sb) and air pore content (Cv). The appropriate equation according to Heukelom and Klomp regarding the modulus of stiffness of the asphalt mixture presented in equation (5)[15]:

$$S_{me} = S_b\left[1 + \left(\frac{2{,}5}{n}\right)\left(\frac{C_v}{1 - C_v}\right)\right]^n \quad (5)$$

$$n = 0{,}83 \log \frac{4 \times 10^4}{S_b}$$

where: Sme = stiffness mixture of asphalt concrete (MPa); Sb = bitumen stiffness (MPa); Cv = aggregate volume concentration (%); n = the stiffness constant of an elastic mixture based on bitumen stiffness.

$$C_v = \frac{V_A}{V_A + V_B} \quad (6)$$

where: Cv = aggregate volume concentration (%); VA = volume of aggregate (%) and VB = volume of binder (%).
Based on equation (5) a curve of the relationship between asphalt mix stiffness and vehicle speed can be made, where the stiffness modulus value is the slope of the straight line curve.

3. Results and discussions

3.1. Effects of vehicle speed on the stiffness modulus

This analysis is carried out on variations in vehicle speed which refers to fluctuations in loading time (10, 20, 30, 40, 50 and 60 kph) using the Nottingham design method (equation 1 to 6) at temperatures of 20 and 40°C to the modulus stiffness. The analysis was carried out on two types of mixtures namely damar asphalt and conventional asphalt. Data from the results on damar asphalt and conventional asphalt based on variations in the vehicle speed at 20°C are presented in tables 5 and 6.

Table 5. Damar asphalt stiffness based on variations in vehicle speed at 20°C

No	v	T	Sb[a] (MPa)			Sme[b] (MPa)		
	(kph)	(°C)	AC – WC[c]	AC – BC[d]	AC – Base[e]	AC – WC[c]	AC – BC[d]	AC – Base[e]
1	10	20	38,551	38,879	38,551	5364,635	5391,952	5364,635
2	20	20	48,997	49,414	48,997	6187,109	6218,116	6187,109
3	30	20	56,375	56,854	56,375	6718,341	6751,684	6718,341
4	40	20	62,273	62,803	62,273	7119,194	7154,274	7119,194
5	50	20	67,271	67,843	67,271	7444,382	7480,857	7444,382
6	60	20	71,650	72,260	71,650	7719,617	7757,262	7719,617

[a] Bitumen stiffness
[b] Asphalt mix stiffness
[c] Asphalt concrete wearing coarse
[d] Asphalt concrete binder coarse
[e] Asphalt concrete base coarse

Table 6. Conventional asphalt stiffness based on variations in vehicle speed at 20°C

No	v	T	Sb[a] (MPa)			Sme[b] (MPa)		
	(kph)	(°C)	AC – WC[c]	AC – BC[d]	AC – Base[e]	AC – WC[c]	AC – BC[d]	AC – Base[e]
1	10	20	24,444	24,652	23,908	5262,794	5289,004	6583,082
2	20	20	31,067	31,331	30,386	6050,891	6080,564	7519,845
3	30	20	35,745	36,049	34,961	6558,899	6590,761	8120,064
4	40	20	39,485	39,821	38,619	6941,768	6975,256	8570,688
5	50	20	42,653	43,016	41,718	7252,102	7286,897	8934,899
6	60	20	45,430	45,817	44,434	7514,595	7550,485	9242,258

[a] Bitumen stiffness
[b] Asphalt mix stiffness
[c] Asphalt concrete wearing coarse
[d] Asphalt concrete binder coarse
[e] Asphalt concrete base coarse

Based on tables 5 and 6 it is known that increasing speed affects the decrease in the loading time value and when the vehicle speed decreases the loading time becomes high. High loading time causes the strain that occurs to be even greater and result in reduced value of material stiffness [16]. The highest stiffness of the mixture drops at a speed of 10 to 20 kph. This is confirmed by the opinion of Pandey et al through his research on fatigue and rutting, from his research it is known that both fatigue and rutting critical strains increase with decrease in vehicle speed and they resembles the static loading conditions if the vehicle speed decreases to 20 kph and less time is found to have more detrimental effect on the flexible pavement [17]. The analysis was then continued at 40°C to determine the effect of vehicle speed on modulus stiffness. Data from the results on damar asphalt and conventional asphalt based on variations in the vehicle speed at 40°C are presented in tables 7 and 8.

Table 7. Damar asphalt stiffness based on variations in vehicle speed at 40°C

No	v	T	Sb[a] (MPa)			Sme[b] (MPa)		
	(kph)	(°C)	AC – WC[c]	AC – BC[d]	AC – Base[e]	AC – WC[c]	AC – BC[d]	AC – Base[e]
1	10	40	1,248	1,259	1,248	568,398	571,780	568,398
2	20	40	1,587	1,600	1,587	671,885	675,852	671,885
3	30	40	1,826	1,841	1,826	740,490	744,842	740,490
4	40	40	2,017	2,034	2,017	793,151	797,795	793,151
5	50	40	2,178	2,197	2,178	836,425	841,310	836,425
6	60	40	2,320	2,340	2,320	873,437	878,526	873,437

[a] Bitumen stiffness
[b] Asphalt mix stiffness
[c] Asphalt concrete wearing coarse
[d] Asphalt concrete binder coarse
[e] Asphalt concrete base coarse

Table 8. Conventional asphalt stiffness based on variations in vehicle speed at 40°C

No	V	T	Sb[a] (MPa)			Sme[b] (MPa)		
	(kph)	(°C)	AC – WC[c]	AC – BC[d]	AC – Base[e]	AC – WC[c]	AC – BC[d]	AC – Base[e]
1	10	40	0,445	0,449	0,422	394,946	397,282	535,419
2	20	40	0,566	0,571	0,536	466,414	469,153	628,787
3	30	40	0,651	0,657	0,617	513,784	516,789	690,377
4	40	40	0,719	0,725	0,681	550,142	553,349	737,506
5	50	40	0,777	0,783	0,736	580,020	583,392	776,147
6	60	40	0,827	0,834	0,784	605,573	609,087	809,136

[a] Bitumen stiffness
[b] Asphalt mix stiffness
[c] Asphalt concrete wearing coarse
[d] Asphalt concrete binder coarse
[e] Asphalt concrete base coarse

Based on Table 7 and 8 the effect of vehicle speed on modulus of stiffness at 40°C is not different from the temperature of 20°C it is known that increasing vehicle speed impacts on increasing the modulus stiffness value. What makes it different is the modulus stiffness value which decreases at all speed variations because the temperature increases from 20 to 40°C.
Asphalt mixture stiffness modulus values generated from the Nottingham design method calculations in tables 5 to 8, then entered into the Cartesian diagram to find the relationship between vehicle speed and stiffness modulus for temperatures 20 and 40°C are distinguished between damar asphalt and conventional asphalt divide into AC- WC, AC-BC and AC-Base. The relationship of the modulus of stiffness and vehicle speed at temperatures of 20 and 40°C is illustrated in Figures 1 to 6.

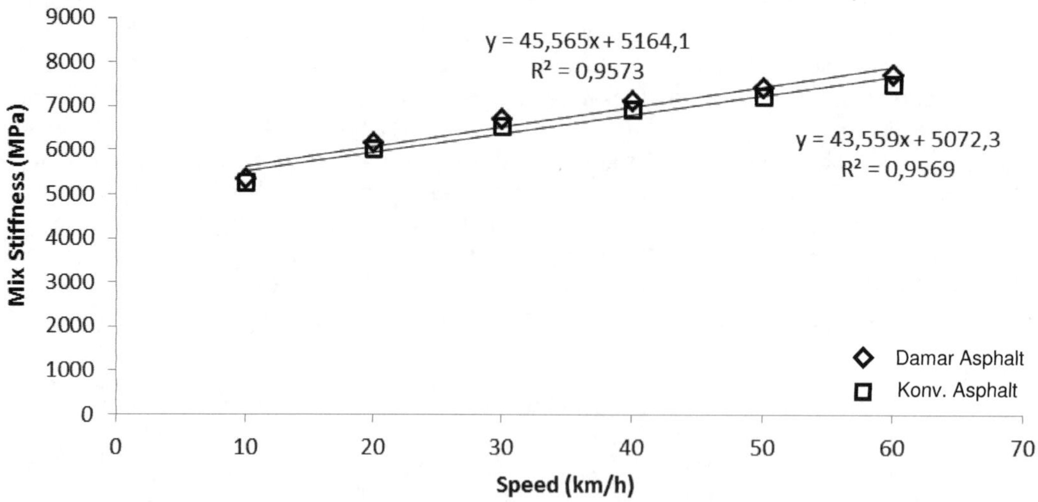

Figure 1. Comparison of stiffness of damar asphalt mixture with conventional asphalt based on variations in vehicle speed at a temperature of 20°C in the AC-WC layer.

Figure 1 shows that vehicle speed has a linear relationship with modulus stiffness in the AC-WC layer at 20°C. The correlation value between the speed and stiffness of damar asphalt modulus (0.9573) and conventional asphalt (0.9569) is high, so it can be said that they have a strong correlation.

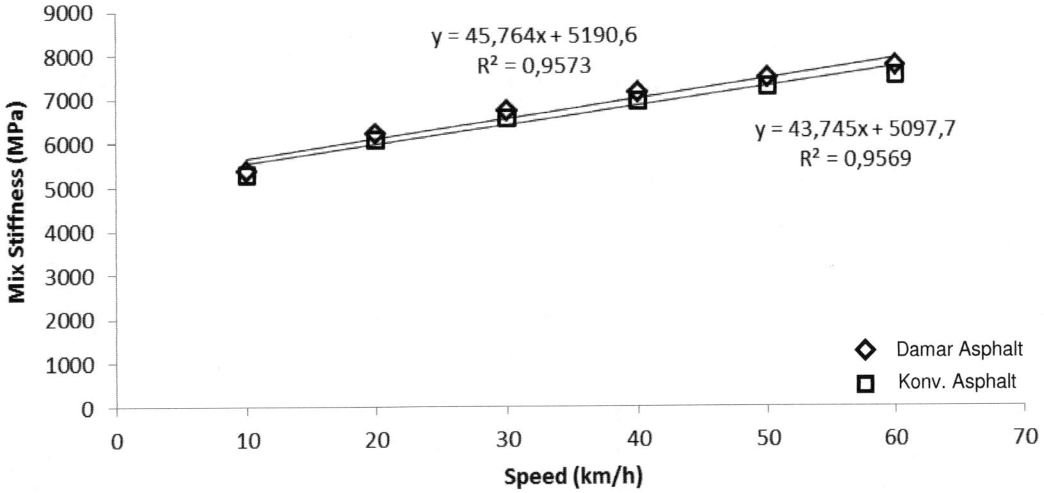

Figure 2. Comparison of stiffness of damar asphalt mixture with conventional asphalt based on variations in vehicle speed at a temperature of 20°C on the AC-BC layer.

Figure 2 shows that vehicle speed has a linear relationship with modulus stiffness in the AC-BC layer at 20°C. The correlation value between the speed and stiffness of damar asphalt modulus (0.9573) and conventional asphalt (0.9569) is high, so it can be said that they have a strong correlation.

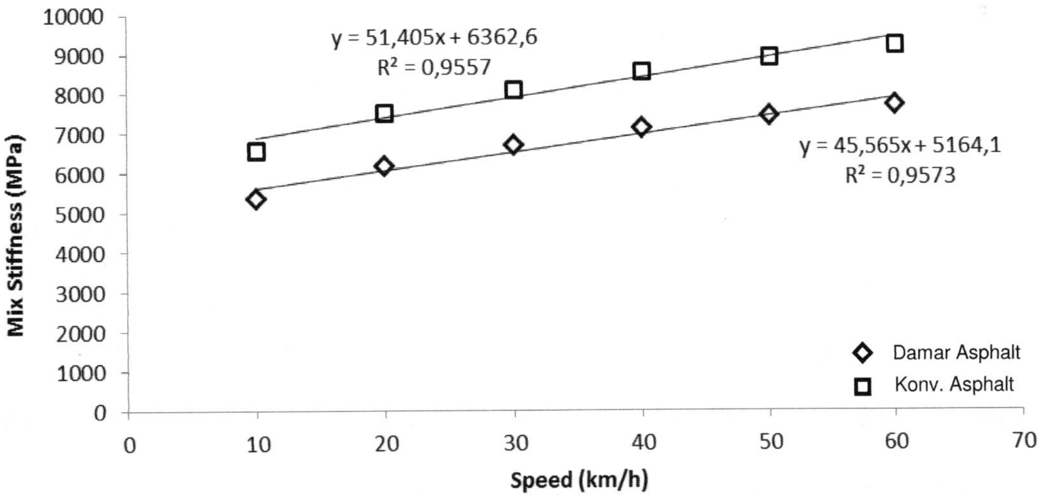

Figure 3. Comparison of stiffness of damar asphalt mixture with conventional asphalt based on variations in vehicle speed at a temperature of 20°C on the AC-Base layer.

Figure 3 shows that vehicle speed has a linear relationship with modulus stiffness in the AC-Base layer at 20°C. The correlation value between the speed and stiffness of damar asphalt modulus (0.9573) and conventional asphalt (0.9557) is high, so it can be said that they have a strong correlation.

IOP Conf. Series: Materials Science and Engineering **858** (2020) 012002 doi:10.1088/1757-899X/858/1/012002

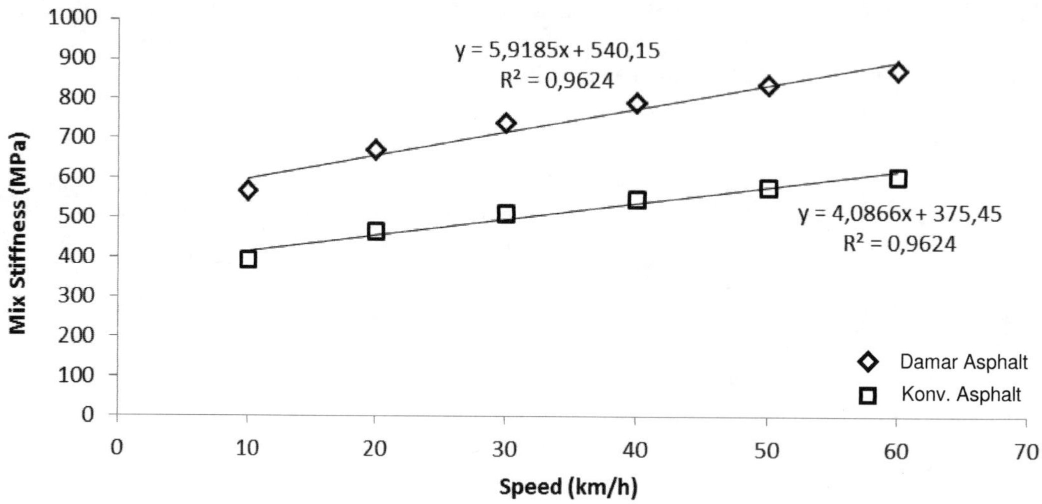

Figure 4. Comparison of stiffness of damar asphalt mixture with conventional asphalt based on variations in vehicle speed at a temperature of 40°C in the AC-WC layer.

Figure 4 shows that vehicle speed has a linear relationship with the modulus stiffness in the AC-WC layer at 40°C. The correlation value between the speed and stiffness of damar asphalt modulus (0.9624) and conventional asphalt (0.9624) is high, so it can be said that they have a strong correlation.

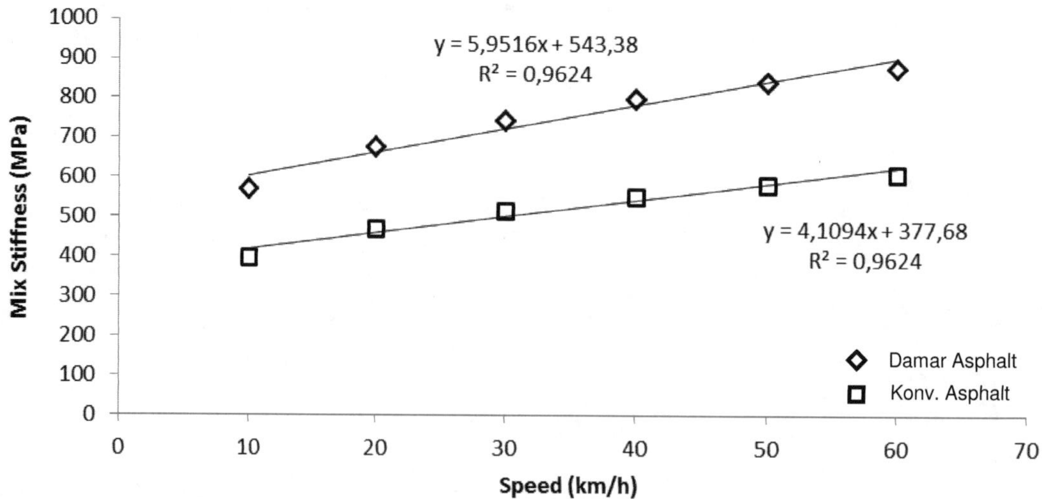

Figure 5. Comparison of stiffness of damar asphalt mixture with conventional asphalt based on variations in vehicle speed at a temperature of 40°C on the AC-BC layer.

Figure 5 shows that vehicle speed has a linear relationship with the modulus stiffness in the AC-BC layer at 40°C. The correlation value between the speed and stiffness of damar asphalt modulus (0.9624) and conventional asphalt (0.9624) is high, so it can be said that they have a strong correlation.

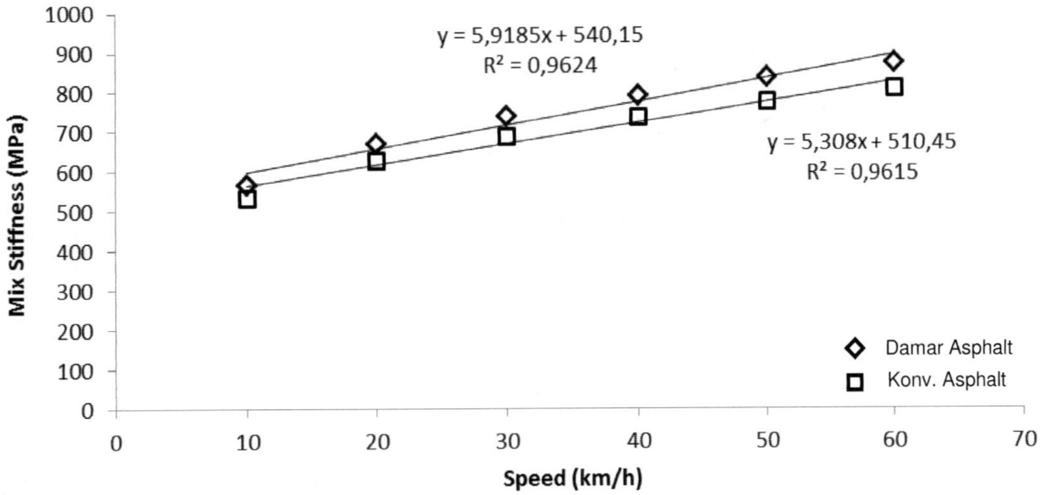

Figure 6. Comparison of stiffness of damar asphalt mixture with conventional asphalt based on variations in vehicle speed at a temperature of 40°C on the AC-Base layer.

Figure 6 shows that vehicle speed has a linear relationship with the modulus of stiffness in the AC-Base layer at 40°C. the correlation value between the speed and stiffness of damar asphalt modulus (0.9624) and conventional asphalt (0.9615) is high, so it can be said that they have a strong correlation.

Based on figures 1 to 6, if the modulus stiffness value between damar asphalt and conventional asphalt in all layers at 20°C (Figure 1 to 3) is compared with conventional asphalt and asphalt modulus stiffness values in all layers at 40°C (Figure 4 to 6), there will be a shift in the line equation that widens the range of modulus stiffness values between damar asphalt and conventional asphalt due to the influence of temperature. The decrease in modulus stiffnes value from damar asphalt is smaller compared to conventional asphalt. So that the ability of damar asphalt to withstand the weight of moving vehicles according to variations in speed is better than conventional asphalt.

4. Conclusion

Based on the analysis of the effect of vehicle speed and pavement temperature on the elastic mixture stiffness using the Nottingham Design Method it is known that the vehicle speed increased from 10 to 60 kph resulting in decreased loading time and an increase in elastic mixture stiffness in all types of asphalt mixtures. The stiffness of damar asphalt mixture due to variations in speed from 10 to 60 kph in AC-WC and AC-BC layers at a temperature of 20°C is higher than in conventional asphalt but in the AC-Base layer is lower than in conventional asphalt. Entering at 40°C all types of damar asphalt mix have a greater stiffness than conventional asphalt at all types of vehicle speed variations. Decreasing the stiffness of all types of damar asphalt mixture is slower than conventional asphalt mixtures from 20 to 40°C so that the damar asphalt mixture is more resistant to changes in temperature and vehicle speed than conventional asphalt.

Acknowledgment

The authors would like to thank the support and discussion given by Ir. Ary Setyawan, M.Sc., Ph.D and Dr. Florentina Pungky Pramesti, ST., MT from Sebelas Maret University of Surakarta to carry out this paper. Last but not least, I would like to thank my family and friends who have been a constant source of encouragement.

References

[1] Tenriajeng A T 2002 *Rekayasa Jalan Raya 2* (Jakarta: Gunadarma)
[2] Mazurek G and Iwański M 2016 *Modern Building Materials, Structures and Techniques* (Poland: Elsevier) p 731 – 738
[3] Van der Poel C 1954 Representation of rheological properties of bitumen over a wide range of temperature and loading times *Proc. of 1st International Congress of Rheology* vol 2 (London: Butterworth's Scientific Publication) p 331-337
[4] Aschuri I 2004 The stiffness modulus of asphalt concrete for design life Prediction of pavement under tropical condition *Proc. The 1st PhD Seminar on Civil Engineering* (Indonesia: *Civil Eng. Journal*)
[5] Sousa J B, Lysmer J, Chen S and Monismith C L 1988 *Transportation Research Record 1207* **145-168**
[6] Taherkhani H, Nejad F M, Khodaii A and Ahari A S 2017 *International J. of Transportation Engineering* Vol.5 1 2017 **1-16**
[7] Lubis A S, Muis Z A and Iskandar T D 2018 *IOP Conf. Ser.: Mater. Sci. Eng.* **309** (1) 012113
[8] Sjahdanulirwan 2009 Pengaruh waktu pembebanan terhadap kekuatan beton aspal lapis permukaan *J. Jalan Jembatan* vol 28 No 2 (2011) (Bandung: Jurnal Jalan Jembatan)
[9] Riwibowo N 2017 *Analisis Pengaruh Suhu Perkerasan dan Loading Time Terhadap Umur Pelayanan Jalan Menggunakan Metode Analisis (Studi Perbandingan Jalan Tol dan Non Tol)* (Surakarta: Muhammadiyah University Press)
[10] Samad E 2011 *Civil Engineering Forum* **1163-1174**
[11] Premana H K 2012 *Analisis Pengaruh Kecepatan Kendaraan Terhadap Umur Rencana Perkerasan Jalan Dengan Menggunakan Metode Analitis (Studi Kasus Ruas Jalan Tol Semarang)* (Surakarta: Muhammadiyah University Press)
[12] Brown S F, Brunton J M and Stock A F 1985 The Analytical Design of Bituminous Pavements *Proc. Institution of Civil Engineers* vol 79 (UK: *Institution of Civil Engineers*) p 1-31
[13] Dai X, Hu X and Gao Y 2019 *Advances in Materials Science and Engineering* vol 2019, ed Victor M Castaño (China: Hindawi) p 1-11
[14] Ullidtz P 1979 A fundamental method for prediction of roughness, rutting and cracking of pavements *Proc. Association of Asphalt Paving Technologists* vol 48 (US: *Association of Asphalt Paving Technologists*) p 557-586
[15] Heukelom W and Klomp A J 1964 Road design and dynamic loading *J. Assoc. Asphalt Paving Technologists* vol 33 p 92-125
[16] Nugroho A 2012 *Analisis Pengaruh Kecepatan Kendaraan Terhadap Umur Rencana Jalan Dengan Menggunakan Metode Analitis (Studi Kasus Ruas Jalan Rembang-Bulu)* (Surakarta: Muhammadiyah University Press)
[17] Pandey S, Chowdary T H and Rokade S 2017 Effect of higher axle loads on service life of bituminous pavement *Proc. World Conf. on Pavement and Asset Management* (Italy: ResearchGate)

IOP Conf. Series: Materials Science and Engineering **858** (2020) 012003 doi:10.1088/1757-899X/858/1/012003

The Assessment of Physical Condition of Delingan Dam in 2019 as an Evaluation on Dam Maintenance

Erhar Augusto[1], Cahyono Ikhsan[1], Rintis Hadiani[1]
[1]Civil Engineering Program, Faculty of Engineering, Sebelas Maret University of Surakarta, Surakarta, Indonesia

E-mail: erhar.augusto@gmail.com, cahyono1970@yahoo.co.id, and rintisha@gmail.com

Abstract. For operating and maintaining dam activities, there is an activity serving the main function, The Assessment of Dam Condition to make it the main building well-monitored and maintained. In addition to having many advantages, dam also hides a big potential hazard. For the dam to function maximally, good maintenance and evaluation are required. Delingan Dam in Karanganyar Regency was constructed by Dutch Government since 1920. This research employed case study approach to evaluate the physical condition of dam for Delingan Dam at 2019. This research aimed to excuse the physical condition of dam, viewed from the component dam body so that component weight is assumed to be 100%. The method employed in this research was descriptive qualitative one. The result of physical assessment of Delingan Dam body obtained score of 91.63%, indicating that the physical condition of Delingan Dam in 2019 is good (condition score o >90% - 100% and damage level < 10%). To maintain physical condition of Delingan dam, it is recommended to operate and to maintain the dam maximally.

1. Introduction

Human beings have developed dam since thousands of years ago to meet the need for water to many interests. Since 50s, dam development initiated by public and private sectors throughout world has grown very rapidly along with population and economic growth. About 45,000 large dams have been constructed for the purposes above. There are about 120 dams in Indonesia, managed by Ministry of Public Work and People Housing, PLN (State Electricity Company) and private in mining sector and in other groups. Viewed from its distribution, most dams are located in Java Island, Bali, Lampung, Sulawesi Selatan (South Sulawesi), Kalimantan Selatan and Timur (South Borneo and East Borneo), Nusa Tenggara Barat (West Nusa Tenggara) and Nusa Tenggara Timur (East Nusa Tenggara). Many dams have been constructed, beneficial to store water, and functioning to control flood. However, the risk of dam failure can also occur, and even the small dam can potentially take property and life toll, when all aspects related to the dam maintenance is considered inadequately. In Dam Operation and Maintenance Activity, there is an activity serving main function, the Assessment of Dam Condition to keep the condition of dam as the main building well-monitored and maintained. In Karanganyar there is a dam constructed by Dutch Government since 1920 and completed in 1923, Tirtormarto Dam or called Delingan Dam. The assessment of dam condition is required to find out the safety level of dam and as the matter of consideration in the dam rehabilitating activity. Considering the problem, an assessment should be conducted on the physical condition of Delingan Dam.

2. Experimental

Some researchers, such as Herlen (2009), explain that the objective of dam management is to prevent the risk of dam damage from occurring [1]. Marzulian Ramli and Bambang E. Yuwono (2015) state "to minimize the failure of dam operation, maximum dam management is required" [2]. Yulius Heryant, et.al. (2014) suggest that in operation and maintenance of dam, an assessment is required as included in the Guideline of Assessment of Dam's Physical Condition referring to Government Regulation Number 37 of 2010 about dam [3].

2.1. Dam operation

The Guideline of Dam Operation and Maintenance issued by Directorate General of Water Resource in 2003 mentions that before the operation of dam, the water capacity of dam and the chart of the correlation between dam water surface elevation, puddle width, and water volume stored have been calculated and known. However, the correlation is affected by water storage rate and sedimentation rate in the bottom of dam. These factors should be considered in the Guidance of Dam Operation and Maintenance. The elements related to dam operation are:

- Dam's storage capacity,
- Design flood and Flood Tracking,
- Estimated Water Flow,
- Charging Schedule and Water Releasing Procedure, and
- Operational instruction.

2.2. Dam maintenance

The Guideline of Dam Operation and Maintenance issued by Directorate General of Water Resource in 2003 mentions that Maintenance is any attempt or measure aimed at ensuring the preserved utilization of water resource and dam function, complementary buildings, and other irrigation infrastructure including its environment corresponding to its planned service life [4]. Two basic activities conducted in maintenance are routine maintenance and repairing work. The Guidance of Dam Operation and Maintenance should confirm that the two activities should be planned well including dam maintenance or repairing system, cost need, personnel availability, and equipment need. The result of maintenance implementation should always be evaluated, recorded, and documented chronologically.

2.3. The assessment of dam condition

Regulation of Dam characteristic issued by Government Regulation Number 37 of 2010 about Dam has issued a Guideline of Dam and its Reservoir's Physical Condition Assessment issued or published by *Balai Bendungan* in 2010 [5]. The assessment of dam and its reservoir's physical condition was conducted on some components: 1) Reservoir including: a) sedimentation, b) Reservoir Border Area, c) Green Belt Area, and d) Reservoir Inflow; 2) Dam body including a) downstream slope, b) dam peak, c) upstream slope, d) supporting field, and e) leakage; 3) Complementary buildings including a) intake, b) outlet, and c) spillway. The weighing of each component and subcomponent in the assessment of dam's physical condition serves as the reference in estimating damage index in each component and subcomponent of dam. Component and scoring in each dam as well as reservoir, according to Standard Guidelines of Dam's Physical Condition Guidelines in 2010, are as follows.

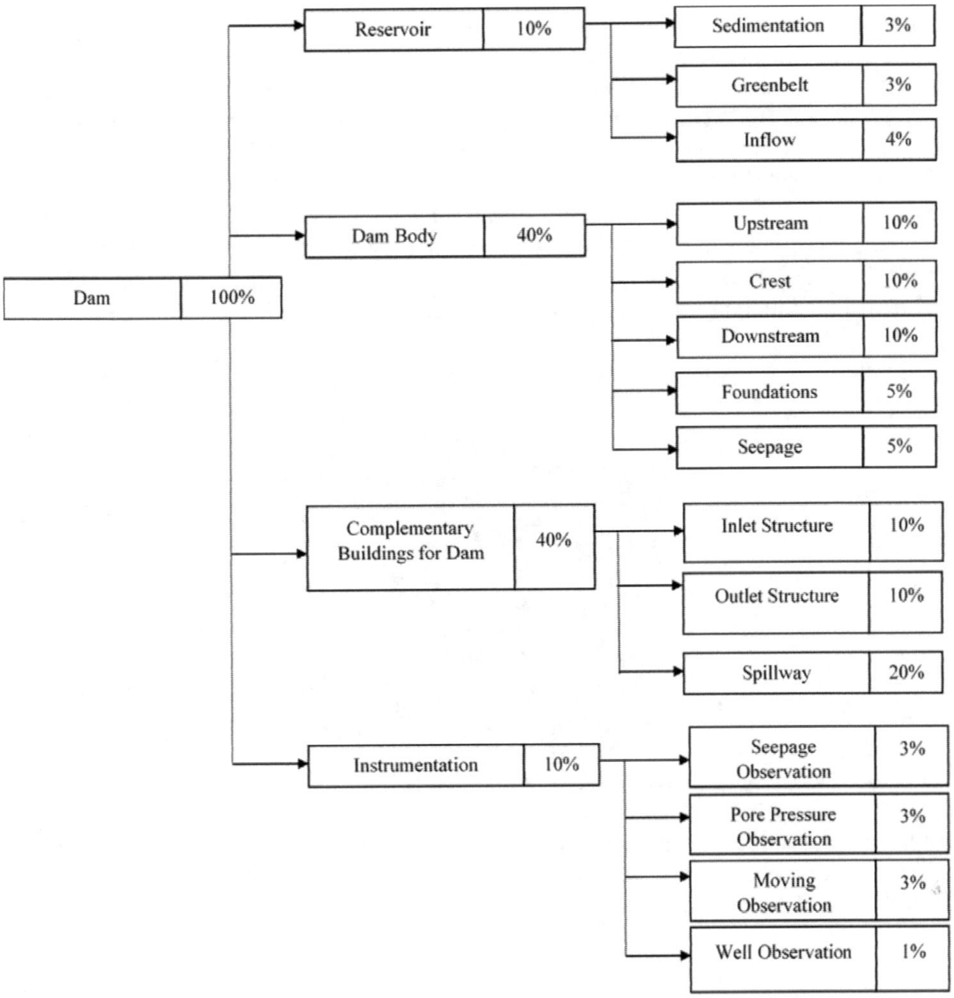

Figure 1. Scheme of Component and Scoring of Dam and Its Reservoir's Physical Condition Assessment.

The classification of condition assessment is determined based on the problem and the damage level of dam and its reservoir. The classification of condition assessment can be seen in table 1.

Table 1. Condition value of reservoir component

No.	Condition	Explanation
1	Good	The condition score is >90% and the damage level is <10% compared with the early condition of building.
2	Mildly damaged	The condition score is 80-89% and the damaged level is 10-20% compared with the early condition of building.
3	Moderately damaged	The condition score is 60-79% and the damaged level is 21-40% compared with the early condition of building.
4	Severely damaged	The condition score is < 60 and the damaged level is >40% compared with the early condition of building.

Source: [1] *Balai Bendungan (Anonymous, 2010)*

IOP Conf. Series: Materials Science and Engineering **858** (2020) 012003 doi:10.1088/1757-899X/858/1/012003

3. Result and Discussion

In Karanganyar there is a dam constructed by Dutch Government since 1920 and completed in 1923, Tirtormarto Dam or called Delingan Dam. Geographically, Delingan Dam is located on coordinates 7° 35'14, 60" SL and 110° 59'22, 60'' EL. Considering the river area system, Delingan Dam belongs to Bengawan Solo River Area, and Bengawan Solo River Flow Area. The management of Delingan Dam is under BBWS Bengawan Solo today. Delingan Dam is an upstream slope homogeneous type-landfill dam with declivity of 1:1.5 with concrete slab layer, and declivity of 1: 2.0 in the downstream slope with thick grass, and fishbone drain, and toe drain on the leg of dam body. This dam is ± 21.00 tall, and ± 885.00 m long, with average peak width of 4,5 m. The assessment of Delingan Dam's physical condition in 2016 on dam body component is as follows.

3.1. The Assessment of upstream slope condition

Figure 2. Upstream Slope Condition in 2019

Considering the result of survey on body dam in a good condition, slope protection consists of riprap (rock), regular or stable declivity condition. There is no extraordinary condition. This upstream slope condition has score of 90 (good).

3.2. The assessment of dam crest condition

Figure 3. Condition of dam crest in 2019

The condition of crest seems to be good, with no surface crack, no deflection, no decline, and with linearity as planned. There is no extraordinary condition. The crest condition has score of 91 (good).

IOP Conf. Series: Materials Science and Engineering **858** (2020) 012003 doi:10.1088/1757-899X/858/1/012003

3.3. The assessment of downstream slope condition

Figure 4. Downstream Slope Condition in 2019

Generally, the condition of dam is still good, the slope protector consists of riprap (rock), the declivity of slope seems to be organized/stable, with no sign of movement, no erosion plot and ruin, weathering beginning to appear, wild plant consisting of short grasses only, no subsidence, no animal hole/nest, and no leakage and wet area. There is no extraordinary condition. The score of downstream slope condition is 94 (good).

3.4. The appearance of right and left support condition

Figure 5. Condition of Right and Left Support in 2019

Considering the result of survey or observation, the dam is generally in good condition with no leakage, crack, wild plant, and sign of movement. The score of right and left support condition is 91 (good).

3.5. Seepage condition

Figure 6. Condition of Seepage and V-notch in 2019

The condition of seepage is good, with clean environment, no perennial plant, the V-notch (seepage measurement tool) still functioning well, and the seepage water is clear. This condition obtains score of 92 (good).
These assessments can be seen in table 2 presenting physical condition and weight scores of dam body component.

Table 2. List of assessment for physical condition of dam with dam body component

No	Name of Building	Weight Score (%)	Performance Score	
			NKF (%)	NKB (%)
(1)	(2)	(3)	(4)	(5) = (3) x (4)
	Dam Body			
1.	Upstream Slope	25.00	90.00	22.50
2.	Crest	25.00	91.00	22.75
3.	Downstream Slope	25.00	94.00	23.50
4.	Left and Right Support	12.50	91.00	11.38
5.	Seepage	12.50	92.00	11.50
	Total Score	100.00		91.63

4. Conclusion

The conclusion of research was physical condition assessment on Delingan Dam body obtains score of 91.63%, indicating that Delingan Dam physical condition is good in 2019 (condition score >90% - 100% and damage level <10%).

Acknowledgment

The author would like to say thank you as much as possible to Ir. Ary Setiawan, M.Sc. (Eng)., Ph.D. as Head of Master Program of Civil Engineering, Sebelas Maret University Surakarta; Prof. Dr, Ir. Sobriyah, M.S., as Supervisor I; Dr. Cahyono Ikhsan, S.T., M.T. and_Dr. Ir. RR. Rintis Hadiani, MT., as Supervisor II; Dr. Ir. Mamok Suprapto, M.Eng., as a Lecturer Examiner; all staffs of the Master Program in Civil Engineering Sebelas Maret University Surakarta for helping many authors during lectures.
Especially for my beloved family for praying continuously, in supporting me both morally and materially to complete this education; fellow students of Master Program in Civil Building Rehabilitation and Maintenance Engineering; and all of those having assisted the author in completing this thesis which the author cannot mention one by one.

References

[1] Herlen, "Pengelolaan Bendungan PLTA PB. SOEDIRMAN," in *Seminar Komite Nasional Indonesia untuk Bendungan Besar (KNI-BB)*, Yogyakarta, 2009.

[2] M. M. Ramli and B. E. Yuwono, "Identifikasi Faktor Dominan Pengelolaan Waduk Yang Berberpengaruh Terhadap Kinerja Waduk," in *Seminar Nasional Teknik Sipil V Tahun 2015 – UMS*, Surakarta, 2015.

[3] Y. Heryan, "Pengembangan Penilaian Kondisi Fisik Bendungan Untuk Penentuan Penanganan Pemeliharaan Bendungan study Kasus: Waduk Lodan," *Jurnal Teknik Sipil Magister Teknik Sipil Universitas Sebelas Maret,* vol. II, no. No. 1, March, 2014.

[4] Anonymous, Keputusan Direktur Jendral Sumber Daya Air No. 199/KPTS/D/2003, Tentang

Pedoman Operasi, Pemeliharaan dan Pengamatan Bendungan, Jakarta: Direktur Jendral Sumber Daya Air, 2003.

[5] Anonymous, Peraturan Pemerintah Nomor 37 Tahun 2010, Tentang Bendungan, Jakarta: Kementerian Pekerjaan Umum, 2010.

IOP Conf. Series: Materials Science and Engineering **858** (2020) 012004 doi:10.1088/1757-899X/858/1/012004

Retention Pond and Pump Station as an Alternative To Flood Management In Bengawan Jero Irrigation Area

Sabdoyono Wiyasa Harimurti [1],
1Student of Master of Civil Engineering Program, Faculty of Engineering, Sebelas Maret University of Surakarta, Indonesia

E-mail: sabdoyono.w@gmail.com

Abstract. Water is the source of life to all living organisms. But in a very large volume, water can be problematic to life. One of problems generated by a large volume of water is flood disaster. River, as a water source, is one of natural resources with multifunction to human beings. Some areas in Indonesia have increasing flood intensity as the rainy season comes. Retention pond and pump station is one solution to dealing with flood puddle occurring in area. One of areas experiencing flood routinely annually is Bengawan Jero Irrigation Area (D.I. Bengawan Jero). The design of retention pond and pump station in DI Bengawan Jero was calculated using flood flow rate with repeated period of 10 years. From the result of rainfall data processing with repeated period of 10 years using Nakayasu method, it could be found $Q_{planned}$ of 7.37 m³/s at hour 1. The storage volume needed to deal with flood in the Retention Pond in Bengawan Jero Irrigation area is 3.909 m³, using 2 pumps with capacity of 2.5 m³/s. From hydrograph analysis, it could be concluded that the use of retention pond and pump station can control the peak flow rate by suppressing or cutting the flood peak to occur.

1. Introduction

Bengawan Jero Irrigation Area is located in Lamongan Regency area and a little in Gresik Regency, East Java Province. Flood occurs in this area in certain period. Considering topographical condition and land use, Bengawan Jero area can be classified into three parts: 1) Bengawan Jero Barat with medium declivity and relatively smooth gravitation flow is appropriate to agricultural purpose. This location lies in the west part of Sukodadi-Karanggeneng Street. 2) Bengawan Jero Dalam with lower declivity is allocated for *tambak* (embankment) and a littlepart for farming in MT 3. Puddle occurs in this area in certain period. This location lies in the east of Sukokadi-Karanggeneng Street through the south of Kuro bordered with low dike from Kuro and K. Glagah through K. Clanggang. 3) Bengawan Jero Luar is located in the east part of Bengawan Jero Dalam, exactly in the east of K. Glagah- K. Clanggang. As aforementioned, flood problem in Bengawan Jero Dalam area needs special attention and thereby is interesting to study, so that the author

IOP Conf. Series: Materials Science and Engineering **858** (2020) 012004 doi:10.1088/1757-899X/858/1/012004

wants to give recommendation to the related authority concerning the flood management scenario in Bengawan Jero Irrigation Area.

2. Experimental

2.1 Hydrological Analysis

Evy Harmani[1] found that the hydrograph analysis shows that the use of retention pool in fact can control the peak flow rate by suppressing or cutting the flood peak to occur. Doddy Yulianto's [2] study also concluded that the result of analysis showed that the combination of retention pool and infiltration well can store runoff volume with flood flow rate repeated period f 10 years. This system infiltrates again the water into aquiver layer all at once. Hydrological analysis is needed to find out the hydrological characteristics of Bengawan Solo River irrigation area. Data of rainfall was processed into the planned rainfall data to be analyzed later into the planned flood flow rate: a) Data of regional rainfall used based on rainfall data recording existing in Bengawan Jero River Flow Area. b) Hydrographic analysis on flood based on annual rainfall, using Q 5 year and Q 10 year, with Nakayassu Synthetic Unit (HSS) Method.

2.2 Hydraulic Analysis

After finding out the flow rate of flood, simulation was conducted using HEC-RAS 4.1.0 2010 software to find out the profile of water surface and river's storage capacity. The result of simulation constituted layout of flood distribution occurring along the channel, so that the location affected with flood in Bengawan Jero River can be found.

3. DISCUSSION AND CONCLUSION

3.1. Hydrological Analysis

3.1.1 Rainfall Analysis

To determine regional rainfall, Thiessen Polygonal method was used between the closest stations in Bengawan Jero River Flow Area (DAS Bengawan Jero). The closest station of DAS Bengawan Jero plotted for this method was Sembung Rain Station. Data of rain refers to Investigative Survey and Rehabilitation Design report for Bengawan Jero Irrigation area in drainage (continuation) in 2016. hydrographic chart used in planned flow rate Q 5yr and Q 10 yr can be developed. Hydrograph chart of Planned Flood can be seen in Figure 1.

IOP Conf. Series: Materials Science and Engineering **858** (2020) 012004 doi:10.1088/1757-899X/858/1/012004

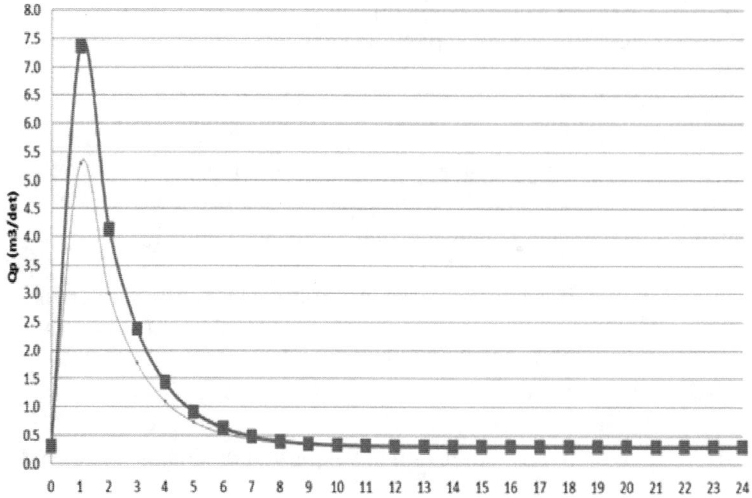

Figure 1. Chart of Planned Flood Hydrograph

Figure 1 shows the peak flow rate of flood at Q 5 years, 5.298 m3/s at hour 1. Meanwhile the peak flow rate at Q 10 yr is 7.37 m3/s at hour 1.

3.2. Hydraulic Analysis using HRC-RAS Modeling

To find out the hydraulic behavior phenomenon in a river/stream flow, simulation or numerical analysis is required to describe both existing and planned conditions of channel. The analysis is conducted using HEC RAS 4.1.0 2010 numerical modeling. HEC RAS is designed to simulate one-dimension steady flow and unsteady flow. The profile of water surface during flood can be found using HEC RAS software. In this research the analysis was conducted using unsteady flow; the analysis involved existing and planned ability of channeling the planned flood flow rate. The procedure of modeling is as follows:

1. Starting HEC RAS
2. Developing new projects including:
   a. On HEC –RAS main window, select menu *File*, then *New Project*
   b. Select Directory and folder wanted or make new folder by clicking Create Folder, writing folder name, click OK (to save all HEC-RAS files).
   c. Then name the Project/title and file name, click OK
   d. Select SI (simulation) unit and continue the steps of input geometric data, input background image of Blawi River, develop river plot, input cross section, defined border condition, and run the modeling program

3.3. Hydraulic Analysis on Existing Condition of Blawi River in DI Bengawan Jero Irrigation Area

The analysis on Capacity of Blawi River plane in Bengawan Jero Irrigation Area is conducted on the existing condition of river aiming to find out the maximum flowing capacity in each of river segments. This analysis was conducted using planned flow rate Q 5yr (Q5) and Q10 yr (Q10). The result of analysis on Bengawan Jero River condition can be seen from the profile of water surface:

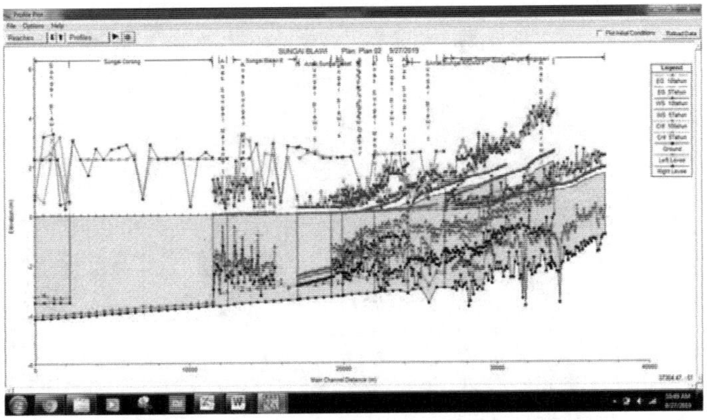

Figure 2. Existing Longitudinal Plane of Blawi River in DI Bengawan Jero with planned flow rate Q 2 year (Ramification of Blawi River – Estuary Downstream)

Figure 3 shows that the elevation of Blawi River in Bengawan Jero Irrigation Area in some stationing develops flood puddle. This research conducts a more specific review on Bengawan Jero Dalam River (in the analysis, it is called Blawi 2). The river shows the raise of water flow rate of 0.4m. This flow rate is planned periodically with Q10 years in the attempt of coping with flood occurring in Blawi River in long term. From the result of simulation on planned flow rate Q10 year, it can be found the difference of flood depth, indicating that the longer (more year) the simulation is done, the deeper and the wider is the flood puddle. The difference can be seen in the Table 2.

Table 1. Puddle Width

NO	PLANNED FLOW RATE	PUDDLE WIDTH (M^2)
1	Q10 Yr	766.11

3.4. A Study on Flood in Bengawan Jero Dalam Irrigation Area

Having found out the existing result of HEC-RAS, a study should be conducted on the flood occurring in Bengawan Jero Irrigation Area because the location often encounter flood almost annually. It can be seen from the Map of Flood-Vulnerable Area as shown in Figure 3.

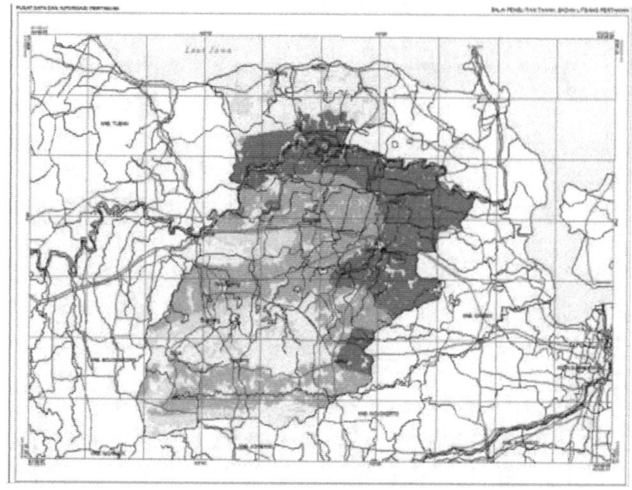

Figure 3. Map of Potentially Vulnerable-Flood Area

Figure 4 shows that the flood occurring in Bengawan Jero Dalam is due to high Flood flow rate and too low land elevation compared with river elevation, with runoff of 0.4 m. So, Bengawan Jero Irrigation Area will encounter flood annually if no attempt is taken such as repairing the channel, retention pool and pump. Therefore, an alternative to flood management is needed in the area. The design of retention pond and pump house can be seen in figures 4, 5, and 6.

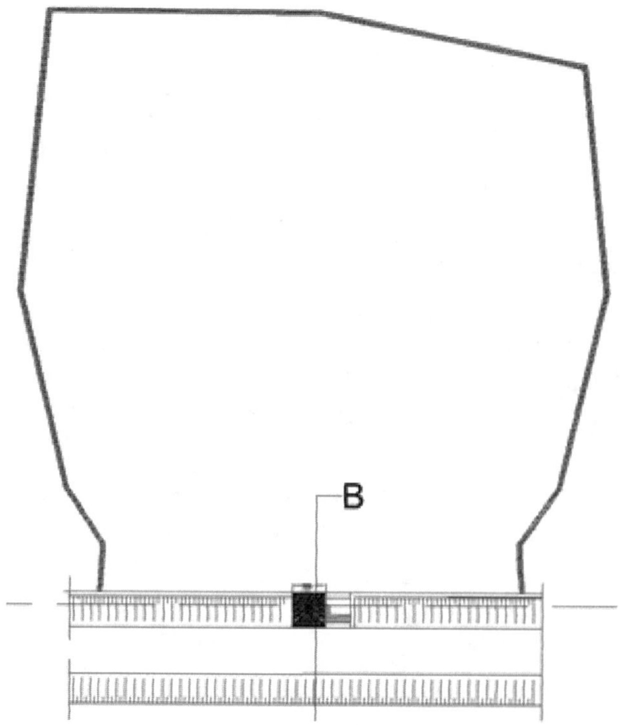

Figure 4. Top View of Retention Pool

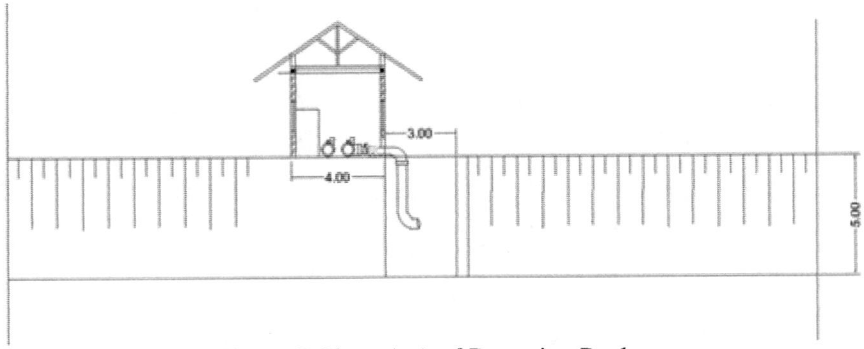

Figure 5. Piece A-A of Retention Pool

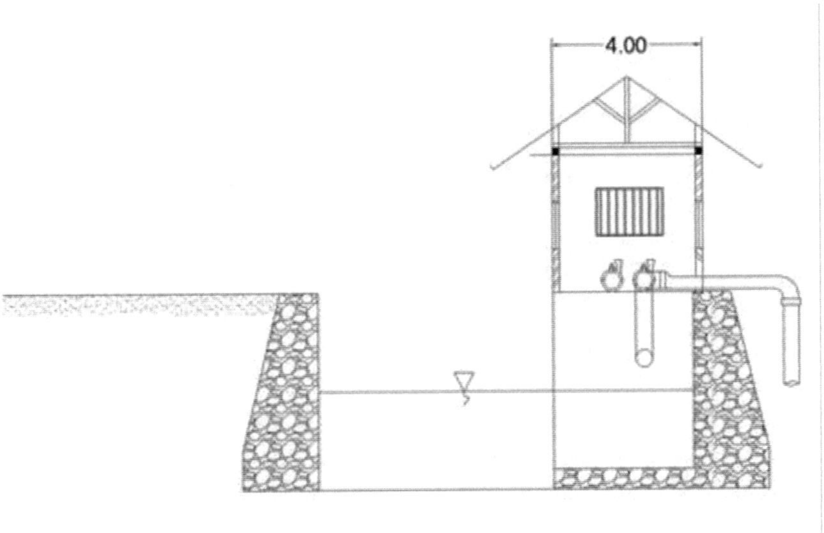

Figure 6. Piece B-B of Retention Pool

4. CONCLUSION

The conclusion of research is Storage volume needed in Retention Pond to cope with flood in Bengawan Jero Irrigation Area is 3,909 m^3, using 2 pumps with capacity of 2.5 m^3/s.

References

[1] Azhar Husain, dkk, 2017, Flood Modelling by usung HEC-RAS. International Journal of Engineering Trends and Technology (IJETT), Volume 50 No 1 Agustus 2017

[2] Darwin Pakpahan, dkk, 2014, Kajian Optimalisasi Sistem Irigasi Rawa (Studi Kasus Daerah Rawa Semangga Kabupaten Merauke Propinsi Papua), Jurnal Media Kominikasi Teknik

Sipil Donny Adzan. M, dkk, 2008, Kritisnya Kondisi Bendungan Di Indonesia, Seminar Komite Nasional Indonesia untuk Bendungan Besar (KNI-BB), Surabaya 2-3 Juli 2008.

[3] Doddy Yulianto, dkk, 2009, *Pemanfaatan Kolam Retensi dan Sumur Resapan Pada Sistem Drainase Kawasan Padat Penduduk. Jurnal Teknik Sipil Volume 5 Nomor 2,* Oktober 2009: 93-169

[4] Ichsan Saputra, 2015, Kajian Hidrologi dan Analisa Kapasitas Tampang Sungai Krueng Langsa Berbasis HeC-HMS dan HEC-RAS, Jurnal Teknik Sipil Unaya Volume 1 Nomor 1, Januari 2015

[5] Nur Azizah Affandy, dkk, 2011, Pemodelan Hujan Debit Menggunakan Model HEC-HMS di DAS Sampean Baru. Jurnal Teknik Sipil Institut Teknologi Sepuluh Nopember, Surabaya

[6] M. Baitullah Al Amin, 2016, Analisis Genangan Banjir di Kawasan Sekitar Kolam Retensi dan Rencana Pengendaliannya (Studi Kasus: Kolam Retensi Siti Khadijah Palembang), Journal Of Regional and City Planning, Vol 27 No.2, PP 69-90, Agustus 2016

[7] Muhammad Dwi Prayogo, dkk, 2015, Perencanaan kolam Retensi dan Stasiun Pompa pada Sistem Drainase Kali Semarang, Jurnal Sipil Universitas Diponegoro

IOP Conf. Series: Materials Science and Engineering **858** (2020) 012005 doi:10.1088/1757-899X/858/1/012005

Using Baking Powder as Additional Concrete Material

FK Budi Nugroho[1], S Sumarni[1], AG Thamrin[1], Roemintoyo [1], FD Isnantyo[1]
[1] Department of Civil Engineering Education, Faculty of Teacher Training and Education, Sebelas Maret University
Sebelas Maret University, Jl. Ir. Sutami 36A Surakarta 57126 Indonesia

E-mail: Fajarkus09@gmail.com
Srisumarni@staff.uns.ac.id
Agthamrin2@yahoo.com
Roemintoyo@staff.uns.ac.id
Isnantyo@staff.uns.ac.id

Abstract. Baking Powder is a useful ingredient for developing which basically used in increasing food volume. It is made from alkaline sodium bicarbonate elements, acid salts and corn starch. Baking powder is a chemical compound consisting of $NaHCO_3$ formula. This compound belongs to the salt group and has been used for a long time. The main added ingredients of retarders are divided into several categories according to their chemical composition, including: 1) Lignosulphonic Acid and other salts, ie., Na, Ca or NH4, 2) Hydro-carboxylic acids and their salts. In this study we used baking powder as an added ingredient. This study aims at determining the setting time, compressive strength and the freshness of concrete with variations in the addition baking powder are 0.00%; 0.15%; 0.30%; 0.45% of the weight of cement. This is an experimental research. Test samples for compressive strength in the form of a concrete cylinder. The results showed that maximum value of compressive strength of concrete is in the addition of 0.45% baking powder with an increase in compressive strength of 6.43% of the nominal concrete.

1. Introduction

Concrete is one of the materials used for construction such as building structures, highways, piers, and various other structures. The development of concrete requires an increase in quality. To obtain a qualified concrete, there are several factors to be considered, among others: material, mixture proportions and work processes. Among the materials that can improve the quality of concrete include: additives from chemicals, materials with high density, silica-containing materials, and others. Among the materials used to improve the quality of concrete have been previously studied. [1-15].

[16] one of the additive ingredients to improve the quality of concrete is water - Reducing Admixtures need ingredients can reduce the need for water mixture to produce a highly consistence concrete, Water - Reducing Admixture is, therefore, used among others to avoid reducing the cement content and slump value from producing concrete with a comparative value or a low cement water factor ratio, all of which will increase the concrete compressive strength.

Material used for reducing water may be in the forms of both organic or inorganic ingredients for non-air-entrained concrete or with air in reducing the mixed water content. In addition, this added material can be used to modify the time of binding of concrete or mortar as a result of changes in the cement water factor. The mixed composition of these added ingredients is generally distinguished into 5 classes: 1. Lignosulfonic acid and the content of salts, 2. Modification and derivatives of lignosulfonic acid and salts. 3. Hydroxylated carboxylic acids and their salt content. 4. Modification of hydroxylated carboxylic acids and their salt content. 5. Other materials such as: Inorganic materials such as zinc, salts, barracks, phosphates, chlorides; Amino acids and their derivatives: Carbonic acid, polysaccharin and acid sugar; Polymer mixtures, such as ethers, melamic, naptan, silicon, hydrocarbon-sulfate derivatives. Baking powder is a chemical compound comprising of NaHCO3 formula, the mention of which is frequently abbreviated as dense. This compound belongs to the salt group and has been used for a long period of time. This compound is also called baking soda, sodium bicarbonate, sodium hydrogen carbonate, and others.

In addition to concrete quality, setting time considerably needs attention. One of the jobs which demands slowing down the initial cement binding is the transportation process. The fresh concrete with a long distance from the batching plant to the concrete casting location while maintaining its workability. The initial cement binding is the time required for the cement paste to change its properties from liquid into solid. To cope with this problem, we need additional substances that function to slow the binding of cement (retarder admixture). Retarder mixtures are chemical aids that function to slow down the initial binding time (time setting), i.e., due to hot weather conditions or prolong of the hardening time intended to avoid cold connections, which enables the mixture work for longer periods. Additives that can be used as inhibitors include sugar, sucrose, sodium gluconate, glucose, citric acid and tartaric acid.[17]

Baking powder is sodium bicarbonate mixed with other substances of slightly acid. By referring to building materials and civil engineering No. 22 / SE / M / 2015 concerning guidelines for the use of chemical additives in concrete, baking powder (NaHCO3) is a chemical containing salt, which is Sodium (Na). Thus, it enables to be used as additional material in concrete, for the Na includes an example of Lignosulphonic Acid.

Potassium bitartate in case reacted with sodium bicarbonate will establish a compound called baking powder. Under controlled conditions, this reaction will produce sodium potassium bitartate salt, which is often referred to as rochelle salt of class A. A. [18]

Among the researches on setting time of concrete, among others: the addition of sodium silicate delayed the hydration of tricalcium aluminate and thus prolonged the setting process of oil-well cement.[19] that polymer latexes perform good water-reduction effect and delayed setting behavior on CSA cement mortar. [20]. The accelerating-retarding effect of triethanolamine on the initial setting time is caused by the different intensity of formed ettringite, which is governed by the triethanolamine dosage. This finding provides information about the role of triethanolamine and ettringite formation in the initial setting. [21]

In this research we use baking powder as an additive to determine the time of adjustment, compressive strength and freshness of concrete.

2. Theoretical Underpinnings

2.1. Cement Setting Time (Vicat Test)

The initial binding time of cement is the time needed to harden, calculated from the reaction of the cement with water and becoming cement paste. So that the cement is stiff enough to hold the load. Setting time occurs when the hardening or binding of the cement fibers. The initial binding of cement according to the standard is a minimum of 45 minutes. The initial binding of this cement must occur / proceed slowly. It is intended that there is a gap between stirring with other construction installations so that the workmanship has no difficulty in working. Final binding of cement according to the standard after 8 hours. After the cement is attached to the construction it should not be affected by interference, because it will damage the construction bond. If it is disturbed, it is better if the construction is torn down. [22]. The initial binding time can be determined when the vicat needle gets penetration as deep as 25 millimeters, and the final binding time, when the vicat needle cannot penetrate. [23]

2.2. Slump Test

Concrete slump is a decrease in height at the center of the upper surface of the concrete measured immediately after the slump test mold is removed. Slump test is an empirical test / method used to determine the consistency / stiffness (easy to do or workable) of a mixture of fresh concrete (fresh concrete). Rigidity in a concrete mixture shows how much water is used. Slump value is proportional to the value of the concrete mixture's moisture content, but inversely proportional to the strength of the concrete. In a concrete mixture / mix, water content is very important because it determines the level of workability. Concrete mixture that is too liquid will cause low concrete quality, and long to dry. While the concrete mixture that is too dry causes the mixture to be uneven and difficult to print. [24]

2.3. Compressive Strength of Concrete

Concrete compressive strength identifies the quality of a structure. The higher the level of structural strength desired, the higher the quality of the concrete needed.

3. Experiments

3.1 Materials

Materials used in this study include:

Portland cement. The cement used is semen gresik with 40 kg packaging. The coarse aggregate used is broken stone with a maximum size of 20 mm taken from PT Pancadarma Puspawira, Solo, Central Java. Has fulfilled the requirements of the Indonesian National Standard [25]. Fine aggregate used is using sand from Kaliworo River, Klaten, Central Java. Has fulfilled the requirements of the Indonesian National Standard [26-27]. The water used in the clean water is in the PTB FKIP UNS laboratory.

The added retarder used in this research is baking powder produced by PT Gunacipta Multirasa, Tangerang, Indonesia.Experimental Procedure

Examination of Material

The material tested was fine agregat (sand) and course agregat (gravel) which had to meet the prescribed requirements of Indonesia standard. which included, among others:

Fine Agregat testing include:

a) Testing of sludge levels.
b) Testing levels of organic substances.
c) Testing specific grafity.
d) Grading testing.
e) Testing the water content.

Course aggregate testing.

a) Testing for abrasion.
b) Grading testing.
c) Testing specific grafity.

3.2. *Mix design concrete*

At the calculation stage the planning is to mix concrete with or design a concrete mixture adding baking powder with additional variations of 0%, 0.15%, 0.30%, 0.45% of the weight of cement. The concrete mix plan calculation was carried out to determine the need for gravel, sand, cement and water. The mixed method used in this study applied a standard Indonesian mixed design method [28].

3.3. *Setting time test*

Testing step of the initial cement setting time used the Vicat apparatuses test. This test is carried out using American standard. [29] The following is Vicat Apperature or Vicat Tool used to know the time of binding of cement.

3.4. ***Workability Test***
Testing is done using Abrams cone and steel rod as crusher, testing is conducted to know the value of slump concrete (workability) based on Indonesian Standard.

3.2.5. *The Compressive Strength Test of Concrete*
The test sample is in the form of a concrete cylinder with a diameter of 15 cm and a height of 30 cm, tested using a Compressing Testing Machine, so that the maximum load value is obtained when the sample is cracked.

The compressive strength was calculated by using equation as follows:

$$\sigma = P/A \quad (1)$$

where σ= compressive strength, P= load, A= area of specimen

4. Results and Discussion
The result of fine agregat material test is shown in Table 1. Based on the results of fine aggregate (natural sand) testing, the values of sludge, organic matter and bulk specific gravity have met the standard requirements.

Table 1. Properties of fine agregat (Sand)

Properties	Values
Water Content	0.53 %
Specific Gravity	2.573
Fineness Modulus	3.063
Zone	1
Absorption	1.091 %

Based on the results of fine aggregate (sand) testing, the values of sludge, organic matter and bulk specific gravity tests have met the standard requirements. [30].

Table 2. Properties of course agregate (Gravel)

Properties	Values
Abration	9.32 %
Specific Gravity	2.55
Fineness Modulus	3.385
Zone	2
Absorption	1.07 %

Based on the results of course rough aggregate testing (gravel), the value of gradation, specific gravity, and abrasion tests have met the standard requirements [31]. The data of the results of initial setting time test is shown in Figure 1.

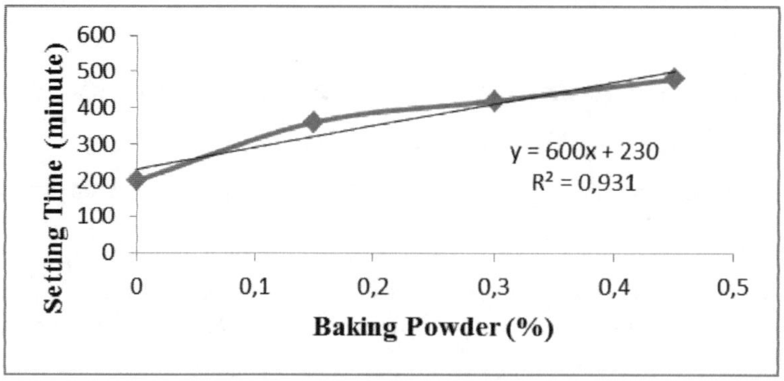

Figure 1. Result of Setting Time

Based on the results of setting time testing, concrete by adding baking powder with a percentage, among others: 0.00%; 0.15%; 0.30%; 0.45% of the weight of the cement affects the bonding time of the concrete and forms a linear equation Y = 600X + 230, with variable X being the variation of baking powder percentage and Y variable is the binding time of the concrete. The regression value is 0.91.

The graph shows that the greater the addition of baking powder, the higher the binding time of the cement, which means that the more baking powder percentage added to the concrete the longer the drying process. This shows that baking powder has chemical contents that inhibit the interaction process between cement and water. The increase in concrete binding time for each variation of baking powder addition compared to normal concrete binding time without the addition of baking powder is shown as in Figure 2.

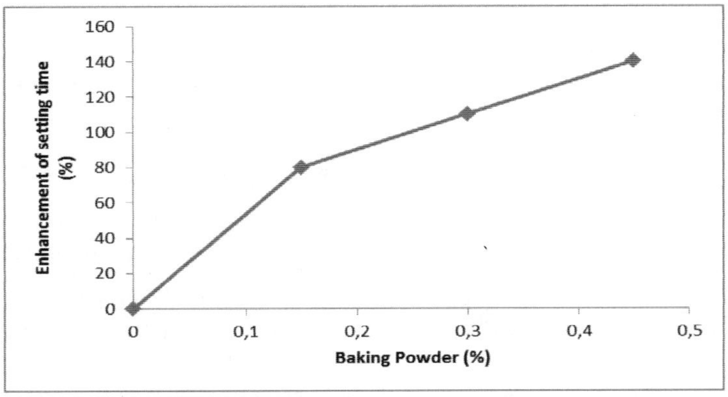

Figure 2. The increased binding time from binding time without baking powder addition.

In figure 2, the binding time in the variation of baking powder 0.45% cement increase 140% of the cement without any additional baking powder.

Concrete slump test data results are shown as in Figure 3.

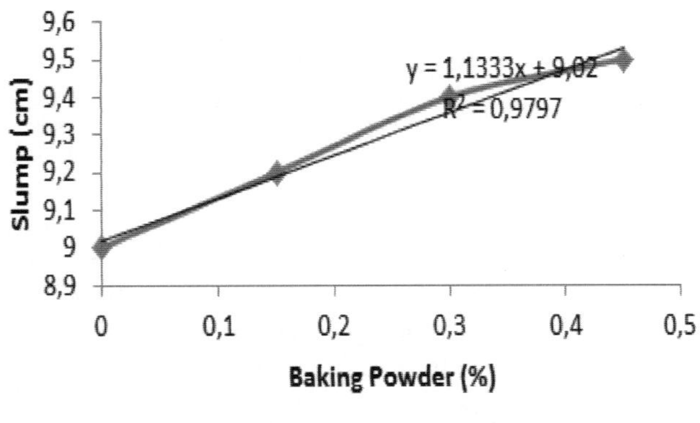

Figure 3. Slump Test

Based on the results of slum testing, concrete by adding baking powder with a percentage include: 0.00%; 0.15%; 0.30%; 0.45% of the weight of the cement affects the nature of the dilution (workability) and forms a linear equation that is Y = 1.1333X + 9.02, with the variable X is the variation in the percentage of baking powder and the variable Y is the workability. The regression value is 0.98.

The graph shows that the greater the addition of baking powder, the higher the workability, which means that the more baking powder percentage is added, the concrete will be thinner. This shows that baking powder when mixed with water and cement becomes runny. The increase in the value of slum for each variation of baking powder added compared to concrete slum without the addition of baking powder is shown as shown in Figure 4.

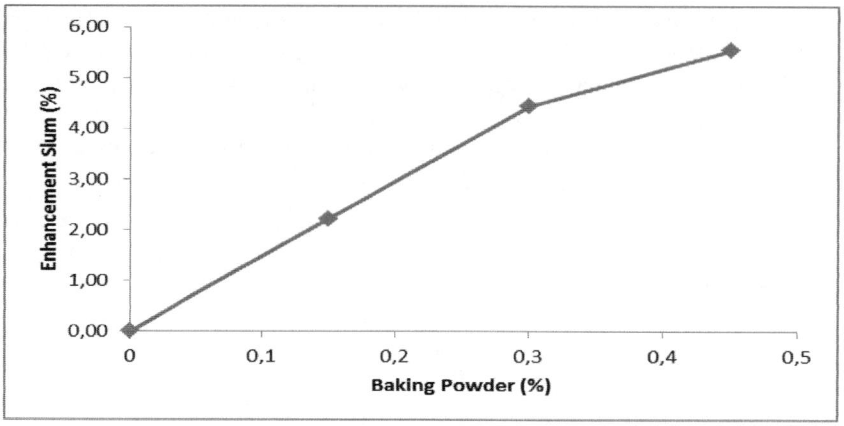

Figure 4. Increasing the slum value of the slum value without the addition of baking powder.

From the results of an increase in the slump value with a variation of the addition of baking powder results in a small increase, this shows that the interaction between baking powder and water has no effect of dilution. Referring to the observations during the slum test, the slum test is carried out shortly after the concrete is finished mixed, and does not wait any longer, it is possible to increase the slum value if the test is carried out by giving a time lag of testing for the concrete constituent reacting especially water, cement and baking powder.

The result of compression strenght test of concrete is shown in Figure 5.

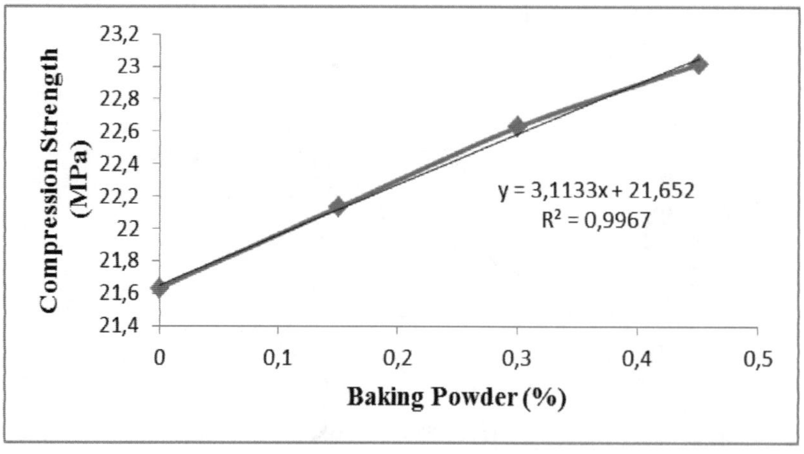

Figure **5.** Variations of baking powder versus compressive strength

Based on the results of the concrete compressive strength test that the concrete by adding baking powder with a percentage include: 0.00%; 0.15%; 0.30%; 0.45% of cement weight affects the compressive strength of concrete and forms a linear equation Y = 3.1133X + 21.652, with variable X is the variation of the baking powder percentage and Y variable is the compressive strength. The regression value is 0.99.

The graph shows that the greater the addition of baking powder the higher the compressive strength, which means that the more baking powder percentage is added, the stronger the concrete.

The increase in compressive strength value of each variation of the addition of baking powder compared to the compressive strength of concrete without the addition of baking powder is shown as shown in Figure 6.

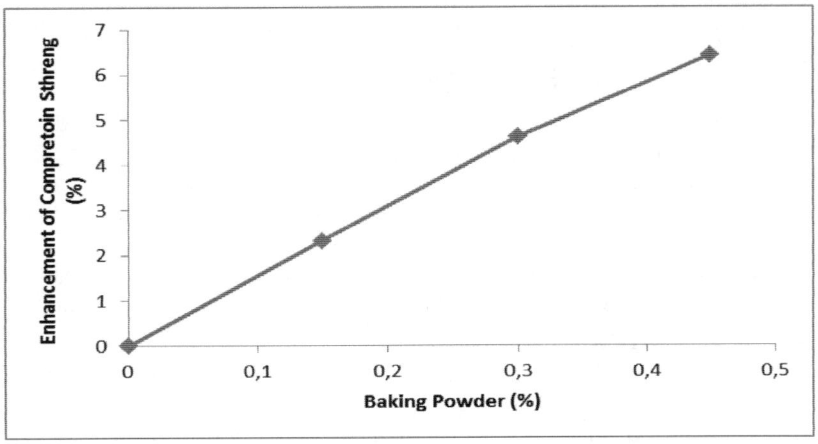

Figure 6. Increasing the compressive strength value of concrete from the compressive strength value of concrete without the addition of baking powder.

The highest concrete compressive strength of 23.03 MPa, occurs in concrete with the addition of 0.45% baking powder, this value indicates that there is an increase in compressive strength of 6.43% of the concrete without the addition of baking powder. This shows that baking powder has properties as a superplasticizer which is an added material to increase the compressive strength of concrete. The highest compressive strength value of 23.03 MPa. Based on the Indonesian national standard, this concrete is still a normal type of concrete.

5. Conclusion

This study concludes that the greater addition of the baking powder used, the longer the hardening and working capacity are found to be slightly increased. The maximum value of concrete compressive strength is the application of 0.45 per cent baking powder with an improvement of 6.43 per cent of gross concrete compressive strength.

References

[1] Xiao, J., Ma, Z., Sui, T., Akbarnezhad, A., & Duan, Z. 2018. Mechanical properties of concrete mixed with recycled powder produced from construction and demolition waste, *Journal of Cleaner Production, 188*, , pp.720-731.

[2] Mazloom, M., Ramezanianpour, A. A., & Brooks, J. J. 2004. Effect of silica fume on mechanical properties of high-strength concrete, *Cement and Concrete Composites, 26,* (4), pp. 347-357.

[3] Chen, S. D., Hwang, C. H., & Hsu, K. C. 1999. The effects of sulphonated phenolic resins on the properties of concrete. *Cement and concrete research, 29,* (2), pp.255-259.

[4] Ramachandran, V. S., & Malhotra, V. M. 1996. Superplasticizers. In *Concrete Admixtures Handbook William,* Andrew Publishing, pp. 410-517.

[5] Sitorus, L. R. 2018. Analisis Kuat Tekan Terhadap Umur Beton dengan Menggunakan Admixture *Superplasticizer Viscocrete-3115* N.

[6] Janowska-Renkas, E. 2015. The influence of the chemical structure of polycarboxylic superplasticizers on their effectiveness in cement pastes. *Procedia Engineering, 108*, pp. 575-583.

[7] Irawan, I. 2017. Pengaruh Silica Fume Terhadap Beton Mutu Tinggi Self Compacting Concrete, Doctoral dissertation, Universitas Pendidikan Indonesia.
[8] Chen, H.J. and Wu, C.H. 2018. Influence of Aggregate Gradation on the Engineering Properties of Lightweight Aggregate Concrete, *Appl. Sci.* 2018, 8, 1324; doi:10.3390/app8081324.
[9] Siˇcáková, A. & Špak, M. 2019. The Efect of a High Amount of Micro-Fillers on the Long-Term Properties of Concrete, *Materials* 2019, 12, 3421; doi:10.3390/ma12203421.
[10] Zahiri, F., Eskandari-Naddaf, H. Optimizing the compressive strength of concrete containing micro-silica, nano-silica, and polypropylene fibers using extreme vertices mixture design. Front. Struct. Civ. Eng. 13, 821–830 (2019). https://doi.org/10.1007/s11709-019-0518-
[11] Wang Y, Hu S, He, Z.2019. Mechanical and Fracture Properties of Fly Ash Geopolymer Concrete Addictive with Calcium Aluminate Cement. Materials (Basel, Switzerland), 15 Sep 2019, 12(18), DOI: 10.3390/ma12182982
[12] Rajanna, S.K. Vinjamur, M. & Mukhopadhyay, M. 2017, Robust Silica Aerogel Microspheres from Rice Husk Ash to Enhance the Dissolution Rate of Poorly Water-Soluble Drugs, *Chemical Engineering Communications, 204* (2): pp. 249-253, DOI: 10.1080/00986445.2016.1263618.
[13] Rasoni, Y., & Yurisman, Y. 2014. Penelitian Pembuatan Beton Mutu Tinggi Dengan Semen PCC Menggunakan Sikafume dan Viscocrete-10 Sebagai Bahan Tambah. Abstract of Undergraduate Research, Faculty of Civil and Planning Engineering, Bung Hatta University, 1(2).
[14] Abolpoura, B., Mehdi Afsahia, M., Hosseini, S.G. 2015. Statistical analysis of the effective factors on the 28 days compressive strength and setting time of the concrete, *Journal of Advanced Research,* 6, (5), September 2015, pp. 699-709, DOI: https://doi.org/10.1016/j.jare.2014.03.005
[15] Alqadia, A.N.S,. Bin Mustapha, K.N., Naganathan,S. & Al-Kadic, Q.N.S.(2013). Development of self-compacting concrete using contrast constant factorial design, *Journal of King Saud University - Engineering Sciences, 25*, (2), July 2013, pp.105-112, https://doi.org/10.1016/j.jksues.2012.06.002
[16] American standart ASTM. 1982. Standard Specification for Chemical Admixture for Concrete Type F. American Society for Testing Materials, ASTM C 494-82, Philadelphia.
[17] Otunyo,A. W., Onwusiri, S. C. & Nwaiwu, N. (2015), Effect of Sugar Cane Juice on Slump Values, Setting Times and Strength Of Concrete, *Nigerian Journal of Technology (NIJOTECH), 34*, (2), April 2015, pp. 254 – 258, DOI: http://dx.doi.org/10.4314/njt.v34i2.6
[18] Ar, Indah. 2009. Persenyawaan dari Golongan I A, Natrium Bikarbonat dan Kalium Bitartart sebagai bahan baku pembuatan Garam Rochelle, Artikel dalam https://indbongolz.files.wordpress.com/2010/06/garam-rochelle.pdf.
[19] Guo S, Zhang Y, Wang K, Bu Y, Wang C, Ma C, Liu H. 2019. Delaying the hydration of Portland cement by sodium silicate: Setting time and retarding mechanism. Construction and Building Materials, 205:543-8. https://doi.org/10.1016/j.conbuildmat.2017.08.168
[20] Li L, Wang R, Lu Q. 2018. Influence of polymer latex on the setting time, mechanical properties and durability of calcium sulfoaluminate cement mortar, Construction and Building Materials, 169:911-22.
https://doi.org/10.1016/j.conbuildmat.2018.03.005
[21] Yaphary YL, Yu Z, Lam RH, Lau D. 2017. Effect of triethanolamine on cement hydration toward initial setting time, Construction and Building Materials, 141:94-103.
https://doi.org/10.1016/j.conbuildmat.2017.02.072
[22] Indonesian product standardization SNI 03-6827-2002
[23] American standart ASTM C191-92
[24] Indonesian product standardization SNI 03-1972-2008
[25] Indonesian product standardization SNI S-04-1989-F
[26] Indonesian product standardization SNI 1969-2008
[27] Indonesian product standardization SNI 2417-2008
[28] Indonesian product standardization SNI 03-3449-2002

[29] Indonesian product standardization SNI 1969-2008
[30] Mulyono, Tri. 2004. *Concrete Technology*. Yogyakarta: Publisher Andi.
[31] American standart ASTM C187

Utility of local quartz powder and silica fume to produce high early strength of self compacting high strength concrete.

A Lisantono[1] and J B Susanto[2]
[1,2] International Civil Engineering Program, Department of Civil Engineering, Universitas Atma Jaya Yogyakarta, Yogyakarta, Indonesia, 55281.

E-mail: ade.lisantono@uajy.ac.id; josuabernard@gmail.com.

Abstract. Generally concrete needs 7 to 14 days to reach minimum adequate compressive strength. Sometimes it is needed to get high early compressive strength of concrete for some reason. This research conduct an experimental program to make high early strength of self compacting concrete which is using local quartz powder and silica fume. Twenty one cylinder specimens were made and tested to get the compressive strength at age of 1 day, 3 days, 7 days, 14 days, 21 days, and 28 days. While the modulus elasticity of concrete were tested only at the age of 28 days. The cylinder specimen was using standard size of (150 × 300) mm^2. The experimental program shows that the average compressive strength of concrete in 1 day, 3 days, 7 days, 14 days, 21 days, and 28 days were 10.00 MPa, 27.16 MPa, 36.41 MPa, 40.93 MPa, 50.36 MPa, and 48.66 MPa, respectively. While the average modulus elasticity of the concrete at the age of 28 days was 18,703.45 MPa.

1. Introduction

Concrete becomes famous materials for building construction in the past decades up to now. If the cast in situ concrete is using for material of building construction, the age of the concrete to reach the maximum compressive strength is very important. Generally, the concrete needs 28 days to reach the maximum compressive strength. Normally, concrete needs 7 to 14 days to achieve minimum compressive strength of 0.7 to 0.85 fc' as a minimum strength. If the minimum strength can be achieve, the scaffolding and formwork can be removed and the work can be proceed to the next step. It can be said that the project of building construction needs high early strength of concrete to make a faster project construction.

Experimental study to produce high early strength of concrete were conducted by Soni et al. [1]. This study is using plasticizer to achieve high early strength and the results showed that the concrete could get 40% high early strenth using the plasticizer. Yasin et al. [2] used local materials of calcite stone from Tuban East-Java and sand from Bengawan Solo River, Bojonegoro East-Java to produce high early strength of concrete. The results showed that the high early strength of concrete can be achieved using the local materials. However, both studies were not intended for high early strength of self compacting concrete.

In several decades ago up to now, self compacting compacting concrete was developed to solve the problem for complex form work of concrete structure such as thin element of structure or curved element of structure, especially in the element of structures which has congested reinforcement [3, 4]. So that, to get a high early strength of self compacting concrete using local quartz powder and silica fume, this research was conducted.

IOP Conf. Series: Materials Science and Engineering **858** (2020) 012006 doi:10.1088/1757-899X/858/1/012006

2. Experimental

2.1 Materials

Portland Pozzolana Cement (PPC) used in this project. Portland Pozzolana Cement is a hydraulic cement consisting of a homogeneous mixture between portland cement with fine pozzolan, which is produced by grinding portland cement and clinker pozzolan together, or mix evenly with portland cement powder with pozzolan powder, or a combination of grinding and mixing, where the levels of pozolan are 6% to 40% of the mass of Portland cement. Crushed stone with size of 0.5 – 1 cm used as coarse aggregates. The crushed stone is taken from local material in Yogyakarta Province. Quartz sand is taken from Bangka Belitung. In addition, nano quartz which has the finer mesh was also used as filler. The quartz sand with size 0.3– 0.8 mm used as fine aggregates and additional quartz powder which has 200 meshes is also used as additional filler in the concrete. Silica Fume used in this project to encourage the pozzolana activity to begin earlier and continuing the hydration process of concrete so the concrete keep gaining its strength as conducted by Laura et. al. [5]. HRWR (High Range Water Reducer) is used in this research to improve the workability of the concrete in low water to cement ratio.

2.2 Mix Design and Specimens

Several trial and errors of mix design were conducted to carry out the optimum mix design. The mixing was done by several steps. Before getting the optimum design, the initial mix design followed the mix design proposed by Ma and Dietz [6], where the concrete uses the full quartz sand and quartz powder. However, the workability of concrete was not good due to the low water cement ratio. Therefore, the mix design was changed and increased the water to cement ratio by 0.01 and balanced with proper the use of SPC until it forms a homogenous mixture. After the last trial, the optimum mix design of self compacting high strength concrete was gotten.

All the dry materials were mixed for around 2 minutes, and then added by water and HRWR slowly by periods until it forms a homogenous mixture of concrete. After that, the setting time and behavior of the mixtures were studied within one day. The cylinder specimens were cured into the water to keep the moistures of the specimens. The mix design of the self compacting high strength concrete is shown in Table 1.

Table 1. The mix design of high strength concrete per m^3.

No	Properties	Amount (kg)
1	Cement	510
2	Silica Fume	76.5
3	Water (WCR 0.35)	178.5
4	Total Aggregates	1443.3
5	Fine Aggregates (Quartz Sand)	485.52
6	Quartz Powder	208.08
7	Coarse Aggregates	749.7

Twenty one cylinder specimens with the size of (150×300) mm^2 were made and tested in this study. The testing's of concrete were conducted at 1 day, 3 days, 7 days, 14 days, 21 days, and 28 days. In addition, the specimen at age of 28 days was also tested to investigate the modulus of elasticity of the concrete.

IOP Conf. Series: Materials Science and Engineering **858** (2020) 012006 doi:10.1088/1757-899X/858/1/012006

2.3 Testing Specimen

The requirement test for self compacting concrete (SCC) were slump flow test and L-Shape box test. These testing were conducted in this study to carry out the filling ability and passing ability of the fresh concrete mixture EFNARC [7]. The slump flow test was depicted in Figure 1.

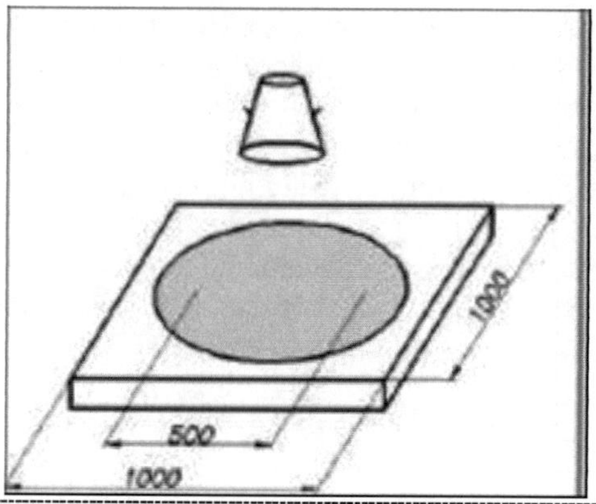

Figure 1. Slump flow test

The L-shape box testing is using the standard L shape box size as shown in Figure 2.

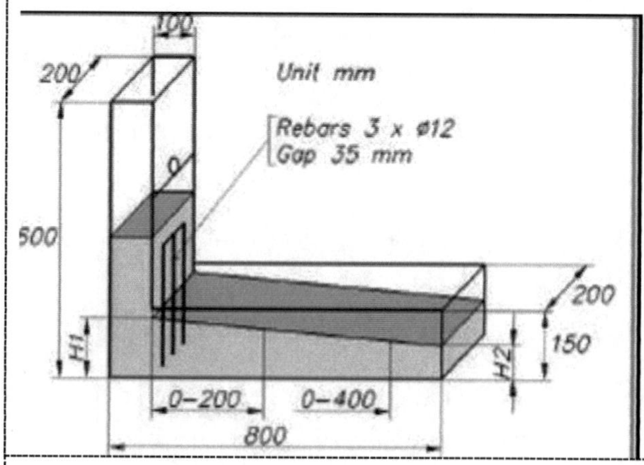

Figure 2. L-Shape box test

Universal Testing Machine (UTM) with the capacity of 30,000 kgf as shown in the figure 3 was used to conduct the compressive strength and modulus of elasticity testing of concrete.

IOP Conf. Series: Materials Science and Engineering **858** (2020) 012006 doi:10.1088/1757-899X/858/1/012006

Figure 3. Universal Testing Machine

3. Results and Discussions

3.1 Slump flow and L-shape box test

Slump flow test result of the concrete can be seen in Table 2. The slump flow results of the self compacting concrete were in the range of EFNARC requirement. The requirement for the slump flow is 600 to 800 mm, the result of slump flow of this research were in the range of EFNARC requirement.

Table 2. Slump flow test result

	Mixing for testing of 1 Day	**Mixing for testing of 3 Day**	**Mixing for testing of 7 Day**	**Mixing for testing of 14 Day**	**Mixing for testing of 21 Day**	**Mixing for testing of 28 Day**	**EFNARC**
Flow 1	680 mm	670 mm	680 mm	690 mm	700 mm	680 mm	
Flow 2	680 mm	680 mm	680 mm	680 mm	750 mm	680 mm	600 up to 800 mm
Flow 3	690 mm	650 mm	650 mm	700 mm	720 mm	650 mm	
Flow 4	700 mm	680 mm	630 mm	700 mm	720 mm	630 mm	
Average	688 mm	670 mm	660 mm	693 mm	723 mm	660 mm	
T50	4 sec.	4 sec.	5 sec.	3 sec.	3 sec.	5 sec.	2-5 sec

Another testing of fresh concrete in this study was L-shape box test. The result of L-shape box test was shown in Table 3. It can be seen that the result of L-shape box test of the concrete follows the requirement of EFNARC. The requirement of L-shape box test for self compacting concrete according to EFNARC was in the range of 0.8 to 1.0. The result of L-shape box test of this research was in the range of 0.8 to 1.0. So, it can be said that the concrete of this study can be classified as self compacting concrete.

Table 3. L-shape box test result

	Mixing for testing of 1 Day	**Mixing for testing of 3 Day**	**Mixing for testing of 7 Day**	**Mixing for testing of 14 Day**	**Mixing for testing of 21 Day**	**Mixing for testing of 28 Day**	**EFNARC**
Blocking Ratio	0.83	0.86	0.842	0.87	0.82	0.86	0.8<H<1

The testing of the concrete was conducted at the age of concrete of 1 day, 3 days, and 7 days to investigate the early strength of the concrete and also for the concrete at the 14 days, 21 days, and 28 days to investigate the development strength of the concrete. The compressive strength result of the concrete can be seen in Table 4, while the development of the compressive strength can be seen in Figure 4. It can be seen from Table 4 and Figure 4 that the average compressive strength at the early strength at 1 day; 3 days; and 7 days were 10.00 MPa; 27.16 MPa; and 36.41 MPa, respectively. The compressive streength at 1 day was not so satisfied due to the quality of quartz sand which was taken from local material. The quartz sand from local material was not as good as the super fine grains size.

While the compressive strength at 14 days; 21 days; and 28 days were 40.93 MPa; 50.36 MPa; and 48.66 MPa, respectively. The concrete which has a compressive strength more than 41.4 MPa can be categorized as high strength concrete [8]. So, it can be said that the self-compacting concrete in this research can be classified as high strength concrete, because the compressive strength of the concrete at 21 days and 28 days had the value more than 41.4 MPa.

Table 4. The compressive strength test result

Age	Specimen	Load (kN)	Strength (MPa)	Average (MPa)
1 Day	1	120	6.79	10.00
	2	200	11.32	
	3	210	11.88	
3 Days	1	450	25.46	27.16
	2	510	28.86	
	3	480	27.16	
7 Days	1	650	36.78	36.41
	2	590	33.39	
	3	690	39.05	
14 Days	1	680	38.48	40.93
	2	700	39.61	
	3	790	44.7	
21 Days	1	870	49.23	50.36
	2	920	52.06	
	3	880	49.8	
28 Days	1	790	44.71	48.66
	2	880	49.8	
	3	910	51.49	

IOP Conf. Series: Materials Science and Engineering **858** (2020) 012006 doi:10.1088/1757-899X/858/1/012006

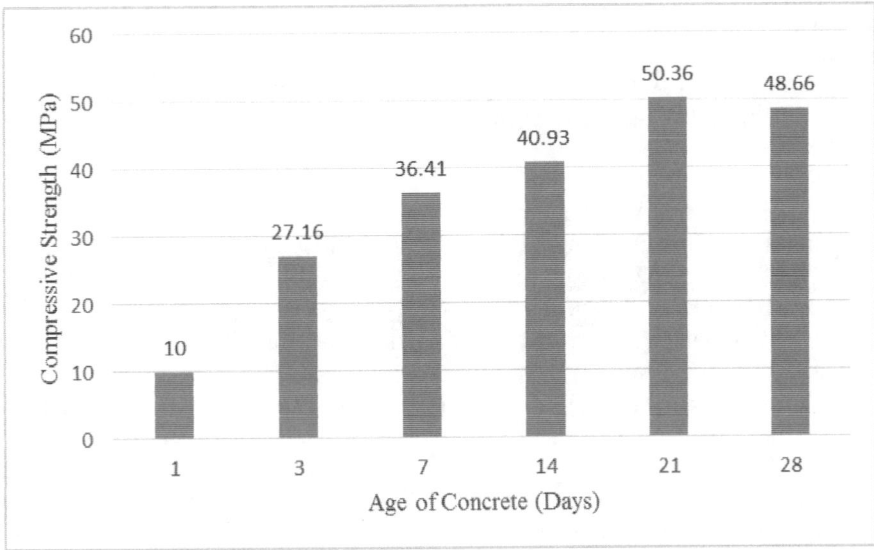

Figure 4. The development strength of concrete

The modulus of elasticity of the concrete was conducted when the concrete has reached 28 days. The modulus of elasticity of the concrete was depicted in Table 5. It can be seen from Table 5 that the average of modulus of elasticity was 18,703.45 MPa. The modulus of elasticity of self-compacting concrete in this research has the value less than the normal concrete which usually has the modulus elasticity more than 20,000 MPa. This indicated that the self-compacting concrete in this study was not as stiff as the normal concrete.

Table 5. The modulus of elasticity.

Cylinder	**Modulus of elasticity (MPa)**	**Average (MPa)**
1	17,512.87	
2	19,525.82	18,703.45
3	19,071.68	

4. Conclusion

Based on the result of experiments program, several conclusions can be drawn as follow:

1. The fresh properties testing of self compacting high early strength of concrete fulfils the requirements of Self Compacting Concrete. It is indicated by the average of slump flow and L-Shape Box testing in 1 day; 3 days; 7 days; 14 days; 21 days; and 28 days.
2. The early strength of self compacting high strength concrete in one day, three days, and seven days were 10 MPa; 27.16 MPa; and 36.41 MPa, respectively. This indicated that the early high strength of self compacting concrete which is more than 20 MPa can be achieved at three days.
3. The compressive strength of self compacting concrete in 14 days; 21 days; and 28 days were 40.93MPa; 50.36 MPa; and 48.66 MPa, respectively. The compressive strength of the self-compacting concrete at 21 days and 28 days had the value more than 41.4 MPa. It means that the self compacting concrete in this research can be classified as high strength concrete.

Acknowledgments
The Authors are wishing to acknowledge to the head and staff of the Laboratory of Structures and Materials, Department of Civil Engineering, Faculty of Engineering, Universitas Atma jaya Yogyakarta for the facilities, so the research can be conducted.

References
[1] Soni, V.A., Chauhan, S.A., Tandei, U.U., and Vatsal, P. 2015 An Experimental Study to Produce High Early Strength Concrete *International Journal for Innovative Research in Science & Technology* 2 05, **pp 167-171**.
[2] Yasin, A. K., Bayuaji, R., Susanto, T. E. 2017 *A Review in High Early Strength Concrete and Local Materials Potential* Department of Civil Infrastructure Engineering, Institut Teknologi Sepuluh November Surabaya.
[3] Okamura, H. and Ouchi M., 2003, Self-Compacting Concrete, *Journal of Advanced Concrete Technology* **1** (1) 5-15.
[4] Ozawa, K., Maekawa, K., Kunishima, M., and Okamura, H., 1989, Development of performance concrete based on the durability design of concrete structures. *Proceeding of the Second East Asia and Pasific Conference on Structural Engineering and Construction* (EASEC-2), 1, **pp. 445-450.**
[5] Laura D., Diana B., Ina P., Laura V. 2017 *Procedia Engineering* 172, **pp 202 – 210.**
[6] Ma, J. and Dietz, J 2017 *Ultra High Performance Self Compacting Concrete* Universitat Leipzig, German.
[7] EFNARC 2002 *Specification and Guideline for Self Compacting Concrete* EFNARC Association.
[8] SNI 03-6468, 2000, *Guidance of High Mixed Design with Portland-Cement and Fly Ash* (in Indonesian), Badan Standardisasi Nasional.

Effect of Water Content on the Characteristics of Bagasse Ash-Calcium Carbide Residue Stabilized Organic Soil

J Hatmoko[1] and L Handoko[2]
[1,2] Department of Civil Engineering Universitas Atma Jaya Yogyakarta, Jl. Babarsari No. 44 Yogyakarta, Indonesia

E-mail : john.trihatmoko@uajy.ac.id

Abstract. Research about chemically stabilized organic soil to improve shear strength has been done by many researchers. However, they have never undertaken the research about the the effect of water content on physical and chemical characteristics of stabilized organic soil. A set of experimental program performed in this research were chemical composition of organic soil (OS), calcium carbide residue (CCR), and bagasse ash (BA); also the physical characteristics of OS. The admixture (60% CCR + 40% BA) was added to the soil with proportion 5, 10, 15, 20, 25 and 30% with various water content , and the samples were cured in 7, 14, 21, and 36 days- curing time. Then, the experiment of physical properties of the stabilized soil was undertaken. The physical characteristics of the stabilized soil change with respect to the water content, quantity of admixture and the curing times. On the high water content, the changes of the physical properties are more significant compared to that on the lower water content. The unit weight, specific gravity, and the degree of acidity increase whereas void ratio and organic content decrease according to the proportion of the admixtures and curing time.

1. Introduction

Organic soil is found all over the world including in Indonesia. The soil is categorized as soft soil having a low shear strength, high compressibility, low degree of accidity, low specific gravity, and high void ratio. Therefore, if it is used as the construction materials there will be some problems such as high settlement, creep and foundation failure. Practically, the soil is cut and replaced by the soil having better physical and mechanical properties such as sand, gavel or crushed stone. However, the problem is going to happen if the substitution materials is far away form the construction site. Then, soil improvement program is better alternative. Some research effort to improve engineering properties of organic soil with ordinary portland cement, lime, fly ash, calcium carbide have been sucesfully done [1-4]. Chemical stabilization by admixtures to improve engineering properties soft clay has been also perfectly done by many researchers [5-8], because clay is rich of mineral and silica (SiO_2) content that generate pozzolanic reaction. Even though organic soil has a lack of mineral, silica, and high water content the pozzolanic reaction still occur. However, formation of gel obtained from pozzlanic reaction fill the pore and cover the organic fiber [9]. It improves physical and mechanical properties of stabilized organic soil but the improvement is not as high as on clay stabilized soil. The chemical agent usually used to stabilize organic soil are cement, lime, fly ash, rice husk ash, bagasse ash or combination of those admixtures. Some research about using fly ash as an admixture for soft clay has been perfectly performed [10-14]. However, less research was done on organic stabilized soil. To stabilize organic soil is more difficult compared to that of soft clay [15,16], because organic soil does not have enough mineral, and has a low degree of accidity. Therefore, because of the existence of fly ash is very limited , the use of the other admixture having similar chemical properties to fly ash to improve engineering properties of organic soil is interesting.

Bagasse ash (BA) is a waste residue of sugar factory. It contains high proportion of silica (SiO_2), alumina (Al_2O_3), ferrit (Fe_2O_3), however, it has low proportion of lime (CaO). Therefore other material containing high CaO such as calcium carbide is required to generate pozzolanic reaction. Calcium carbide residue (CCR) contains high calcium oxide (CaO). Reaction between calcium in the CCR and silica and alumina in BA derive calcium silicate hydrate (C-S-H) and aluminum silicate hydrate (C-A-H). And during long period of curing, there is formation of calcium-silicate-aluminate hydrate (C-S-A-H). Those hydrates are in the form of gel that improve physical and mechanical characteristics of stabilized soil. Stabilization of an organic soil with lime, cement, bagasse ash, fly ash, rice husk ash, fly ash + cement, bagasse ash + lime has been [17-19]. However, the research about bagasse ash-calcium carbide residue to alter the physical and chemical properties of organic soil suc as water content (w), specific gravity (G), unit weight (γ), organic content (OC) and the degree of accidity (pH) was never performed. Therefore, the aim of this research are to study the changes of physical and chemical properties of organic soil stabilized by bagasse ash and calcium carbide residue.

Both organic and an organic soil are formed directly by physical or chemical weathering of rock. Structure of organic soil is derived from the mixing of organic matter and mineral, and it is generated from compaction of degradation of organic materials. Organic soil is usually lack of mineral and silica then a pozzolanic reaction due to addition of admixture on organic soil is not as fast as on clay. The pozzolanic reaction on organic soil delays due to organic content and a low degree of acidity. Pozzolanic reaction mechanism on organic soil has not known yet, however the mechanism has been proposed by [9, 16]. The pozzolanic reaction does not occur between admixture with the mineral in the soil but it happens between the materials in the admixtures. For example, reaction between SiO_2 in the fly ash and CaO in the CCR. The proposed pozzolanic reaction mechanism in organic soil are as follow: organic materials change composition and structure of gel calcium-silicate-hydrate, cementous materials forming soil particle chain, and the new hydration that will be formed. Organic matters contains humic acid that delay pozzolanic reaction. In the water, the weight of organic materials is more than 10 times of its dry unit weight delaying the hydration process. Organic materials construct complicated surfaces with alumina (Al_2O_3), silica (SiO_2) and ferrit (Fe_2O_3), then the complicated surfaces delay the hydration process. In this research, CCR and bagasse ash is used as admixture agents [9].

Calcium carbide residue (CCR) is a waste material derived from production of acetyline gas. Calcium carbide (CaC_2) is generated from chemical reaction between cacium carbide and charcoal . CaC_2 react with water (H_2O to generate the acetyline gas (C_2H_2). The compelete reaction can be reviewed on [20-23]. The use of CCR and rice husk ash improves significantly the unconfined compression strength of clay [7,8]. Bagasse ash (BA) is the residue generated from production of sugar. Original bagasse ash has low composition of SiO_2 + Al_2O_3 + Fe_2O_3, however, when the BA is burned to 300 °C, it contains of SiO_2 increases to 62.75%, and 85.21% for (SiO_2) + (Al_2O_3) + (Fe_2O_3) [25].

2. Experimental

2.1. Materials

2.1.1. Organic Soil

Organic soil was taken from Ketapang Regency, The Province of West Kalimantan, Indonesia. The soil is fibrous peat with 95% organic content, and it is considered as soil with high organic content. The water content is 498% due to the properties of fiber that absorb water. The Laboratory tests of physical properties of the soil use the standard ASTM-1984 (Peat Testing Manual), and the result is presented on Table 1. The soil sample is still on the range of the previous research [25-26]. Based on ASTM D-4427-92, the soil is classified as moderate fibrous peat with high degree of acidity.

2.1.2. Admixture

The admixture agents used in this research is the mixed of calcium carbide residue (CCR) containing high calcium (CaO), and bagasse ash (BA) that contains high pozzolanic compound SiO_2 + Al_2O_3 +

Fe_2O_3 = 73.19% (Table 2). The high content of CaO on CCR, pozzolanic materials on BA, and the present of water, the pozzolanic reaction occurs showed by the form of the gels calcium silicate hydrate, aluminum silicate hydrate or calcium-aluminum-silicate hydrate. The formed gels covers the fiber on the peat, then there will be improvement of the properties of organic soil.

Table 1. Physical properties of Organic Soil

Soil Parameters	Sample	Previous Research [25,26]
Spec. Gravity(G)	1.47	1.4 – 1.7
Void Ratio (e)	7.1	6.89 – 11.09
The Unit weight (γ),(t/m^3)	1.08	0.9 – 1.25
Degree of acidity (pH)	4.6	3 – 7
Water content (w), %	498	450 – 1500
Organic content (OC),%	95	62.5 – 98
Fiber content (Fc), %	53.1	39.5 – 61.3

Tabel 2. Chemical content of CCR and BA

Chemical content	CCR	BA
SiO_2	8.43	**45.28**
Al_2O_3	1.67	**15.15**
Fe_2O_3	0.91	**12.76**
SiO_2 +Al_2O_3 +Fe_2O_3= 73.19%		
CaO	**78.76**	11.54
Na_2O	0.00	0.00
K_2O	0.67	0.00
MgO	2.98	0.00
MnO	0.09	0.00
LOI	0.16	0.21

2.2. Procedure
Before mixed to the soil, calcium carbide residue (CCR) was oven dried and the bagasse ash (BA) was burned on 300 °C to get the higher SiO_2 + Al_2O_3 + Fe_2O_3 content. The admixture was prepared to the proportion of 40% BA + 60% CCR. The admixture, then, was mixed with the dry organic soil with the proportion of 5, 10, 15, 20, 25 and 30% with 498, 548 and 598% water content. To wait the pozzolanic reactions occurred, the stabilized soil then cured in 7, 14, 21, and 36 days of curing period. Finally, the specific gravity (G), the unit weight (γ), void ratio (e), degree of acidity (pH), organic content (OC) of stabilized soil were tested.

3. Results and Discussion
3.1. Water Content (w)
Figure 1 and 2 are presentation of the relation between water content and binder proportion with various curing times. With respect to admixture proportion, the average decrease of water content is 300%. For example at 36 days curing time, at the initial water content 498% and 548%, the water content decrease to 150 and 155%. The decrease of water content is probably due to water absorbent during pozzolanic reaction that generate cementous agents filling out the pore in organic soil and covering peat-fibers [10]. The decrease of water content, however, tends to be constant at 20% admixture and 21 days curing period because the ending of pozzolanic reaction. The trend is different from the stabilized an organic clay, the longer the curing period and the more admixture content, the pozzolanic reaction more intensively occur [6].

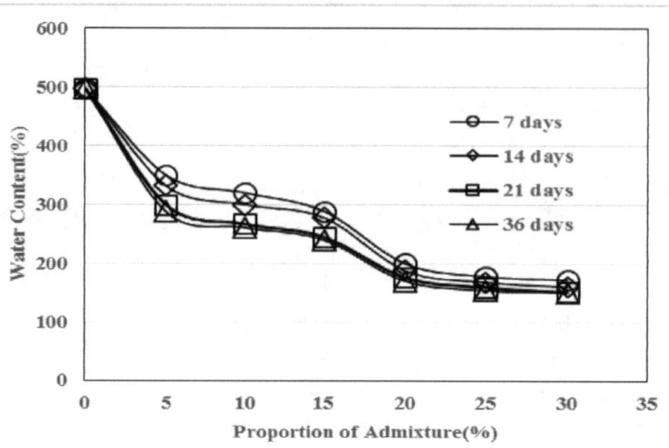

Figure 1. Proportion of Admixture vs. water content (w = 498%)

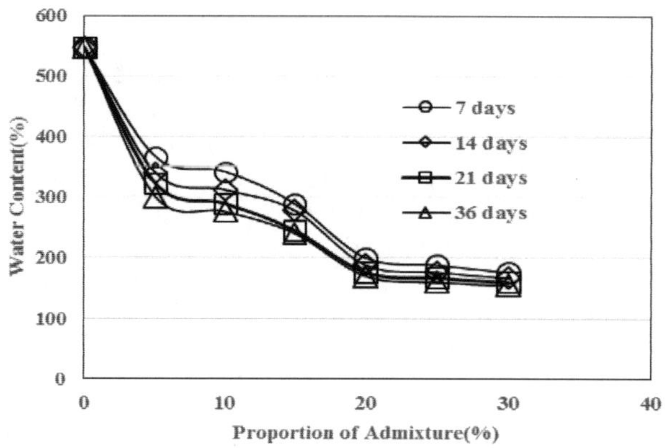

Figure 2. Proportion of Admixture vs. water content (w = 548%)

3.2. The Unit Weight (γ)

Due to addition of admixtures in the organic soil, the unit weight of stabilzed soil improves parallel to the proportion of binder quantity (Figure 3). This improvement is due to the pozzolanic reaction generating calcium-silicate-hydrate or aluminum-silicate-hydrate that fill out the pore and cover the soil fibers. However, the improvement of the unit weight of stabilized soil goes slowly, and tends to be constant after 20% proportion of admixture and 21 days curing period. It is due to lack of mineral in organic soil. With respect to the water content, the unit weight of stabilized soil significantly improve on 21 days curing time for both 5% and 20% admixtures (Figure 4). On 36 days curing period, on the other hand, there is no improvement of the unit weight on both 5% admixture and slightly improvement on 20% admixture. On 5% of admixture, the pozzolanic reaction does dot effectivelly occure , it results on small absorbtion of water for the reaction. On 20% admixture, however, the pozzolanic reaction intensivelly happpen that result on the improvement of the unit weight of stabilized organic soil.

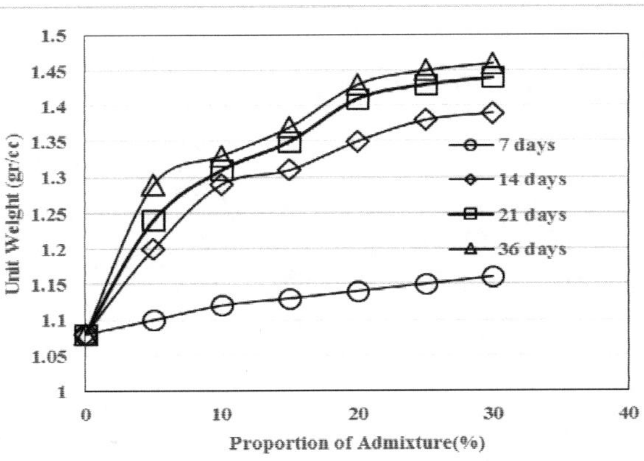

Figure 3. Proportion of admixture-Unit weight (w = 498%)

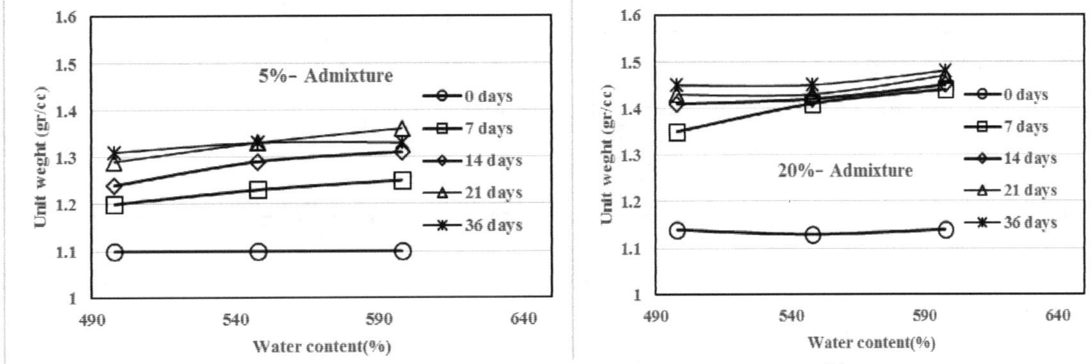

(a) (b)

Figure. 4. Water content – Unit weight (a: 5% Admixture, b: 20% Admixture)

3.3. Specific Gravity (G)

Specific gravity of original soil is 1.47 (Figure 5) that is far below the specific gravity of normal soil such as clay having G more than 2.5. For 0 days curing time with any initial water content, there is almost no changing of specific gravity due to lack of mineral on organic soil, than no pozzolanic reaction occur. It is different from that of on an organic soil. In the early period, an organic soil contain Na^+ and K^+ that can be changed by Ca^{2+} or Mg^{2+} in the admixture during reaction of cation changes, than specific gravity of soil improve on 0 day curing period [6]. On 7 days curing time and 498% water content, the specific gravity of stabilized soil improve to 1.95 with 20%, then small improvement and constant up to 30% of admixture agents. This trend is similar to the sample with 548 and 598%, however, the improvement of specific gravity is bigger compared to that of on 498% water content. The specific gravity goes to 2 and 2.1 for 548 and 598% water content.

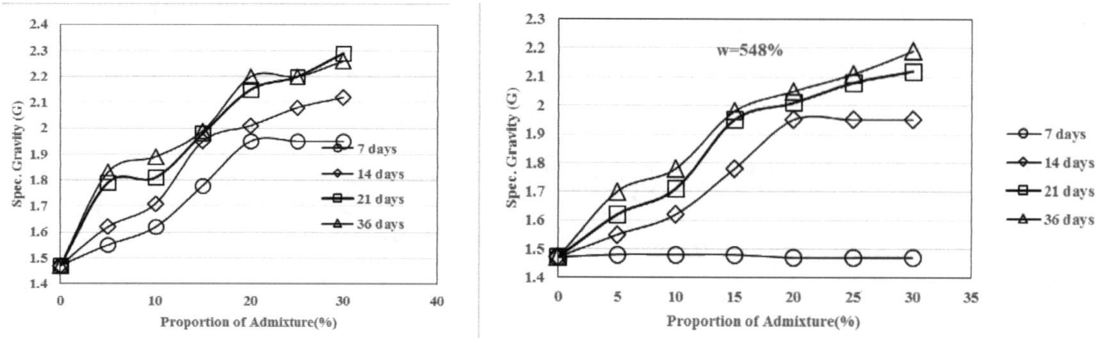

(a) (b)

Figure 5. Proportion of Admixtures – Specific Gravity (a : w =498%; b : w=548%)

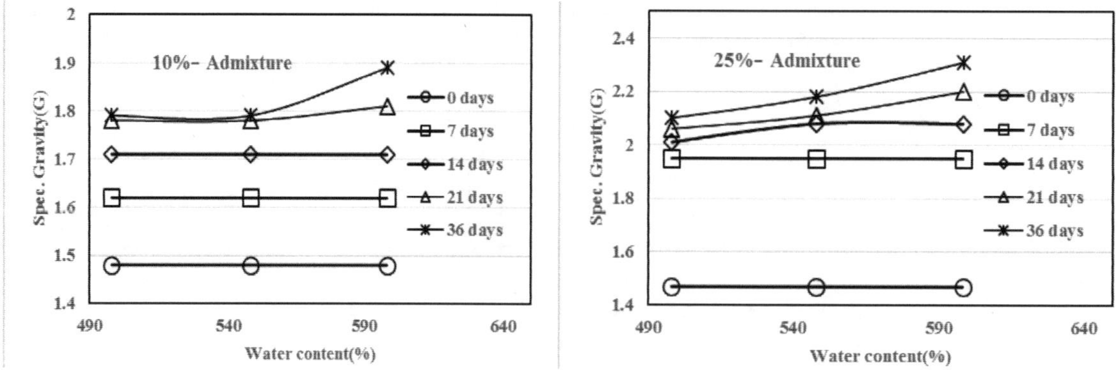

(a) (b)

Figure 6. Water content – Spec. Gravity (a: 10% Admixture, b: 25% Admixture)

Figure 6 is the presentation of the relation between water content and specific gravity. The behavior of the curves of 10% and 25% admixtures are similar. The significant improvement occur on 598% water content and 36 days curing time, because the pozzolanic reaction needs more water, and more curing time to get the best reaction.

3.4. Void Ratio (e)

Void ratio of original soil is 7.1. It decreases almost liniarly with the binder proportion. After 20% proportion of admixture, there is no more significant decrease on void ratio (Figure 7). It means that 20% is optimum proportion of admixture. With respect to the water content, the decrease of void ratio on 598% water content is bigger compared to those on 498%. For example, at 36 days curing time, void ratio decreases to 3.4 on 498%, and to 2.9 on 598% water content. It is the fact that the water is required on pozzolanic reaction. Basically, the decrease of void ratio is because of filling out the pore by the gels formation as the product of pozzolanic reaction. However, the decrease of void ratio from 21 days to 36 days curing period is not significant, and tends to constant. This phenomenon indicates that after 21 days curing period, there is no poxzzolanic happpens.

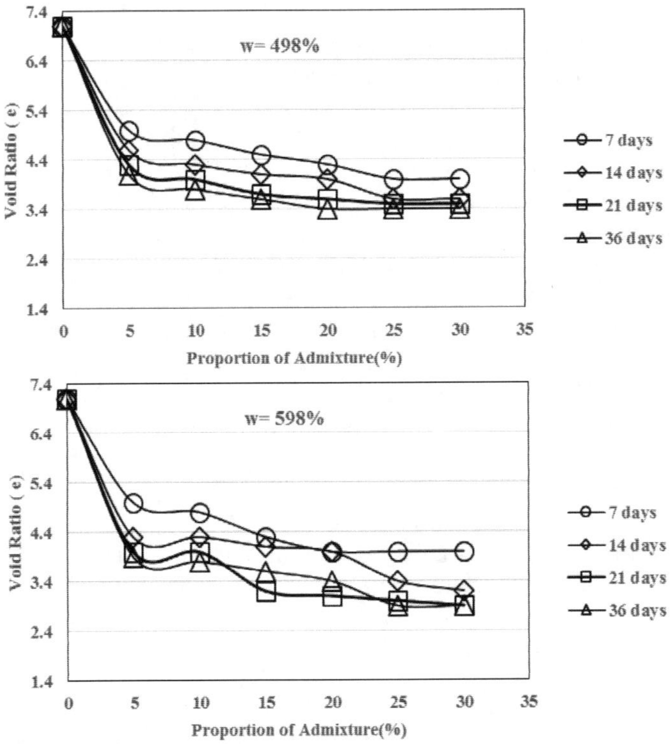

Figure 7. Proportion of Admixtures – Void Ratio

3.5. Organic Content (OC) and Degree of Acidity (pH)

Similar to void ratio, organic content decreases with respect to the water content, binder quantity, and time curing (Figure 8).With respect to binder quantity, on 36 days curing period and on 498% water content, OC decrease from 95% to 51%. With respect to curing period, on 30% binder proportion and 598% water content, OC decrease from 60% (7 days) to 45% (36 days). Moreover, OC decreases with respect to the water content, for example, for 30% water content, 36 days curing time, OC decreases from 51% on 498% water content to 45% on 598% water content. All those phenomena show that cementous reaction require admixture, water and curing period.

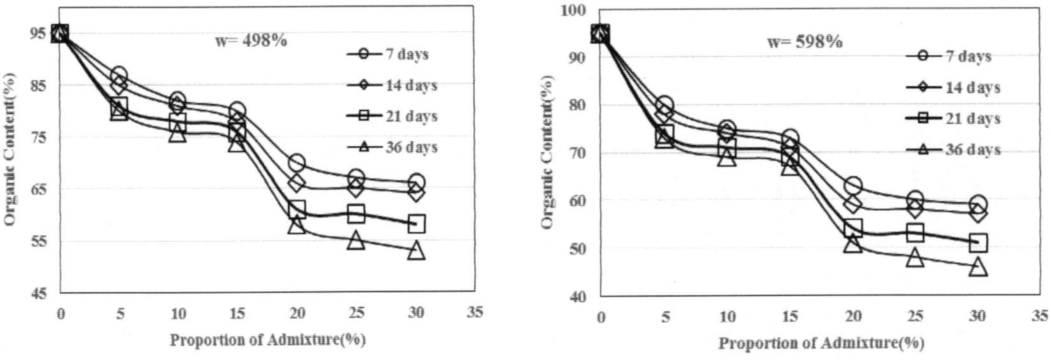

Figure 8. Proportion of Admixture – Organic Content

Beside improving the unit weight, specifif grvity, decreasing void ratio and organic content, the degree of acidity (pH) of stabilized organic soil increases with respect to the water content, binder proportion and curing period. Initial degree of acidity is 4.7%, it improves to 7.1 on 30% binder proportion, 36 days curing time and 598% water content. The improvement is because of characteristics of binder (40% CCR + 60%AAT) that can change the acidity of organic soil to neutral.

However, at 498% water content the improvement of pH is smaller than that of at 598% water content, that indicates the effect of water content to pH improvement.

4. Conclusion

A series of experimental program and analysis have been done, then the following conclusion can be drawn. The Ketapang Peat has high organic content, around 95%, that is classified as high organic content-peat. The binder was (40% bagasse ash + 60% calcium carbide residue). The bagasse ash has high content of silica, alumina and ferrit whereas calcium carbide has high CaO content, therefore combination of both materials with additional water generate the pozzolanic reaction. The addition of water content improves the specific gravity, unit weight, and the degree of acidity of organic soil. However, it decreases the void ratio and organic content of soil. Similarly, binder proportion and curing period improve the physical and chemical properties of stabilized organic soil. The optimum admixture proportion were found on 20%, and curing period were found on 21 days.

Acknowledgement

The authors would like to deliver special thanks to The minister of Research and Higher Education Republic of Indonesia for the funding of the research through the Fundamental Research of Higher Education Grant. Thank is also addressed to Research Institute and Public Affairs of Universitas Atma Jaya Yogyakarta.

References

[1] Keshawarz M S, and Dutta U 1993 *Fly ash for soil improvement* (GSP 36), ASCE Geotechnical Special Publication No. 36, New York, pp. 30 – 42
[2] Sridharant A, Prashant J P, and Sivapullaiah P V 1997 *Proc. of ICE Ground Improvement* **1**(3), 169-175.
[3] Kanieaj S R, and Havanagi V G 1999 *Cemment Conccrete Research* **29**, 673 – 677
[4] Bin-Shafique S, Edil T, Benson C, and Senol A 2004 *Proc. of ICE Geotech. Eggrg* **157**(4), 239 – 249.
[5] Parsons R.L, and Kneebone, E 2015 *Ground Improvement* **9**(1), 33-38.
[6] Thali M, Bankole G.M.2011 *Jour.of Geotech. Geoenviron. Eggrg*. **16**, 983-996
[7] Horpibulsuk S, Phetchuay C, Chinkulkijniwat A 2012 *Jour. of mtls. in Civil Eggrg.*@ASCE **24**(2) 184-193.
[8] Yadu L, and Triphati R K 2011 *Intl. Jour. Of Earth. Sci. and Eggrg*. **4**(6) 42-45
[9] Cetin B, Aydilek A H, and Guney Y 2012 *Resource and Conserv. Recycle* **58**, 8-17
[10] Venda O J P, Correia A S, and Lopes J S 2014 *Jour. of Geotech. and Geoenviron. Eggrg.* @ASCE **140**(9) 04014053
[11] Ferguson G 1993 *Fly Ash for Soil Improvement* ASCE New York
[12] Acosta H A, Edil T B, and Benson C H 2003 *Geo Engineering Report, No. 03-03*, Dept. Of. Civil and Environ. Eggrg. Univ. Of Wiscounsin, Madison, WI
[13] Prabakar J, Dendorkar N, and Morchale R K 2004 *Construction Building Materials* **18**(4) 263-267
[14] Trzebiatowski B D, Edil T B, and Bensson C H 2015 *Recycled Mtls. in geotech.*(GSP 127), ASCE, Reston, VA, 123-136.
[15] Hampton M B, and Edil T B 1998 *Soil improvement for big digs* (GSP 81), ASCE, Reston, VA, 135 – 148.
[16] Janz M, and Johansson SE 2012 *Report 9, Sweedish Deep Stabilization Research Center, Sweden.*
[17] Thalib M, and Bankole G M 2011 *Jour.l of Geotech. Geoenviron. Eggrg* **16** 983 – 996.
[18] Yadu L, and Triphati R K 2011 *Int. Jour. of Earth Sci. and Eggrg* **4**(6) 42 – 45.
[19] Correia A A S, Venda O P J, and Lemons L J L 2013 *Proc. 18th Int. Conf. on Soil Mech. and Geotech. Eggrg.* Presses des Ponts, Paris, 2457 – 2460.

[20] Jaturapitakkul C, and Roongreung B 2003 *Jour. of mtls. in Civil Eggrg. @ASCE* **15**(5) 470-475.
[21] Makarkarat N, Jaturapitakkul C, and Laosamathikul T 2010 *Jour. of mtls. in Civil Eggrg.* @ASCE **22**(11) 1164-1170.
[22] Somna K, Jaturapitakkul, C. and Kajivichyanukul, P 2011 *Jour. of mtls. in Civil Eggrg. @ASCE* **23**(3) 298-30
[23] Kampala A, and Horpibulsuk S 2013 *Jour. of mtls. in Civil Eggrg. @ASCE* **25**(5) 632-644.
[24] Horpibulsuk 2013 *Jour. of mtls. in Civil Eggrg., ASCE* **125**(5) 470 -475.
[25] Mochtar N.E, and Indrasurya B 2010 *Proc. Of the 1st Makasar Inter.Conf. on Civ. Eggrg.(MICCE2010)*, March-9-10
[26] Yulianto F.E, and Mochtar N.E 2010. *Proc. Of the 1st Makasar Inter.Conf. on Civ. Eggrg.(MICCE2010)*, March-9-10

Constructing hydrotalcite Mg-Al-CO_3 by utilizing CO_3^{2-} from natural resource

L Mahardiani[1], S Saputro[1], R Arinda[1] and Y F Astuti[1]
[1]Department of Chemistry Education, Faculty of Teacher Training and Education, Universitas Sebelas Maret, Jl. Ir. Sutami 36A, Kentingan, Surakarta, Indonesia

E-mail: lina80_ssa@yahoo.com and mahardiani.lina@staff.uns.ac.id

Abstract. This research aims to construct hydrotalcite Mg-Al-CO_3 by tuning the pH synthesis and molar ratio. The tuning was carried out by applying different synthesis pH. In addition, utilization of snail shell waste is expected to replace CO_3^{2-} ion that needed for anion component in hydrotalcite production. For this study precipitation and hydrothermal methods were applied for synthesis. The precipitation had been carried out in three steps, namely isolation of ion CO_3^{2-} from snail shell, preparation of Na_2CO_3 solution using snail shell powder as ion CO_3^{2-} source and synthesizing hydrotalcite, where hydrothermal method was carried out at 150 °C. Confirmation of hydrotalcite obtained was done using SEM and FTIR.

1. Introduction

Belongs to class of anionic clays, hydrotalcite owns interesting properties, such as high surface area, basic properties, thermally stable, abundant ion exchange surface sites, small crystal size and memory effect [1-3]. Hydrotalcite as a functional interlayer material is composed from metal cations as layer with interlayer spaces consist of anion and water molecules as building blocks [1, 4-6]. With general formula of $[M^{2+}_{(1-x)}\, M^{3+}_x\, (OH)_2]\, (A^{n-})_{x/n}.zH_2O$, the M^{2+} and M^{3+} as divalent and trivalent, respectively are changeable. There are a lot of studies on synthesizing hydrotalcite-like materials by substituting Mg^{2+} and Al^{3+} as main metal of the framework [2, 3, 7, 8]. This changeable condition also applicable for anion, where CO_3^{2-} are typical ion existing in the interlayer of hydrotalcite [9, 10]. This is suggesting that by varying cations, their ratios and interlamellar anions hydrotalcite can be formed [11]

Designing and constructing hydrotalcite have been conducted in several methods, including precipitation and coprecipitation [12], sol-gel [13], urea hydrolysis [14], and thermal decomposition [15]. Most of studies were carried out by applying other metal ion for replacing M^{2+}, M^{3+} and A^{n-} in the network using chemical source which is expensive. Therefore, some studies have been conducted on the potential starting materials for preparing hydrotalcite, like dolomite, aluminum slag, fly ash [16]. This fly ash, for example, was confirmed its chemical constituents consisted of SiO_2, Al_2O_3, Fe_2O_3 with small amount of CaO. About 97.5% of calcium carbonate ($CaCO_3$) composed the snail shell [17-19] making snail shell is promising material for functional material, namely hydroxyapatite [20], perovskite [21] and catalyst [22].

In this study, we are reporting the construction of hydrotalcite Mg-Al-CO_3 where CO_3^{2-} was obtained from snail shell waste. The effect of molar ratio of Mg-Al-CO_3 and pH were investigated.

IOP Conf. Series: Materials Science and Engineering **858** (2020) 012008 doi:10.1088/1757-899X/858/1/012008

2. Experimental

2. 1 Chemicals and materials

Snail shell was obtained from local farmer. $Mg(NO_3)_2.6H_2O$, $Al(NO_3)_3.9H_2O$, NaOCl, HCl, Na_2SO_4 were purchased from PT. Merck Chemicals and Life Sciences, Indonesia. All chemicals were used with no further purification.

2.2 Methods

2.2.1 Preparation of CO_3^{2-} from snail shell

The snail shell was washed and soaked in NaOCl 0.01% for 2 h and washed again. Followed with soaking snail shell in HCl 0.6% for 30 min then washed again for several times. Cleaned snail shell was dried for 1 h at 110 °C prior for grinding and sieving for uniform size. The snail shell powder then calcined at 400 °C for 90 min. Calcined carbonate powder was obtained.

2.2.2 Formation of Na_2CO_3

Calcined carbonate powder was dissolved in distilled water. Sodium sulphate was added to the solution and kept for 30 min. The solution filtered and filtrate then was tested for sodium carbonate formation using phenolphthalein test.

2.2.3 Synthesis of Mg-Al-CO_3 hydrotalcite

The hydrotalcite was synthesized by precipitation and hydrothermal methods. The molar ratio of three solutions were varied with composition = 3:1:1 and 3:1:3 have been prepared as follow. Solution A was prepared by dissolving Mg and Al nitrate salts with different molar concentration, namely $[Mg(NO_3)_2]$ = 0.048 M, $[Al(NO_3)_3]$ = 0.050 M for 3:1:1 (hereafter known as A_1) and $[Mg(NO_3)_2]$ = 0.016 M, $[Al(NO_3)_3]$ = 0.05 M for 3:1:3 (known as A_2). Solution A_1 and A_2 then slowly added to solution B, which is sodium carbonate (Na_2CO_3) with concentration 0.016 M. The mixture, solution C_1 and C_2 was obtained. The pH of the solution C_1 and C_2 was maintained at 12 and 14 using NaOH, solution D_1 and D_2 was obtained. Solution D was kept for 24 h under constant stirring. Then placed into Teflon lined stainless steel autoclave for hydrothermal analysis for 48 h at 150 °C. The precipitate obtained by filtration was dried at 110 °C resulting hydrotalcite with different molar ratio and pH, namely H_1-12, H_1-14, H_2-12 and H_2-14.

2.2.4 Characterization

The surface functional groups of catalyst were detected by a Fourier transformed infrared (FTIR) spectrometer (Shimadzu IR-Prestige21). The samples (0.1 wt%) were mixed with KBr and analyzed in the 4000 – 400 cm^{-1} wavenumber range, with resolution of 2 cm^{-1}, by averaging 45 scans. Scanning electron microscopy (SEM) was obtained with a Hitachi SU 3500 microscope using an accelerating voltage of 5 kV.

3. Results and discussion

3.1 Infra-red Spectroscopy

Confirmation on functional groups of hydrotalcite samples were carried out by infrared spectroscopy. Figure 1 exhibits all spectra of H_1-12, H_1-14, H_2-12 and H_2-14. All samples show similar IR spectra where a broad band was observed at 3500 cm^{-1} assigned for –OH stretching vibration of brucite-like layer, the lattice H_2O and the interlayered water, as reported by others [3, 6, 10]. In addition, peak for deformation of –OH group was also observed at 1669 cm^{-1}. Other study reported that intensity of –OH deformation was varied depending on interlayer anions [23]. In this study, the anions for all samples were same, namely CO_3^{2-} and NO_3^-, therefore the intensity for all spectra at this wavenumber was same.

Another broad but weak intensity peaks at about 3026 cm^{-1} was detected and was attributed to H–bonding between water molecule and anions in the interlayer [1, 24]. In this case, carbonate ion-water interaction. Meanwhile, at wavenumber 1379 cm^{-1}, sharp and strong band indicating presence of NO_3^- ion [23, 25-27], free carbonate ion and disordered nature of the interlayer [1]. Bending mode of carbonate was observed at wavenumber 1123 cm^{-1} [28]. Bands at fingerprint area, 702 and 563 cm^{-1}, were suggesting interaction between Al and Mg with lattice oxygen (M – O) [29] and asymmetric vibration of carbonate ion [1, 30].

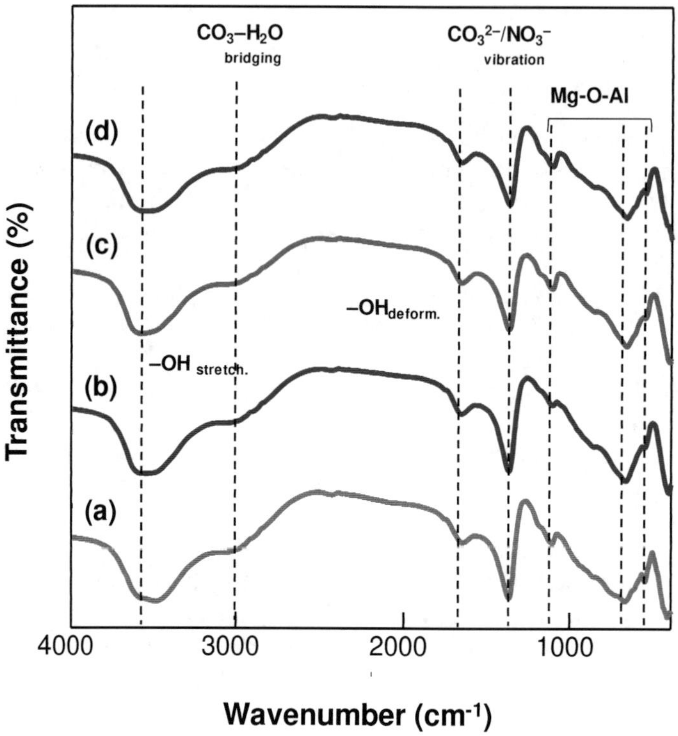

Figure 1. The infrared spectra of (a) H_1-12, (b) H_1-14, (c) H_2-12, and (d) H_2-14, where H_1: molar ratio of Mg:Al:CO_3 = 3:1:1, H_2: molar ratio of Mg:Al:CO_3 = 3:1:3 and pH = 12 and 14.

3.2 Scanning Electron Microscopy

The morphology of the hydrotalcite synthesized in this study is shown in figure 2. There is no difference on the crystal morphology as the effect of molar ratio and pH. All hydrotalcite has well-crystallized mix of hexagonal and plate-like shape. These results are in the accordance with some studies although the method of synthesis was different, namely urea method [4], co-precipitation and hydrothermal [5, 27]. However, the size of crystal as seen in figure 2 has irregular size which opposite to the results of study of Hibino et.al. and Wang et.al which has identical crystal particles. In addition, the crystal was agglomerated and interconnected with each other and formed nano particles aggregates. This is also similar with the previous study [5] where hydrotalcite synthesize at pH $\geq$ 11 formed mesoporous Mg_3-Al_1-CO_3 from accumulation of uniform nano particles aggregates. This is because according to isoelectric point (IEP), the more basic the solution, the growth of Mg_3Al_1-CO_3 was repressed as result of negative charge of the hydrotalcite surface [5].

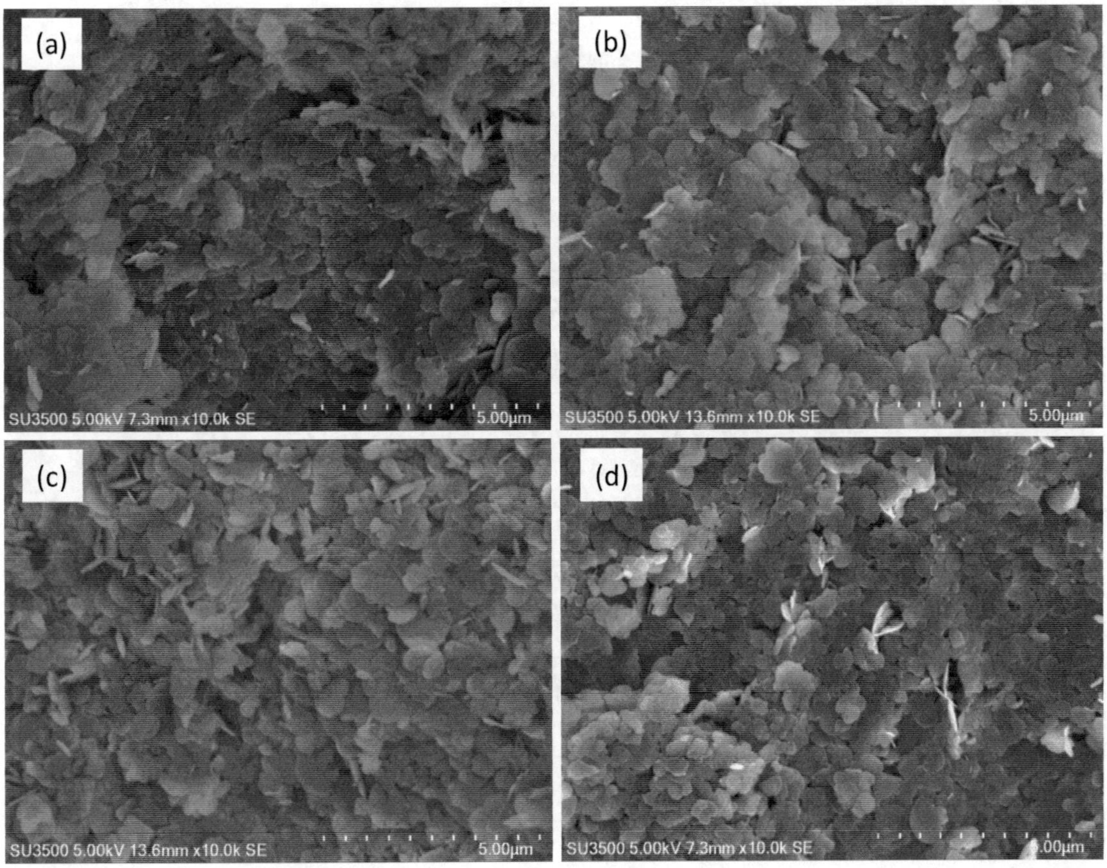

Figure 2. The image of SEM result of (a) H_1-12, (b) H_1-14, (c) H_2-12, and (d) H_2-14, where H_1: molar ratio of Mg:Al:CO_3 = 3:1:1, H_2: molar ratio of Mg:Al:CO_3 = 3:1:3 and pH = 12 and 14.

4. Conclusion

The constructing of hydrotalcite with promote the carbonate ion (CO_3^{2-}) from natural resource had been done. CO_3^{2-} from snail shell is successfully incorporated by precipitation and hydrothermal methods with variation of molar ratio and pH of solution. The results were confirmed by FTIR and SEM analysis. From FTIR results, peaks assigned for CO_3^{2-} were observed at 1379 and 3026 cm^{-1}. As for SEM analysis, the morphology of hydrotalcite with different pH showed no different that mesoporous Mg_3Al_1-CO_3 particles was formed.

Acknowledgment

This research was funded by Universitas Sebelas Maret, Surakarta – Indonesia via Mandatory Grant of PNBP UNS with contract no. 516/UN27.21/PP/2019.

References

[1] Cavani F, Trifiro F and Vaccari A 1991 *Catal. Today* **11** 173
[2] Chubar N, Gilmour R, Gerda V, Micusik M, Omastova M, Heister K, Man P, Fraissard J and Zaitseu V 2017 *Adv. Colloid Interface Sci.* **245** 62
[3] Zhou H, Jiang Z and Wei S 2018 *Appl. Clay Sci.* **153** 29
[4] Hibino T and Ohya H 2009 *Appl. Clay Sci.* **45** 123
[5] Wang Q, Tay H H, Guo Z, Chen L, Liu Y, Chang J, Zhong Z, Luo J and Borgna A 2012 *Appl. Clay Sci.* **55** 18

[6] de Sa F P, Cunha B N and Nunes L M 2013 *Chem Eng J.* **215-216** 122
[7] Zubair M, Daud M, McKay G, Shehzad F and Al-Harthi M A 2017 *Appl. Clay Sci.* **143** 279
[8] Jawad A, Peng L, Liao Z, Zhou Z, Shahzad A, Ifthikar J, Zhao M, Chen Z and Chen Z 2019 *J. Clean. Prod.* **211** 1112
[9] Tamura H, Chiba J, Ito M, Takeda T and Kikkawa S 2004 *Solid State Ion.* **172** 607
[10] Zhang B, Luan L, Gao R, Li F, Li Y and Wu T 2017 *Colloid Surf A Physicochem Eng Asp* **520** 399
[11] Bukhtiyarova M V 2019 *J. Solid State Chem.* **269** 494
[12] Sharma U, Tyagi B and Jasra R V 2008 *Ind. Eng. Chem. Res.* **47** 9588
[13] Tichit D, Lorret O, Coq B, Prinetto F and Ghiotti G 2005 *Microporous Mesoporous Mater.* **80** 213
[14] Rao M M, Reddy B R, Jayalakshmi M, Jaya V S and Sridhar B 2005 *Mater. Res. Bull.* **40** 347
[15] Kim Y, Yang W, Liu P K T, Sahimi M and Tsotsis T T 2004 *Ind. Eng. Chem. Res.* **43** 4559
[16] Muriithi G N, Petrik L F, Gitari W M and Doucet F J 2017 *Powder Technol.* **312** 299
[17] Srinivaskannan C and Abu Baker M Z 2004 *Bio-mass and Bio-energy* **27** 89
[18] Oladoja N A and Aliu Y D 2009 *J. Hazard. Mater.* **164** 1496
[19] Zhang Y, Liu S and Wu H 2014 *BTAIJ* **9(8)** 303
[20] Asimeng B O, Fianko J R, Kaufmann E E, Tiburu E K, Hayford C F, Anani P A and Dzikunu O K 2018 *J. Asian Ceram. Soc.* **6(3)** 205
[21] Fatimah I, Rahmadianti Y and Pudiasari R A 2018 *IOP Conf. Ser.: Mater. Sci. Eng.* **349** 012026
[22] Kaewdaeng S, Sintuya P and Nirunsia R 2017 *Energy Procedia* **138** 937
[23] Kloprogge J T, Wharton D, Hickey L and Frost R L 2002 *American Mineral.* **87** 623
[24] Iyi N, Matsumoto T, Kaneko Y and Kitamura K 2004 *Chem. Mater.* **16** 2926
[25] Aramendia M A, Borau V, Jimenez C, Luque J M, Marinas J M, Ruiz J R and Urbano F J 2001 *Appl. Catal. A: Gen.* **216** 257
[26] Abdolmohammad-Zadeh H and Kohansal S 2012 *J. Braz. Chem. Soc.* **23** 473
[27] Wiyantoko B, Kurniawati P, Purbaningtias T E and Fatimah I 2015 *Procedia Chem.* **17** 21
[28] Zhang W H, Guo X D, He J and Qian Z Y 2008 *J European Ceramic Society* **28** 1623
[29] Heriyanto, Heraldy E, Nugrahaningtyas K O 2018 *IOP Conf. Ser.: Mater. Sci. Eng.* **333** 012006
[30] Zhang P, Qian G, Xu Z P, Shi H, Ruan X, Yang J and Frost L L 2012 *J. Colloid Interface Sci.* **367** 264

Simple synthesis of rGO (reduced graphene oxide) by thermal reduction of GO (graphene oxide)

R Hidayat[1], S Wahyuningsih[2] and AH Ramelan[2]
[1]Chemistry Master Department, Graduate School, Sebelas Maret University
[2]Inorganic Material Research Group, Faculty of Mathematic and Natural Sciences, Sebelas Maret University

Email: sayekti@mipa.uns.ac.id

Abstract. The rGO has been synthesized successfully by combination of modified Hummers method and thermal reduction. First, the GO was synthesized by modified Hummers method (chemical exfoliation) of graphite flake. This reaction obtained graphite oxide (GrO) as intermediate material. The materials were characterized by FTIR, XRD, TG/DTA and SEM. The FTIR data shows the functional group of graphite, GrO, GO and rGO, while XRD data shows crystal parameter of them. The TG/DTA was used to study the rGO formation at various of temperature (200, 300 and 400 °C). The rGO had carbonyl, carboxyl, hydroxyl and alkene functional groups and 2 theta peak at around 25° corresponds to [002] plane. The rGO layer by thermal reduction of GO had wrinkle sheets structure from SEM image appearance. In this result, the simple thermal reduction strategy of rGO is possible to control functional group and layer number depends on its application.

1. Introduction

Graphene, two-dimensional (2D) sheet material from graphite is arranged of sp^2-hibridized carbon atoms that has unique characteristic [1,2]. This material has attracting characteristic, high thermal conductivity, good elasticity, high specific surface area and good mechanic stability [3]. Firstly, graphene was discovered in 2004 by mechanical exfoliation of graphite with adhesive tape [4]. Today, graphene becomes a popular material and be applied in many applications. Graphene can be applied in nano-technology, electronic devices, agriculture, foods technology and etc [5,6].

The synthesis of graphene attracts researcher to obtain graphene at higher amount and easier strategy than mechanical exfoliation. The Hummers and its modified method are the most popular method to synthesis graphene. This method was introduced by Hummers and Offeman in 1958, before Novoselov *et al* (2004), named by graphene oxide (GO). Hummers and its modified method refer to chemical exfoliation process on graphite flake [7]. This method usually occurs in strong acid and oxidizing agent solution mixture. Therefore, it is difficult to determine the mechanism of formation and oxygen functional group of GO clearly [8]. Controlling the oxygen functional group content can be approached by reduction strategy of GO. The reduction of GO results rGO which has more similar structure with pure graphene. The GO can be reduced via thermal, chemical, solvo/hydro-thermal, electrochemical, photo and multistep reduction [5,9]. The thermal reduction has advantages on high reduction degree, environmentally friendly and lower cost. This reduction method had been done in previous study, but mechanism of the influence of thermal to GO structure is not reported yet [6].

In this work, GO was synthesized by chemical exfoliation (modified Hummers method) of graphite powder. This reaction resulted GrO as intermediate of GO which is obtained by ultrasonication of GrO. GO was reduced by simple thermal reduction then obtained rGO. All of materials were characterized by Fourier transform infrared spectroscopy (FTIR), X-ray diffraction (XRD), thermal analysis (TG/DTA) and scanning electron microscopy) SEM to know the formation of rGO.

2. Experimental

2.1. Materials

Graphite powder as graphene precursor, sulphuric acid (H_2SO_4) and hydrogen peroxide (H_2O_2) were obtained from Merck. Potassium permanganate ($KMnO_4$) was obtained from SIGMA-ALDRICH and aquadest was supplied by Laboratory of MIPA Terpadu UNS.

2.2. Procedure

Chemical exfoliation method was used to synthesize GO from graphite powder that refers to Hummer's modified method [6,7,9]. A mixture of 2 g graphite powder and 80 mL H_2SO_4 was stirred continuously until dark homogenous mixture was obtained. The oxidizing agent ($KMnO_4$) 8 g was added slowly to the mixture with constant stirring. The temperature of mixture was maintained with ice flakes to avoid excessive raising temperature oxidation reaction. This mixture changed from dark to dark green homogenous suspension then maintained for 24 hours at room temperature. The H_2O_2 0.4 mL was added to dark brown suspension so the oxidation process was terminated (brown suspension). The brown suspension was added by aquadest to reach pH of 7. Then, the brown suspension was concentrated by eliminated the water content. The brown condensed was dried under vacuum freeze drying for 3 days until brown powder was obtained.

The dried brown powder was known as graphite oxide (GrO). A mixture of GrO 200 mg and 200 mL of aquadest was ultrasonicated for 3 hours to exfoliate graphene sheets in GrO powder. Subsequently, the mixture was condensed and dried under vacuum freeze drying for 3 days so dark brown powder (GO) was formed. The reduced graphene oxide (rGO) was obtained by thermal reduction of GO powder. GO powder was reduced by annealing at 200, 300 and 400 °C. The dark brown powder was changed to dark powder after thermal treatment as rGO powder.

2.3. Characterization

The FTIR was used to investigate functional group was recorded by Shimadzu with IRPrestige-21 type, The XRD was performed by Rigaku Miniflex 600 Benchtop with CuKα radiation ($\lambda = 0.15406$ nm), The TG/DTA was carried out by STA Linseis at 30 - 600 °C temperature (10 °C/min), The morphology imaging was observed by SEM from Vega 3 Tescan with 15.0 kV resolution under high vacuum condition.

3. Result and Discussion

The rGO was synthesized successfully by thermal reduction of GO. While, The GO was obtained from chemical exfoliation of graphite powder, it was known as Hummers and its modified method. The graphite powder was oxidized by $KMnO_4$ in extremely sulphuric

IOP Conf. Series: Materials Science and Engineering **858** (2020) 012009 doi:10.1088/1757-899X/858/1/012009

acid condition. The oxidation process caused a color change from dark to dark green suspension and it was maintained for 24 hours. The oxidation reaction was stopped by adding of H_2O_2, so the dark green suspension become a dark brown suspension as GrO suspension. The GrO powder was ultrasonicated with aquadest for 3 hours to exfoliate graphene sheets in GrO then the dark brown suspension was dried to obtain GO powder. The alteration of functional group of graphite, GO and rGO can be seen in Figure 1.

The FTIR spectra of GrO and GO show the peak at 1050, 1310, 1630, 1740 and 3320 cm^{-1}. Respectively, these peaks correspond a functional group of C-O, C-OH, C=C, C=O and OH. The existence of alkene, carbonyl and hydroxyl functional group ensure that GO was synthesized successfully [6,9]. The clear alteration of FTIR pattern can be seen from graphite to GrO and GO. There is no difference between GrO and GO FTIR spectra because GO just exfoliated sheets from GrO so both of them have same FTIR pattern.

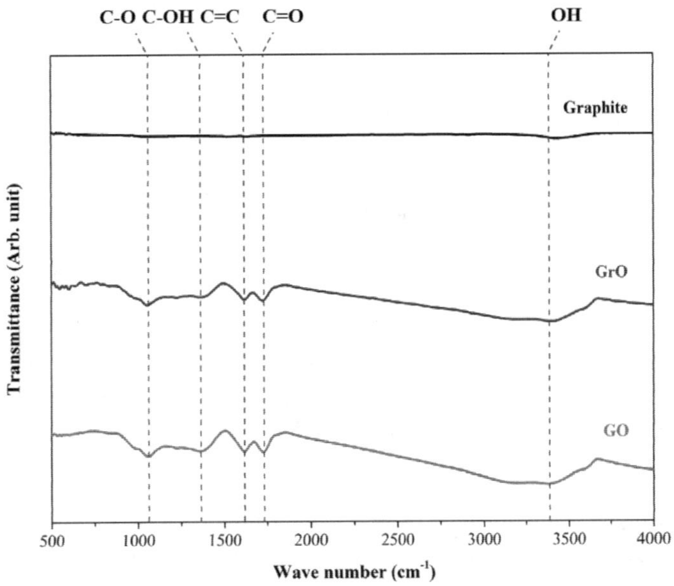

Figure 1. FTIR spectra of graphite, graphite oxide (GrO) and graphene oxide (GO).

Then, rGO was obtained via simple reduction treatment of GO powder. The GO powder was annealed at various temperatures (200, 300 and 400 °C) in oven. The annealing process of GO caused a dark brown powder changed to dark powder as rGO material. The alteration is caused the decreased of oxygenated functional group on GO sheets. The decreased of carbonyl and hydroxyl functional group can be seen in Figure 2. The carbonyl and hydroxyl have higher absorbance (in Figure 1) than Figure 2, so reduction process of GO have occurred. The rGO-200 and rGO-300 have same FTIR pattern that show peaks at 1220, 1630 and 1740 cm^{-1}, respectively correspond a functional group of C-OH, C=C, and C=O. But, the difference phenomenon is showed by rGO-400. The rGO-400 has significant peak at 610, 1120, and 3410 cm^{-1} correspond as C-H, C-O and O-H functional group. The existence of C-H shows that the functional group of GO is reduced by adding of H atom. The carbon atom bonding also was broken and formed O-H functional group then the structure rearrangement occurred. This reason is correlated by thermal analysis data in Figure 3.

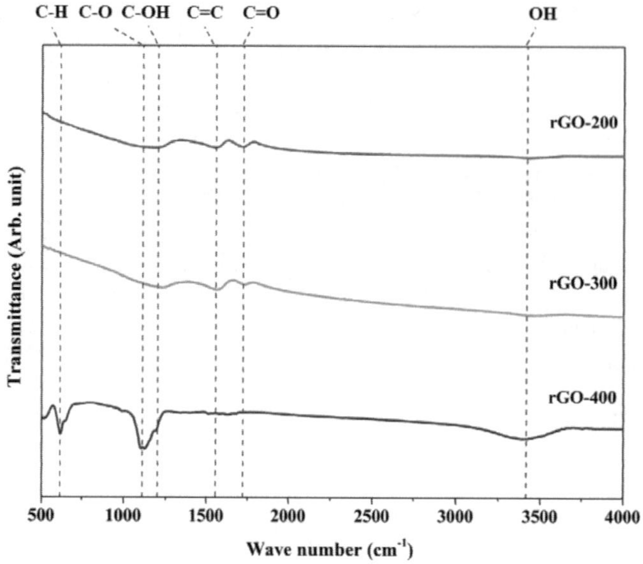

Figure 2. Effect of annealing temperature on functional group of rGO.

The figure 3 is TG/DTA graph of GO that shows structure decomposition of GO against temperature increasing. Figure 3 shows the decomposition steps of GO that corresponds with functional group on GO sheets. First step (I), GO was decomposed (21%) at 30 °C until 200 °C, shows as the elimination of water molecule and unstable oxygen functional group. Second step (II), the decomposition was occurred at 200 °C until 430 °C (30%). This decomposition is caused by elimination of more stable oxygen functional group. Then, third step (III), the weight loss is 32% that was occurred by decomposition of unstable carbon (CO and CO_2) started at 430 °C [9]. Afterwards, DTA curve (blue line) shows the structure transformation of GO sheets become rGO sheets. There are exothermic peak at 230, 430 and 450 °C as transition phase of GO sheets that is correlated with FTIR data (Figure 2) [10]. The FTIR data of rGO-200 and rGO-300 are significantly different with rGO-400 because from DTA data was occurred a high decomposition (% weight loss) of GO sheets at temperature more than 400 °C. This decomposition also leads structure rearrangement of rGO sheets. Furthermore, the structure of rGO sheets is shown in Figure 5.

Figure 4 shows diffractogram of crystal structure transformation of graphite, GrO and GO. The diffraction of graphite shows an intense peak at 2 theta of 26.08° as [002] plane. This peak is disappeared at diffraction pattern of GrO as intermediate compound of GO. The highest peak shifts from 26.08° to 10.0° which corresponds to [001] plane. The interlaminar space (d-space) value also changes from 0.35 nm to 0.89 nm, respectively for graphite and GrO (Table 1). This is can be proof that oxygen functional group from oxidation process is entrapped in interspace of graphite layers [6,9]. The GrO diffraction pattern also shows a peak at 42.28° corresponds to [100] plane. The GrO diffraction shows a broad peak around 20° - 40° indicating that GrO is an intermediate state between graphite and GO.

IOP Conf. Series: Materials Science and Engineering **858** (2020) 012009 doi:10.1088/1757-899X/858/1/012009

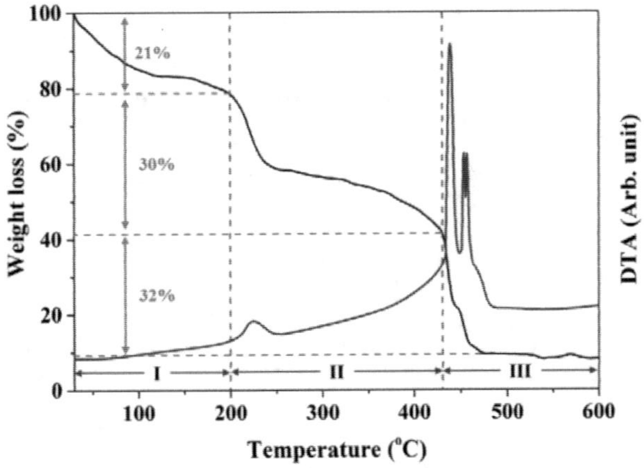

Figure 3. Thermal graph (TG/DTA) of graphene oxide (GO).

The crystal height (L_c) of GrO is lower than graphite (13.26 to 48.61 nm) also indicates that carbon bonding in graphite layers is partially broken. So, The GrO contains a mixture compound of graphite with interspace inserted oxygen functional group and partially interrupted. Whereas, the GO has similar the highest peak with GrO diffractogram. The [001] and [100] plane of GO appear at 2 theta of 10.54° and 42.61°, then interlaminar space value and crystal height of GO are close with GrO (respectively, 0.84 nm and 11.47 nm). But, there is no broad peak (at 20° - 40°) in GO diffractogram, which means that GO sheets totally formed without intermediate compound remaining. The GO sheets are arranged by 14 of graphene layers (Table 1), this value is drastically decreased from the number of graphene layers of graphite (131 layers). The decreasing of the number graphene layers indicates that chemical exfoliation of graphite is successfully applied.

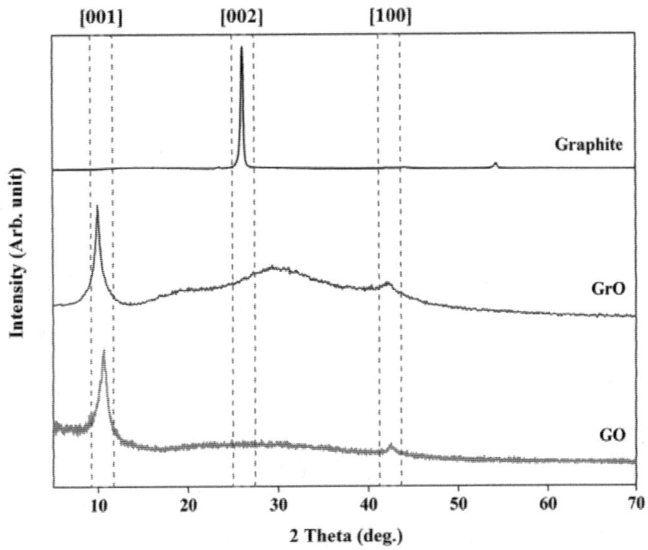

Figure 4. Diffractogram of graphite, GrO and GO.

Table 1. Crystallographic characteristic parameter of graphite, GrO, GO and rGO.

Material	2 theta (deg.)		(d_{002})	(L_c)	(L_a)	(N_c)
	[002] or [001]	[100]	(nm)	(nm)	(nm)	
Graphite	26.08	-	0.34	44.82	-	131
GrO	10.00	42.28	0.88	13.11	20.72	15
GO	10.54	42.61	0.84	11.32	14.72	14
rGO-200	24.93	43.57	0.36	1.78	4.93	5
rGO-300	24.72	43.74	0.36	1.85	5.12	5
rGO-400	26.18	42.58	0.34	1.06	4.68	3

d_{002} : Interlaminar space, L_c : Crystal stack height, L_a : In-plane crystallite size,
N_c : Number of graphene layers

The diffraction pattern of rGO is shown in Figure 5. The highest peak of diffraction shifts from 10.54° (GO) to around 25° (rGO) corresponds to reappearance of [002] plane [6]. The rGO-200 and rGO-300 have similar diffraction pattern that corroborates with the FTIR data of both. So, rGO-200 and rGO-300 are indicated in the similar crystal structure and supported by crystallography data in Table 1. However, this appearance is not shown by rGO-400 that has broadening peak at around 2 theta of 20 - 40°. The broad peak shows lower crystallinity that indicates the defect of rGO structure. The defect of structure is caused by functional group elimination, so rearrangement of compound occurs at 400 °C, but not perfectly. The number of graphene layers of rGO-400 is also fewer than rGO-200 and rGO-300 (Table 1) which shows decomposition of functional group not only impacts on weight loss but also breaks the graphene interlayers bonding, and the crystal height of rGO-400 is decreased.

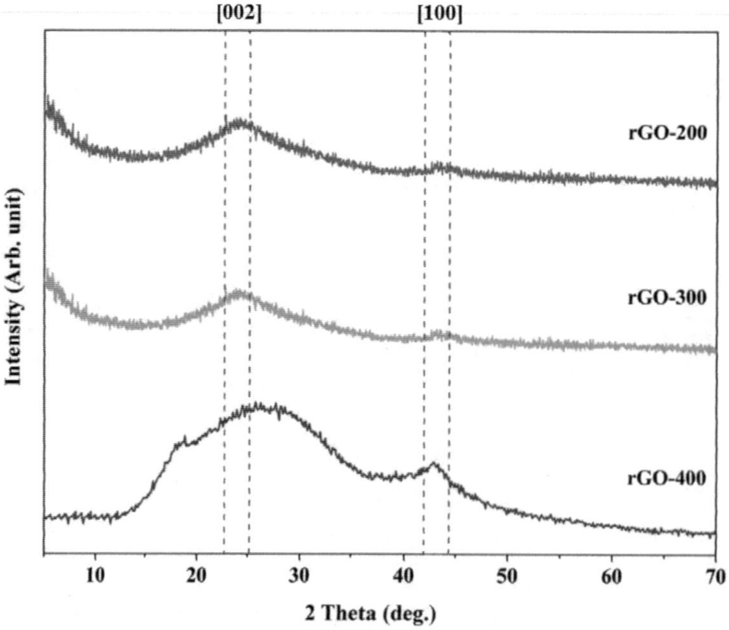

Figure 5. Effect of annealing temperature on crystallite characteristic of rGO.

The morphology of graphite, GO and rGO is known in Figure 6. Graphite has three dimensional (3D) structure which is seen in Figure 6(a). Graphite is arranged by many graphene layers, that can be known in Table 1 (131 layers). So, from SEM image, graphite shows crystal stack. While, GO and rGO are 2D material and can be seen in Figure 6 (b and c). The arrangement of GO causes highly reduction of graphene layers. Moreover, the morphology of material is not like crystal stack but sheets look that indicates GO is successfully synthesized with modified Hummers method. The different appearance is shown by rGO image, which is seen wrinkled sheets. The thermal reduction of GO occurs rapidly, so it impacts on structure rearrangement of graphene layers. But, the formation of rGO by simple thermal reduction is successfully arranged from SEM images.

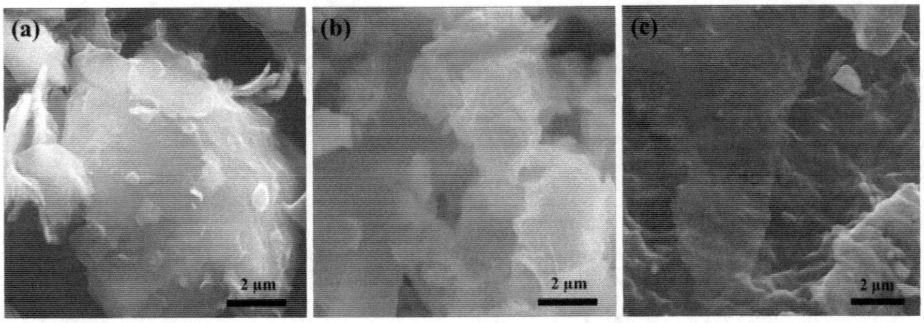

Figure 6. SEM images of (a) graphite, (b) graphene oxide (GO) and (c) reduced graphene oxide (rGO).

4. Conclusion

The reduced graphene oxide has been successfully synthesized by chemical exfoliation (modified Hummers method) of graphite powder and simple thermal reduction. The chemical exfoliation of graphite obtains graphene oxide with graphite oxide intermediate. The rGO formation can be known from FTIR and XRD characterization, then corroborated with TG/DTA and SEM image data. The different temperature of thermal reduction results a different functional group and crystal structure. So, the controlling functional group and crystal structure of rGO is possible by temperature control.

Acknowledgement

This research was funded by Sebelas Maret University through the Hibah Pascasarjana PNBP 2019 program (No. 516/UN27.21/PP/2019).

Reference

[1] Lv K, Fang S, Si L, Xia Y, Ho W and Li M 2017 *Appl. Surface Sci.* **391** 218-227.
[2] Wang Y, Wang X, Zhang M, Fang L, Jin L, Gao J, Zhang Y, Yang B, He G and Zun Z 2019 *J. Alloys Compounds* **770** 243-251.
[3] Peng T, Zhang J, Ray S, Fakhouri H, Xu X, Arefi-Khonsari F and Lalman JA 2018 *Environ. Sci. Pollution Res.* **25** 17553-17564.
[4] Novoselov KS, Geim AK, Morozov SV, Jiang D, Zhang Y, Dubonos SV, Grigorieva IV and Firsov AA 2004 *Science* **306** 666-669.
[5] Gunda R, Madireddy BS and Dash RK 2018 *Appl. Nanosci.* **8**.
[6] Saleem H, Haneef M and Abbasi HY 2018 *Mater. Chem. Phys.* **204** 1-7.
[7] Hummers WS Jr and Ofeman RE 1958 *J. Am. Chem. Soc.* **80** 1339.

[8] Yi M, Zhang W, Shen Z, Zhang X, Zhao X, Zheng Y and Ma S 2013 *J. Nanopart. Res.* **15**.

[9] Lavin-Lopez MP, Paton-Carrero A, Sanchez-Silva L, Valverde JL and Romero A 2017 *Adv. Powd. Technol.* **28** 3195-3203.

[10] Niu Y, Fang Q, Zhang X, Zhang P and Li Y 2016 *Phys. Let. A* **380** 3128-3132.

IOP Conf. Series: Materials Science and Engineering **858** (2020) 012010 doi:10.1088/1757-899X/858/1/012010

Freundlich Adsorption Isotherm in the Perspective of Chemical Kinetics (III): Isolation Method Approach

Patiha, M Firdaus[1], S Wahyuningsih[1], F Rahmawati[1], T Kusumaningsih[1]
Department of Chemistry, Faculty of Mathematic and Natural Sciences,
Sebelas Maret University, Jl. Ir. Sutami 36A, Surakarta, 57126, Indonesia
Tel. / Fax. +62 271 669376

E-mail: maulidan@mipa.uns.ac.id

ABSTRACT. Determination of *F*reundlich *A*dsorption *I*sotherm (*FAI*) fitting-constant is mostly carried out using different amounts of adsorbent with fixed amount of adsorbates, calculated for 1 g or 1 mgr, and measured at the same contact time. This technique is used together with *L*angmuir *A*dsorption *I*sotherm (*LAI*). It is considered empirical but, kinetically, it is similar to the *I*solation *M*ethod (*IM*). The purpose of this study is to provide a theoretical basis for the use of the *IM* approach in conjunction with *LAI* based for *FAI* fitting-constant determination and provide suggestions for data analysis techniques. The study was carried out theoretically, while the validity test was carried out by applying the findings to the literature data which were supplemented by statistical tests. The results show that the present technique is somewhat dubious. The technique is not suitable for every adsorbent-adsorbate pair (violating the basic assumptions of the *LAI* kinetics aspect) and the non-linear analysis technique give an order-reaction that is not always consistent with the adsorption condition. A new technique has been introduced. This new technique gives a value of $b = n$ (*P*ercentage *D*ifference, *PD*, is less than 5%).

1. Introduction

The *F*reundlich *A*dsorption *I*sotherm (*FAI*), one of the first isotherms equations introduced to describe the adsorption phenomena, is given in equation:

$$S = KC^b \tag{1}$$

where S is mass adsorbed per adsorbent mass, C is concentration in solution, and K and b are fitting-constants. S is an intensive properties. The fitting-constant b is dimensionless.

EQ. (1) has no theoretical foundation [1, 2]. It is empirical and have nearly no, or limited in its usefulness for its ability to fit data or to interpret the physical significance of the coefficients. The value of K is obtained from the intercept and b from the slope of the double-logarithmic plot of log S versus log C of equation:

$$\log S = \log K + b \log C \tag{2}$$

The kinetic approach has also been introduced [3]. It is based on equation:

$$-dC/dt = k_1 C^{n_1} - k_2 S^{n_2} \tag{3}$$

At equilibrium the derivative is equal to zero. Then, after some steps *EQ*. (3) becomes:

$$S = (k_1 / k_2)^{1/n_2} C^{n_1/n_2} \tag{4}$$

where n_1 and k_1 are the order and fractal rate constant for adsorption and n_2 and k_2 for desorption. *EQ.* (4) is identical with *EQ.* (1) if $(k_1/k_2)^{1/n2} = K$ and $(n_1/n_2) = b$. However, it is unlikely that the rate of adsorption is independent of S_o.

A more acceptable derivation has been given. It is based on equation:

$$-dC/dt = k_1(S_o - S)C^{n_1} - k_2 S \tag{5}$$

where S_o is the maximum adsorbed. If $S << S_o$, and at equilibrium, after some steps, *EQ.* (5) becomes:

$$S = \left[\frac{k_1}{k_2}\right] S_o C^{n_1} \tag{6}$$

where n_1 is the fractal dimension for adsorption. Here, the fractal dimension with respect to adsorption (S_o) and desorption (S) are assumed to be *first*-order for *EQ.* (6) = *EQ.* (1) if $(k_1/k_2)S_o = K$ and $n_1 = b$ [4].

EQ. (1) is usually used together with the Langmuir Adsorption Isotherm (*LAI*):

$$S = \frac{K_{ads} C}{k + C} \tag{7}$$

or with its fractal form

$$S = \frac{K_{ads} C^b}{k + C^b} \tag{8}$$

to find the equation that works better for a data, *LAI* or *FAI*. It is important to note that both *EQ.* (7) and *EQ.* (8) are derived by kinetics approach [5, 6].

EQ. (1) can also be derived from *EQ* (8). If $C^b << k$, *EQ.* (8) becomes

$$S = \frac{K_{ads} C^b}{k} \tag{9}$$

EQ. (9) = *EQ.* (1) if $Kads/k = K$. From kinetics point of view, the reaction is of the *b*th-order with respect to adsorbate *C*.

Based on *EQ.* (6) and *EQ.* (9), *FAI* is *LAI* at low adsorption and $n = b$. Despite the claim that the value of $0 < b < 1$ [7, 8], it has been proved that the fitting constant b represents the same meaning with reaction order n [9, 10]. So, there is no need to cheque whether an adsorption process works better for *FA*I or *LAI* [11].

The *k* value can be determined from the intercept of the plot of:

$$\log S = \log (K_{ads}/k) + b \log C \tag{10}$$

The value of K_{ads} is determined from the condition of $C^b >> k$. In this condition,

$$S = K_{ads} \tag{11}$$

By *EQ.* (11), the reaction is of the *zeroth*-order with respect to adsorbate concentration [12].

The determination of the *FAI* fitting-constant is mostly carried out by using different amounts of adsorbent with the same amount of adsorbate, calculated for 1 g or 1 mgr, and measured at the same contact time. This technique is considered empirical but, kinetically, is similar to the isolation method.

There are 2 interesting problems arise. First, as has been discussed, dividing the equilibrium concentration with the amount of adsorbent used is not valid for every condition. It is depend on the reaction order with respect to adsorbent. Theoretically will only valid for *first*-order reaction. This facts can be related to isolation method.

Second, there is no definite requirement regarding the ratio of the amounts of adsorbates and adsorbents used; in the condition of $C^b << k$, or $C^b >> k$, or in both conditions. But, by *EQ.* (9), *FAI* is valid only if $C^b << k$. Then, the 3 cases can cause confusion in analytical techniques, linear or non linear.

Based on the above discussion, there are two objectives of this study.

1. To identify the possible error made by the method carried out by using different amount of adsorbents under fixed amount of adsorbates.
2. To clarify the use of data analysis techniques by linear and non linear method.

2. Experimental

2.1. Data

This research is a literature study. The data used are presented in Table 1 [13], Table 2 [14], and Table 3 [15]. Discussion will be focused on the determination of *FAI* fitting-constant that based on the *LAI* and or *fractal-LAI* model. More specifically on the effect of the amount of adsorbent used.

Table 1. Effect of the amount of Adsorbent on the adsorption of Acetic Acid at 298 °*K* and Contact Time 90 Minutes

Amb. IRA-67 *G*	C_o $g.\ L^{-1}$	C_e $g.\ L^{-1}$	S_e $g.\ mg^{-1}$	C_e M	v $M\ min.^{-1}$	v $M\ min.^{-1}g^{-1}$
0.25	105.60	94.20	0.0456	1.570	2.11x10^{-3}	8.44x10^{-3}
0.50	105.60	85.50	0.0402	1.425	3.72x10^{-3}	7.44x10^{-3}
0.75	105.60	74.40	0.0416	1.240	5.78x10^{-3}	7.71x10^{-3}
1.00	105.60	69.60	0.0360	1.160	6.67x10^{-3}	6.67x10^{-3}
1.25	105.60	61.20	0.0355	1.020	8.22 x10^{-3}	6.58x10^{-3}
1.50	105.60	54.60	0.0340	0.910	9.44x10^{-3}	6.29x10^{-3}
1.75	105.60	46.20	0.0340	0.770	1.10x10^{-2}	6.29x10^{-3}

PS:

1. In the original data, the contact time is not given. The time 90 minutes is taken from the data of the effect of contact time of Acetic Acid.
2. In other to have the same condition with other, only 7 of the 8 data used.
3. The results show that the *LAI* fits the experimental data but give negative value for K_L (= $/K_{ads}$).

Table 2. Effect of the amount of Granular Activated Carbon, GAC, on the adsorption of Soluble Organic Nitrogen at 293 °*K* and Contact Time 2 Hours

GAC *Mg*	C_o $mg.\ L^{-1}$	C_e $mg.\ L^{-1}$	S_e $mg.\ mg^{-1}$	C_e M	v $M\ hour.^{-1}$	v $M\ hour.^{-1}g^{-1}$
0.2	0.90	0.77	0.65	1.283x10^{-5}	1.08x10^{-6}	5.42x10^{-6}
0.5	0.90	0.65	0.500	1.083x10^{-5}	2.08x10^{-6}	4.17x10^{-6}
2.0	0.90	0.32	0.290	5.33x10^{-6}	4.83x10^{-6}	2.42x10^{-6}
5.0	0.90	0.19	0.142	3.17x10^{-6}	5.92x10^{-6}	1.18x10^{-6}
10.0	0.90	0.14	0.076	2.33x10^{-6}	6.33x10^{-6}	6.33x10^{-7}
20.0	0.90	0.09	0.041	1.50x10^{-6}	6.75x10^{-6}	3.38x10^{-7}
50.0	0.90	0.06	0.017	1.00x10^{-6}	7.00x10^{-6}	1.40x10^{-7}

PS: The results of calculation show that the *FAI* fits the experimental data but give negative value for q_m (= K_{ads}) and b = 0.32.

Table 3. Effect of the amount of Activated Carbon, AC, on the adsorption of Methylene Blue at 298 °*K* and Contact Time 72 Hours

AC *mg*	C_o *mg.* L^{-1}	C_e *mg.* L^{-1}	S_e *mg. mg*$^{-1}$	C_e *M*	*v* *M hour.*$^{-1}$	*v* *M hour.*$^{-1}$ *mg*$^{-1}$
1	25	19.9	5.1	6.22x10^{-2}	2.21x10^{-4}	2.21x10^{-4}
5	25	10.8	2.84	3.38x10^{-2}	6.16x10^{-4}	1.23x10^{-4}
10	25	4.2	2.08	1.31x10^{-2}	9.03x10^{-4}	9.03x10^{-5}
12.5	25	3.5	1.72	1.09x10^{-2}	9.33x10^{-4}	7.46x10^{-5}
25	25	1.0	0.96	3.13x10^{-3}	1.04x10^{-3}	4.16x10^{-5}
30	25	0.70	0.81	2.19x10^{-3}	1.06x10^{-3}	3.53x10^{-5}
100	25	0.20	0.248	6.25x10^{-4}	1.08x10^{-3}	1.08x10^{-5}

PS: The stress of this article is not for determining the better isotherm that fits for the experimental data. But the results show *LAI* fits better and linear and non-linear method give different values for the fitting-constant.

2.2. Framework and execution technique

2.2.1. First case.

Determination of the reaction order for each reagent whose reaction rate depends on the concentration of two (or more reagents) is difficult to do simultaneously. The *I*solation *M*ethod (*IM*) is the solution.

For simplest case:

$$v = k\,[A_o]^{\alpha}[B_o]^{\beta} \tag{12}$$

If $[A]_o$ is made much smaller than $[B]_o$: say one hundredth, because $[B]$ is practically unchanged, the rate law becomes

$$v = k^{'}\,[A_o]^{\alpha} \tag{13}$$

where

$$k^{'} = k\,[B_o]^{\beta} \tag{14}$$

The reaction has pseudo order α. The value of $k^{'}$ is obtained from the intercept of the plot of log $[A_o]$ versus log v [of *EQ*. (13)].

There are 2 ways to obtain β. But, to the extent used in this study, $[B]_o$ is still made higher, say, two fold. Then, the rate law becomes

$$v = k^{''}\,[A_o]^{\alpha} \tag{15}$$

where

$$k^{''} = k\,[2B_o]^{\beta} \tag{16}$$

The reaction also has pseudo order α. The value of $k^{''}$ is obtained from the intercept of the plot of log $[A_o]$ versus log v [of *EQ*. (15)]. Then, the value of β is obtained using *EQ*. (14) and *EQ*. (16). That is to say, 2 data are enough for β determination. As for this study, because $[A_o]$ are made all same and α does not change, then the value of β is determined from the slope log $[B_o]$ versus log v of *EQ*. (12). However, because basically in *EQ*. (6), *EQ*. (7), and *EQ*. (8), the value of β is 1 (*first*-order), this value is then used as the standard for evaluating the β value obtained. Also, because the reaction-order can only be integers or half-integers [16, 17], the values obtained are then rounded to the closest order. The result of calculation is presented in Table 4.

2.2.2. Second case.

Based on above discussion, there are 2 points to be discussed. 1. The validity of non-linear analysis. 2. As can be seen later, by the New Technique, *LAI* or *fractal-LAI* and *FAI* are valid for Acetic Acid but not for both Soluble Organic Nitrogen and Methylene Blue. The result of calculation is presented in Table 5.

3. Results and discussions

3.1. Results

3.1.1. First case

Table 4. Effect of the Amount of Adsorbent on the Reaction Order, *n*, with Respect to Adsorbent for Acetic Acid, Soluble Organic Nitrogen, and Methylene Blue by Conventional and Improved Technique

Adsorbate	Theory	Conventional	New Technique	
		All Data	First 3 Data	Last 3Data
Acetic Acid	1	$0.84 \approx 1$	$0.91 \approx 1$	$0.84 \approx 1$
Soluble Organic Nitrogen	1	$0.34 \approx 0.5$	$0.65 \approx 0.5$	$0.06 \approx 0$
Methylene Blue	1	$0.35 \approx 0.5$	$0.62 \approx 0.5$	$0.02 \approx 0$

As seen, in Table 4, there are 2 terms, Conventional and New Technique. Term Conventional is used for representing the conventional method while New Technique for method that dividing data into 2 part; at the out-set and near-end. This is because, adsorption data that fits the *LAI* or *fractal-LAI*, takes place in two different reaction mechanisms. If the adsorbed is still small ($S<<So$), adsorption is of the *bth*-order with respect to adsorbate concentration (and so fit *FAI*). But if if all surfaces covered, than the adsorption is of *zeroth*-order [18]. Of the the 3 adsorbates, only the Acetic Acid that meets the criteria ($n = 1$), both by Conventional or New Technique.

3.1.2. Second case

Table 5. Results of Statistical Analysis for b and n by Conventional and New Technique for Acetic Acid, Soluble Organic Nitrogen, and Methylene Blue

	Conventional		New Technique			
	All Data		First 3 Data		Last 3 Data	
Acetic Acid						
b	0.409	≈ 0.5	0.345	≈ 0.5	0.142	≈ 0.0
n	0.410	≈ 0.5	0.339	≈ 0.5	0.149	≈ 0.0
Soluble Organic Nitrogen						
b	1.373	≈ 1.5	-	-	-	-
n	-	-	-	-	-	-
Methylene Blue,						
b	0.601	≈ 0.5	-	-	-	-
n	-	-	-	-	-	-

As seen in Table 5, there is no result given for both Soluble Organic Nitrogen and Methylene Blue. As for Acetic Acid, the New Method gives the acceptable value $b = n$ (Table 6) for the fitting-constant of *FAI* (the first three data). The value of *b* is obtained from the slope and *K* from the intercept of *EQ*. (2). To have the same condition in *b* determination, the value of *n* is determined from the slope of the plot of log C_e (column 5) versus log v (column 7) of the data.

The proof of compatibility of *K* and the rate constant ***k*** value can also be obtained using equation:

$$\Delta t \, \boldsymbol{k} \, [MW]^{1-n} \, w_{ads}^{-1} = K \qquad (17)$$

where Δt, $\boldsymbol{k}$, *MW*, *n*, w_{ads}, and *K* are the time, rate constant, molecular weight of the adsorbate, reaction-order, (mean) weight of the adsorbent, and *FAI* fitting constant (19). Results of calculation is presented in Table 6.

Table 6. Values of *b* and *n* and *K* and *k* pairs using the New Method and their Percentage Difference (*PD*) for Acetic Acid

	b	*n*	*K*	***k***	$K_{EQ.\ (17)}$
	0.345	0.339		$7.00\text{x}10^{-3}$	
			$9.19\text{x}10^{-3}$		$9.43\text{x}10^{-3}$
PD %	1.74			2.61	

Relatively each pair of *b* and *n* and *K* and ***k*** has the same value. These meant that *b* and *n* concepts represent the same idea.

3.2. Discussion

There are not enough facts to explain why Acetic Acid meets the criteria and two other not. Probably depend on the structure and nature of each adsorbate and adsorbent; not on the ratio of adsorbent use. The facts that both Soluble Organic Nitrogen and Methylene Blue are not appropriate for the New Technique does not have to mean that the isolation method is not suite for *FAI* fitting-constant determination. The amount of reactants reacted or products obtained at any time, can always be calculated using the rate constant, *k*, value. Moreover, the facts that the Last 3 Data show that adsorption is independent of the amount of adsorbent and can be used to determine maximum adsorption.

It is of the first interest to discuss the facts presented in Table 5 and Table 6. In Table 5, for Acetic Acid, the First 3 Data gives *half*-order, which meant that Acetic Acid dissociated before being adsorbed. Because, the Last 3 Data gives *zeroth*-order (so as in column 4 (Table 1) than these three data give the value of K_{ads} (of *LAI*). These 2 facts at once proofed that non-linear analysis technique is not appropriate for *FAI* fitting-constant determination in its conjunction with *LAI*. The non-linear analysis give *b* or *n* which is always the same with that using data at the out-set and near end of adsorption. (There are 3 possible situation. If both First 3 Data and Last 3 Data give $n \neq 0$ -integers or half-integers- then the data only fit for *FAI*. If First 3 Data gives $n = 1$ and Last 3 Data gives $n \neq 0$, then the data fit for *FAI* and *LAI*. And, if both First 3 Data and Last 3 Data give $n = 0$, then the data fit only for *LAI*.). Facts in Table 6 prove that the concepts of *b* and *n* representing the same idea.

4. Conclusion

Kinetically, the method with the use of different amounts of adsorbent under fixed adsorbate conditions is not always appropriate for *fractal-LAI* based of *FAI* fitting-constant determination. Non-linear data analysis technique is not suited for *fractal-LAI* based of *FAI* fitting-constant determination. A New Technique has been introduced.

Acknowledgement

We thank the Dean and Chairman of the Department of Chemistry, FMIPA UNS, the committee of ICAMBF UNS 2019, data sources, and all those who have helped in writing / improving the article and its presentation.

References

[1] Spark D L 2003 *Environmental Soil Chemistry* 2, ed (San Diego: Academic Press) p 151
[2] Essington M E 2004 *Soil and Water Chemistry: An Integrative Approach* (Boca Raton, FL: CRC Press) p 343
[3] Skopp J 2009 *J. Chem. Ed.*, **86** p 1341

[4] Patiha, Firdaus M, Wahyuningsih S, Nugrahaningtyas K D, Hidayat Y 2018. I*OP Conf. Ser.*: *Mater. Sci. Eng.* **333** Number 1
[5] Sposito G 1980 *Soil Sci. Soc. Am. J.* **44** pp 652-654
[6] Goldberg S and Sposito 1984 *G. Soil Sci. Soc. Am. J.* **48** pp 772-778
[7] Gregory T, Karns C L, Shimizu K D 2005 A Critical Examination of the Use of Freundlich Isotherm in Characterizing Molecularly Imprinted Polymers (MIPS) *Analytica Chimica Acta* **52** pp 107-113
[8] Sawyer C N, McCarty P L, Parking G N 2003 *Fundamental of Chemistry for Environmental Engineering and Science* 5, ed (New York: McGraw-Hill Companies, Inc.) pp 100-105
[9] Patiha, Firdaus M, Wahyuningsih S, Nugrahaningtyas K D, Hidayat Y 2018 *IOP Conf. Ser.*: *Mater. Sci. Eng,* **333** Number 1
[10] Patiha, Firdaus M, Rahmawaty F, Wahyuningsih S, Kusumaningsih T 2019 Freundlich Adsorption Isotherm in the Perspective of Chemical Kinetics (II); Rate Law Approach 14, th, *J. C. C.*
[11] Patiha, M Firdaus, S Wahyuningsih, K D Nugrahaningtyas, Y Hidayat 2018. *IOP Conf. Ser.*: *Mater. Sci. Eng.* **333** Number 1
[12] Patiha, E Heraldy, Y Hidayat, M Firdaus 2016. *IOP Conf. Ser.*: *Mater. Sci. Eng.* **107** issue 1
[13] Uslu H, Inci I, Bayazit S S 2010. *J. Chem. Eng. Data* 55 pp 1295-1299
[14] Sawyer C N, McCarty P L, Parking G N 2003 *Fundamental of Chemistry for Environmental Engineering and Science* 5, ed (New York: McGraw-Hill Companies, Inc) pp 100-105
[15] Potgieter J H 1991 *J. Chem. Ed.* **68** pp 349-350
[16] Levine I N 2003 *Physical Chemistry* 6, ed. (New York: McGraw-Hill Companies, Inc.) p 517
[17] Espenson J H 1995 *Chemical Kinetics and Reaction Mechanisms,* 2, Ed. (New York: McGraw-Hill, Inc.) p 90
[18] Patiha, Firdaus M, Rahmawaty F, Wahyuningsih S, Kusumaningsih T 2019 Freundlich Adsorption Isotherm in the Perspective of Chemical Kinetics (II); Rate Law Approach, 14, th, *J. C. C.*
[19] Patiha, Heraldy E, Hidayat Y, Firdaus M 2016. *IOP Conf. Ser.*: *Mater. Sci. Eng.* **107** issue 1
[20] Patiha 2013.: *Alchemy* **9** pp 145-155
[21] Patiha and Firdaus M 2016 *Bulgarian Journal of Science Education* 25 (5) pp 687-694

Phase characterization and electrical properties of fluorhydroxyapatite : influence of molarity variation and microwave irradiation

Z S Kubro[1,2]*, K Dahlan[1,2] and S T Wahyudi[1]
[1] Division of Biophysics, Department of Physics, IPB University, Bogor, Indonesia 16680
[2] Advanced Research Laboratory, IPB University, Bogor, Indonesia 16680

E-mail: zudah_simaatul@apps.ipb.ac.id

Abstract. Hydroxyapatite is one of the calcium phosphates that is most often used to repair tissue hard as teeth and bones due to its excellent osteoconductive properties and bioactivity, but hydroxyapatite has low chemical stability at high temperatures, high dissolution values in the biological environment and low resistance to damage in acidic environments. The addition of fluorine to hydroxyapatite can make hydroxyapatite have superior mechanical properties because of its higher thermal stability. It also can increase calcium phosphate mineralization and crystallization in the process of forming teeth and bones. This research was carried out by synthesis of fluorhydroxyapatite using microwave irradiation-assisted precipitation method, by varying the molarity ratio and the microwave irradiation used. The XRD results show that samples at high microwave irradiation (100P) both in molarity variations of [P]/[F] 4 and 6, fluorhydroxyapatite has the least impurity intensity and the highest fluorhydroxyapatite intensity among the others. FTIR analysis also shows that all samples have the same group and fluorhydroxyapatite has been formed which is shown in the OH...F functional group at wave number 3539-3544 cm^{-1}. Microwave irradiation and molarity ratio [P]/[F] showed no relationship with conductivity value of fluorhydroxyapatite.

1. Introduction

Recent decades, several recent studies have shown an increase in dental caries globally. This increase occurs in children and adults, both in primary and permanent teeth [1]. WHO estimates that DMFT (Decayed, Missing and Filled Teeth) globally occurs in 12-year-old children in 209 countries included in their database. Of the total 3.681.724.312 teeth, around 244.571.686 teeth had caries. This is based on data available in 2015 from the WHO Oral Health Database, Country / Area Profile Program (CAPP) [2].

Dental caries is a scientific term for tooth decay or cavities. This is caused by certain types of bacteria that produce acids that destroy tooth enamel and dentin [3]. Initially, the remnants of food in the teeth decay which then makes the teeth become porous, hollow, and even broken. The process of tooth decay occurs due to microorganism activity in carbohydrates, which makes carbohydrates gradually become acidic due to the decay process. The acidic environment (pH below 5.5) [4] is what makes demineralization in hard tooth tissue [5]. This happens because the concentration of hydrogen ions increases (which indicates an increase in acidity) and calcium ions decrease. This demineralization process will then form small pores on the enamel surface which for a long time allow it to become a hole so as to allow the bacterial invasion to the inside of the teeth [6].

Hydroxyapatite [$Ca_{10}(PO_4)_6(OH)_2$] is one of the most commonly used calcium phosphates to repair hard tissues [7] such as teeth and bones due to osteoconductive properties [8] and bioactivity [9] which is very good. Hydroxyapatite is a special mineral component found in tissues, specifically dental and bone tissue [7]. In addition, synthetic hydroxyapatite has the ability to bind directly with teeth and bones because of its chemical composition which is the same as teeth and bones [10].

Behind the advantages of hydroxyapatite, it turns out there are some deficiencies found in its use as a biomaterial. Hydroxyapatite has low chemical stability at high temperatures as indicated by the formation of other phases such as tricalcium phosphate which often pollutes the results [10]. In addition, hydroxyapatite also has a high dissolution value in the biological environment and low resistance to damage in the acidic environment [7].

Fluorine has a role in the formation of tooth enamel. The presence of fluorine ions makes teeth more resistant to demineralization by acids from bacteria, and influences bacterial metabolism, and influences the formation of polysaccharides in carbohydrates in cells that are used as reserves to produce acids [11]. When fluorine (F^-) substitutes the hydroxyl group (OH^-), the solubility of bone mineral minerals decreases because some fluorine substitutes hydroxyapatite (commonly called fluorhydroxyapatite) which is less soluble than hydroxyapatite [10]. Substitution of fluorine into hydroxyapatite material will increase stability and prevent second caries by inhibiting bacterial metabolism [12]. Fluorine also has superior mechanical properties when sintered at high temperatures because of its higher thermal stability than hydroxyapatite. The presence of fluorine also increases calcium phosphate mineralization and crystallization in the process of the tooth and bone formation [13] and influences cell addition, proliferation, differentiation and morphology of osteoblast cells [14].

The research that will be carried out in the synthesis of fluorhdroxyapatite based on chicken eggshells using precipitation method assisted by microwave irradiation. The use of microwaves in the synthesis process is expected to help increase the degree of crystallinity, reduce energy consumption and produce nano-sized particles, because in several previous studies such as those conducted by Hassan et al. (2015) the use of microwaves has succeeded in increasing the degree of crystallinity, reducing energy consumption and producing nano-sized particles [15].

2. Experimental

2.1. Materials and tools

The material used in this study was CaO powder from chicken eggshells, distilled water, H_3PO_4, and NH_4F. The tools used in the study include sample containers, measuring cups, Erlenmeyer, goblets, crussible, dropper pipettes, magnetic stirrers, plastic wrap, Vulcan furnaces, Nabertherm furnaces, digital balance sheets, funnels, mortar, spatula, infusion tools, glass cups measuring, drip flask, Perkin Elmer Spectrum One FT-IR Spectrometer, and x-ray diffraction (XRD) Shimadzu XRD-7000, x-ray diffraction (XRD) Rigaku MiniFlex and HiTESTER 3522 -50 LCR Meters.

2.2. *Methods*

2.2.1. Synthesize of fluorhydroxyapatite with precipitation method. Synthesize fluorhydroxyapatite with varying the molarity ratio and the microwave irradiation were synthesized through precipitation method. The molarity ratio [P]/[F] were 6 and 4 in order to have corresponding compositions of $Ca_{10}(PO_4)_6F_1OH_1$, $Ca_{10}(PO_4)_6F_{1.5}OH_{0.5}$ (by replacing the OH group with F ions in molar ratios of 0.5 and 0.75 [15]. The varying microwave irradiation are 30P, 50P and 100P (30P or 30 percent means the microwave oven is producing microwaves 30 percent of the time and 70% other times, microwaves are turned off, neither does the 50P and 100P microwave irradiation).

0.5 M of CaO powder was dissolved into 100 ml distilled water, 0.3 M H_3PO_4 was dissolved into 100 ml of distilled water and also NH_4F was dissolved into 100 ml of distilled water following a variation of [P]/[F] molarity that is 6 and 4. A total of 50 ml of 0.3M H_3PO_4 was mixed into 50 ml of NH_4F solution to obtain a mixed solution of H_3PO_4 and NH_4F as much as 100 ml then drops into a CaO suspension accompanied by a stirring process during precipitation process at a speed of 300 rpm. Then the suspension is put into the microwave oven to be given microwave irradiation. At this stage, the samples were treated in the form of variations of microwaves irradiation used, i.e 30P, 50P and 100P for 45 minutes. After the sample is irradiated with microwaves, the sample is dried using a furnace at 100 °C for 45 minutes then sintered at 600 °C for 2 hours.

2.2.2. X-ray diffraction analysis. The fluorhydroxyapatite powder were characterized with X-ray diffraction (XRD) Shimadzu XRD-7000, and Rigaku MiniFlex) to determine the phase. Data obtained from XRD analysis then matched with Joint Committee on Powder Diffraction Standard (JCPDS) with numbers 15-0876 (fluorapatite) and number 09-0432 (hydroxyapatite).

2.2.3. Fourier transform infrared spectroscopy (FTIR). The fluorhydroxyapatite powder was analysis using Perkin Elmer Spectrum One FT-IR Spectrometer, It was prepared by means of 1 mg of the sample mixed with 300 mg KBr and then formed into transparent pellets. These pellets are then analyzed using FTIR with a wavenumber range of 350 to 4000 cm^{-1}

2.2.4. LCR-meter characterization. Electrical conductance of samples was measured by using an HiTESTER 3522-50 LCR Meter. The measurement is carried out using a frequency interval of 50 Hz to 50 kHz. The sample to be tested is placed between two electrodes and connected to the LCR meter for scanning.

3. Results and discussions

3.1. Phase characterization

This research was carried out by the method of precipitation assisted by microwave irradiation by mixing 0.5M CaO suspension and 0.3M H_3PO_4 solution to obtain a Ca/P molarity ratio of 1.67. Henceforth varying the molarity ratio [P]/[F] and the irradiation of the microwaves used. The resulting sample will be adjusted to the fluorhydroxyapatite database, namely with the Joint Committee on Powder Diffraction Standards (JCPDS) literature number 15-0876 belonging to fluorapatitite and number 09-0432 of hydroxyapatite since the literature data for fluorhydroxyapatite is not yet available. Figure 1 and figure 2 are the results of the XRD characterization of fluorhydroxyapatite samples with each treatment.

The X-ray diffraction pattern in fluorhydroxyapatite samples with molarity variations [P]/[F] 4 with different microwave irradiation is shown in figure 1 From the three graphs with each microwave variation, all of them indicate the presence of the fluorhydroxyapatite phase, only the difference is the number of impurities accompanying it and the high intensity of fluorhydroxyapatite obtained. The highest intensity of fluorhydroxyapatite was found in samples with the highest variation of microwave irradiation, which is 100P, and the intensity of fluorhydroxyapatite decreased with decreasing irradiation of the microwaves used. The $Ca(OH)_2$ phase appeared in all three samples with different intensities. Samples with 30P microwave irradiation had the highest $Ca(OH)_2$ intensity between the other two samples. The intensity of $Ca(OH)_2$ also seems to decrease with increasing microwave irradiation used. This shows that the higher the microwave irradiation used, the more the fluorine ion will react to, replacing the hydroxyl group. In addition, the CaO phase also appeared in all samples as well as the $Ca(OH)_2$ phase, which was supported through the FTIR results which showed the presence of CaO functional groups in all three samples. But the results obtained, only seen in samples with 100P of microwave irradiation.

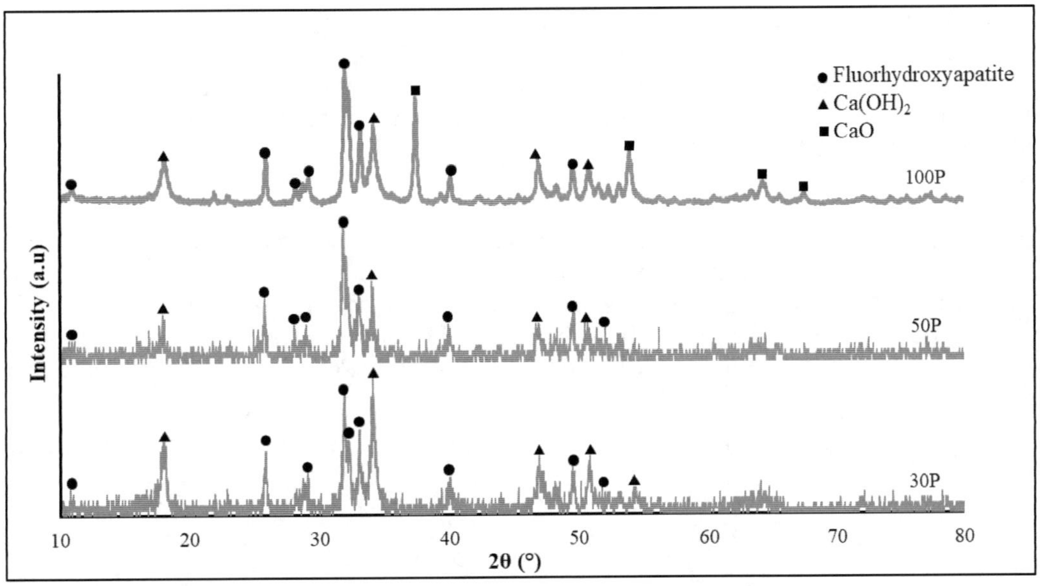

Figure 1. The X-ray diffraction pattern of fluorhydroxyapatite powder with variations in molarity of [P]/[F] 4 with variations in microwave irradiation 30P, 50P and 100P.

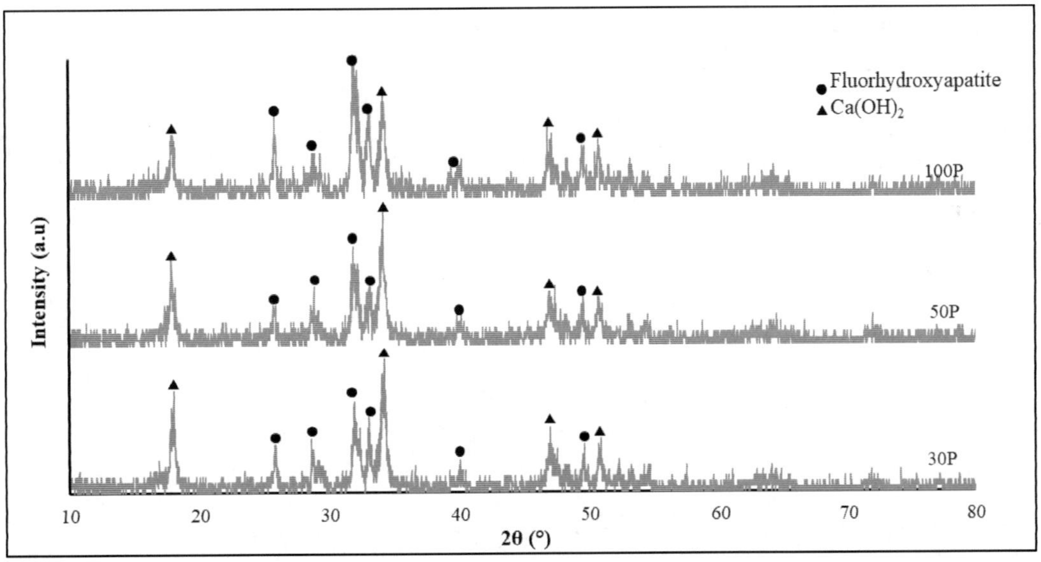

Figure 2. The X-ray diffraction pattern of fluorhydroxyapatite powder with variations in molarity of [P]/[F] 6 with variations in microwave irradiation 30P, 50P and 100P.

Figure 2 shows the pattern of X-ray diffraction of samples with variations in molarity of [P]/[F] 6 with variations in microwave irradiation. Fluorhydroxyapatite phase appeared in all samples with different intensities, and the highest intensity was owned by samples with a variation of 100P microwave irradiation. The $Ca(OH)_2$ phase also appeared in all samples, but the intensity was lowest at 100P microwave irradiation. It can be seen that the greater the microwave irradiation used, the intensity of $Ca(OH)_2$ that appears will also be smaller.

X-ray diffraction pattern from the two pictures above with the variation of molarity and microwave irradiation used, it can be seen that the intensity of fluorhydroxyapatite obtained is higher in molarity variations in [P]/[F] 4 than [P]/[F] 6. Results what is seen is the $Ca(OH)_2$ phase which is found to be not so significant in [P]/[F] 6 decreases when adding microwave irradiation, unlike in [P]/[F] 4 which shows a significant difference in the $Ca(OH)_2$ phase appears on the sample. This is appropriate because the concentration of fluorine in molarity variations of [P]/[F] 4 is greater than [P]/[F] 6 so that the existing fluorine will also replace more hydroxyl groups as the microwave irradiation is used, as evidenced by reduced intensity of $Ca(OH)_2$.

3.2. Functional groups

Observations using FTIR testing were carried out to determine the functional groups contained in the sample and to know the fingerprint of the sample that cannot be seen through XRD testing. Figures 3 and 4 show the FTIR spectra of samples with molarities variations [P]/[F] 4 and 6 with variations in the microwave irradiation of 30P, 50P and 100P. The same functional groups are shared by all samples namely phosphate (PO_4^{3-}), hydroxyl (OH^- and OH...F), carbonate (CO_3^{2-}) and CaO. The wavenumbers of each functional group are in table 1.

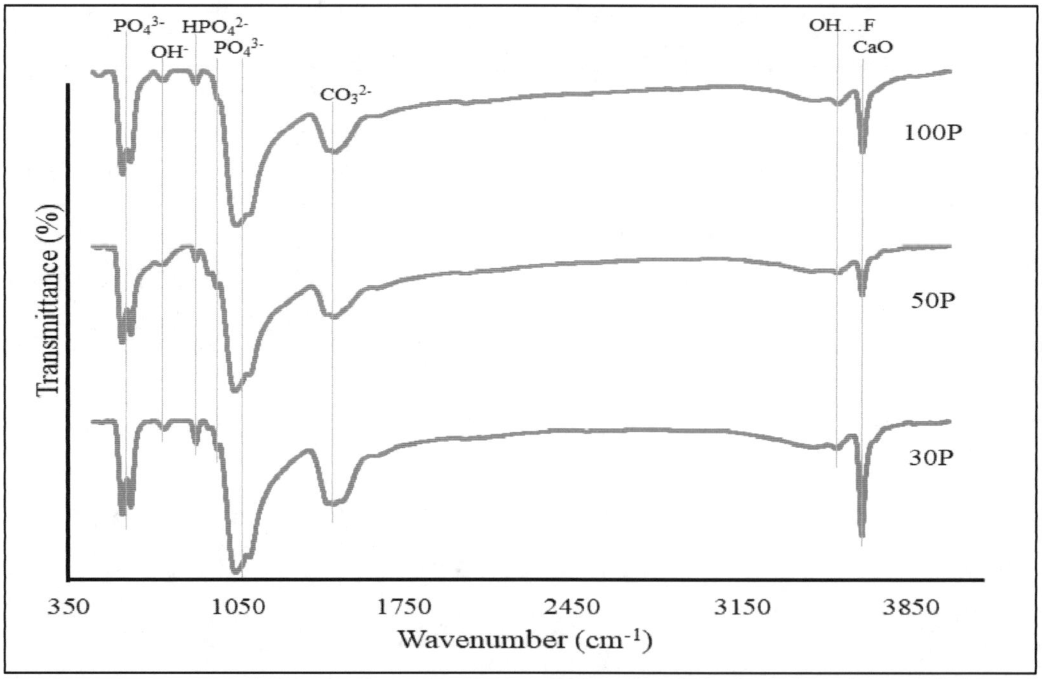

Figure 3. FTIR spectra for fluorhydroxyapatite with a molarity variations of [P]/[F] 4 and with variation of microwave irradiation 30P, 50P and 100P.

IOP Conf. Series: Materials Science and Engineering **858** (2020) 012011 doi:10.1088/1757-899X/858/1/012011

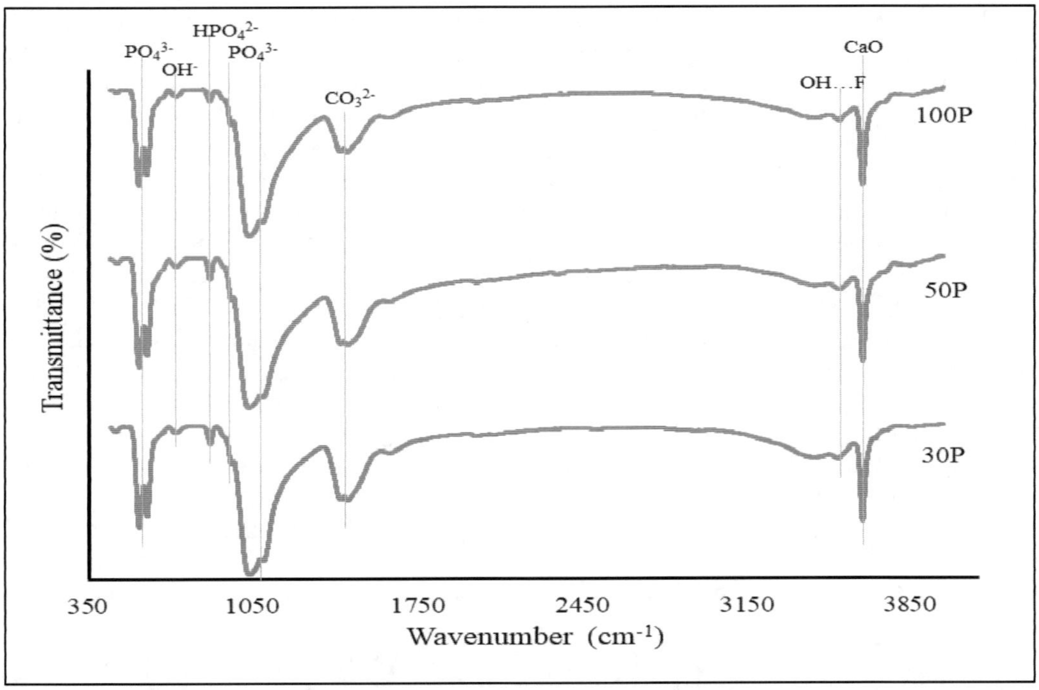

Figure 4. FTIR spectra for fluorhydroxyapatite with a molarity variations of [P]/[F] 4 and with variation of microwave irradiation 30P, 50P and 100P.

Table 1. The fluorhydroxyapatite FTIR assignments and wavenumber of each variations

Assignment	Literature	Wavenumber (cm-1)					
		[P]/[F] 4			[P]/[F] 6		
		30P	50P	100P	30P	50P	100P
OH^-	744[1]	747	744	744	746	746	746
v_3 PO_4^{3-}	1048[1]	1042	1045	1052	1038	1038	1039
v_1 PO_4^{3-}	970[1]	966	966	965	965	965	965
CO_3^{2-}	1473[1]	1466	1467	1466	1460	1461	1460
v_4 PO_4^{3-}	568[1]	573	572	573	571	572	572
v_4 PO_4^{3-}	605[1]	607	609	607	609	609	609
HPO_4^{2-}	870[1]	876	877	874	875	875	874
CaO	3640[2]	3646	3647	3645	3644	3645	3645
OH...F	3546[1]	3539	3541	3543	3542	3547	3546

[1] Sumber : Eslami *et al.* 2009
[2] Sumber : Nasrazadani *et al.* 2008

IOP Conf. Series: Materials Science and Engineering **858** (2020) 012011 doi:10.1088/1757-899X/858/1/012011

The HPO_4^{2-} function group appears at wave numbers about 870 cm^{-1} in all samples. The CaO functional group appears in all samples at wavenumbers at around 3645-3646 cm^{-1} [16]. This justifies the data shown in XRD testing. This CaO functional group arises because it has not bonded with other compounds during the process of precipitation and sintering. In addition, it is also possible to increase the volume of CaO, so that it cannot be paired with other functional groups as expected.

Characteristics of PO_4^{3-} functional groups have four different asymmetric vibration modes including v_1, v_2, v_3 and v_4. Vibration modes v_1 and v_2 are observed respectively at wavenumbers 961 and 471 cm^{-1}. Vibration mode v_3 is the main vibrational mode of the PO_4^{3-} functional group which is at wavenumbers between 1000-1100 cm^{-1}, and finally is the vibration mode v_4 at wave number 560-610 cm^{-1} which occupies two crystal lattice locations, 603 and 574 cm^{-1} respectively [7]. The PO_4^{3-} function group found in the sample is only vibration mode v_1, v_3, and v_4. Vibration mode v_1 PO_4^{3-} of all samples is around the wave numbers 966 and 965 cm^{-1} [16]. For vibration mode, v_3 PO_4^{3-} of all samples are around wave numbers 1038 to 1052 cm-1 and vibration mode v_4 are at wavenumbers between 571-573 cm^{-1} and 607-609 cm^{-1} [16].

Wave numbers around 744-747 cm^{-1} appear in all samples that are owned by hydroxyl (OH^-) functional groups. When this hydroxyl group is partly replaced by a fluorine ion, it will switch to a new bond to an OH ... F bond [17]. This functional group appears in all samples around the wave number 3542-3546 cm^{-1}. This is a unique characteristic possessed by fluorhydroxyapatite which is not possessed by other types of calcium phosphate. Based on this, it is clear that all the samples produced were fluorhydroxyapatite.

The carbonate (CO_3^{2-}) functional group also appeared in all fluorhydroxyapatitesamples, which were present in wavenumbers around 1460-1467 cm^{-1}. This indicates the presence of apatite carbonate. This carbonate apatite can arise due to absorption of fluorhydroxyapatite samples of carbon dioxide from the atmosphere [16].

3.3 Conductivity evaluation

Conductivity can occur due to ion jumps. In fluorhydroxyapatite, the conduction mechanism is related to the release of fluoride ions along the c axis from the cell unit to the lattice in the interstitial state and then back again. Fluoride ions must move to other positions by forming defects, where these defects can occur through thermal activities such as Schottky defects which require high activation energy due to the nature of the fluorhydroxyapatite material [18].

The conductivity of the sample can be seen in figure 5 and figure 6 It can be seen that the conductivity value of all samples increased with increasing frequency, in other words, the conductivity value is linearly related to frequency. This can be explained through theory, where the frequency value corresponds to the R-C (resistor-capacitor) model. The ion contained in a material is a complex conductive and capacitive state [19]. Meanwhile, the irradiation used in the process of fluorhydroxyapatite synthesis does not show any relationship with the conductivity value of fluorhydroxyapatite. It can be seen that the variation in molarity of [P]/[F] 4 and [P]/[F] 6 conductivity values are different and are not consistent with increasing frequency. Fluorhydroxyapatite with 30P of microwave irradiation at [P]/[F] 4 has the lowest conductivity among other samples, while at [P]/[F] 6 the lowest conductivity is owned by 100P of microwave irradiation. This shows that the irradiation of the micro irradiation microwave used does not affect the conductivity value of fluorhydroxyapatite.

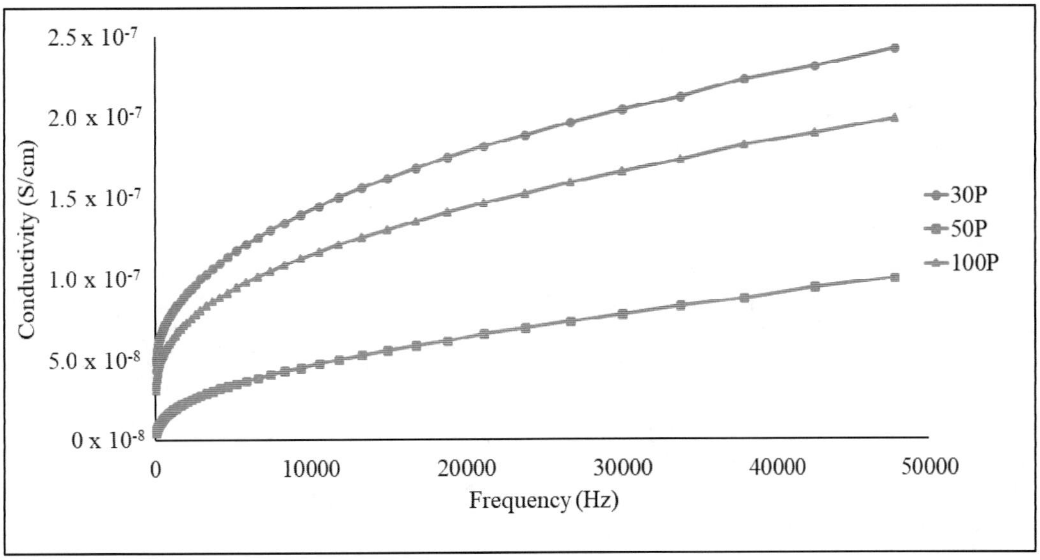

Figure 5. The conductivity of fluorhydroxyapatite with a molarity variation of [P]/[F] 4 and variation of microwave irradiation 30P, 50P and 100P.

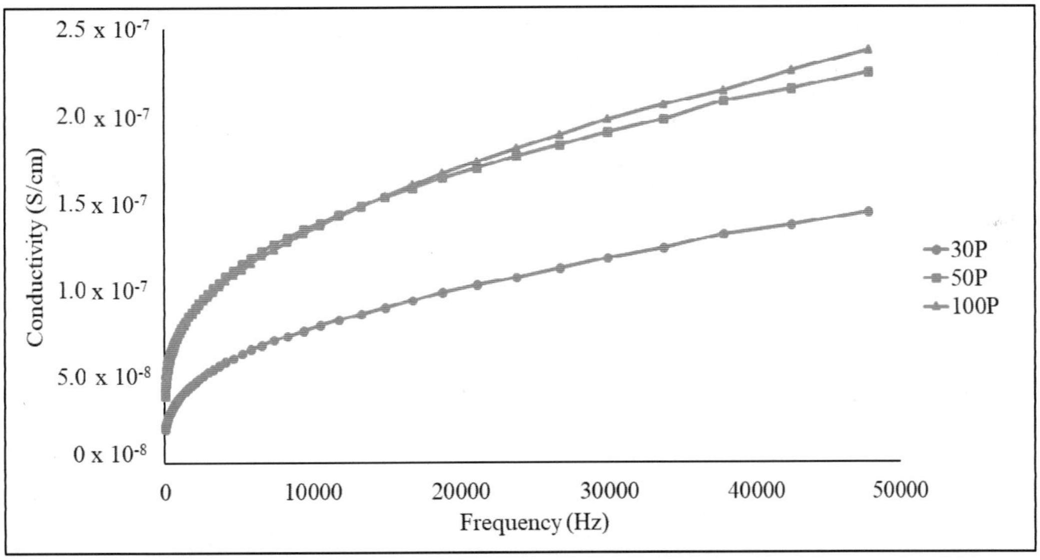

Figure 6. The conductivity of fluorhydroxyapatite with a molarity variation of [P]/[F] 6 and variation of microwave irradiation 30P, 50P and 100P.

4. Conclusion

Fluorhydroxyapatite synthesis based on chicken eggshells was carried out using precipitation method assisted by microwave irradiation, by varying the ratio of molarity and microwave irradiation used. The XRD results show that samples at 100P irradiation both in [P]/[F] 4 and 6 molarity variations, fluorhydroxyapatite has the least impurity intensity among the others and the highest fluorhydroxyapatite intensity among the others. FTIR analysis also shows that all samples have the same functional groups. The FTIR analysis also shows that fluorhydroxyapatite has formed, which is shown by the emergence of the OH...F functional group at wavenumber 3539-3544 cm^{-1}. Microwave irradiation and molarity ratio [P]/[F] showed no relationship with conductivity value of fluorhydroxyapatite

References

[1] Bragramian R A, Godoy F G, and Volpe A R 2009 *Am J Dent.* **22** 3-8
[2] Nodstrom M 2018 Global DMFT for 12-years-olds: 2015
[3] American Dental Association 2014 Dental Caries
[4] Heintz S D, Bastos J R M and Tomita N 1996 *J. Dent. Res* **75**:192
[5] Lussi A, Jaeggi T and Ucharer U J 2008 *J. Dent. Res* **87**:429-30
[6] Jarvinen V K, Stabholz A and Wilkinson G 1990 *J. Dent. Res* **79**(6):942-7
[7] Fathi M H and Zahrani E M 2009 *J. Alloys and Compd* **475**:408–414
[8] Rameshbabu N, Kumar T S S and Rao K P 2006 *Bull. Mater. Sci* **29**(6):611–615
[9] Wei M, Evans J H, Bostrom T and Grondahl L 2003 *J. Mat Sci: Mat Med* **14**:311-320
[10] Eslami H, Hashjin M S and Tahriri M 2008 *J. Ceram. Process Res.* **9**(3):224-229
[11] Ferreira J M, Aragao A K, Rosa A D, Sampaio F C and Menzes V A 2009 *Braz.Oral Res* **23**:446-451
[12] Taheri M M, Kadir M R A, Shokuhfar S, Hamlekhan A, Shirdar M R and Naghizadeh F 2015 *Mater. Des* **82**:119–125
[13] Kim H W, Kim H E and Knowles J C 2004 *Biomaterials* **25**:3351–3358
[14] Tredwin C J, Young A M, Neel E A A, Georgiou G and Knowles J C 2014 *J. Mat Sci: Mat Med* **25**:47–53
[15] Hassan M N, Mahmoud M M, El-Fattah A A and Kandil S 2015 *Ceram. Int* **42**:1-10
[16] Nasrazadani S and Eureste E 2008 *Application of FTIR for Quantitative Lime Analysis* (Texas : University of North Texas)
[17] Eslami H, Solati-Hashjin and Tahriri M 2009 *Mater. Sci. Eng C* **29**:1387–1398
[18] Shafiei F, Behroozibakhsh M, Moztarzadeh F, Haghbin-Nazarpak M and Tahriri M 2012 *Micro Nano Lett* 7:109-114
[19] Laghzizil A, Herch N E, Bouhaouss A, Lorente G and Macquete J 2001 *J. Solid State Chem* **156**:57-60
[20] Gitting J P, Bowen C R, Dent A C E, Turner I G, Baxter F R and Chaudhuri J B 2009 *Acta Biomater* **5**:743–754

Characterization and performance of low cost amendments to immobilize lead in contaminated soil

Himawan[1,2], P Setyono[1], S Suntoro[1], Sentot B Rahardjo[1]
[1]Postgraduate Program of Sebelas Maret University, Surakarta Indonesia
[2]Department of Chemistry Cenderawasih University, Jayapura Indonesia

E-mail: himawanhim.11@gmail.com

Abstract. Immobilization is a method of remediation to reduce lead (metals) mobility by soil amendments to prevent from uptaking of plant roots and ground water contamination. This research aims to produce, characterize, and assess the efficacy of low cost solid amendments in immobilizing lead in naturally contaminated soil. Biochars [produced from chicken manure (CM), chicken bone (CB), farmyard manure (FM)] and bioslurry (BS) were characterized with FTIR, SEM, and XRF. Amended soils were incubated and sequentially extracted prior to Pb measurement with AAS. Chars were made by pyrolysis at 450°C for 90 minutes. Characterization data from FTIR showed that all amendments had similar spectra showing the existence of hydroxyl, carbonyl, C=C, C-H, C-O-C groups in the amendments. SEM images showed porous structure of the chars having pores diameter between 1.95 and 13.4 μm. XRF spectra revealed that dominant elements possessed by amendments were Ca, Na, Si (BS & FM), and P (TA & CM). The immobilization study showed that performance order from the most effective was CB > CM > FM > BS. Chicken bone biochar (CB) was the best amendment which immobilized up to 79.71% Pb and remediated soil from very high contamination-moderate risk into low contamination-low environmental risk status.

1. Introduction

Lead contaminated agricultural soil has become a global problem because of its risks to the health of living organism through food chain. Lead in soil can also contaminate groundwater. Human activities such as mining, smelting, agriculture and industry have played an important source of Pb in soil. Heavy metal is non-biodegradable and possesses long-term persistence in soils [1], therefore effective remediation measures are needed [2].

Remediation of Pb contaminated soil by conventional techniques such as excavation, soil washing and landfilling are costly and not suitable for agricultural soil in developing countries [2]. A sustainable method with low energy input, improving soil quality, and adding economic benefits is being sought [3]. Phytoexraction and immobilization are the two methods of in situ soil remediation which are considered the most appropriate because they are cheap, potentially maintain the structure and function of soil, and are most socially accepted. Phytoexraction and immobilization are contradictory *in situ* methods in remediation of lead-contaminated agricultural soil. Phytoexraction exploits plant ability in taking up Pb, so that efforts to increase Pb solubility in soil solution are needed. On contrary, immobilization focuses on reducing lead mobility by soil amendments to prevent from up taking of plant roots and ground water contamination.

Soil lead immobilization with low cost amendments is promising soil remediation for developing country. Biochar, carbonaceous porous material produced from pyrolysis of biomass, has widely applied due to its alkaline and good sorptive properties for metals and organics [4, 5]. Besides the wide application as soil ameliorant in agricultural cultivation [6], alkaline and nutrient-rich slurry biogas has been used to remediate heavy metals from polluted soils and reduce Pb levels in soil extraction and test plant seeds [7]. Bioslurry also detoxifies Hg demonstrated by increased activity and weight of microbial biomass [8].

Environmental risk assessment of metal contamination in soil can be better performed by sequential extraction procedure (SEP) proposed by the Community Bureau of Reference (BCR) than conventional metal total concentration [9, 10]. In SEP, metal concentration in the most mobile fraction is distinguished from other more stable fractions. In BCR procedure, metal in soils is defined operationally into four geochemical fractions / phases i.e, acid extractable (F1), bound to Fe/Mn oxides (F2), bound to organic matter (F3, and residual fraction (F4). Acid extractable is the most mobile fraction and percentage of F1 determines the environmental risk of the metal [9, 11-13].

Surface properties, pH, and chemical composition of amendments are often associated with the ability of metal sorption and organic pollutants [14, 15]. Even so, there is still not enough information on the main properties that determine the effectiveness of Pb immobilization due to the level of contamination and soil properties [16]. Characterization of amendment material was carried out by FTIR, SEM, and XRF. The study was conducted by soil incubation techniques and continued by three-stage extraction using the modified Community Bureau of Reference (BCR) method and measurement of Pb concentrations by AAS. Objectives of this research are (i) to explore the properties of amendments, (ii) assess efficacy of low-cost amendments in influencing soil-Pb mobility, (iii) evaluate probable environmental risk of Pb in soil after immobilization.

2. Experimental

Amendment materials used in this research were collected from District Jaten and District Gondangrejo in Karanganyar Regency. Contaminated soil was collected from Cinangka, Bogor Regency. We prepared 4 low cost amendments from agricultural byproduct to minimize input of the immobilization process. Chars were made from chicken manure (CM), chicken bone (CB), farmyard manure (FM) by pyrolysis at 450 °C for 90 min in laboratory scale muffle furnace. Solid amendments were pulverized in a wooden mortar and filtered through 0.5 mm plastic filter and characterized with Fourier transform infrared (FTIR) spectroscopy to evaluate functional groups and XRF to perform elemental analysis. SEM was utilized to examine surface properties of the solid. Amended soils were incubated, measured of pH in water (1:2.5), and analyzed for loss of ignition (LOI) as a proxy to organic content. Incubation was conducted in 150 mL plastic container. Each 25 g of air-dried soil was admixed homogenously with 5% amendments, moistened, and incubated for 5 weeks in the dark at room temperature [17]. The incubated soil were checked and moistened weekly.

SEP with modified three steps BCR [9, 10] was carried out for studying redistribution of Pb into the geochemical phases after incubation of amendments. This method distinguishes metals into 4 fractions and is summarized in Table 1.

Table 1. Three step modified BCR sequential extraction procedure

	Fraction	Procedure
F1	Acid extractable	Add 40 mL of 0.11 M acetic acid o 1.00 of soil sample. Shake for 16 h
F2	Reducible	Add 40 mL of 0.5 M $NH_2OH.HCl$ (adjusted pH2 with HNO_3) to F1 residue. Shake for 16 h
F3	Oxidizable	Add 10 mL of H_2O_2 8.8 M, incubate at 85 °C for 1 h (Repeat procedure twice and cold). Add 50 mL CH_3COONH_4 (adjusted pH2 HNO_3). Shake for 16 h

Residual Fraction (F4) remaining in the last step was extracted with aqua regia [1, 18, 19]. Solid solution separation between steps was done by centrifugation at 2500 rpm for 20 min. Solid was then washed with 20 mL deionized water, shaken for 15 min and centrifuged at 2500 rpm for 20 min. All metal quantification in each step was measured with atomic absorption spectrophotometer (AAS) Shimadzu AA 6500.

Impact to the environment was approached with the values of individual contamination factor (ICF) and Risk Assessment Code (RAC). ICF was the measure of retention time of the metal before reaching the environment and determined by dividing the sum of non-residual fractions by residual fraction (F4) [9, 13]; also called as ratio of secondary to primary phase, RSP [20]. RSP can be used to distinguish anthropogenic sources of heavy metal from natural sources. The value of CF expresses contamination degree. RAC (%F1) expresses risk or mobility of the metal/pollutant and is determined with the proportion of metal in F1 from the total metal concentration in soil. Immobilization efficiency (E) was employed to evaluate the effectiveness of the methods [21] and formulated as follow:

$$E = 100\,(M_o - M_e) / M_o$$

M_o : metal extract concentration in soil without amendment
M_e : metal extract concentration in amended soil .

3. Results and discussions

3.1. Characteristics of soil and amendments

Table 2 shows the important physical-chemical properties of soil samples. Soil sample had acidic pH, loam texture, and rather low organic compound (LOI). Pb content is extremely high for agricultural soil in China, USA, and Australia [12]. Ceiling concentrations in China Soil quality standard for grade B and grade C are 350 and 500 mg/kg. Grade C includes upper limits to ensure agricultural production and plant growth and Pb content in contaminated soil exceeded 8 times of the standard. Also, It exceeded 10 times of U.S. EPA (400 mg/kg). Similarly, Australia has Health investigation level (HIL) A and C standard for residential with garden/accessible soil and open space respectively. Compared to Values of HIL A and HIL C for lead (300 and 600 mg/kg); the soil exceeded 13 times and 6 times of Australian standard.

Table 2. Physical-chemical properties of soil sample

Texture (%)			Soil Texture	Pb total	EC	pH	CEC	LOI
Sand 2 – 0,05	Silt 0,05-0,002	Clay <0,002	(USDA)	ppm	mS/cm		me/100 g	(%)
25.54	49.70	24.76	LOAM	4029	0.13	5.5	17	8.45

FTIR spectra and SEM micrograph of the biochars were shown in Figure 1 and 2. FTIR data showed 9 peaks in FM spectra and 13, 14, 15 peaks in the spectra of CM, CB BS respectively. Some similar important peaks were tabulated in Table 3. Characterization data from FTIR showed that solid amendments had similar spectra indicating the existence of hydroxyl, carbonyl, carbonate, phosphate C=C, C-H, and C-O groups in the amendments [22]. Peak of hydroxyl group appeared at frequency of 3500 cm^{-1} corresponded to OH bond stretching. Aromatic C=C and C=O peaks observed 1613 cm^{-1} [15]. The peaks observed at 1420-1450 and 1034 cm^{-1} showed that biochar also had abundant CO_3^{-2} and PO_4^{-3}. These functional groups can act as precipitation agent and sorption sites for Pb in soil. Carbonates and phosphates compound of Pb have low solubility. For example, hydroxypyromorphite, $Pb_{10}(PO_4)_6(OH)_2$ $K_{sp} = \sim 10^{-78}$ will be effectively reduce lead concentration in soil solution down to order of μg/kg.

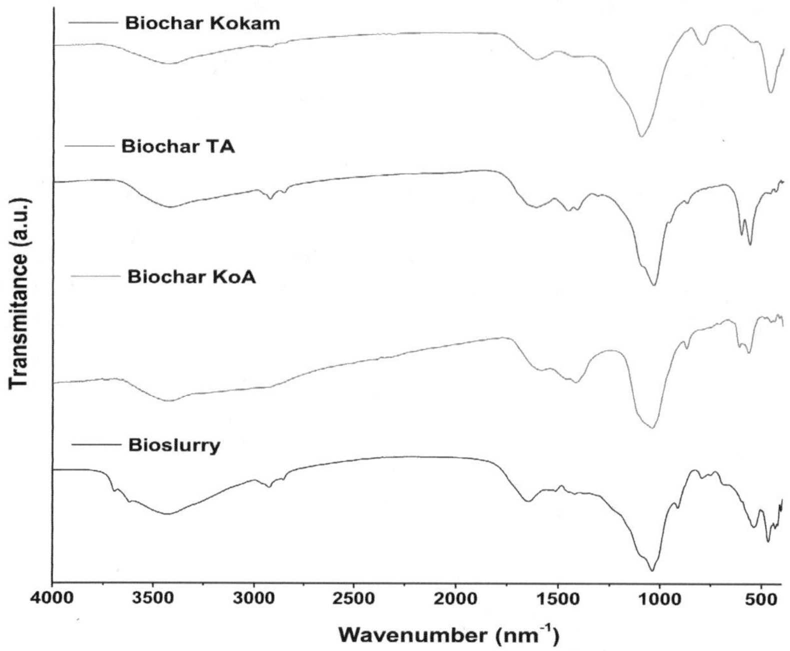

Figure 1. FTIR spectra of chicken manure biochar (CM)

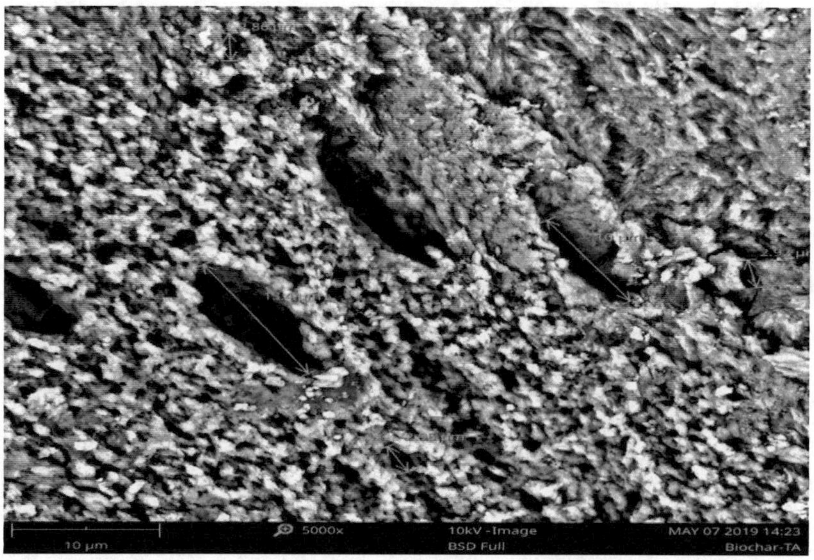

Figure 2. SEM photomicrograph of chicken bone biochar (CB)

Table 3. Peaks, area, and functional groups in the FTIR spectra of amendment

Peak (cm^{-1})	CB	CM	FM	BS	Note
1000-1050	1034.85(31.18)	1040.64(23.17)	-	1036.78(23.30)	Phosphate; PO_4^{3-} C-O alcohol
1410-1465	1415.81(0.393); 1458.2(0.688)	1417.74(-1.04); 1463.07(0.062)	1448.6(0.008)	1421.6(0.049)	Carbonate; $CO_3^{=}$ C=C aromatic
1600-1700	1613.52(5.865)	-	1613.52(0.184)	1643.42(0.039)	C=O carboxyl or ketones
2850-3000	2853.81(0.144) 2925.17(0.997)	2929.03(0.061)	2855.73(0.004) 2926.14(0.083)	2852.84(0.02) 2923.25(0.691)	C-H alkane
3420-3550	3421.87(20.09)	3435.37(0.196)	3439.23(0.031)	3428.62(0)	-OH

SEM images showed porous structure of the chars having pores diameter between 1.95 and 13.4 µm. CM and FM had comparable pore size of 1.95 – 3.76 and 1.49 – 4.95 µm respectively whereas CB had the largest pore size of 2.86 – 13.4 µm. Compared to other biochars produced from woody material (*Acacia magnium, Eucalyptus grandis, Gmelina arborea*), our biochar (CM, CB, and FM) had more irregular pore size and shape [15]. The relatively low temperature (450oC) and particularly pyrolysis time (90 min) might cause incomplete carbonization process. In spite of poor pore properties, lower temperature pyrolysis was favorable in Pb immobilization because resulting higher available phosphate [23] and increasing available P, K, and Ca associated with effective Pb stabilization [24]. Elemental analysis data from XRF is summarized in Table 4.

Table 4. Elemental % composition of amendments and soil

Material \ element	O	K	Ca	S	P	Si	Na
CM	32.41	18.86	17.91	2.25	7.08	1.96	9.33
CB	35.12	1.82	34.16	0.67	10.54	0.94	10.83
FM	43.33	4.30	6.82	0.97	2.63	24.70	10.19
BS	43.36	2.86	7.88	2.74	2.26	19.16	BDL
C	34.14	1.98	2.26	BDL	0.74	9.04	17.74

Table 4 showed that CM and CB had relatively high mineral content. These minerals had the potential to enrich plant nutrients and increase soil fertility. Si element is prominent for FM and BS as much as 24.70 and 19.16%. The high calcium and phosphorus (Ca and P) content was observed for CM and CB. The values were 17.91 and 7.08% for CM and 34.16 and 10.4% for CB. Chicken manure biochar, CM also had the highest K Content as much as 18.86%. Content of available P, K, and Ca in the amendment was essential in effective Pb stabilization [24]. Because of limitation in instrumental sensitivity, sulfur content in soil control and sodium content in BS were below the detection limit of measurement (BDL). The high content of P in CB and CM confirmed FTIR spectra data at 1046 cm^{-1} showing phosphate peak.

3.2. Efficacy of amendments in influencing soil-pb mobility and fraction redistribution

The effect of amendments on soil properties, fraction distribution of Pb, efficacy of immobilization, and risk is summarized in Table 4. Amendments cause changes of soil pH from 5.5 to (6.6 – 8.3) and F1-lead from 17.1 to (3.47–14.72) ppm and induces immobilization efficiency (%) E between 13.92 and 79.71%. Residual fraction (F4) and reducible fraction (F2) also changed significantly but organic fraction (F3) were relatively constant. Residual fraction (F4) changes between 4.69 and 36.23 ppm whereas F2 changes between -3.21 and -34.5 ppm.

The immobilization order is CB > CM > FM > BS. The low cost immobilizing amendments can be applied, either individually or in combination, as phytostabilization agent. The use of amendments together can provide multi benefits for plants such as decreasing Pb solubility, increasing soil nutrients, and improving physical and biological properties of the soil [20, 25, 26].

All chars rises soil pH and immobilizes Pb. This fact is in good agreement with previous reseach [10, 20, 23]. In general, chars are more effective than bioslurry. The immobilization order of CB > CM > FM > BS is followed with the increasing trend of F4 and decreasing of F2 and F1 (Figure 3).

Table 5. Soil properties, BCR fractions Pb, immobilization efficiency (%E), and risk assessment after incubation with amendments

Effect / AMD	pH	LOI	F1	F2	F3	F4	F1-4	**E (%)**	CF	RAC and **RISK**	
C	5.5	8.78	17.1	69.3	11.0	13.7	111.1	**0.00**	7.1	15.39	**M**
CM	8.0	9.88	6.15	49.35	13.04	38.23	106.8	**64.03**	1.79	5.76	**L**
CB	8.3	9.47	3.47	34.8	13.0	49.93	101.2	**79.71**	1.03	3.42	**L**
FM	8.3	9.82	12.3	58.7	12.1	19.5	102.6	**28.07**	4.26	11.99	**M**
BS	6.6	10,48	14.72	66.09	10.4	18.39	109.6	**13.92**	4.96	13.43	**M**

Remarks:

C	: control soil
CM	: chicken manure biochar
CB	: chicken bone Biochar
FM	: farmyard manure biochar
BS	: bioslurry
E	: efficiency of immobilization
CF = RSP	: Concentration factor; CF = (F1 + F2 + F3) / F4
RAC	: Risk assessment code = %F1.

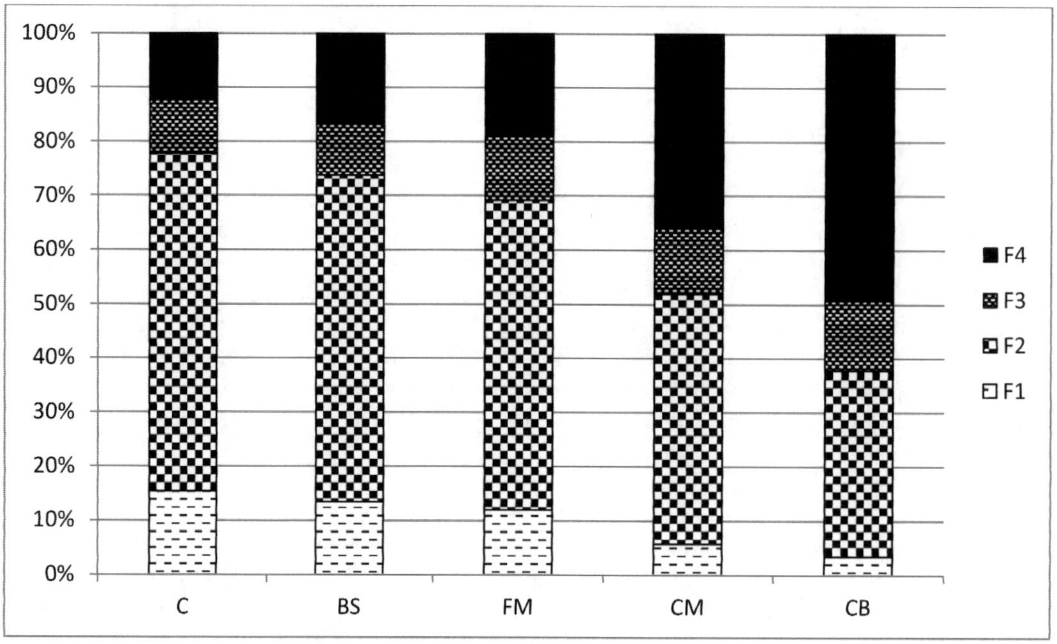

Figure 3. Redistribution of Pb geochemical fraction after amendment

3.3. Environmental risk of Pb after immobilization

Values of contamination factor (CF) and Risk Assessment Code (RAC) are employed to assess the risk. The relationship is: CF $\leq$ 1 low contamination ; 1 $\leq$ CF $\leq$ 3 moderate contamination; 3 $\leq$ CF $\leq$ 6 considerable contamination; and CF > 6 very high contamination [13]. RAC (%F1) expresses risk or mobility of the metal / pollutant and is determined with the proportion of metal in F1 from the total metal concentration in soil. The potential risk is: % RAC < 1 (no risk); 1 < % RAC < 10 (low risk); 10 < % RAC < 30 (moderate risk); 31 < % RAC < 50 (high risk); and % RAC > 50 (very high risk). Table 5 shows that polluted soil has CF and RAC values of 7.1 and 15.39 respectively. It means that the soil has very high Pb-contamination with moderate environmental risk. In this work, CF values of soil after amendment vary from 1.03 to 4.96 and RAC values vary from 3.42 to 13.43%. CB (CF 1.03 ; RAC 3.42) and CM (CF 1.79 ; RAC 5.76) were the two most effective immobilizing amendments which remediating soil into low contamination, low environmental risk status.

The changes of environmental risk may be associated with redistribution of lead geochemical fractions during incubation with amendments. Reducible fraction (F2) is dominant in soil before incubation (69.3 ppm). Effective immobilizing amendments (CB & CM) significantly decrease F2, F1 and increase residual fraction (F4). The decrease of F2 is 34.5 and 19.95 ppm while the decrease of F1 is 13.63 and 10.95 ppm for CB & CM respectively. It is, therefore, compensated by the F4 increase of 36.23 and 24.43 ppm.

Toxicity characteristic leaching procedure (TCLP) test is other method in assessing ecological risk of material and waste from USEPA. The test were performed by shaking 1.00 g of soil (material) with 20 mL of 0,1 M acetic acid (solution #2; pH 2.88 ± 0.01) for 16 h in a plastic centrifuge tube [17, 23, 27]. Supernatant was then filtered through quantitative filter paper and measured for leachate concentration. This procedure was very similar to F1 procedure in modified BCR employed in this research work. The only differences were concentration and volume of acetic acid solution; 20 mL of 0,1 M in TCLP versus 40 mL 0.11 M in BCR procedure; hence he result should be comparable. Concentration of acid extractable Pb fraction (F1) of the soil was 17.1 ppm and therefore can be classified as hazardous waste (exceeded regulatory level of 5 ppm). Only CB reduced successfully Pb concentration in acid fraction below critical level; which was 3.47 ppm.

4. Conclusion

FTIR data showed the existence of hydroxyl, carbonyl, C=C, C-H, C-O-C, carbonate, and phosphate groups in the amendments. SEM images showed irregular porous structure of the chars having diameter between 1.95 and 13.4 μm. XRF spectra proved that dominant elements possessed by amendments were Ca, K, Si (for BS & FM), and P (for CB & CM). All amendments immobilized lead in the order of CB > CM > FM > BS. Chicken bone biochar (CB) was the best which immobilized up to 79.71% Pb and might redistribute it from reducible and acid soluble into residual fraction. CB and CM remediated soil from very high contamination-moderate risk into low contamination-low environmental risk status.

Acknowledgment

Thanks to Lembaga Pengelola Dana Pendidikan (LPDP) Kementerian Keuangan Republik Indonesia for funding of this research work on Contract Number PRJ-4940 / LPDP.3/2016.

References

[1] Mahar A, Wang, P, Li R H, and Zhang Z Q 2015 *Pedosphere* **25**(4) 555–568
[2] Khalid S, Shahid M, Niazi N K, Murtaza B, Bibi I, and Dumat C 2016. *Journal of Geochemical Exploration*, http://dx.doi.org/10.1016/j.gexplo.2016.11.021
[3] O'Connor D, Peng T, Zhang J, Tsang D C W, Alessi D S, Shen Z, Bolan N S, and Hou D 2018. *Science of the Total Environment* 619-620: 815-826
[4] Puga A P, Abreu C A, Melo L C A, and Beesley L 2015 *Journal of Environmental Management* **159** 86-9.
[5] Ahmad M, Lee S S, Lee S E, Al-Wabel M I, Tsang D C W, and Ok Y S 2016 *Journal of Soil and Sediments* DOI 10.1007/s11368-015-1339-4

[6] Groot L and Bogdanski A 2013 *Bioslurry = Brown Gold? A Review of Scientific Literature on the Co-product of Biogas Production* (Rome: Food and Agricultural Organization of the United Nations)
[7] Sanchez M G, Romera I G, Szakova J, Kaplan L, and Tlustos P 2015 *Environ Sci Pollut Res* **22** 14325-14336.
[8] Sanchez M G, Klouza M, Holeckova Z, Tlustos P, and Szakova J 2016 *Environ Sci Pollut Res* **23** 14254-14268.
[9] Nemati K, Abu Bakar N K, Abas M R, and Sobhanzadeh E 2011 *Journal of Hazardous Materials* **192** 402-410
[10] Lu K, Yang X, Gielen G, Bolan N, Ok Y S, Niazi N K, Xu S, Yuan G, Chen X, Zhang X, Liu D, Song Z, liu X, and Wang H 2017 *Journal of Environmental Management* **186** 285-292.
[11] Zhang L, Liao Q, Shao S, Zhang N, Shen Q, and Liu C 2015 *International Journal of Environmental Research and Public Health* **12** 14115-14131.
[12] Yang K, Zhang T, Shao Y, Tian C, Cattle S R, Zhu Y, and Song J 2018 *International Journal of Environmental Research and Public Health* **15** 613.
[13] Tytla M 2019 *International Journal of Environmental Research and Public Health* **16** 2430. Doi: 10.3390/ijerph 16132430
[14] Zhang X, Wang H, He L, Lu K, Sarmah A, Li J, Bolan N S, Pei J, and Huang H 2013 *Environ Sci Pollut Res* **20** 8472-8483. DOI 10.1007/s11356-013-1659-0
[15] Hernandez L S, Ardila-A A N, and Zapata R B 2017 *Revista Facultad de Ingenieira* **26**(46) 123-130. DOI http://doi.org/10.19053/01211129 v26n46.2017.7324
[16] Nartey O D, and Zhao B 2014 *Advances in Materials Science and Engineering* 2014. http://dx.doi.org/10.1155/ 2014/715398.
[17] Huang G, Su X, Rizwan M S, Zhu Y, and Hu H 2016 *Environmental Science Pollution Research* **23** 16845-16856. DOI 10.1007/s11356-016-6885-9.
[18] Lim J E, Ahmad M, Lee S S, Shope C L, Hashimoto Y, Kim K R, Usman A R A, Yang J E, and Ok Y S 2013 *Clean – Soil, Air, Water* **41**(12) 1235-1241.
[19] Rodriguez l, Gomez R, Sanchez V, and Azcarate J A 2015 *Environ Sci Pollut Res* **23** 6046-6054.
[20] Yang X, Liu J, McGrouther K, Huang H, Lu K, Guo X, He L, Lin X, Che L, Ye Z, and Wang H 2016 *Environ Sci Pollut Res* **23** 974-984. DOI 10.1007/s11356-015-4233-0
[21] Wuana R A, and Okieimen F E 2011 *International Scholarly Research Network ISRN Ecology* **2011** 1-20.
[22] Cao X, Ma L, Liang Y, Gao B, and Harris W 2011 *Environmental Science & Technology* Dx DOI.org/10.1021/es103752 u
[23] Rajapaksha A U, Ahmad M, Vithanage M, Kim K R, Chang J Y, Lee S S, and Ok Y S 2015 *Environ Geochem Health* http://DOI 10.1007/s10653-015-9694-z
[24] Bolan N, Kunhikrishnan, A, Thangarajan R, Kumpiene J, Park J, Makino T, Kirkham M B, and Scheckel K 2014 *Journal of Hazardous Materials* **266** 141-166. http://dx.doi.org/10.1016/ j.jhazmat.2013.12.018
[25] Kabas S, Faz A, Acosta J A, Arocena J M, Zornoza R, Martinez-Martinez S, and Carmona D M 2014 *Environ Geochem Health* **36** 41-54. DOI 10.1007/s10653-013-9517-z
[26] Herath I, Kumarathilaka P, Navaratne A, Rajakaruna N, and Vithanage M 2014 *J Soils Sediments* DOI 10.1007/s11368-014-0967-4
[27] Rizwan M S, Imtiaz M, Huang G , Chhajro M A, Liu Y, Fu Q, Zhu J, Ashraf M, Zafar M, Bashir S, and Hu H 2016 *Environ Sci Pollut Res*.DOI 10.1007/s11356-016-6695-0

IOP Conf. Series: Materials Science and Engineering **858** (2020) 012013 doi:10.1088/1757-899X/858/1/012013

FTIR, SEM and Micro-CT Scan results of biocomposite scaffold characteristics

K Dahlan[*1,2], **D Y Asra**[1] **and Y W Sari**[1]
[1]Division of Biophysics, Department of Physics, IPB University, Bogor, Indonesia
[2]Advanced Research Laboratory, IPB University, Bogor, Indonesia

[*]Email: kdahlan@apps.ipb.ac.id

Abstract. CHA/PVA/Alginate based biocomposite scaffolds have been successfully synthesized by using freeze drying methods. Carbonated hydroxyapatite or CHA powder was firstly synthesized using calcium from chicken eggshells, phosphate acid, and sodium carbonates as starting materials. Polyvinyl alcohol (PVA) has been chosen in this experiment to reinforce the mechanical properties of the CHA/alginate scaffolds. The alginate concentrations of 3%, 4%, and 5% were used as porosifiers and $CaCl_2$ was added as a crosslink agent. Properties of the biocomposite were characterized by fourier transform infrared (FTIR) spectrophotometer, scanning electron microscope with energy dispersive X-ray (SEM-EDX), and micro computed tomography (Micro-CT) Scanner. FTIR results of the scaffolds show spectra corresponding to functional groups of CHA, PVA, and alginate which include O-H, C-O, C = O, C = C, O-Na, P-O and P = O. The morphology of the scaffolds was evaluated by SEM and clearly shows many interconnectivity pores with the values range from 4.92 to 28.48 μm which are sufficient for osteoblast cells to grow. The EDX results show that the scaffolds contain Ca, P, C, O, Na, Cl, and S with the Ca/P ratio ranges from 1.81 to 2.67. Porosity of the scaffolds was measured by micro-CT scanner presenting the range of 25.86% to 42.74%. These scaffolds reveal good prospects for bone substitute application.

1. Introduction

The need of biomaterial is increasing with the number of bone damage cases due to the increasing number of motor vehicle accidents. Every day thousands of surgical procedures are performed to replace or repair tissue that has been damaged. Bone is a highly dynamic connective tissue, which provides adequate mechanical strength and structural support to the body. At macroscopic level, it is classified as cortical or compact bone and cancellous or trabecular bone [1]. The inorganic substance of bone forms a structure that approximates the structure of calcium phosphate apatite carbonation. The main inorganic elements in bone are Ca^{2+} (24.5%) and P^{3-} (11.5%). In addition, bones also contain trace elements such as Na^+ (0.7%), Mg^{2+} (0.55%), Cl^- (0.1%), K^+ (0.03%), F^- (0.02%), and CO_3^{2-} (5.8%) [2]. The mineral composition in human bones differs slightly from hydroxyapatite (HA) stoichiometry due to the presence of impurities in human bones which include carbonate ions, chlorides, fluorides, magnesium, and sodium [3]. Apatite biology in bone has the characteristics of low crystallinity and non-stoichiometry [4]. Bone apatite contains 8% of carbonate ions in apatite crystal structure which HA does not have. The content of carbonate ions in the apatite structure plays an important role in the process of resorption by osteoclasts.

Apatite carbonate (CHA) can be synthesized by various synthesis methods such as precipitation, hydrothermal, ultra-sonication, sol-gel, mechanochemical techniques, and microwave. Microwave is an electromagnetic wave which has frequencies range from 300 MHz-300GHz and may provide a rapid and shorter synthesis time because the waves can penetrate and excite water and lipid molecules evenly. Microwave irradiation may produce nano size material, faster temperature rise, lower energy use, high yield and purity [5]. Synthesis of CHA will produce bone-like material that can stimulate bone growth. But CHA has a weakness that is fragile when it is used as a single element. For this reason, there is a need to mix CHA with elastic compounds such as polyvinyl alcohol (PVA) and alginate. PVA is a non-toxic, water-soluble, biocompatible, and biodegradable synthetic polymer, which is widely used in biochemical and biomedical applications [6]. Alginate is derived from brown algae which has hydrophilic and biocompatible properties, and is relatively economic [7].

In the applications, biomaterials needed by orthopaedic doctors are in the form of scaffolds (porous scaffolds). For bone tissue engineering, therefore, it is necessary to control the porosity and pore characteristics, mechanical strength, biodegradation properties, and bioactivity of the scaffolds. The use of ceramics/polymer-based scaffolds will make the scaffolds osteoconductive and reinforce the scaffolds [8]. The scaffold pores can be produced using freeze drying method. Freeze drying is a simple method for producing scaffolds because it does not need high temperatures which may damage the polymer structures. The water contained in the scaffolds will leave the pores and make the scaffolds become soft and elastic. Based on these matters, in this study, CHA/PVA/alginate biocomposite scaffolds were developed through freeze-drying method. Calcium source used was chicken egg shell (CES) waste which compose of about 95% calcium carbonate, 3.5% glycoproteins, and proteoglycans [9]. Biocomposite scaffolds made from CHA, PVA, and alginate are expected to be one of biomaterial candidates that have potential to be developed for human bone application [10].

2. Experimental

CES, H_3PO_4, and $NaHCO_3$ were used as starting materials for CHA synthesis. CES were collected and washed using distilled water and their membranes were peeled off to remove macro dirts. The cleaned and dried CES were then calcined at 900°C for 8 hours to remove the organic components. $CaCO_3$ will release CO_2 and other organic compounds to produce CaO. CaO was then ground to fine powder using mortar and pestle. 50 ml of 1M $Ca(OH)_2$ was mixed with 50 ml of 0.6M H_3PO_4 and 50 ml of 0.3M $NaHCO_3$ at molar ratio of $Ca^{2+}:PO_4^{3-}:CO_3^{2-}$ = 1.67:1:0.5. The mixing processes were stirred using a magnetic stirrer and the pH was maintained at 13. The resulted solution was then immediately placed into a microwave oven 400 W for 30 minutes.

The initial stage of making scaffold was dissolving 7.5 g PVA powder into a 100 mL distilled water at a temperature of 60°C until homogenous. CHA powder was added to the PVA solution and alginate (3%, 4%, and 5%) was added slowly to the solution until gel was formed. After the solution is stirred, 2 ml of 0.03 M $CaCl_2$ was added as a crosslink agent. The mixtures of CHA/PVA/alginate were then put in a multi-plate well. The quenching gel process was carried out in a freezer for 24 hours, and then followed by freeze drying processes using a freeze dryer until the sample was dried at a temperature of -50°C. The scaffolds obtained were then characterized by using FTIR, SEM-EDX, and micro-CT scanner. FTIR measurements were carried out to determine the functional groups of samples. Morphology analysis, pore size, and element content were observed by using SEM-EDX with magnification 50x-1000x. Porosity analysis was obtained from micro-CT scan results.

3. Results and Discussion

3.1. Biocomposite Scaffolds of CHA/PVA/Alginate

CHA powder has hard and brittle properties. Meanwhile, scaffolds needed in medical applications are biodegradable amorphous scaffolds. Therefore CHA needs to be composited with polymers, such as PVA and alginate, to improve the mechanical properties and form pores. The Micro CT scan images of the CHA/PVA/alginate biocomposite scaffolds is shown in Figure 1. The results of mixing the three ingredients (CHA-PVA-alginate) form a gel mixture and after freeze drying processes produce a scaffold that is elastic like a sponge. Pores in the scaffold appear as a side effect of the disappearance of water (ice crystals) through the sublimation processes at low temperature of about -50°C for 48 hours.

Variation in temperature and time during freeze drying processes greatly affects the performance of the scaffolds. Some experiments suggested that, before freeze drying, it is necessary to consider some factors describing the characteristics of the scaffold such as material composition, mold size, and density.

Figure 1. Micro-CT scan images of the CHA/PVA/alginate biocomposite scaffolds

3.2. FTIR Spectroscopy Analysis

Na-alginate compound is one of the natural polymers with R-O-R groups containing carboxyl groups (C-O), alkana (C-H), carbonyl (C=O), sodium isomers (O-Na), and hydroxyl groups (O-H) [11]. The FTIR spectra identified by the functional groups of alginate compounds are shown in Figure 2a. The absorption peak of the C-O group was detected at wave number 1127 cm^{-1}, the C-H group was identified at wave number 2925 cm^{-1}, the C=O group at wavenumber 1645 cm^{-1}, the O-Na group at wave number 1423 cm^{-1}, and the group O-H at wave numbers 615 cm^{-1} and 3447 cm^{-1}. Meanwhile FTIR spectra for polyvinyl alcohol are shown in Figure 2b which shows the functional groups of O-H, C-H, C = C, and C-O, respectively, at wave numbers 3447 cm^{-1}, 2924 cm^{-1}, 1632 cm^{-1}, and 1035 cm^{-1}.

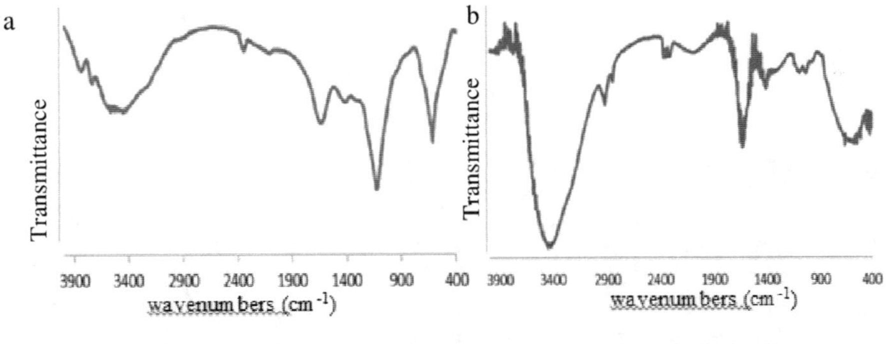

Figure 2. FTIR spectra of Na-alginate and polyvinyl alcohol

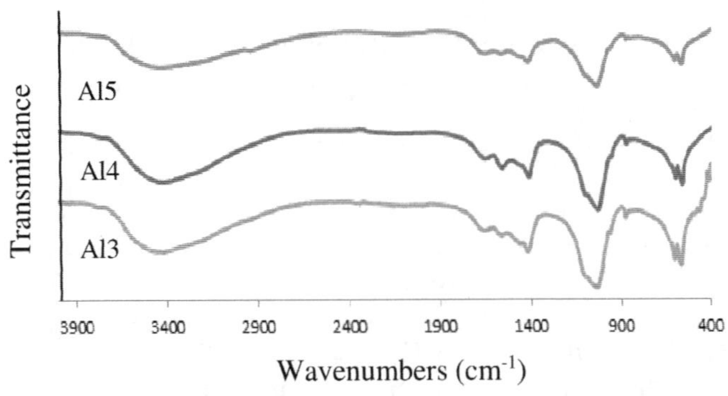

Figure 3. FTIR Spectra of Al3, Al4, and Al5 samples

Figure 3 shows the FTIR spectra of the CHA/PVA/alginate biocomposite with the different concentrations of alginate (3%, 4%, and 5%). The spectra show that the functional groups identified increase in number due to the combination of the functional groups of CHA, PVA and Na-alginate. Based on the FTIR test results, it is obtained that the spectrum with the same patterns and wavenumbers with absorption rates are not too much different. The groups identified by this FTIR spectrometer include carboxyl (C-O), carbonyl (C = O), isomer sodium (O-Na), hydroxyl, and phosphate. The O-H group arises because it is the main group of CHA, alginate and PVA. The C=O, C= C, and C-O bonds confirm the presence of alginate and PVA in the sample, and the P-O and P = O groups are characteristic groups of calcium phosphate. Sodium O-Na isomer shows the presence of sodium alginate in the sample. The results of this characterization do not show the emergence of new functional groups in the scaffold due to the addition of PVA and alginate. In other words the addition of PVA and alginate does not change the composition of compounds chemically, but only changes the properties of the material physically.

3.3. Scanning Electron Microscopy (SEM) Analysis

The results of pore size measurements using ImageJ software reveal the average pore size of the scaffolds as presented in Table 1. The pore size found in these scaffolds tends to be small. This is because the mass ratio of CHA used was large. However, the pore size in some samples is sufficient for the growth of osteoblast cells. Interconnected pore structure is important to bone in-growth and interface support. The micro-porous structure is beneficial to capillary in-growth, nutrient transport and biological properties. Figure 4 shows the SEM images of the morphology of the scaffold surface.

Table 1. Pore size of scaffold biocomposites

Alginate Content (%)	Pore Size (μm)	
	Min (μm)	Max(μm)
3	6.69	11.92
4	4.92	18.35
5	8.78	28.48

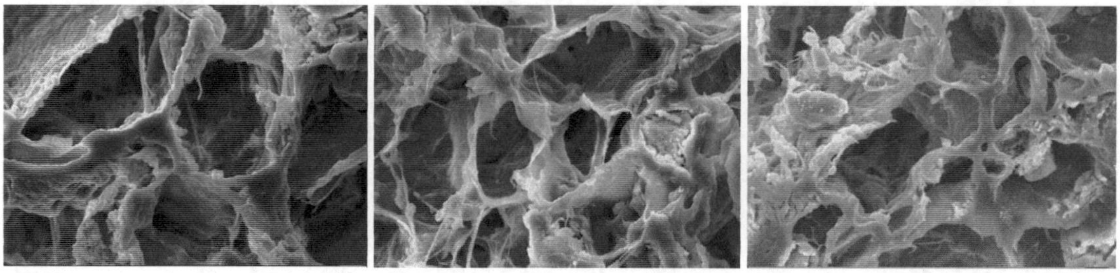

Figure 4. Morphology and pore shapes of the scaffolds

3.4. Analysis of Elemental Composition of the Scaffolds

The EDX results show that the scaffolds compose of carbon, oxygen, phosphorus, sodium, calcium, and other minor elements. The distribution of the elements contained in the composite is not evenly distributed in all composite parts. Based on the percentage of elements contained in the composite, the Ca/P molarity ratio was calculated and the result is shown in Table 2. This result indicates that the Ca/P composite molarity ratio is greater than the stoichiometry of HA molarity ratio 1.67. The increase of Ca/P molarity ratio may be due to the presence of $CaCl_2$ used as a crosslink agent, so that the amount of Ca contained in the composite slightly increases. However, this result clearly shows the relation between alginate content and Ca/P ratio which drives a conclusion that the greater the alginate content in the sample, the smaller the Ca/P ratio produced.

Table 2. Ca/P ratio of the scaffolds

Alginate Content (%)	Ca/P Ratio
3	2.67
4	2.65
5	1.81

3.5. Porosity Measurements

Porosity of the samples for different alginate concentrations was measured by using a Micro-CT scan. Figure 5 shows the relation between the porosity and the alginate concentration. Porosity obtained at each scaffolds of 3%, 4%, and 5% alginate, respectively, are 37.17%, 25.86%, and 42.74%. The values indicate that the porosity is low. Macropores hydroxyapatite for bone filler in spongious femur bone requires porosity of about 70% [12]. However, scaffold porosity greater than 80% can reduce the mechanical strength of the scaffolds. Meanwhile, commercial scaffolds made by Sigma Aldrich have an average porosity value of about 55%.

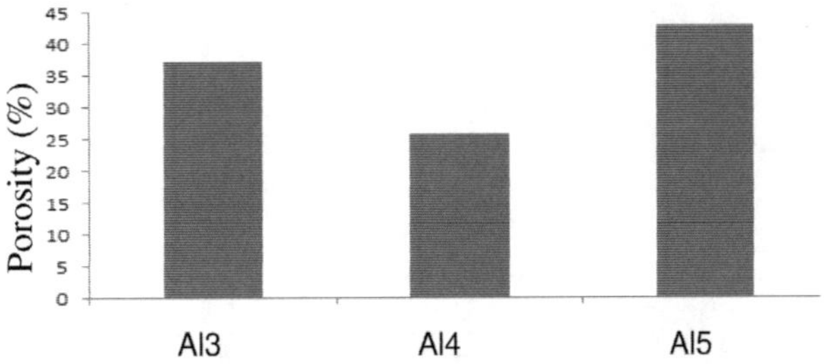

Figure 5. Scaffold porosity of Al3, Al4, and Al5 samples

4. Conclusion

CHA/PVA/Alginate biocomposite scaffolds, with Na-alginate concentration variation of 3%, 4%, and 5%, have been successfully synthesized. FTIR results confirm the presence of O-H, C-O, C = O, C = C, O-Na, P-O and P = O spectra which correspond to the functional groups of CHA, PVA, and alginate. SEM images show the surface morphology of the scaffolds with the interconnected pore structure. The size of the pores range from 4.92 to 28.48 μm. EDX results reveal the Ca/P ratios range of 1.81 to 2.67. The value of the pore sizes found is directly proportional to the concentration of the alginate used in this research. Micro-CT analysis presents that the porosity of the scaffolds obtained is less than 50%. From this experiment, it is found that the sample with alginate content of 5% is the most optimal scaffolds to be used as a bone replacement material since it has large pore size, high porosity and Ca/P ratio close to pure HA.

Acknowledgments

The authors show their appreciation and gratitude to Direktorat Riset dan Pengabdian Masyarakat, Direktorat Jenderal Penguatan Riset dan Pengembangan, Kementerian Riset, Teknologi, dan Pendidikan Tinggi, Republik Indonesia for their support to this research.

References

[1] Rosa I M, Mazzoni E, Manfrini M, D'Agostino A, Trevisiol L, Nocini R, Trombelli , Barbanti-Brodano G, Martini F and Tognon M 2019 *Int. J. Mol. Sci.* **618** 1
[2] Weiner S, Wolfie T, Wagner H D 1999 *J. Struct. Biol.* **126** 241
[3] Wang H L, Zuo Y, Zhang L, Yang W H, Zou Q, Zhou S and Li Y B 2010 *Mater. Res. Innov.* **1** 375

[4] Zhoou W Y, Wang M, Cheung W, Guo B C, Jia D M 2008 *J. Mater. Sci.* **119** 103
[5] Fan L, Du Y, Wang X, Huang R, Zhang L and Hu L 2005 *J. Macromol. Sci. A* **42** 41
[6] Kamoun E A, Kenawy E S, Tamer T M, El-Meligy M A and Eldin M S M 2015 *Arab. J. Chem.* **8** 38
[7] Aly A F, Agameia A, Eldesouky A S and Sharaf M A 2010 *J. Appl. Sci. Res.* **C** 1
[8] Zhoou W Y, Wang M, Cheung W, Guo B C and Jia D M 2008 *J. Mater. Sci.* **119** 103
[9] Wahyudi S T, Dewi S U, Anggraeni A, Dahlan K, Akhmaloka, Zulfikar M A and Hertadi R 2014 *Adv. Mater. Res.* **896** 276
[10] Noviana D *et al.* 2011 *Proceedings-2nd International Conference on Instrumentation, Communication, Information Technology, and Biomedical Engineering* 403
[11] Sopyan I, Mel M, Singh R and Khalid K 2007 *Sci. Technol. Adv. Mater.* **8** 116
[12] Martin B, Suchy T, Tonar Z, Foltan R, Hubalek M and Kalbacova 2018 *Ceram-Silikaty* **62** 194

Adsorption of Pb(II) ion using pectin membrane

B Hastuti[1] **and S Hadi**[2]

[1] Department of Chemistry Education, Faculty *of* Teacher Training and Education, Universitas Sebelas Maret
[2] Department of Pharmacy, Faculty of Mathematics and Natural Sciences, Universitas Sebelas Maret

E-mail: Budihastuti@staff.uns.ac.id

Abstract: Disposal of heavy metal ion into body waters may result in the disruption of aquatic biota ecosystem, thus needs to be overcome. In this study, a cellulose-based adsorbent membranes was developed as adsorbents of Pb ion. The membrane was prepared by mixing chitosan gel in 2,5% acetic acid solution with pectin gel in water for 3 h. The solution then was printed out into a polypropylene container and dried. The membrane characterization was performed using FTIR for functional groups identification, and XRD and SEM for physical characteristic analysis. The results showed the stability and resistance of the pectin membrane in acidic environment. FTIR data confirmed main functional groups that participated in the adsorption process included carboxyl and hydrogen. The XRD data show that the adsorbent has an amorf phase. Furthermore, a solid structure of the adsorbent was showed by SEM data. Adsorption of pectin membranes to Pb (II) ions reaches a maximum at pH 4, an optimum time of 60 min with an adsorption capacity of 92.68% and 92%, respectively. Furthermore, the pectin membrane in adsorbing Pb (II) ions follows the Freundlich isotherm.

1. Introduction

Heavy metal contamination in aquatic ecosystems become a serious environmental problem, due to heavy metal toxicity and its accumulation in aquatic habitats [1]. Some of heavy metal are naturally occurring elements and are needed in small amounts as micronutrients by human body, w*hile* certain heavy metals *have possibility to* accumulated in the body and food chain, and bring a risk to human health.

Lead (Pb) is heavy metal which is associated with the damage to the central nervous system. Lead ions in the blood act as heavy metal poisons that affect almost all hemoglobin synthesis and porphyrin metabolism. Pb metal ion contaminants cause mental retardation [2]. Lead can also damage the kidneys, liver and reproductive system, basic cellular process and brain function[3]. Symptoms of poisoning are characterized by anemia, insomnia, headaches, dizziness, irritability, muscle weakness, hallucinations and kidney damage [4].

Several technologies were developed for heavy metal Pb removal and remediation from the aquatic environment [5]. Current treatment technology that is commonly used and is being developed is filtration, sedimentation, and adsorption. Compared to other methods, adsorption is one method that does not require high costs. Adsorption is the absorption of a substance (molecule or ion) on the surface of the adsorbent. The absorption mechanism can be divided into two, namely, physical absorption (physical absorption) and chemical absorption (chemisorption) [6].

One of the adsorbent biomaterials that has the potential to adsorb heavy metal ions is pectin[7]. Pectin is a polysaccharide that forms cell walls in plants [8]. Potential sources of pectin come from fruit and vegetables or waste. Pectin has a carboxyl group that can interact with heavy metal ions

[9,10]. In this study, pectin will be synthesized into a membrane which is applied as an adsorbent for the metal ion Pb (II).

2. Experimental

2.1 Materials

All the primary reagents used were analytical grade purchased from Merck, Germany. Standard of Pb(II), NaOH, HNO_3, Acetic Acid 5%, Aquadest, Pectin from Surabaya.

2.2 Preparation of adsorbent.

0,2 g pectin was dispersed in 10 mL of acetic acid 5 % and then stirring until 1 h to form a gel. The pectin gel was then poured into a membrane casting and dried. After drying, the formed pectin was then characterized using FTIR to analyze functional groups, XRD to analize membrane crystallinity, and SEM to analyze surface morphology.

2.3 Effect of acidity (pH) of the solution on metal adsorption

10 mg of pectin membrane was put into a glass containing 10 mL of 50 ppm Pb (II) solution which had pH variation of 2, 3, 4, 5, 6 and 7 with 0.1 M NaOH, 1 M NaOH and 0.1 M HNO_3 The mixture was stirred for 3 h at a speed of 40 rpm. Then the mixture was filtered. The filtrate and the blank solution were diluted with 0.1 M HNO_3 solution then analyzed by Atomic Absorption Spectroscopy (AAS).

2.4 Effect of contact time on metal adsorption

10 mg of pectin membrane was put into a glass containing 10 mL of 50 ppm Pb (II) solution which has been adjusted to the optimum pH then stirred with different time variations with variations in contact time of 15, 30, 45, 60, 75, 180, 360, 720 and 1440 min. Then the mixture was filtered, the filtrate was diluted with 0.1 M HNO_3 solution then analyzed with AAS.

2.5 Effect of solution concentration on metal adsorption

10 mg of pectin membrane was put into a glass containing 10 mL of Pb (II) solution with varying concentrations that have been adjusted to the optimum pH as the results of previous observations. The initial concentrations of Pb (II) solution used were 25, 50, 75, 100, 200, 300, 400 and 500 ppm. The mixture is stirred for 1 h. The mixture was filtered then the filtrate was diluted with 0.1 M HNO_3 solution then analyzed with AAS)

3. Results and discussions

3.1 FTIR characterization

Fourier Transform Infrared (FTIR) Spectroscopy were used to analyze the characteristic of the functional active group of the membrane. From figure 1, pectin membran has three peak in the region 1600 cm^{-1} and broad peak at in the region 3000-3700 cm^{-1} indicating of the presence of hydrogen-bonded OH groups on 3464 cm^{-1} and vibration of C=O on 1600 cm^{-1}. Spectra, 1300-1072 cm^{-1} is C=O stretching and 1380 cm^{-1} is C-H bending.

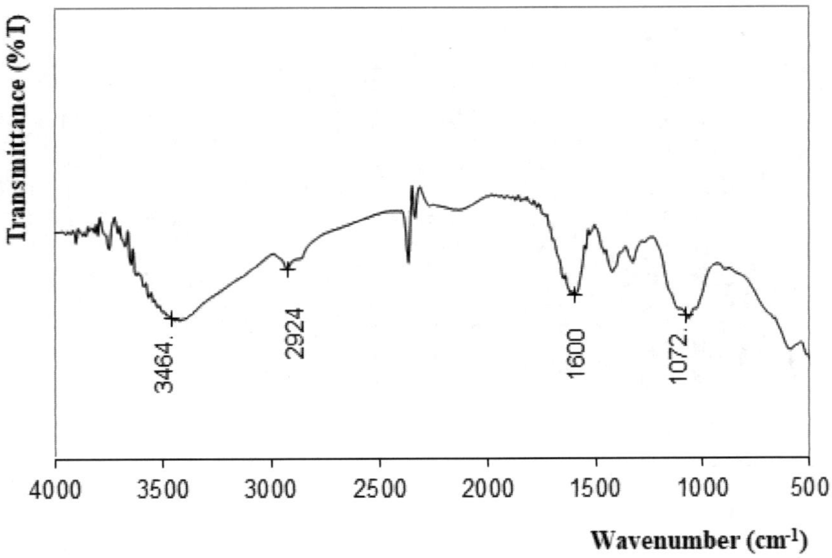

Figure 1. FTIR spectra of pectin membrane

3.2 XRD characterization

Pectin membrane was analyzed using X-Ray Diffraction (XRD) to understand the phase of the material. The XRD spectra of pectin membrane (Figure 2) show that the properties of pectin is amorphous. It can be seen characterized by the peak at $2\theta = 20°$. The peak at $2\theta = 20°$ shows that the membrane have an amorf phase.

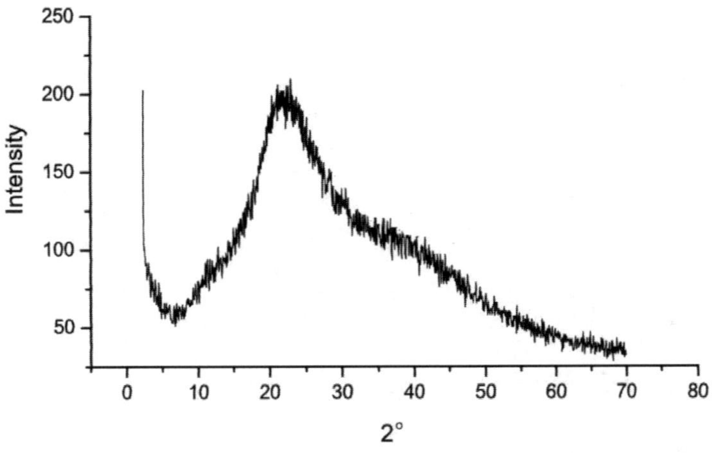

Figure 2. XRD spectra of pectin membrane

3.3 SEM characterization

The surface morphology of pectin membrane was informed by Scanning Electron Microscope (SEM). SEM image shows that the pectin membrane have small and homogeneous pores (Figure 3). The surface and cross section of membrane showed in the 500X magnification. The cross-section image of pectin membrane shows the pores are smoother and smaller. This surface characteristic is an indication the membrane as an efficient absorbent for Pb(II) metal ion.

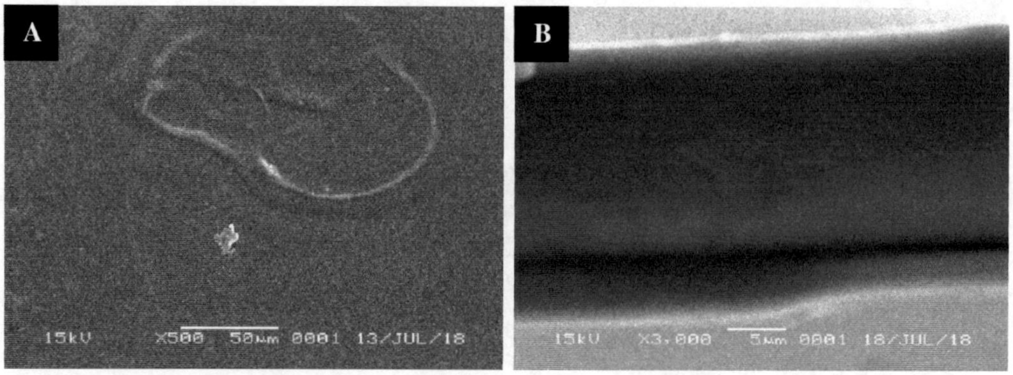

Figure 3. The SEM images of pectin membrane: **A.** surface of the membrane; **B.** Cross section of the membrane (500X magnification)

3.4 The Adsorption test

The effect of acidity (pH) of the solution on Pb(II) adsorption

The acidity of the solution is one important factor that determine the ability of metal ion adsorption. The pattern of adsorption of Pb (II) ions with variations in the pH of the solution is shown in figure 4. At pH 2, the adsorption of pectin membranes to Pb (II) ions has an absorption of 20%, and significantly increase and each maximum at pH 4, with absorption capacity of 90.45%. At pH 5 and higher, the adsorption tend to decrease.

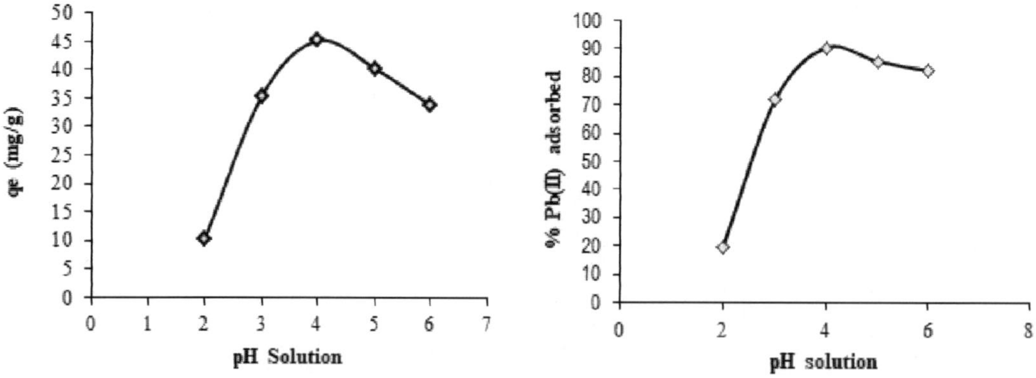

Figure 4. The effect of pH on the pectin membrane adsorption

This maximum adsoption of the pectin membrane on pH range of 4 to 6 showing considerable potential the adsorbent for real applications, particularly for environmental water and industrial wastewater which tends to be having acidic pH.

3.5 The effect of contact time on Pb(II) adsorption

The profile of the effect of contact time on Pb (II) adsorption is shown in figure 5. Pb (II) adsorption increases rapidly as soon as it is contacted and reaching equilibrium after 60 min. By the time increasing, the adsorption tend to decreases. In the variation of the contacting of the pectin membrane against the Pb (II) ion, the optimum contact time was obtained at the contact time of 60 minutes with an absorption of 90.11%.

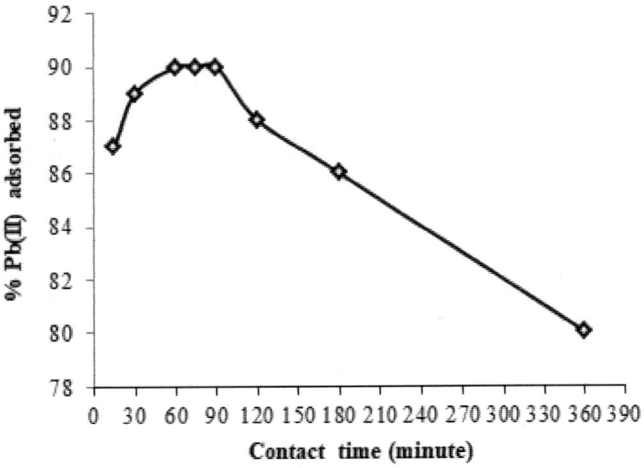

Figure 5. The effect of contact time on the adsorption of Pb(II) onto pectin membrane

3.6 Effect of initial Pb (II) ion concentration

The results of the analysis of the ability of Pb (II) ion adsorption by Pectin adsorbents are shown in figure 6. The amount of metal ions adsorbed by the adsorbent is affected by the concentration of metal ions in solution. The optimum concentration of Pb (II) ions adsorbed on the pectin membrane has increased to a concentration of 300 ppm with an adsorption capacity of 209.57 mg/g. This indicates that the membrane pectin is able to adsorb maximum Pb (II) ions at a concentration of 300 ppm. Increasing the concentration of metal ions increases the number of adsorbent active sites which bind metal ions and decreases the free active sites. At the concentration of Pb (II) 400 ppm ion, the adsorption power of pectin membrane has decreased. This is because the surface of the pectin membrane has saturated with the adsorbed Pb (II) ions.

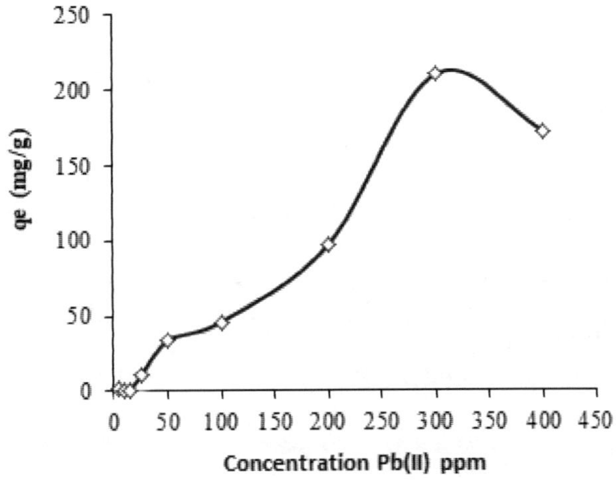

Figure 6. The effect of concentration on the adsorption of Pb(II) onto pectin membrane

3.7 Isotherm adsorption of pectin membrane toward Pb (II) ions

Pb (II) ion adsorption on Pectin adsorbents follows the freundlich isotherm pattern. This shows that the mechanism of pectin membrane adsorption on Pb (II) ions occurs through physical absorption. This shows the interaction between the pectin membrane in adsorbing Pb (II) ions occurs by involving the surface of each adsorbent in a multilayer manner. The pattern of pectin membrane adsorption isotherm in adsorbing Pb (II) ions is shown in figure 7.

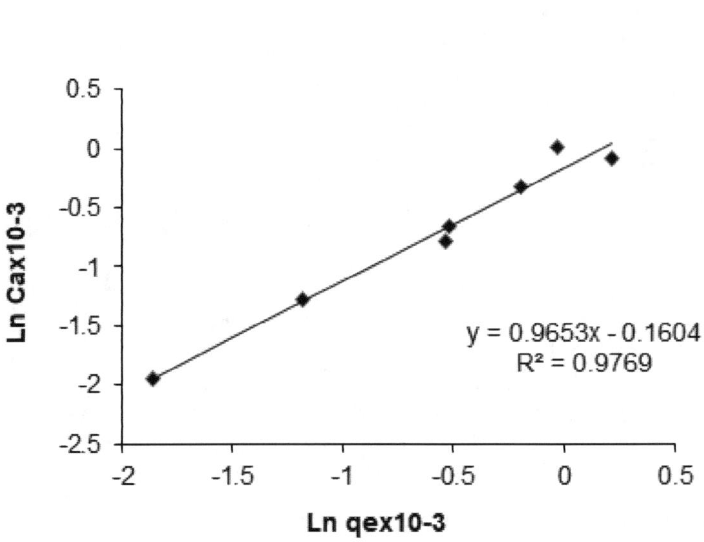

Figure 7. Freundlich adsorption isoterm of Pb(II) ion by pectin membrane.

4. Conclusion

Pectin membran has the ability to absorb metal ion Pb(II), so it can be an alternative to heavy metal Pb bioremoval. Pectin membrane have hydroxyl and carboxyl active groups, and form a microporous membrane. Pectin membrane can adsorp of Pb(II) metal ion at optimum condition of pH 4 and contact time 60 minute with % adsorbed with % Pb adsorbed 92.68 and 90.11%, respectively. The pattern of isotherm adsorption follows the Freundlich isotherm.

References

[1] Muley D V 2013 *Univers. J. Environ. Res. Technol.* **6** 690–94

[2] Ogata F, Kangawa M, Iwata Y, Ueda A, Tanaka Y and Kawasaki N 2014 *Chem. Pharm. Bull.* **62** 247–253

[3] Waghmare VH and Chaudhari U.E. 2014. *Rasayan J. Chem.***7** (4) 375-379

[4] Fu F and Wang Q. 2011, *J. Environ. Manage.* **92** 407–18

[5] Al-Masri M S, Amin Y, Al-Akel B, Al-Naama T. 2010 *Appl. Biochem. Biotechnol.* **160** 976–987.

[6] Gottipati R and Mishra S 2010 *Brazilian J. Chem. Eng.* **27** 357–67

[7] Anwar J, Shafique U, Zaman WU, Salman M, Dar A, and Anwar S 2010, *Bioresource Technology*, **101**, 1752-1755

[8] Mata Y N, Blázquez M L, Ballester A, González F and Muñoz J A 2009 *Chem. Eng. J.* **150**, 289–301

[9] Kupchik L A, Kartel N T, Bogdanov E S, Bogdanov OV and Kupchik M P 2006 *J. Macromol. Chem. Polym. Mater.* **79** 457–60

[10] Wong W W, Alkarkhi A, and Easa A M 2009 *J. Food Sci.* **74** C637–41

IOP Conf. Series: Materials Science and Engineering **858** (2020) 012015 doi:10.1088/1757-899X/858/1/012015

Preparation of Ni/Zeolite as esterification catalyst in biodiesel production based on off-grade crude palm oil

Mazlan[1], **W A Lestari**[2], A **Istiqomah**[2], **D Krisdiyanto**[1], **E Pramono2***, **A W Afrianto**[1]
[1]Department of Chemistry, Faculty of Science and Technology of UIN Sunan Kalijaga Yogyakarta.
[2]Chemistry Department, Faculty of Mathematic and Natural Sciences, Sebelas Maret University Surakarta, Indonesia.

E-mail: edi.pramono.uns@staff.uns.ac.id

Abstract. This aim of this research is preparation to prepare catalyst of 0% Ni/Zeolite, 5% Ni/Zeolite, and 10% Ni/Zeolite. Catalyst of Ni/Zeolite was prepared by impregnation method. Biodiesel production was conducted by 2 steps, the first step was esterification using 3 types of catalyst 0% Ni/Zeolite, 5% Ni/Zeolite, and 10% Ni/Zeolite. The second step was trans-esterification using NaOH catalyst. The catalyst were characterized by FTIR to identify functional group and XRD to identify crystallinity. The acidity of catalyst was determined by ammonia adsorption. The catalyst can reduce FFA content in CPO that influence increasing of yield biodiesel production. The formation of biodiesel was confirmed by the GCMS analysis that identified the methyl ester produced. The highest yield of biodiesel production was obtained using 5% Ni/Zeolite catalyst. Based on physical analysis of biodiesel which includes density, acid number, and water content, biodiesel synthesized from palm oil off grade has qualified the Indonesian National Standard for biodiesel.

1. Introduction

Petroleum fuel based on fossil commodities sustain reduction up to 10% each year, and for the consumption of its fuel increased by 6% [1]. By thus main problem, the presence of petroleum fuel reach the limitation situation. Some researcher try to solve by find some overcome alternative fuels i.e. biodiesel. Biodiesel is an alternative fuel made from renewable natural resources from plants, animals or waste product with several treatments. Palm oil off-grade is one of the basic ingredients which can be used in producing biodiesel. Indonesia is the largest palm fruit producing country in the world [2], but there is some difference category for palm fruit commodities, there is some low-quality palm fruit or also called off grade palm fruit. Off grade Palm is a palm fruit that has a low quality and not fit for the manufacture of cooking oil industries. Off grade Palm oil occurs because of several things such as too fast in harvesting palm fruit, slow in the process of taking oil palm fruit or because it is too long in the mobility of sending fresh fruit to the company, so there are palm fruits that have been rotten before reaching the company. Besides off grade palm oil was also produced from the process of sorting fresh fruit carried out directly by the company and decreasing field data which from the sorting of companies about palm fruit there are 7-10% ware off grade palm fruit [3].

One type of heterogeneous catalyst is a bifunctional catalyst in the form of a catalyst with a metal system. The bifunctional catalyst is chosen because the single metal catalyst does not have high thermal stability and will form a solid metal that can reduce surface area, and facilitate the deactivation of the catalyst

[4-6]. Therefore, metal catalysts need to be dispersed into bifunctional catalysts that are porous, heat-stable and chemical, large surface area, and mechanically strong [7]. Bifunctional catalysts consist of acid and metal sites. Acid site i.e. zeolites and bentonites, while metal sites in transition metals i.e. Nickel, Molybdenum, Cobalt, and can be a metal combination [8].
Zeolites can be modified by the activation process, both physical activation and chemical activation. The zeolite activation process using hydrochloric acid can remove impurities which cover the zeolite pores and can increase the ratio of Si to Al atoms [12]. Addition of Ni content to the zeolite catalyst can increase the acidity of natural zeolite [13, 14]. In addition, an increase of Ni metal content on Ni/Zeolite catalysts with nickel variations of 0, 5, and 10%, shows enhancement in the acidity value of Ni/Zeolite catalysts [15]. Zeolite as a carrier has a role to place the active catalyst metal. Metal is added into zeolite will cause a relatively large surface area, so the reaction will take place more quickly [9]. Metals can be dispersed evenly throughout the pore system to produce an active metal surface [10]. In this way, the catalyst produced is more efficient with several advantages such as large specific surface area, high thermal stability, longer service life, and results in more effective catalysts. There are two things that can cause nickel metal to be impregnated into zeolite. The first is the difference in chemical potential between nickel metal (Ni^{2+}) and aluminium in zeolites which have an excess of 1 electron free pair because they receive 4 electrons in the zeolite framework [11].

2. Experimental

2.1. Materials
Materials in this research are purchased from Merck and Sigma Aldrich without purification and pretreatment that are chloride acid (HCl, Merck), ethanol (Sigma Aldrich), natrium hydroxide (Sigma Aldrich), methanol (Sigma Aldrich), nickel nitrate hexahydrate ($Ni(NO_3)_2.6H_2O$, Sigma Aldrich), silver nitrate ($AgNO_3$, Sigma Aldrich) and aquadest. Natural zeolites obtained from Wonosari, Gunung Kidul.

2.2. Synthesis of Ni/Zeolite Catalyst
Synthesis of Ni/Zeolite Catalyst conducted by dilute of 50 g Natural Zeolite in 200 mL HCl 2M at 90 °C for 5 hours. The catalyst was dried at 120 °C for 2 hours, and the catalyst calcinated at 500 °C for 3 hours. Nickel was loaded into natural Zeolite by impregnation method with varying nickel concentrations (0%, 5%, and 10% (w/w)). The impregnation process was carried out at 80 °C and stirred for other 3 hours. The Ni/Zeolite catalyst was dried at 120°C for 2 hours. The catalysts calcinated at 500 °C for 5 hours. The catalysts was characterized by Fourier-transform infrared spectroscopy (FTIR) and X-ray powder diffraction (XRD). Acidity of catalyst was obtained by ammonia adsorption.

2.3. Catalytic Activity
Catalytic activity was obtained through esterification reaction of off-grade crude palm oil. Esterification reaction was obtained by added methanol into 20 g of off-grade crude palm oil with ratio oil : methanol (1:12). 2% (w/w) catalysts was added into solution and stirred at 50 °C for 2 hours. Quantitative and qualitative analysis of reaction product was obtained by FTIR and GC-MS.

3. Result and Discussion

3.1. Synthesis of Ni/Zeolite Catalyst
Dilution of natural zeolite with HCl 2M is purposed to separate impurities by the ion exchange method (Figure 1). In addition to dissolving impurities, the activation process with HCl solution aims to replace cations in zeolite interlayer spaces or in zeolite canals with hydrogens (H^+). The cations can be dissolved by HCl activation such as Ca^{2+}, Mg^{2+}, Na_+, Fe^{2+}, and Fe^{3+} [16]. HCl activation will increase the surface area and the diameter of zeolite pores due to opened the tip of zeolite layer.

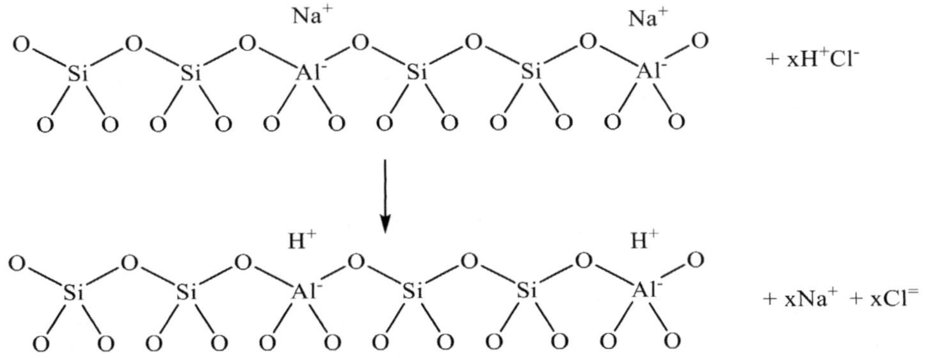

Figure 1. The mechanism of ion exchange in the zeolite activation process.

The crystallinity of natural zeolites is influenced by the ratio between Si/Al. If the Si/Al ratio is higher then its crystallinity will be higher [17]. One of the aims of HCl 2M activation is to reduce the Si/Al ratio in zeolite material. In the activation process, the H^+ atoms will attack the oxygen atoms which are bound to Si and Al to produce a silanol group. Based on the value of bonding dissociation energy, the value of Al-O dissociation energy (116 kkalmol-1) is smaller than Si-O (190 kkalmol-1). When it is activated with 2M HCl, there will be a process of dealumination (Figure 2) on several Al atoms in the zeolite structure.

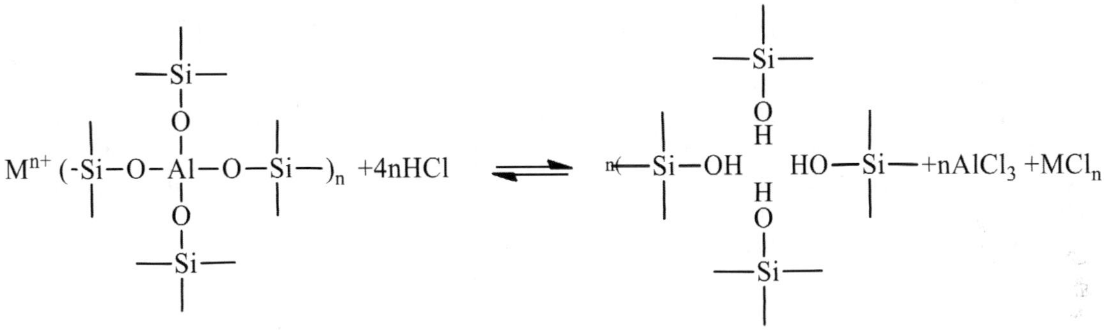

Figure 2. Dealumination reaction in zeolite structure.

FTIR analysis is used to analyze functional groups of zeolite samples. Figure 3. shows the functional groups possessed by 4 zeolite samples namely natural zeolite, acid activated zeolite, 5% Ni/Zeolite, and 10% Ni/Zeolite. Based on FTIR data, there are 4 main peaks of zeolites that can identify the vibrations of the functional groups possessed by Zeolites. The peak absorbance at 482.48 cm^{-1} represents the Si-O-Al functional group and in the peak absorbance at 793.89 cm^{-1}, a peak appears that identifies the bending vibration of Si-O. The peak absorbance at 1043.78 cm-1 represents asymmetry T-O (T= Al or Si), at this peak the wavenumber shifts to a larger one. This indicates that there has been a dealumination process or the process of releasing Al in the zeolite structure to become Al outside the structure. The dealumination process causes decreasing of aluminum in the zeolite structure, decreasing the amount of aluminum causes the distance between Al will be even greater. It causes the vibrations that arise will be greater. In the peak absorbance 1638.00 cm^{-1}, a peak appears which identifies the bending vibrations of O-H from the H_2O molecule. The peak absorbance at 3428.47 cm^{-1}, a peak appears which identifies the O-H stretching vibrations of the physically bound silanol and H_2O groups.

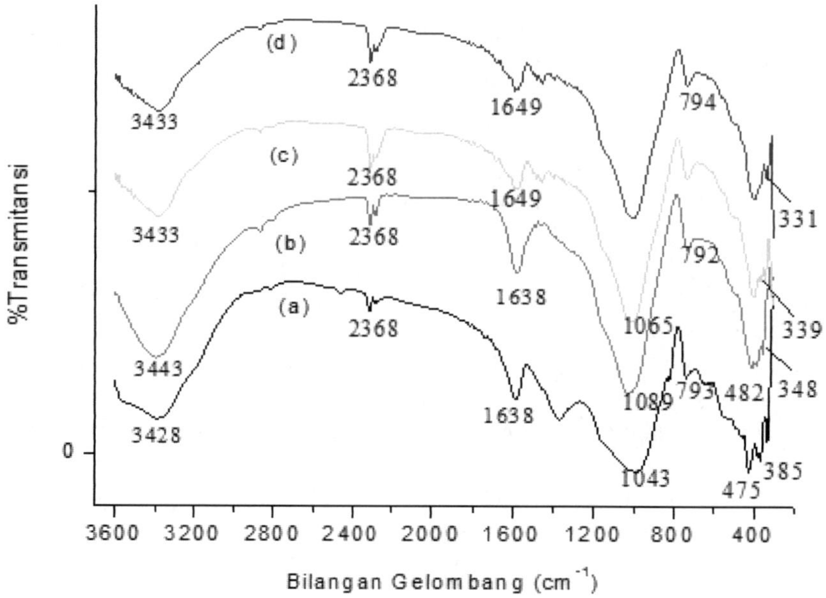

Figure 3. FTIR spectra of (a) natural zeolite, (b) 0% Ni/Zeolite, 5% Ni/Zeolite, (d) 10% Ni/Zeolite.

Vibration due to pore opened by a small cavity. It will give a higher absorption pattern than vibrations in a large cavity. All zeolite samples showed the emergence of a peak at wavenumber 385 cm^{-1} that showed the open pore vibrations of natural zeolites. Whereas on 0% Ni/Zeolite, 5% Ni/Zeolite, and 10% Ni/Zeolite successively peaked at 348 cm^{-1}; 339 cm^{-1} ; and 331cm^{-1}. This diversity indicates that the 5% and 10% Ni/zeolite samples have a non-uniform open pore distribution from the natural zeolite and 0% Ni/zeolite. This indicates the presence of Nickel metal affects the open pore of zeolites.

Based on the XRD spectra (Figure 4) shows that the zeolite catalyst used contains mordenite minerals and several clinoptilolite minerals. The peaks show that the presence of mordenite minerals are in 2θ 13.6°; 19.56°; 25.78°; 26.69°; and 27.83°. The peaks showed the presence of clinoptilolite minerals were 9.86° and 22.39°.it has been reported in recent article [7, 17]. These peaks were compared with JCPDS-International Center for Diffraction Data (2001) for mordenite and clinoptilolite minerals.

IOP Conf. Series: Materials Science and Engineering **858** (2020) 012015 doi:10.1088/1757-899X/858/1/012015

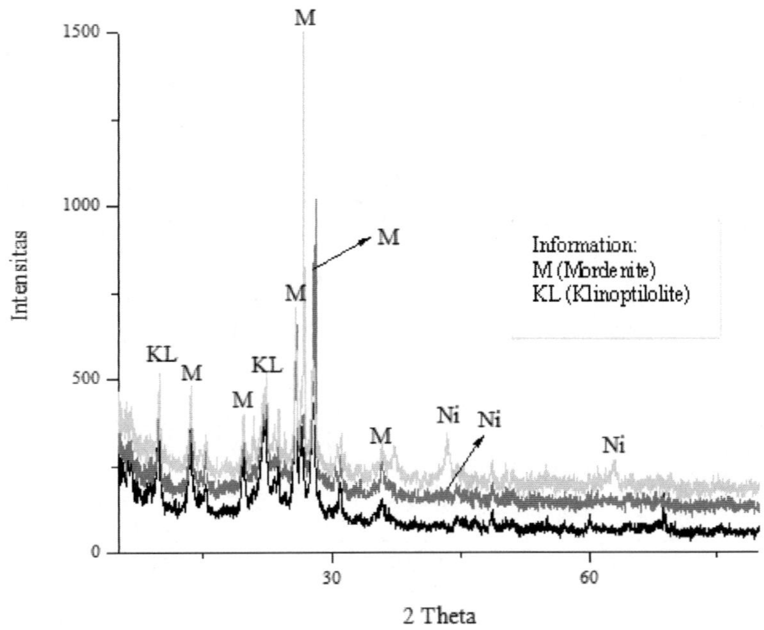

Figure 4. Difractogram of H/Zeolite (black), 5% Ni/Zeolite (red), 10% Ni/Zeolite (green).

Based on XRD analysis, spectra of 5% Ni/Zeolite and 10% Ni/Zeolite catalysts showed that it has similar spectra with 0% Ni/Zeolite spectra. The new peaks that appeared after addition of nickel, it showed the presence of nickel metal with 2θ 43.28° and 63.04°. It means that impregnation of nickel to natural zeolite is successfully obtained. It has been reported in some article and compared with JCPDS-International Center for Diffraction Data for nickel metal [16]. Data 2θ of catalysts can be see in Table 1.

Table 1 The intensity of the 10 largest diffractogram peaks and the crystallinity of acid activated zeolite and balanced by the nickel metal.

2θ	Intensity		
	H/Zeolite	Ni/Zeolite (5%)	Ni/Zeolite (10%)
9.86 (KL)	433	396	515
13.56 (M)	391	383	478
19.72 (M)	282	342	396
22.39 (KL)	385	476	520
25.78 (M)	569	648	699
26.69 (M)	325	381	1615
27.83 (M)	426	555	829
Total Intensity	2811	3181	5052

The catalyst acidity is expressed in milimoles of acidic sites per gram unit of catalyst. The total amount of acid present in the catalyst are both on the external surface and the inner surface of the catalyst. The acidity is determined by the absorption of ammonia gas which is alkaline by zeolites which have an acidic site in its structure. The zeolite acidity can originate from the Lewis acid site, Al and also originate from the Browsted site, H^+. The measurement of acidity from 4 samples aims to determine the effect of Ni metal on the acidity of the acid zeolite activated.

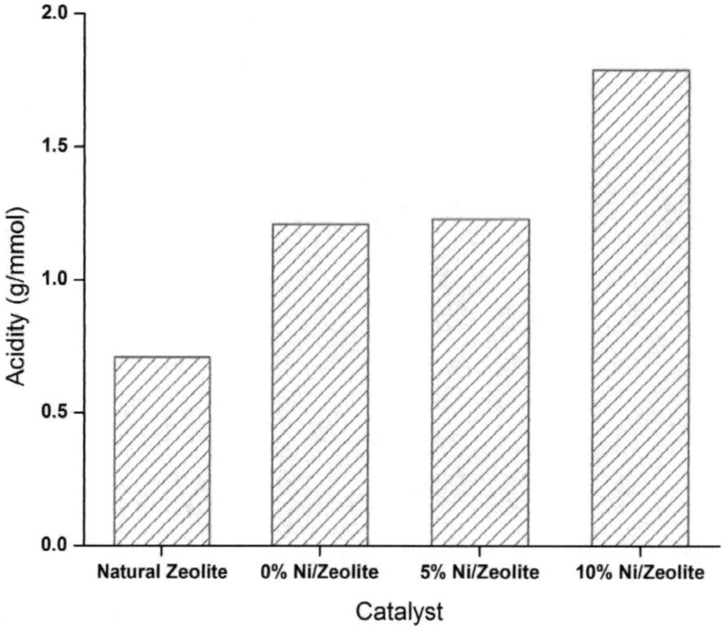

Figure 5. Solids acidity graph of natural zeolite catalyst

Figure 5 shows the solids acidity graph or natural zeolite, 0% Ni/Zeolite, 5% Ni/Zeolite, 10% Ni/Zeolite are 0.71 g/mmol; 1.21 g/mmol; 1.23 g/mmol; and 1.79 g/mmol. These results indicate the amount of solids acidity increases with activation process using HCl 2M and presence of nickel metal. Addition of HCl solution contributes to switch the active site in the zeolite structure become Bronsted acid sites.

3.2. Catalytic Activity

Characteristic of palm oil showed that the levels of free fatty acid was quite high at 6.2813%. This will affect the results of the reaction in the transesterification process. It will triggers the saponification reaction and complicate the separation process of biodiesel and glycerol due to there are layers that produced from saponification reaction. To overcome this problem, it is necessary to reduce free fatty acid (FFA) contain in off-grade crude palm oil.

Based on the esterification production analysis, FFA content in products are 0.2510%; 0.9705%; 1.4557%; and 0.7278% for H_2SO_4, 0% Ni/Zeolite, 5% Ni/Zeolite, and 10% Ni/Zeolite, respectively (Figure 6). It has decreased FFA content in crude palm oil (CPO) with FFA content is 4.02%. Characteristic of CPO can be seen in table 2.

Tabel 2. Characteristic of Off Grade Palm Oil

Characteristic	Unit	Value
Color	-	Reddish yellow
Acid Number	NaOH/g	6.2813
Free Fatty Acid Levels	%	3.8400
Water Levels	%	0.0801
Density	kg/L	0.8678

The best catalyst for esterification reaction was 10% Ni/zeolite (10%) catalyst with % FFA is 0.7278%. H_2SO_4 catalyst is used as a comparison with prepared catalyst. FFA concentration of palm oil with zeolite catalyst are still lower than H_2SO_4 catalyst, but due to insignificant differences and the FFA concentration produced in the esterification reaction is lower than 2%. It can be conclude that Ni/Zeolite catalyst can be used as an alternate to replace H_2SO_4 catalyst in the esterification.

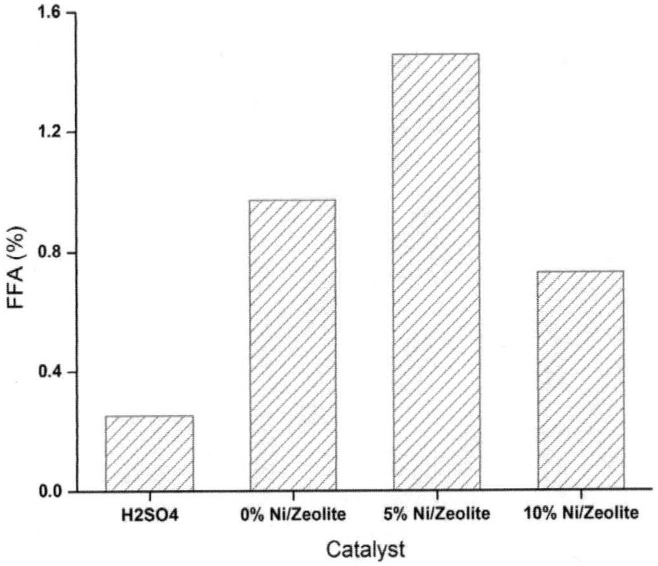

Figure 6. Free fatty acid levels graph of esterified palm oil.

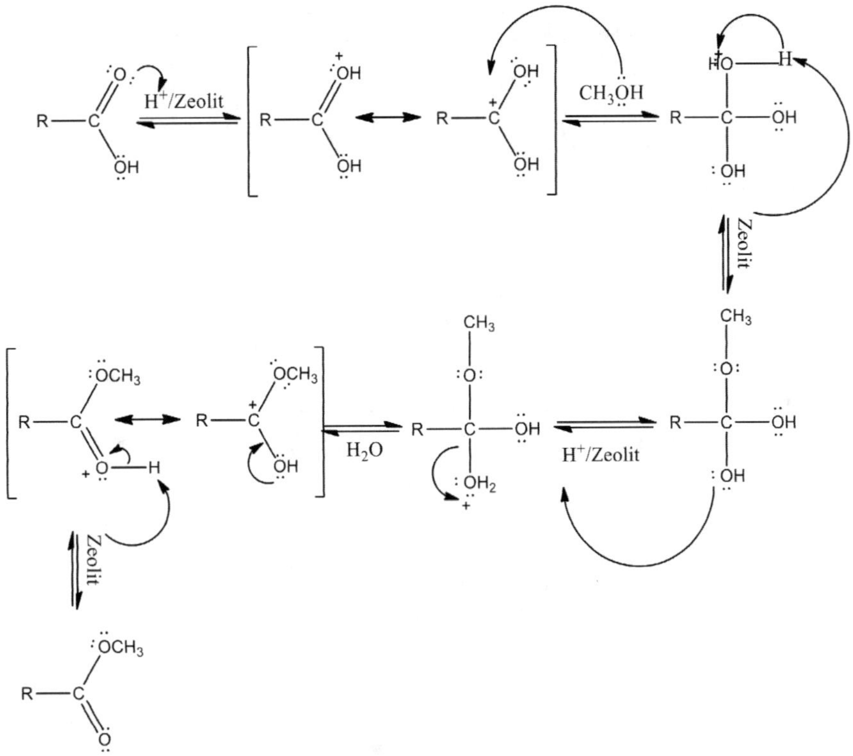

Figure 7. Esterification Mechanism with 0% Ni/Zeolite Catalyst.

Esterification using zeolite have conducted through two steps. The first reaction with 0% Ni/Zeolite catalyst and the second reaction with 5% Ni/Zeolite and 10% Ni/Zeolite catalyst. Both reaction are reversible reaction. In this mechanism, the carbocation is formed first due to the binding (H^+) of 0% Ni/Zeolite to the oxygen atom in the free fatty acid carbonyl group. 0% Ni/Zeolite has H^+ levels greater than natural Zeolite, due to the activation process using HCl 2M. The binding of oxygen atoms by H^+ is a nucleophilic substitution reaction that occurs due to the role of Bronsted acid on the 0% Ni/Zeolite.

Then the process of carbocation attack occurs which is unstable by the electron pair of oxygen atoms in methanol. The last step is the release of H_2O followed by the process of re-forming the catalyst 0% Ni/Zeolite and the formation of methyl esters as a product. The mechanism applies equally when using a 5% Ni/Zeolite catalyst and 10% Ni/Zeolite catalyst. The mechanism of reaction can be seen in Figure 7.

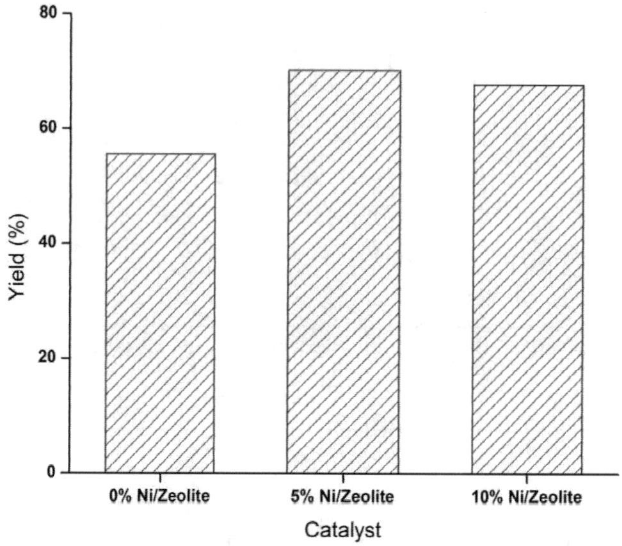

Figure 8. Yield of biodiesel product using variation catalysts.

Figure 8 shows the yield of biodiesel is influenced by the type of catalyst used in the esterification. The type of catalyst used in the esterification were the catalyst without the nickel metal in the form of 0% Ni/Zeolite and with a metal in such as 5% Ni/Zeolite and 10% Ni/Zeolite.yield of biodiesel production can be seen in table 3.

Biodiesel product is characterized by FTIR to determine the functional group. Figure 9 shows that the FT-IR analysis of palm oil biodiesel. The spectra show 8 major peaks that appear. The biodiesel reaction is successfully occur with confirmation of peak at wavelength 1743 cm^{-1}. It is vibration of C=O stretching and the existence of peak at wavelength 1168 cm^{-1} that are vibration of C-O stretching. This is what indicates the existence of a methyl ester group. Biodiesel product peaks can be seen in table 4.

Table 3 The yield of off grade palm oil biodiesel with variations of catalysts type in the esterification reaction

Sample	Catalyst Type	Yield (%)
B01	H/Zeolite	55.60
B02	5% Ni/Zeolite	70.23
B03	10% Ni/Zeolite	67.77

IOP Conf. Series: Materials Science and Engineering **858** (2020) 012015 doi:10.1088/1757-899X/858/1/012015

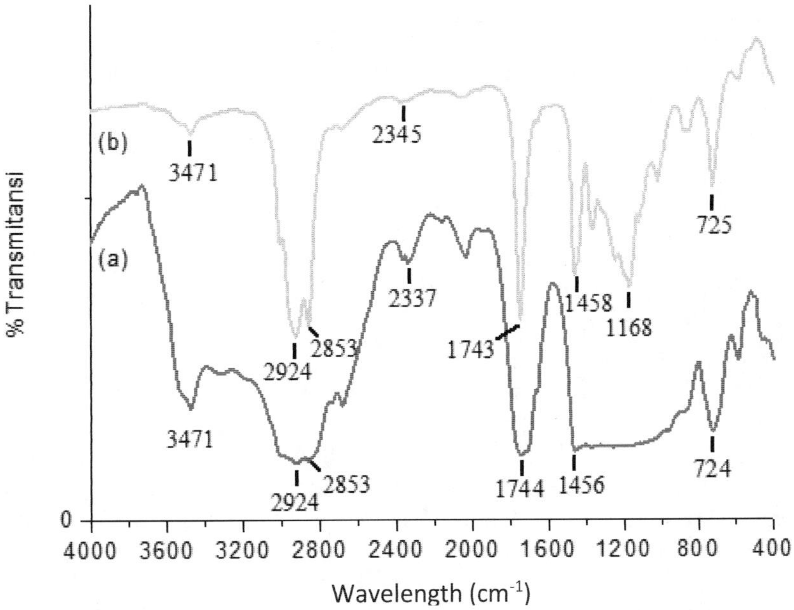

Figure 9. Infrared Spectrum of a. Off Grade Oil Palm b. Biodiesel.

Table 4 Biodiesel functional group from off grade palm oil

Wavenumber (cm^{-1})	Functional group
3464	-OH stretching vibration (H_2O)
3001	C=C bending vibration (aliphatic)
2919	CH stretching vibration (aliphatic) CH_3, CH_2
2855	CH stretching vibration (aliphatic)
1743	C=O stretching vibration (ester)
1458	C-H scissor vibration and bending vibration for methyl
1168	C-O stretching vibration (ester)
725	C-H vibration (aliphatic group)

Qualitative and quantitative analysis is used in determining the amount and type of methyl esters contained in off-grade palm oil biodiesel. The peaks on the GC chromatogram shows the number of methyl esters contained in biodiesel. Each chromatogram peak separated with different retention times. This difference is influenced by the number of carbon chains in each type of methyl ester, where the methyl ester which has the shortest carbon chain will come out first.

Table 5. Retention time and area of the five chromatogram peaks of off grade palm oil biodiesel

Peaks	H/Zeolite catalyst		5% Ni/Zeolite catalyst		10% Ni/Zeolite catalys	
	Retention time (Mins)	Total area (%)	Retention time (Mins)	Total area (%)	Retention time (Mins)	Total area (%)
1	33.029	0.976	33.616	0.530	33.028	0.880
2	37.896	37.568	38.086	38.870	37.855	37.172
3	-	-	41.417	9.540	-	-
4	41.698	54.458	41.616	45.800	41.649	54.789
5	41.999	4.707	42.016	5.270	41.978	5.008
Total		97.707		100.000		97.849

The GC chromatogram showed the existence of five peaks were then analyzed using mass spectrophotometry (MS). Based on MS analysis, it is found that the five peaks that appear that have different areas, ranging from the highest to the lowest area respectively are methyl oleate (45.80%); methyl palmitate (38.87%); methyl linoleate (9.54%); methyl stearate (5.27%); and methyl myristate (0.53%) (Table 5). The appearance of methyl palmitate comes first than methyl oleate and methyl stearate because methyl palmitate has the shortest carbon chain than methyl oleate and methyl stearate. Methyl oleate and methyl stearate have carbon chains with the same length, but because methyl oleate has a lower molecular weight than methyl stearate, it will come out first and have a lower retention time than methyl stearate. The total area of the total methyl ester (biodiesel) chromatogram with 0% Ni/Zeolite, 5% Ni/Zeolite, and 10% Ni/Zeolite catalysts in the mixed esterification are 97.707%; 100%; and 97.849%. The greatest yield of methyl ester (biodiesel) is obtained using a 5% Ni/Zeolite catalyst that is equal to 100% which is strengthened through the yield of biodiesel obtained from off grade palm oil using a 5% Ni/Zeolite catalyst greater than the 0% Ni/Zeolite and 10% Ni/Zeolite catalysts.

Biodiesel obtained from off grade palm oil is analyzed to find out the characteristics of the biodiesel. The results of the analysis compared with the standards in SNI 04-7182-7182-2006. Biodiesel analysis is performed on density of temperatures, acid numbers, and water levels (Table 6).

Table 6. Comparison of Physical Characteristics of Biodiesel [4]

SNI Characteristic of Biodiesel	Standard Biodiesel	Palm Oil Biodiesel
Density of temperature	850-890 kg/m^3	856 kg/m^3
Acid number (mg-kOH/g)	Max. 0.8	0.69
Water levels (%-Vol)	Max. 0.05	Trace (0.05)

Acid number affects the combustion process, if the acid number gets higher, it can deposition the combustion system. The high of acid number for this fuel can be used as a solvent that can improve the quality of the fuel system [18]. Based on Table 6 shows the biodiesel acid number from palm oil grading 0.69 mg-kOH / g. Biodiesel water content based on SNI 04-7182-2006 is a maximum of 0.05%. This is because rising water levels can cause corrosion to the engine and the formation of paraffin crystals at cold temperatures which can clog the flow of fuel. Based on the research results, the water levels of off grade palm oil biodiesel is 0.05 and easily changes depending on the storage process. Table 6 shows that the density, amount of acid, and water levels of biodiesel made from grade palm oil is accordance with the Indonesian National Standard for biodiesel.

4. Conclusion

Biodiesel is the renewable fuel that can replace petroleum fuel. Biodiesel production is conducted through two steps are esterification and transesterification reaction. Esterification reaction is purposed to decrease FFA concentration to increase yield of biodiesel product. The benefit of using heterogeneous catalyst is to separate product and catalyst easily. Ni/Zeolite catalyst showed high convertion from CPO to Biodiesel product and it can reduce FFA content of CPO. It can be conclude that Ni/Zeolite catalyst can be used as an alternative catalyst to replace H_2SO_4 catalyst in the esterification step.

Acknowledgement
The authors wish to thank the chemical laboratory and the Faculty of Science and Technology, Sunan Kalijaga State Islamic University of Yogyakarta for supporting this research.

References
[1] Kuncahyo P, Fathallah A Z M, Semin 2013 Pom. Eng. J. **2** 62-66
[2] Puji S 2015 Study of the Effect of Diethyl Ether Counseling on Biodiesel Synthesis with Heterogeneous CaO Catalysts from Palm Oil Mill Liquid Waste Oil (Yogyakarta: UIN Sunan Kalijaga)
[3] Aziz I, Nurhayati S, Rahman A 2012 Val. J. **2** 511-515
[4] Hadi W A 2009 J. of Reg. Research **8** 2
[5] Musta R 2010 J. of Flux Phys. **7** 149-159
[6] Trisunaryanti W 2015 Catalyst Material and Character (Yogyakarta: Gadja Mada University Press)
[7] Trisunaryanti T, Triwahyuni E, Sudiono S 2005 Indo. J. Chem. **5** 48-53
[8] Yoesepha W, Helwani Z, Saputra E 2016 J. of FTEKNIK **3** 2
[9] Farida I 2018 Modification of Natural Zeolite with Zirconium Oxydechloride PSTA BATAN Products and its Catalytic Test for the synthesis of glycerol-derived compounds (Yogyakarta: UIN Sunan Kalijaga)
[10] Hartono R, Wijanarko A, Hermansyah H 2002 Conf. Ser: Mat. Sci. Eng. **345** 1-4
[11] Istadi 2011 Catalyst Technology for Energy Conversion (Yogyakarta: Graha Science)
[12] Fatimah N F, Utami B 2017 J. C-T. **1** 35-39
[13] Nurhayati N D, Wigiani A 2014 Synthesis of Ni-Cr/Zeolite Catalysts with Separate Impragnation Methods (Surakarta: National Seminar on Chemistry and Chemistry Education VI)
[14] Bijang M C, Aryanto Y, Trisunaryanti W 2002 J. of Tech. **15** 57-69
[15] Ramdhani G D, Fatimah N F, Sarjono A W, Setyoko H, Nurhayati N D 2017 J. of Chem. and Chem Ed. **1** 72-79
[16] Irvantino B 2013 Preparation of Natural Ni / Zeolite Catalysts With Senokimia Method For Catalytic Cracking Polypropylene and Polyethylene (Semarang: Semarang State University)
[17] Wijaya K, Hadi K, Herlina I, Kurnia T A 2016 Nano Material (Application in making biofuels) (Yogyakarta: UGM Press)
[18] Furqon M H, Prasetya A, Wilopo W 2013 J.of Sys Eng, **1** 1

Studies on the Optical and Morphology Properties of DN-F01 Dye Hybrid Black Rice and Ag Metal as Dye-Sensitized Solar Cells (DSSC) Transparent

N E-H Diyanahesa[1], A Supriyanto[1], A H Ramelan[1], M Khairul bin Ahmad[2], F Nurosyid[1], S Wahyuningsih[1], C Cari[1]
[1]Physics Department, Graduate Program, Sebelas Maret University
Jl. Ir. Sutami 36A Kentingan Jebres Surakarta, 57126
[2]Research Management Centre (RMC), Universiti Tun Hussein Onn Malaysia (UTHM) JOHOR, 86400 Parit Raja, Batu Pahat Johor, Malaysia

E-mail: nadiyahesa@gmail.com, agusf22@staff.uns.ac.id

Abstract. Dye-Sensitized Solar Cells (DSSC) is a set of solar cells based on photoelectrocomia, which involves the transfer of electric charge from one phase to another. This research aims to determine the optical and photoelectric properties of DN-F01 dye hybrid black rice and Ag metal as DSSC transparent. TiO_2 (DN-E0P3) was used which was coated on a conductive glass through a spin coating technique and as a photon absorbent was used DN-F01 dye with combination of DN-F01 dye hybrid anthocyanin dye from extracts black rice and DN-F01 dye doped Ag metal. Characterization of glass substrate that has been coated with TiO_2 using FESEM and EDS where with FESEM tools it is seen that TiO_2 on the surface of the glass is more evenly distributed and from EDS it is found that the dominant substance found on the glass surface is Ti. In the absorption of UV-VIS light characterization it is known that combination of DN-F01 dye can absorb the spectrum of light at wavelengths around 400-800 nm. When solar cells are illuminated with sunlight, solar cells can convert solar energy into electrical energy. The voltage, electric current, and highest efficiency produced by DSSC with optimizing the combination of DN-F01 dyes is DN-F01 hybrid Ag metal that is 805 mV, 1,234 μA and 2,145%.

1. Introduction

Solar energy is the largest source of energy, does not contain pollution, is not used up, and is free. Currently, solar energy has not been used as a primary energy source because energy supply still depends on various fossil fuel sources. As of 2002, the availability of fossil energy reserves around the world is running low [1]. To overcome the energy crisis and support human survival, solar energy can be converted into electrical energy. The conversion of solar energy into electrical energy takes place through a photovoltaic cell system. The photovoltaic effect is the event of the discharge of an electric charge in a material as a result of the absorption of light by that material [8]. One mechanism that works based on photovoltaic systems is solar cells. In solar cells, photons from solar radiation are absorbed and then converted into electrical energy. The solar cells that are widely used today are conventional solar cells based on silicon technology [10]. However, the cost of producing silicon is very expensive so that its consumption costs are more expensive than fossil energy sources. Besides the fabrication process of silicon solar cells also use hazardous chemicals [11].

DSSC is one of the non-conventional solar cell technologies that develops in line with the development of nanotechnology. DSSC consists of TiO_2 nanocrystals as photoelectrodes, dyes as light absorbers, and electrolytes as electron donors arranged in a sandwich structure [9]. Basically the working principle of DSSC is the reaction of electron transfer. The first process begins with the excitation of electrons in the dye molecule due to absorption of photons [12]. The second process occurs at the negative electrode (anode), which is the TiO_2 layer where the excited electrons are then injected into the TiO_2 conduction band so that the dye is oxidized. With an electron donor by an electrolyte, the dye molecule returns to its original state and prevents the recapture of the electron by the oxidized dye. In the third process, after reaching the ITO electrode, electrons flow into the counter-electrode which acts as a positive electrode (cathode) through an external circuit [14]. The next process, in the presence of a catalyst at the counter electrode, electrons are received by the electrolyte so that the holes formed in the electrolyte ($I3^-$), due to electron donors in the previous process, combine with electrons to form iodide (I^-) [15,16,17]. The final process, iodide is used to donate electrons to oxidized dyes, to form an electron transport cycle. With this cycle there is a direct conversion from sunlight to electricity [2]. The advantage of DSSC is that it does not require materials with high purity so that the production costs are relatively low [3]. In contrast to conventional solar cells, in DSSC the absorption of light and the transfer of an electric charge occur in separate processes. Light absorption is carried out by dye molecules, while charge transfer is carried out by nanocrystal inorganic semiconductors which have relatively large bandgaps [13]. One of the most commonly used semiconductors is TiO_2 which is relatively inexpensive, commonly found and non-toxic [4].

In DSSC, TiO_2 must have a wide surface so that the dye is absorbed more so that it can increase the current output of solar cells. Dye as a sensitizer includes synthetic dyes and natural dyes. Commercial DSSC uses dye synthesis of ruthenium complex type with efficiency of 9.2% [7]. However, the availability and price are expensive so that an alternative substitute is the natural dye extracted from plant parts [5]. Various types of plant extracts have been used as photosentizers in dye sensitized solar cell systems. Natural dye sensitizers that have been used in DSSC systems include anthocyanins, namely black sticky rice or black rice. The dye material used must be able to absorb a wide spectrum of light for example with using black rice dyes. I-V curve of solar cells using black rice has good efficiency, estimated η value is 0.22% and FF value is 32.6% [23]. The dye material used black glutinous rice for 48 hours soaking obtain the highest efficiency is 0.437×10^{-3} % [24]. The natural dyes are proven to be able to provide photovoltaic effects even though the efficiency is still much smaller than synthetic dyes [6]. Natural dyes that are available when combined with synthetic dye DN-F01 will occur in greater color absorption so it is expected to increase efficiency. So, in this study DN-F01 combined with Ag metal forms have a good structure in the FESEM test and EDS it is found that the dominant substance so that it is expected to produce high solar cell efficiency.

2. Experimental Methodology

2.1. Materials

Fluorine doped Tin Oxide (FTO) glass, TiO_2 (DN-E0P3), Ethanol (C_2H_6O), Polyethylene Glycol ($C_2nH_4n+2On+1$), Citric Acid ($C_6H_8O_7$), and Potassium Iodide (KI), DN-F01, Screen Printable Pt Paste, Ag metal, Black Rice.

2.2. Methods

2.2.1. Preparation of Combination of DN-F01 Dyes

Black rice washed carefully. The finely ground black rice is then stirred with ethanol, acetic acid, and distilled water. 10 grams of black rice and 35 ml of the solvent in the ratio of 30: 4: 16 distirer for 2 hours, then leave it for 24 hours. After 24 hours, the extract was filtered and used as a dye. DN-F01 dye consists of 0.125% wt, 0.5% wt, 1% wt combined with natural dye (black rice). Black rice dye solution is then mixed with DN-F01 solution in a ratio of 1: 1.

2.2.2. Preparation of Photo Anode

FTO glass that has been thoroughly cleaned is deposited with TiO_2 paste with the "Spin Coating" coating technique. For TiO_2 paste (DN-E0P3) mixed with ethanol in a ratio of 1: 1 ie 10 grams DN-E0P3 and 10 ml of ethanol. The protected cell layer is 2 × 2cm^2. The thickness of the subsequent film is around 10-20 μm. After that, the FTO glass that has been coated with TiO_2 is coated with temperature of 550, holding time is 60 minutes, heating rate is 15 minutes. The anneling process is carried out in the carbolite furnace.

2.2.3. Preparation of Electrolyte

0.8 gram of Potassium Iodide (KI) and 10 ml of PEG (Polyetilenglicol) stirrer for one hour until homogeneous then adding 0.127 grams of iodine (I2) mixed until homogeneous [19]. This solution is used as an electrolyte and stored in black bottles coated by aluminum foil.

2.2.4. Cell Preparation

The FTO substrate which was coated by TiO_2 thin film of about 10 cm^2 was immersed in natural dyes, Ag solution and organic dyes DN-F01, each for 24 hours to make a working anode [20]. After that the color particles that have been adsorbed by TiO_2 film are rinsed with ethanol for a few seconds to wash the dye that is not bound. Then the TiO_2 glass is allowed to stand for about 10 minutes to dry. For counter electrodes or cathodes, the FTO glass is coated by a Screen-Printable Pt paste by dripping on a hot plate. **FIGURE 1** shown about chemical structure Anthocyanin, DN-F01, and Ag molecule.

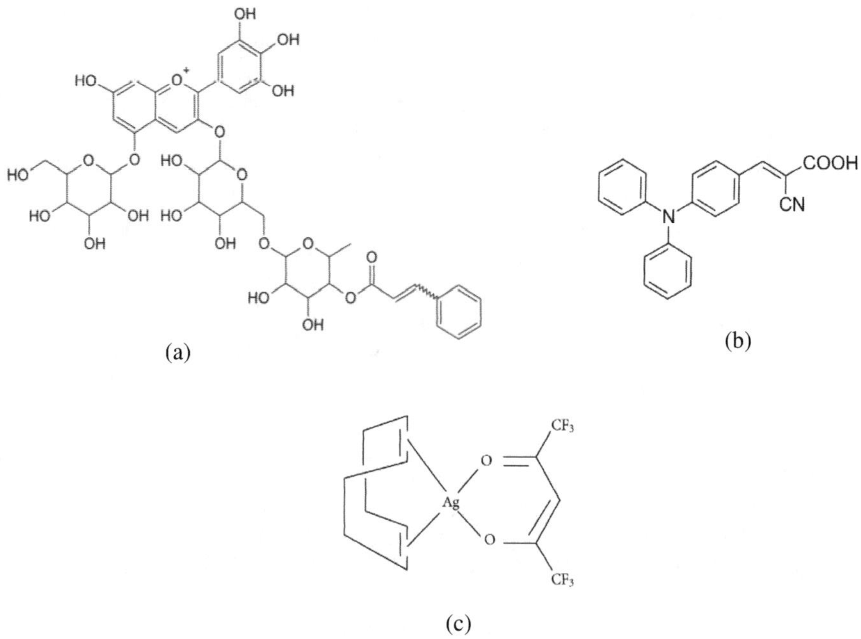

FIGURE 1. Chemical structure (a) Anthocyanin, (b) DN-F01, (c) Ag molecule [13]

The working electrode (anode) and counter electrode (cathode) are then arranged like a sandwich with a paper binder clip and the electrolyte solution is injected into the coated cell of the DSSC component [18].

3. Results and Discussion

3.1. Absorbance of the DN-F01 Combination

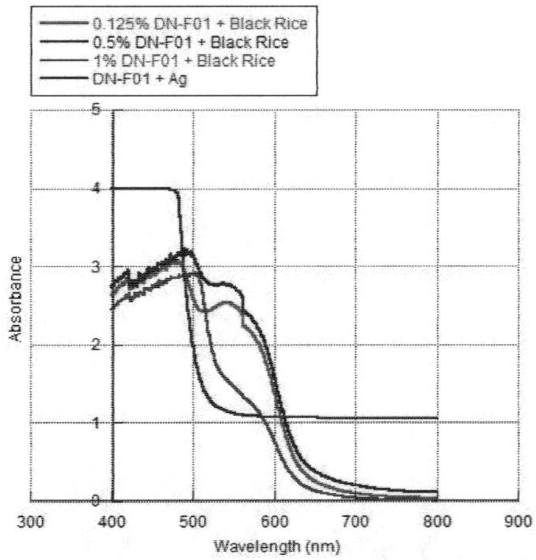

FIGURE 2. Graph of Absorption of the DN-F01 Combination

Absorbance of various extracts black rice, and Ag solution ingredients using a UV Visible shimadzh u 1601 PC Spectrophotometer. **FIGURE 2** shows that the fixed ratio between various natural dyes and solvents produces different absorbances. **FIGURE 2** also shows that the spectrum of dyes extracted from various natural materials has an absorption spectrum that ranges from 400 to 800 nm.

Variation organic dyes (DN-F01) with concentration 0.125% + Black Rice have two peak absorpsion spectrum 420-510 nm and 510- 570 nm. 0.5% DN-F01 + Black Rice has an absorpsion spectrum from 450-580 nm, 1% DN-F01 + Black Rice has an absorpsion spectrum from 400-530 nm, while variation DN-F01 + Ag solution has absorption spectrum that ranges from 480 to 600 nm.

3.2. I-V Performance of DSSC with Combination of DN-F01

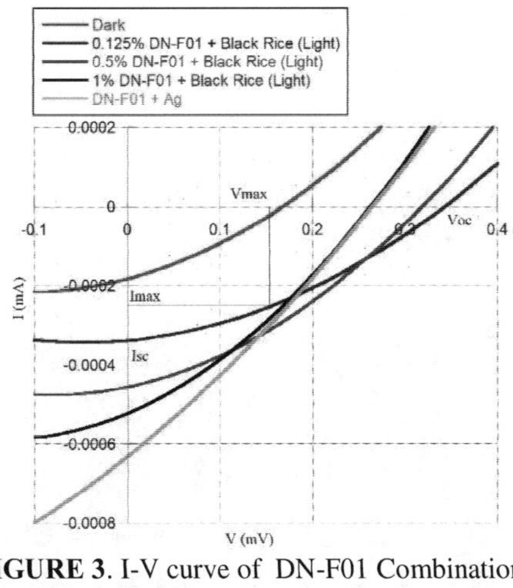

FIGURE 3. I-V curve of DN-F01 Combination

FIGURE 3 shows that the efficiency of 0.125% DN-F01 as much as 2 ml combined with 2 ml of black rice dye obtained 0.996 %. Efficiency of 0.5% DN-F01 as much as 2 ml combined with 2 ml of black rice dye obtained 1.014 %, and 1% DN-F01 as much as 2 ml combined with 2 ml obtained 1.065 %, while DN-F01 as much as 2 ml combined with 2 ml of Ag solution obtained the highest efficiency of all combinations dye, which is 2.145 %.

TABLE 1. Efficiency of DSSC by optimization result

Materials	V_{oc} (mV)	I_{sc} (mA)	*FF*	η (%)
0.125% DN-F01 + Black Rice	364	1.386	0.569	0.996
0.5% DN-F01 + Black Rice	394	1.279	0.579	1.014
1% DN-F01 + Black Rice	378	1.235	0.656	1.065
DN-F01 + Ag	805	1.234	5.396	2.145

TABLE 1 shows that the DN-F01 doped Ag solution obtained the highest efficiency of all combinations dye, which is 2.145 %. So, for the next step is to examine the structure of the working electrodes of DN-F01 dopped Ag solution.

3.4.1. FESEM structure of TiO_2 doped DN-F01

FIGURE 4. FESEM images of TiO2 doped DN-F01

FESEM analysis, results FESEM characterization with a magni fication of 20,000 times was carried out to determine the TiO_2 doped DN-F01 microstructure on the ITO glass substrate. Based on the results of the analysis, it is known that the properties obtained have a nanorod structure but TiO_2 doped DN-F01 particles have not been spread evenly across the surface. The most visible nanorod structure is the nanorod structure between particles. The nanorod structure in the particle and surface profile cannot be known because of the limited resolution of the instrument and research data. The nanorod structure of the TiO_2 layer is an important characteristic of DSSC [21]. More porous layer morphology can increase the number of dye molecules absorbed on the surface of TiO_2 particles. The more nanorod volume formed, the more space can be occupied by dye molecules. This makes photon absorption easier, increasing the number of electrons injected into the TiO_2 surface, which in turn can increase the ability of solar cells. As for the immersion in the dye solution, on the immersion for 24 hours the TiO_2 layer doped DN-F01 has a nanorod structure but has not been spread evenly. The structure of the immersed TiO_2 layer doped DN-F01 can be seen in **FIGURE 4**.

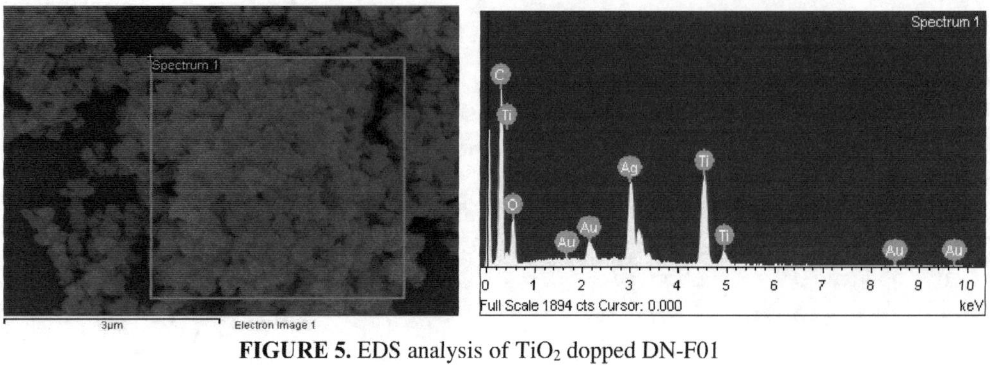

FIGURE 5. EDS analysis of TiO_2 dopped DN-F01

IOP Conf. Series: Materials Science and Engineering **858** (2020) 012016 doi:10.1088/1757-899X/858/1/012016

The EDS spectrum (**FIGURE 5**) exhibits a strong carbon peak which is in agreement with the high percentage of carbon in DN-F01. The EDS spectra also revealed peaks likely to be due to the other materials used or formed during the reduction process. A strong titanium peak shown in the spectrum originates from the copper grid used for FESEM analysis.

3.4.2. FESEM structure of TiO_2 doped DN-F01 + Ag

FIGURE 6. FESEM images of TiO_2 doped DN-F01 + Ag

The results of the sample scan by FESEM provide information related to morphology, including the shape, size and distribution of the sample [22]. Photo of FESEM structure of 50% TiO_2 doped with Ag 50% for 24 hours as shown in **FIGURE 6**. In the picture can be seen TiO_2 doped Ag structure that grows nanowires cross section. This figure also shows the diameter of the 50% TiO_2 doped Ag metal nanowires which is almost the same for three samples, which is between 63.16-210.50 nm. In this sample a high-density homogeneous structure is produced. High density in the sample will increase the amount absorbed, so that the absorption rate of the sample increases.

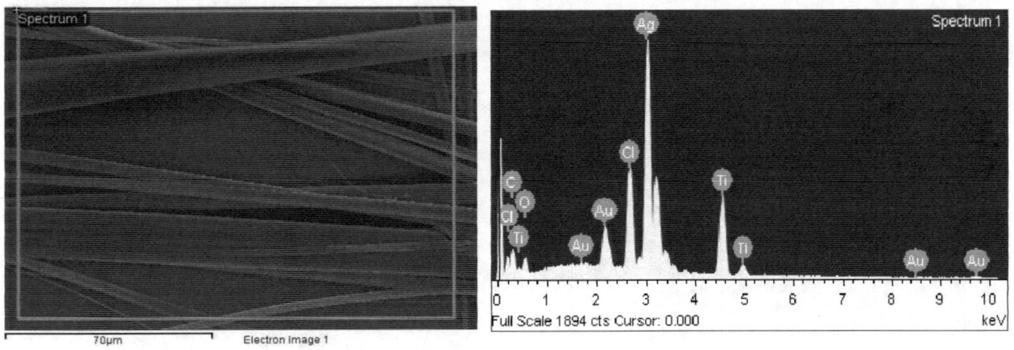

FIGURE 7. EDS analysis of TiO_2 dopped DN-F01 + Ag

The EDS spectrum (**FIGURE 7**) exhibits a strong silver peak which is in agreement with the high percentage of silver in DN-F01 doped Ag. The EDS spectra also revealed peaks likely to be due to the other materials used or formed during the reduction process. A strong titanium and clorin peak shown in the spectrum originates from the copper grid used for FESEM analysis.

3. Conclusion

From the research that has been done, it can be concluded that the combination of syntetic dyes has an influence on the efficiency of DSSC produced, where the highest efficiency is the concentration of synthetic dye DN-F01 doped Ag which is 2.145%. The dye absorption technique on TiO_2 can increase the efficiency of DSSC, where TiO_2 paste combined with DN-F01 doped Ag have higher efficiency than TiO_2 paste which is soaked with natural dye. After knowing the highest efficiency, it can be carried out a morphological structure study as FESEM and EDS of working electrodes that have been soaked DN-F01 doped Ag ie the more volume of nanowires formed, the more space can be occupied by dye molecules. This causes the absorption of high photons so as to increase the ability of DSSC performance.

Acknowledgment
This research was supported by Institute for Research and Community service, Sebelas Maret University by Hibah Penelitian Unggulan Terapan UNS (PUT-UNS) with contract 516/UN27.21/PP/2019 and Research Management Centre (RMC), Universiti Tun Hussein Onn Malaysia (UTHM) JOHOR, 86400 Parit Raja, Batu Pahat Johor, Malaysia.

References

[1] Y. Chiba, A. Islam, Y. Watanabe, R. Komiya, N. Koide, dan L.Y. Han. Jpn. J. Appl. Phys. 45 (2017) L638–L640
[2] R. Buscaino, C. Baiocchi, C. Barolo, C. Medana, M. Grätzel, Md.K. Nazeeruddin, G. Viscardi, Inorg. Chim. Acta 361 (2018) 798–805
[3] H. Zhou, W. Liqiong, G. Yurong, M. Tingli. Journal of Photochemistry and Photobiology A Chemistry 219 (2015) 188–194
[4] S. Hao, W. Jihuai, H. Yunfang, L. Jianming. Solar Energy 80 (2016) 209–214
[5] M.R. Narayan. Renewable and Sustainable Energy Reviews 16 (2018) 208–215
[6] J. K. Lee dan Y. Mengjin. Materials Science and Engineering B 176 (2018) 1142– 1160
[7] Directorate General of Agricultural Product Processing and Marketing, 2006. *"Mangosteen Agroindustry Development Road Map"*
[8] R. A. Wahyuono, *"Dye-Sensitized Solar Cells (DSSC) Fabrication with TiO_2 and ZnO Nanoparticle for High Conversion Efficiency"*. Master Thesis, ITS
[9] A. L. Castro, M. R. Nunes, A. P. Carvalho, F. M. Costa, M. H. Florencio. Solid State Sciences vol. 10, hal. 602 – 606.
[10] M. Z. Asrori, A. Permana, D. Sukma, Darminto, *"Development of Nanocomposite PANi (HCl) – TiO_2 as Anti Corrosion Material"*. Proceeding of 16th National Seminars of Nuclear Power Plant Facilities, Technology and Safety ISSN: 0854 – 2910, hal. 275-281
[11] M. Sardela, *"X-ray Analysis Methods. Advanced Materials Characterization Workshop"*. The Frederick Seitz Materials Research Laboratory – University of Illinois at Urbana-Champaign
[12] Y. Tanaka dan M. Suganuma. J. Sol-gel Sci and Tech. vol. 22, hal. 83 – 89.
[13] K. Varghese, C. A. Grimes. Solar Energy Materials & Solar Cells 92 (2008) 374–384
[14] S. R. Wenham, M. A. Green, M. E. Watt, R. Corkish, *"Applied Photovoltaics"*. Australia:Centre for Photovoltaic Engineering
[15] H. Chang dan Y. J. Lo. Journal of Solar Energy vol. 84, hal. 1833 – 1847.
[16] E. Yamazaki, M. Murayama, N. Nishikawa, N. Hashimoto, M. Shoyama, O. Kurita, *"Utilization Of Natural Carotenoids As Photosensitizers For Dye-Sensitized Solar Cells"*. Solar Energy vol. 81, issue 4, hal. 512 – 516.
[17] W. H. Lai, Y. H. Su, L. G. Teoh, M. H. Hon. Journal of Photochemistry and Photobiology A: Chemistry 195 (2019) 307 – 313.
[18] R. A. M. Ali dan N. Nayan. International Journal of Integrated Engineering (Issue on Electrical and Electronic Engineering) hal. 55 – 62.
[19] K. Hara, T. Horiguchi, Kinoshita, Tohru, K. Sayama, H. Sugihara, H. Arakawa. Solar Energy Materials & Solar Cells 64 (2019) 115 134
[20] P. Wu, J. Shi, J. Chen, B. Wang, L. Guo, *"Graphitis Carbon Nitride Modified by Silicon for Improved Visible Light Driven Photocatalytic Hydrogen Production"*. Nanostructured Materials and Nanotechnology Ceramic Engineering and Science Proceedings vol. 33, issue 7

IOP Conf. Series: Materials Science and Engineering **858** (2020) 012016 doi:10.1088/1757-899X/858/1/012016

[21] S. H. Kang, M. S. Kang, H. S. Kim, J. Y. Kim, Y. H. Chung, W. H. Smyrl, Y. E. Sung. Journal of Power Sources 184 (2018) 331–335

[22] M. Grtzel, Michael. *Photovoltaic performance and longterm stability of dye-sensitized meososcopic solar cells.* C.R. Chimie 9 (2005) 578-583.

[23] A.H. Ramelan, H. Harjana, and L.S. Sakti. *Titanium dioxide nanostructure synthesized by sol-gel for organic solar cells using natural dyes extracted from black and red sticky rice.* Proc. SPIE, Organic Photonics V, (2012) Vol. 8435-84352F-6.

[24] R. Subodro, B. Kristiawan, A.H. Ramelan, S. Wahyuningsih, H. Munawaroh, Q.A. Hanif, M. Z. Saputri. AIP Conference Proceedings, (2017) 1788, 030104-1–030104-8.

Synthesis, characterization and biodiesel conversion of monodispersed sulfonated-silica nanoparticles

W A Lestari[1], O A Saputra[1], E Pramono[1*], A P Chaniago[1]
[1]Chemistry Department, Faculty of Mathematic and Natural Sciences, Sebelas Maret University Surakarta, Indonesia.

E-mail: edi.pramono.uns@staff.uns.ac.id

Abstract. The enhancing need of energy followed by increasing of human population to be main factors in fuels depleting. This motivated researcher over the world to solve and to utilize natural resources as bioenergy for human sustainability in future. Today, there are numerous research related to bioconversion and bioprocessing of natural resources as well as optimizing catalyst in biodiesel conversion. Design of suitable catalyst for biodiesel conversion to be a challenge in this research area. In this paper, we report of preparation, characterization and biodiesel conversion ability of heterogeneous catalyst, so called monodispersion sulfonated-silica nanoparticles (MSN). This nanoparticles was characterized by FTIR, TEM, SEM and SEM-EDAX as well as N2 adsorption-desorption. The result revealed that the catalyst prepared through CTAB templating technique has been successfully synthesized with high monodispersion nano-size. Moreover, its ability in biodiesel conversion up to 13.33% for purity percentage.

1. Introduction

Biodiesel is a fuel that can be derived from biomass sources and can substitute conventional fuel that taken from petroleum sources [1,2] . Biodiesel is one of alternative fuel because it is biodegradable, non-toxic, free of sulfur and organic compound [3,4]. Biodiesel can reduce gas pollution significantly, such as carbon monoxide gas (CO) decrease 4-16%, carbon dioxide gas (CO_2) decrease 50-80% and SOx gas decrease 100% [5]. Basically, biodiesel is produced through transesterification reaction with methanol by alkali catalyst and this step require vegetable oil that contain very low free fatty acid (FFA) [6]. Biodiesel with high FFA, produce through 2 steps reactions, there are esterification and transesterification reaction using homogeneous or heterogeneous catalyst [7].

Homogeneous catalyst is used for oil that contain free fatty acid less than 1%, and heterogeneous catalyst is used for oil that contain free fatty acid more than 1% [7,8]. Using of H_2SO_4 and KOH as catalyst for Biodiesel production from Ceiba pentandra oil produce biodiesel product yield 99.5% [9]. Using of H_2SO_4 and KOH (1.5% : 2%) as catalyst in biodiesel production from castor oil (Karanja) has been reported that has yield 85% [10]. Biodiesel production of crude palm oil with the presence of H_2SO_4 as acid catalyst produce biodiesel product with yield 97% [11]. The weakness of homogenous catalyst is hard to separate from the product and lack of reuseability [12]. Homogeneous acid catalyst can caused environmental problems [13-16]. So heterogeneous catalyst is solution for that problem. That issues have be an trending topic to develop the ideas. Some researcher have synthesis some of heterogeneous catalyst but the main problem is how to enhanced the yield of biodiesel and reuseability.

Mesoporous material can describe as porous separation media without interparticular void [17]. The strength of monolithic are easy to separate, easy to surface modified, and high permeability. Mesoporous catalyst is effective for gas-liquid-solid reaction catalytic. Mesoporous catalyst has high surface area, high efficiency transfer mass and can separate easily [18]. Some researcher have synthesis MSN as catalyst. So in this paper have synthesized successfully MSN material that modified with sulfonate material to enhanced yield of biodiesel.

2. Experimental

2.1. Materials
The employed materials in this research were commercially available from Merck and Aldrich and used without purification except mentioned. The silica sources were tetra ethyl ortho silicate (TEOS, Aldrich). Organosilane functional compound was 3-(trimethoxysilyl)propyl methacrylate (TMPMA, Aldrich). The sulfonate monomer was Sodium-4-vinylbenzene sulfonate (VBS, Aldrich).

2.2. Synthesized of mesoporous silica nanoparticle (MSN)
For synthesized Mesoporous Silica Nanoparticle, 0.789 gram CTAB dilute in 21.6 mL aquades, 3.42 mL ethanol and 0.04 mL ammonium hydroxide solution. The solution stirred 30 minutes in 60°C, 4.83 mL TEOS dropped into solution in 6 minutes and mixed it for 2 hours. The solution has centrifuged and washed by ethanol. The solid phase was dilute in 20 mL methanol and 1.44 mL HCl, stirred for 8 hours in 60°C. The solution has centrifuged and washed by ethanol and dried in 60°C for one night. The powder mention as mesoporous silica nanoparticle (MSN).

2.3. Funcionalized of mesoporous silica nanoparticle (MSN)
Funcionalized of MSN by sulfonate function group, 0.5 gram MSN was added into 2.5 mL TMPMA, 10 mL aquadest and 10 mL ethanol solution. The solution was sonication for 20 minutes, stirred in 110°C for 12 hours. Sulfonate functional group, 0.17 gram VBS and 0.0002 gram BPO as initiator agent was diluted into 10 mL DMF and added into the solution. The solution was stirred 120°C for 20 hours and dried in 60°C for one night. The powder mention as Mesoporous silica nanoparticle sulfonation (sMSN).

2.4. Catalytic Activity
Activity catalytic of sMSN for conversion of palm oil to biodiesel is carried out using the reflux method. Palm oil, methanol and sMSN catalyst are put into a three neck flask with a palm oil: methanol ratio of 1:15. The amount of catalyst used is 5% w / w of palm oil. The mixture is stirred at a stirring speed of 400 rpm in 140 °C for 2 and 4 hours. The formed products were then analysis using GC-MS.

3. Result and Discussion

3.1. Chemical Interaction of MSN Synthesize
Infra-red spectroscopy is one of the simple techniques that can be used to analyze the success of a synthesis that can be seen from the shift, disappear or appearance of a new peak of a functional group. Figure 1 is an infra-red spectra of TEOS precursors and MSN synthesis products. Based on these spectra, MSN has successfully formed which is characterized by the appearance of hydroxyl absorption (-OH) in the region of wave numbers 3500 cm^{-1}. In addition, the disappear of -CH stretching peak in area about 2900-2800 cm^{-1} strengthens the suspicion of the formation of MSN products.

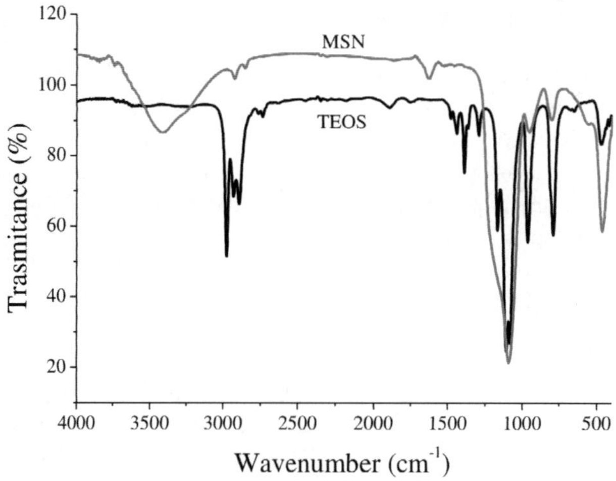

Figure 1. FTIR Spectra of TEOS and MSN.

The chemical illustration of MSN formation is shown in Figure 2.

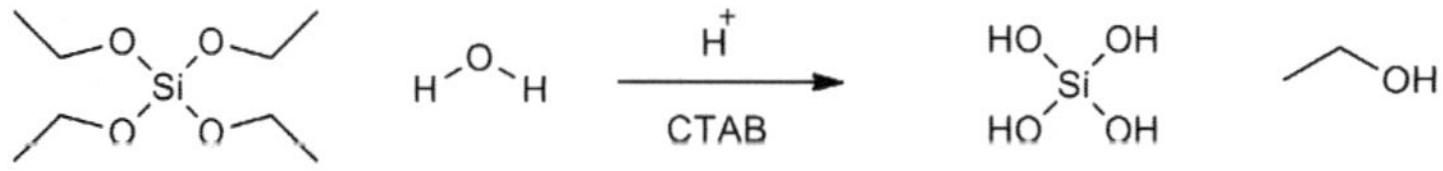

Figure 2. Chemical reaction mechanism of MSN.

The same characterization was also carried out to determine the success of MSN surface modification with sulfonate compounds. Based on the infra-red spectra shown in Figure 3, it can be seen that the appearance of absorbance in the wave number 1690 cm^{-1} which is characteristic of the >C=O group indicates that the TMPMA compound has been successfully attached to the surface of the MSN. In addition, it has also been found that wide absorption in the wave number 1100-1000 cm^{-1} is a sulfonate vibration that overlaps with a Si-O-Si vibration. Based on these data it can be concluded that surface modification has been successful whose chemical reaction mechanism is shown in Figure 4.

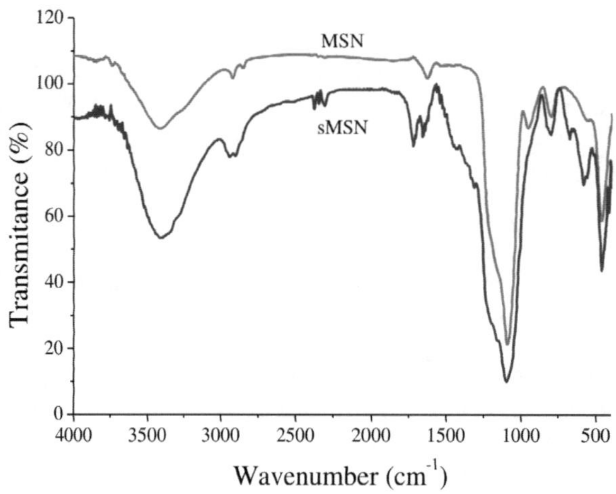

Figure 3. FTIR spectra of MSN and sMSN

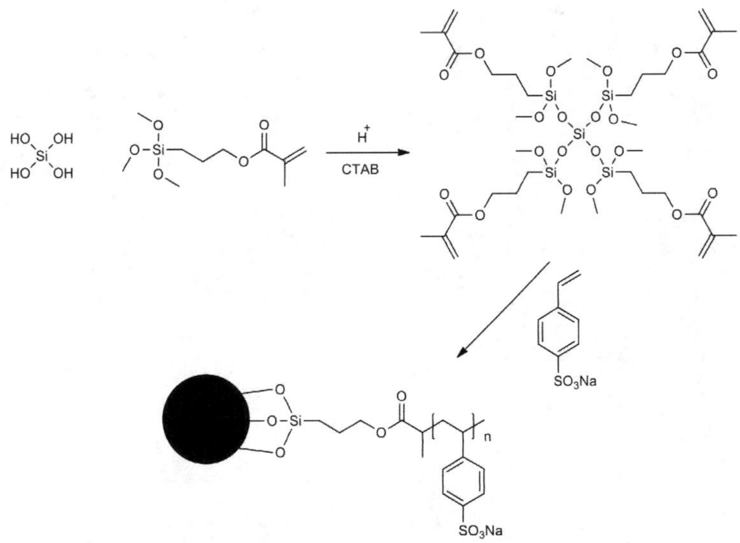

Figure 4. Chemical reaction mechanism of sMSN.

3.2. Morphology Analysis

Morphological analysis by SEM imaging can be used to evaluate and observe the morphological form of the sample. Figure 5a is the SEM morphology of MSN. It is clearly seen that from the morphological structure MSN has a structure similar to a mesoporous, which is a pore that is interconnected. However, after modification with sulfonate compounds, the morphology of the MSN shown in Figure 5b undergoes the union that is made possible by the existence of cross-linking so that one particle with another particle is connected to form a more bulky structure.

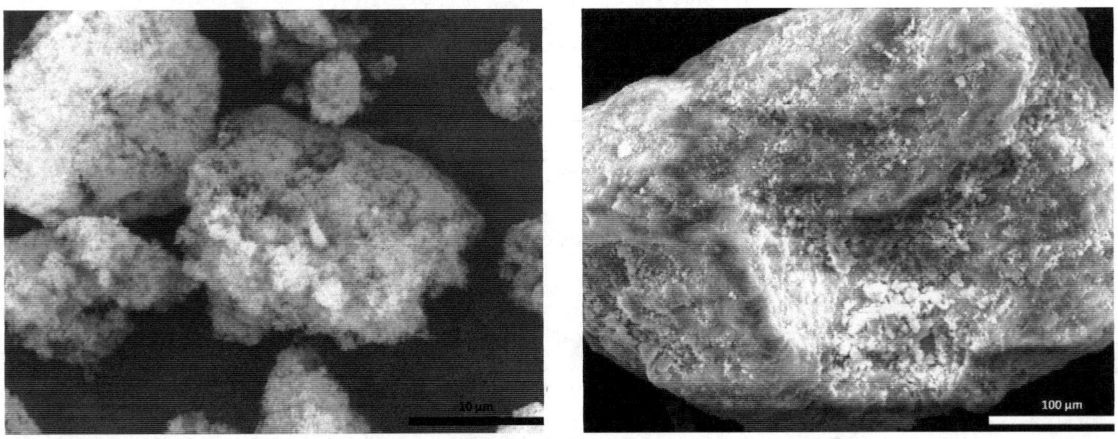

(a) (b)

Figure 5. SEM morphology of (a) MSN, (b) sMSN.

The presence of sulfonate groups on the surface of MSN can be proof by electron mapping and EDX techniques. Figure 6a is an electron mapping of the sMSN sample that the imaging results can conclude the S atom (sulfur, yellow) has distributed on the surface of MSN. The atomic composition of the sMSN is shown in Figures 6b. The S element on the surface of sMSN has a proportion of 1%, while other elements such as C, O, Si and Na are 44.3%, 39.9%, 14.5% and 0.2%, respectively.

IOP Conf. Series: Materials Science and Engineering **858** (2020) 012017 doi:10.1088/1757-899X/858/1/012017

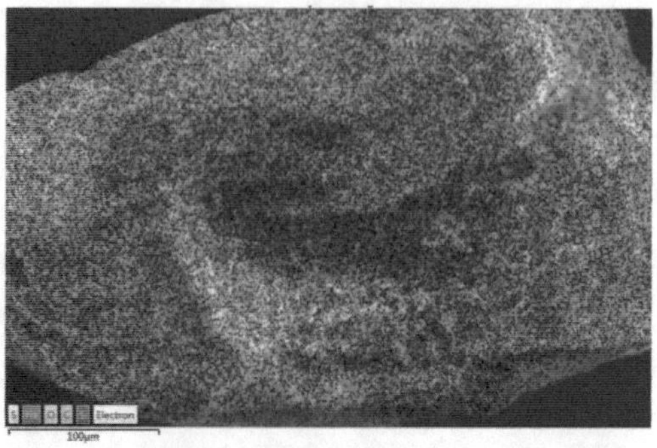

(a)

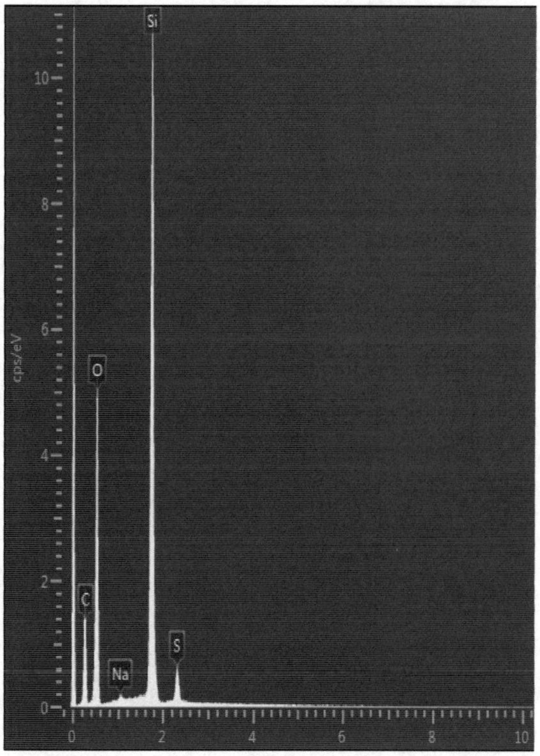

(b)

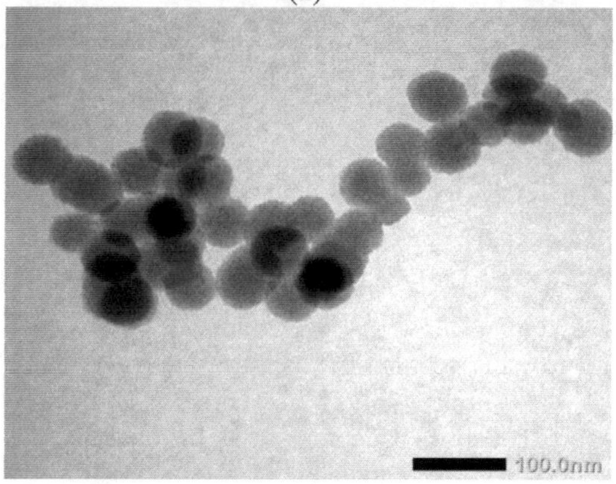

(c)

Figure 6. Imaging electron mapping of sMSN (a) and EDAX (b) of sMSN, TEM Imaging of MSN.

The imaging of nanostructures from MSN was evaluated by TEM. Figure 6c shows that the homogeneity of the nano-particle size distribution of MSN can be said to be evenly distributed. This happened because of the assistance from CTAB which was used as a template in the MSN synthesis process. In Figure 6c, it is clear that one nano-particle is connected to another. In the previous discussion, it has been explained that surface modification causes the structure of MSN to become more bulky, which is possible the connection of MSN particles through TMPMA and sulfonate compounds. The sMSN samples have particle sizes of less than 100 nm, so they can be categorized as nano-materials.

Table 1. Data summaries of physical properties of MSN and sMSN.

Materials	Surface area ($m^2.g^{-1}$)	Pore volume ($cm^3.g^{-1}$)	Average pore radius (nm)
MSN	474.105	8.54×10^{-1}	3.60
sMSN	27.278	8.46×10^{-2}	6.20

Figure 7 show profile N2 adsorption-desorption of MSN and sMSN. It shown that MSN has mesoporous isotherm profile. Isotherm graph of sMSN shown that it has mesoporous isotherm profile. Figure 8 show pore radius and pore volume of MSN and sMSN. It show that MSN and sMSN have same range pore volume is around 8.00 x10-1 – 9.00 x 10-1 cm3g-1. Summary of surface area, pore volume, pore radius can be seen in Table 1. Surface area of MSN is higher than sMSN. It caused sulfonate has been attached into surface of MSN. Modification sulfonate into MSN make MSN surface is covered by sulfonate that caused decreasing surface area of MSN.

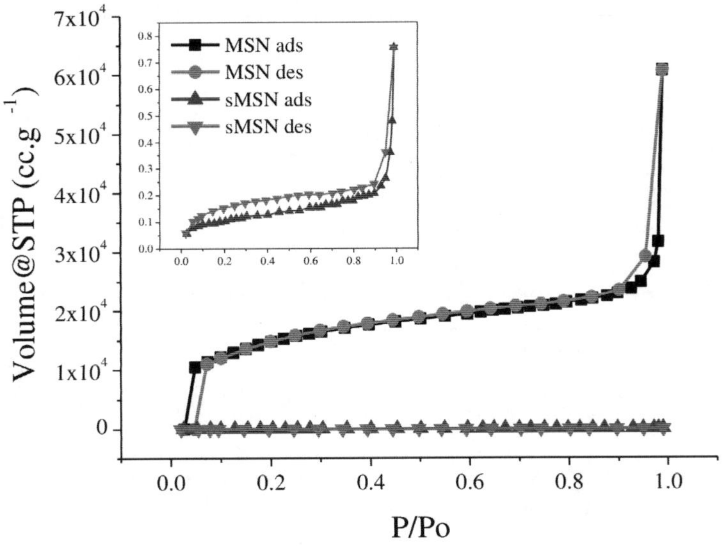

Figure 7. . N_2 adsorption-desorption of MSN and sMSN.

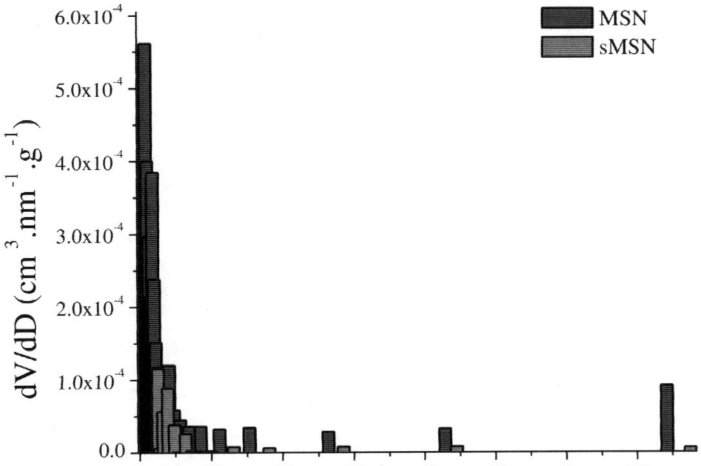

Figure 8. Pore radius distribution of MSN and sMSN.

2. Catalyst Activity of sMSN

Products from the esterification and transesterification process of CPO using sMSN catalyst were analyzed using GC-MS. Figure 9 shows the GC profile of the esterification and transesterification products. FAME product is divided into two products, namely hexadecanoic acid methyl ester (FAME 1) and octadec-9-enoic acid methyl ester (FAME 2), each appearing at a retention time of 16 minutes and 17.8 minutes. In addition to FAME products in the GC profile, free fatty acids and triglycerides were detected. This indicates that biodiesel conversion cannot be done 100%. By setting aside free fatty acid products, it was obtained that the conversion of FAME to triglycerides was 13.33%.

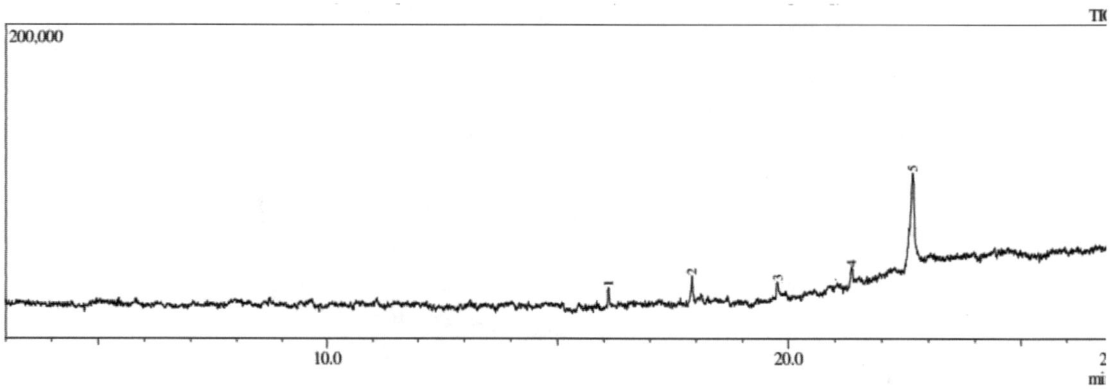

Figure 9. GC chromatogram of biodiesel product.

Figure 10 show the summary of schematic FAME contain in biodiesel product. Result of GC-MS analysis shows that the biodiesel contain of 5 peaks of FAME. Those are hexadecanoic acid methyl ester that change become hentriacontan-16-one if the reaction continuous, octadec-9-enoic acid methyl ester that change become octadec-9-enal if the reaction continuous, and Hexadecanoic acid-2-hexadecanoyloxy-3-hydroxy-prorylester (see table 2). Formed of ketone and aldehyde can be caused of reduction reaction from FAME.

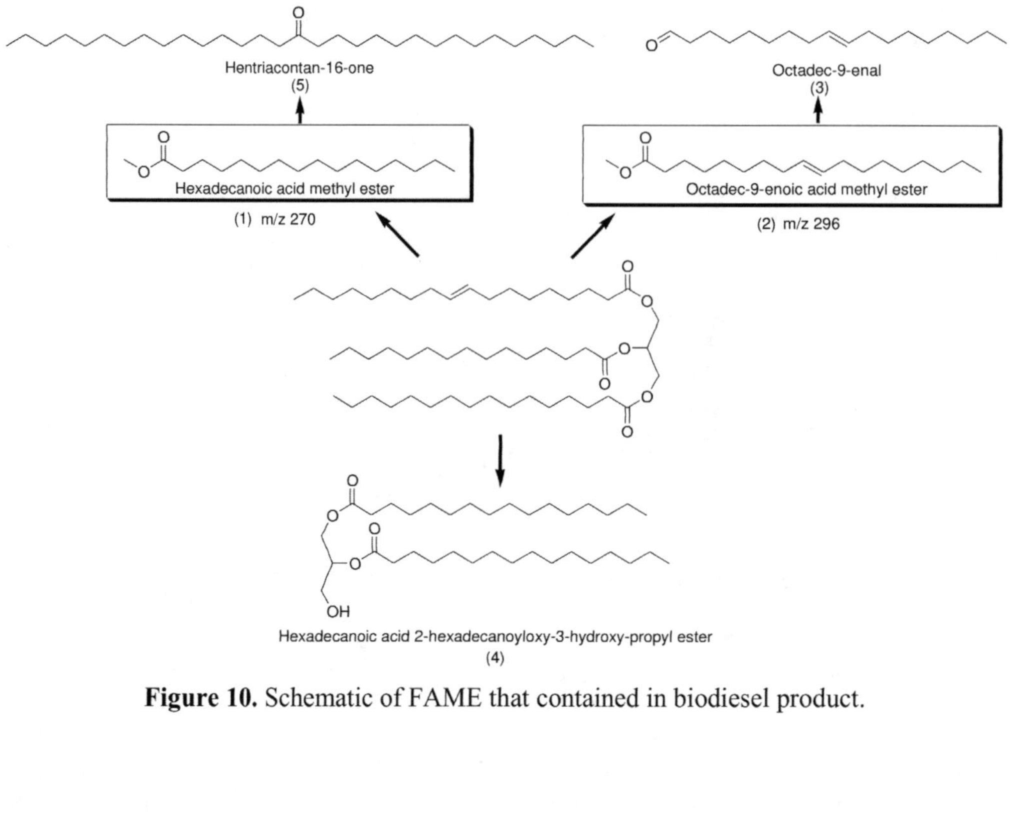

Figure 10. Schematic of FAME that contained in biodiesel product.

Table 2 . List of product from crude palm oil esterification.

No	Compounds	Yields
1	Hexadecanoic acid methyl ester	6.02%
2	Octadec-9-enoic acid methyl ester	7.31%
3	Octadec-9-enal	6.56%
4	Hexadecanoic acid-2-hexadecanoyloxy-3-hydroxy-prorylester	8.74%
5	Hentriacontan-16-0ne	72.37%

4. Conclusion

Biodiesel is renewable fuel that can be produced from biomaterial. The important thing to make biodiesel is transesterification reaction. The reaction is needed base catalyst. The MSN catalyst has been successfully synthesized and modified using sulfonate functional groups. MSN catalysts have a 13.33% conversion activity from palm oil to biodiesel which consists of FAME 1 and FAME 2.

Acknowledgement

Authors would like to acknowledge to Indonesian Directoral General of Higher Education for financial support from Program Kreatifitas Mahasiswa (PKM) 2015.

References

[1] Knothe and Van Gerpen 2005 The biodiesel handbook (Aocs press : Coburg, Germany)
[2] Yusuf N, Salaudeen N and Muhammad F 2011 *Australian Journal of Basic and Applied Sciences* **5** (6) 912-919
[3] Hassan M and Kalam MA 2015 *Procedia Engineering* **56** 39-53
[4] Demirbas A 2009 *Journal Energy convertion and management* **50** 14-34
[5] Xue, Grift and Hansen 2011 *Renewable and Sustainable Energy Reviews* **15**(2) 1098-1116
[6] Jain S, Sharma MP 2010 *Renew. Sust. Energy Rev.* **14** (3) 140-147
[7] Karmakar A, Karmakar S, and Mukherjee S 2010 *Bioresource Technology* **101**(19) 7201–7210
[8] Sharma YC and Singh B 2009 *Renewable and Sustainable Energy Reviews* **13**(6–7) 1646-1651
[9] Sivakumar P, Sindhanaiselvan S, Gandhi N N, Devi S S, and Renganathan S 2013 *Journal of Fuel* **103** 693–698
[10] Patil P D and Deng S 2009 *Journal of Fuel* **88**(7) 1302–1306
[11] Crabbe E, Nolasco-Hipolito C, Kobayashi G, Sonomoto K, Ishizaki A 2019 *Process Biochemistry* **37**(1) 65-71
[12] Semwal S, Arora A K, Badoni R P, and Tuli D K 2011 *Bioresource Technology* **102** 2151–2161
[13] Lee A F and Wilson K 2015 *Catalyst Today* **242** 3–18
[14] Madhu D and Sharma Y C 2017 *Resource Technology* **3** 144–157
[15] Liu Y, Chung J, Jang Y, Mao S, Kim B M, Wang Y, Guo X 2014 *ACS Appl. Mater. Interfaces.* **6** 1887–1892
[16] Tang S, Wang L, Zhang Y, Li S, Tian S, Wang B 2012 *Fuel Process. Technol.* **95** 84–89
[17] Zdravkov B D, Cerm´ak J J, Sefara M, and Jank J 2007 *Central European Journal of Chemistry* **5**(2) 385–395
[18] Zhao Y, Zhou J, Zhang J, Wang, S 2009 *Catalysis Letters* **131**(3-4) 597-605

Effects of HSQ_4 Concentration into dyes of *Carica Papaya* on the Performance of Dye-Sensitized Solar Cell

Agus Supriyanto[1,2], Fernince Ina Pote [1], A. H. Ramelan[2], Fahru Nurosyid[2], Sayekti Wahyuningsih[3]

[1] Department of Physics, Student of Graduate Program, Sebelas Maret University, St. Ir. Sutami 36A Kentingan Surakarta 57126, Indonesia

[2]Department of Physics, Faculty Mathematics and Natural Science, Sebelas Maret University, St. Ir. Sutami 36A Kentingan Surakarta 57126, Indonesia

[3]Department of Chemistry, Faculty Mathematics and Natural Science, Sebelas Maret University, St. Ir. Sutami 36A Kentingan Surakarta 57126, Indonesia

E-mail: agusf22@staff.uns.ac.id, fernince@gmail.com

Abstract. Organic dyes are very important in improving the performance of solar cells, at very low prices and friendly to the environment. DN-F15 transparent green is a squaraine dye with good stability on DSSC or with an alternative name HSQ_4. HSQ_4 is a synthetic dye that is very expensive, but sensitive to light so that it can improve the performance of solar cells. DSSC uses titanium dioxide (TiO_2) as a semiconductor, electrodes are used as substrates for the flow of electric charge. DN-EP01 is a thixotropic terpineol-based screen-printable Pt paste with a viscosity of 4-7 Pa.s at $10s^{-1}$. In furnaces around 450^0C will produce an even and transparent platinum particle layer which has high catalytic activity. The color of the platinum paste is straw or light yellow. Measurement of absorbance dyes solution using a spectrophotometer with a wavelength of 400 nm -800 nm. Characterization of *I-V* DSSC Carica Papaya produced a V_{oc}value of 560 mV, I_{sc} 1.5 x 10^{-4} A, fill factor 0.47 and efficiency of 0.43%. After doping with HSQ_4 the concentration of 1 resulted in a V_{oc} value of 378 mV, I_{sc} 4.0 x 10^{-4} A, fill factor 0.49 and efficiency of 0.14%. An increase in efficiency in variation 3 is 0.24% and variation 5 is 0.35%, this shows the effect of HSQ_4 concentration on the coloring of papaya leaf extract can improve the performance of DSSC.

1. Introduction

Solar energy in Indonesia is the biggest energy potential to be used as an alternative energy source in solar cells. One technology that utilizes solar energy to convert into electrical energy is solar cells. Indonesia can receive more solar heat without cost, which is 4800 W/m^2 every day [1]. Dye-Sensitized Solar Cell (DSSC) has been developed since 1991. This began to emerge as an alternative energy source after the discovery of solar cell materials with the photoelectrochemical principle by Grätzel [2]. Synthetic dyes that are commonly used are Ruthenium (Ru) Complex, this type of dye has excellent performance and can produce high efficiency [3]. Basically, the working principle of DSSC converts light energy to electrical energy on a molecular scale in the form of reactions from electron transfers. The first process starts with electron excitation in the dye due to photon absorption. Where this is one of the roles of the nature of TiO_2. When photons from the sun hit the working electrode on DSSC, the photon's energy is absorbed by the dye attached to the surface of TiO_2. So the dye gets the energy to be excited to carry the energy and is injected into conduction to TiO_2. TiO_2 acts as an electron acceptor. The dye molecules that are left behind are then oxidized. Furthermore, electrons will be transferred through the outer circuit to the opposite electrode. Electrolytes that act as electron mediators so they can produce cyclical processes in solar cells. Electrolytes capture electrons from external circuits with the help of platinum molecules as catalysts. The excited electrons reenter the cell and are assisted by platinum so that they can react with electrolytes in the electrons.

In this study, using synthetic dye HSQ4 which has good absorbance absorption at a wavelength of 703 nm and a maximum of 720 nm [4-5]. Solar cells with synthetic dye HSQ_4 showed very stable work performance in increasing DSSC parameters [6]. Each green plant contains chlorophyll, which is based on each leaf in the plant has the same performance as a solar cell, which absorbs solar energy and converts it into electrical energy [7]. Chlorophyll is the main pigment that is effective as photosensitizer in the process of plant photosynthesis with maximum absorption at 670 nm [8]. In the selection of natural organic material, the chlorophyll must meet the requirements of the characteristics of the active ingredients of solar cells. Where chlorophyll must be able to be a carrier medium for photons absorbed [9]. In this study chlorophyll extract dye used from papaya leaves (Carica Papaya L.). Papaya plants originate from tropical regions in America. This plant has soft and unbranched stems, produces white sap and is filled with leaves that are long and wide [10-11]. The natural chlorophyll dye from Carica Papaya is used as a dye to be hybridized with DN-F15 [12-14]. To obtain conductivity, absorbance and voltage-current characteristics for DSSC performance.

2. Experiments

2.1 Material Preparation

FTO glass is cut into sizes $2\ x\ 2\ cm^2$. This substrate is then put into a clean container and soaked in ethanol for 15 minutes. Ultrasonic Cleaner is used to clean dirt on FTO glass. After the cleaning process is complete, the substrate is dried using a tissue.

2.2 TiO_2 paste, 18-20 nm

DN-EP03 is a paste type TiO_2 which is dissolved in ethanol and stirred for 60 minutes (1: 4). Deposition on the FTO substrate using the spin coating method to get a thin and even layer. This is useful for DSSC as a transparent layer. Layers in furnaces with a temperature of 450°C to 550 °C [15-16].

2.3 Dye extraction

To make synthetic dye HSQ_4, variations in the concentration of *dye*namo transparent green contained 1% wt, 3% wt and 5 % wt then combined with natural dye and dissolved with 5 ml ethanol solvent. Stirred for 30 minutes. To make chlorophyll coloring from papaya leaves, washed with alcohol and then dried, Carica papaya are weighed as much as 10 grams, then crushed with a mortar until smooth and put in 50 ml of ethanol solvent. Stirred for 2 hours.

2.4 Preparation of Electrolyte

The electrolyte solution consisted of three materials that we used in this study. Namely Iodine, Potassium Iodide, and ethylene glycol. Weighed 0.1269 g of pure Iodine and 0.83 g of potassium Iodide and dissolved into 10 ml ethylene glycol [17].

2.5 Characterization

Keithley IV 2602A meters to determine the efficiency of DSSC. To determine efficiency, measurements are made of the current-voltage dense relationship (I-V) that forms a curve. Measurements are usually carried out in bright and dark conditions. I_{SC} current density, open-circuit voltage (V_{OC}), and fill factor FF are the three parameters that determine the quality of a solar cell. Short-circuit current density (I_{SC})

and open-circuit voltage (V_{OC}) are defined as the maximum current and voltage that can be obtained from solar cells. The efficiency of the solar cell is a factor that measures the quality of the performance of a solar cell. The greater the efficiency, the higher the performance of a solar cell that is obtained [18].

Characterization using UV-Vis Spectrophotometer 1601 Pc to determine the wavelength and peak absorbance of the dye. To find out the optical band gap, by applying the Tauc relationship as follows $(\alpha hv)^2 = A\left(hv - E_g\right)$. Where the absorption coefficient is photon energy, A is the constant and the optical band gap is αhvE_g[19]. The electrical conductivity characteristics were tested by Elkahfi 100 /I-V meter.

The DSSC mechanism in Figure 1 shows the DSSC structure consisting of a substrate with transparent electrodes and titanium dioxide (TiO2) [20]. And in Figure 2 is the work cycle of DSSC.

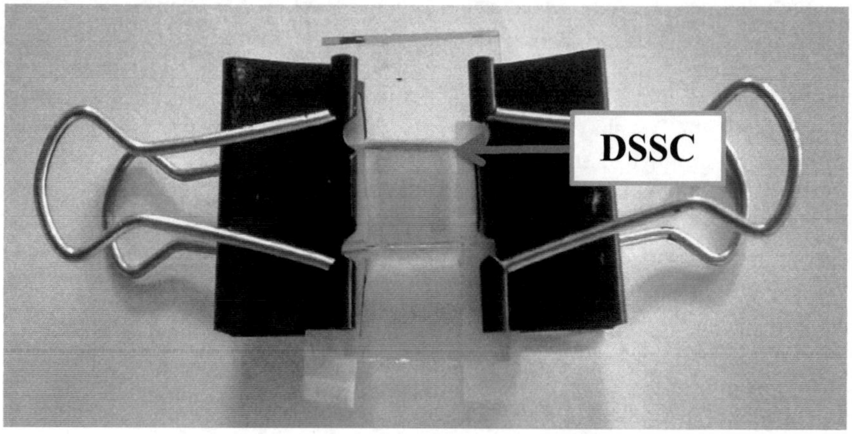

Figure 1. Components of solar cells

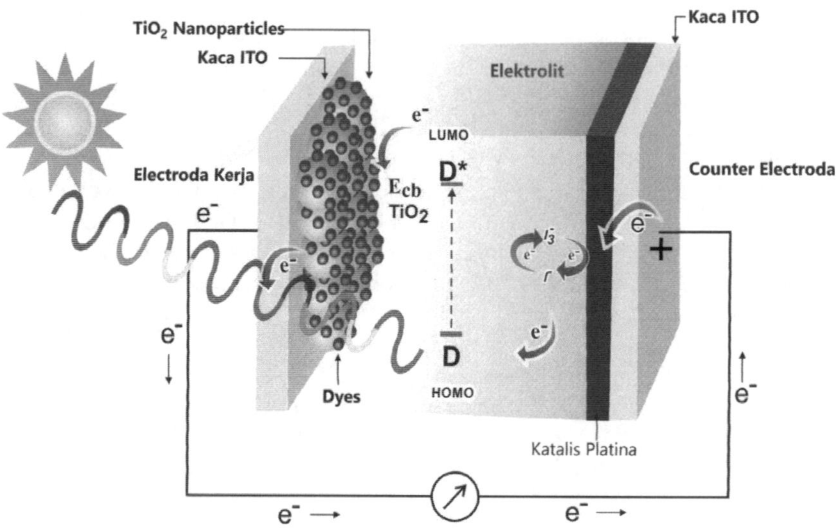

Figure 2. DSSC working principle

3. Results and Discussions

Conductivity characterization is to measure the Carica papaya dye solution, HSQ_4, and natural and synthetic hybrid chlorophyll. The solution is in concentrations of 1, 3 and 5 HSQ_4. Measurement using ELkahfi 100 IV-Meter with halogen lamp beam radiation with an intensity of $1000\ W/m^2$. Electrical conductivity is the ability of a material to conduct electric current if a potential difference is placed at the ends of a conductor, electrons will move through the material. Measurement of Carica papaya dye conductivity and HSQ_4 was carried out using the two-electrode method in the dark and light conditions. Copper is a conductor, whose function can capture electrons, is also able to deliver electrons produced by the dye.

Table 1 shows the fact, the increase in conductivity value is proportional to the increase in the amount of HSQ_4 added to the Carica papaya dye. The conductivity value in the Carica papaya dye is greater, that is 0.003 $\Omega^{-1}m^{-1}$, the conductivity value on HSQ_4 is very small but tends to be the same. Increasing conductivity values on hybrid HSQ_4 with Carica papaya dye, comparable to the many variations of HSQ_4 added.

Table 1. The electrical conductivity of *Carica papaya* with HSQ_4 variations

Material	Conductivity of Dark Condition ($\Omega^{-1}m^{-1}$)	Conductivity of Light Condition ($\Omega^{-1}m^{-1}$)	Δ Conductivity ($\Omega^{-1}m^{-1}$)
Carica Papaya	0,011	0.014	0,003
HSQ_4: 1% Wt	0,00019	0,00021	2 x 10-5
HSQ_4: 3% Wt	0,00011	0,00012	1 x 10-5
HSQ_4: 5% Wt	0,00032	0,00034	2 x 10-5
HSQ_4 1% Wt: Carica Papaya	0,0017	0,0021	0,0004
HSQ_4 3% Wt: Carica Papaya	0,0006	0,0013	0,0007
HSQ_4 5% Wt: Carica Papaya	0,0012	0,0030	0,0018

Table 1 shows the electrical characteristics of the material in dark or light conditions. The graph shows that all materials can conduct electricity in dark or light conditions. The highest conductivity in bright conditions is HSQ_4 5% Wt: Carica Papaya is 0.0018 $\Omega^{-1}m^{-1}$. This shows that the material has good conductivity so that it can help improve DSSC performance.

Table 2. Characterization UV-Visible Spectrophotometer of Carica Papaya with HSQ_4 variations

Material	λ (nm)	Eg (eV)
Carica Papaya	668	2.37
HSQ_4: 1% Wt	706	2.40
HSQ_4: 3% Wt	708	2.41
HSQ_4: 5% Wt	716	2.41
HSQ_4 1% Wt: Carica Papaya	704	2.39
HSQ_4 3% Wt: Carica Papaya	706	2.42
HSQ_4 5% Wt: Carica Papaya	708	2.43

Table 2 shows that the synthetic dye solution obtained from the combination of synthetic dyes and natural dyes was larger. The energy gap is the energy needed by an electron to move to the conduction band. A synthetic dye doping solution will help transfer electrons from the level (LUMO) to the conduction band (CB). From these results, it can be seen that the length of each variation shifts to a larger wavelength, indicating that the mixing of the dyes has been successful. The smaller the material gap energy, the higher the absorption of the sample from incoming light. The high absorption rate of the sample has an impact on the level of electron mobility so that the efficiency of the resulting DSSC increases.

Table 3. Characterization I-V Keithley of Carica papaya with HSQ_4 variations

Material	$V_{oc}(V)$	I_{sc} x 10^{-4}(A)	Fill Factor	Efficiency %
Carica Papaya	560	1,5	0,473	0,43
HSQ_4: 1% Wt	621	3,0	0,62	0,93
HSQ_4: 3% Wt	575	3,5	0,64	1,03
HSQ_4: 5% Wt	530	5,1	0,53	1,14
HSQ_4 1% Wt: Carica Papaya	378	4,0	0,49	0,14
HSQ_4 3% Wt: Carica Papaya	394	2,1	0,47	0,24
HSQ_4 5% Wt: Carica Papaya	454	2,2	0,45	0,35

I-V Keithley characteristics can be seen in Figure 5 and Table 3. From the table there is an increase in solar cells, the synthetic dye HSQ_4 obtained an efficiency value of 1.14% at a variation of 5% Wt. This shows that the combination of natural and synthetic dyes with increasing concentrations can improve DSSC performance [21].

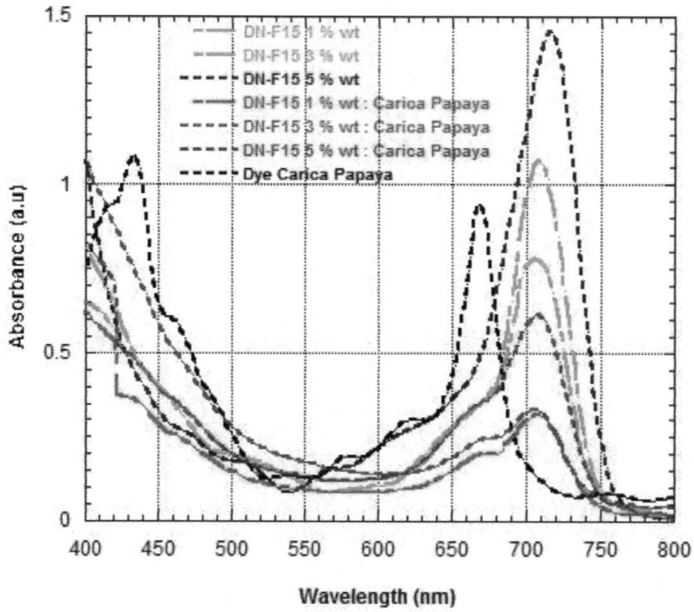

Figure 3. Absorbance Spectrum of HSQ_4 and HSQ_4: Carica Papaya

Figure 3 shows the optical characteristics of the carica papaya dye, HSQ_4, and HSQ_4: Carica papaya. There are seven absorbance peaks located in the wavelength range of 400 nm - 800 nm. From the absorbance results, it can be seen that the increase in HSQ_4 concentration gives a greater shift in wavelength. If more light is absorbed, more electrons can be transferred and DSSC performance increases.

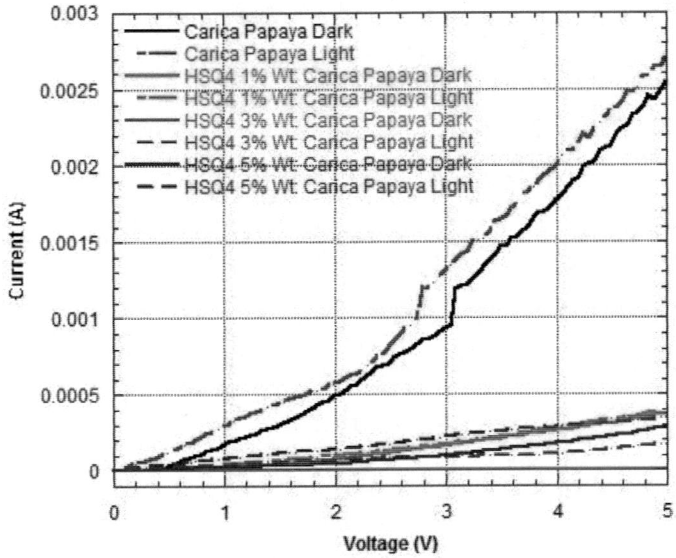

(a)

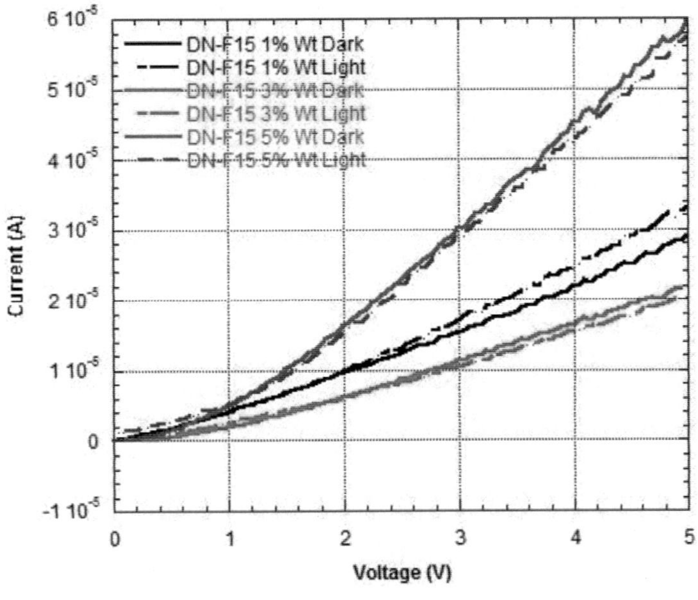

(b)

Figure 4. Electrical conductivity Characteristics for (a) HSQ_4 and (b) HSQ_4: Carica Papaya

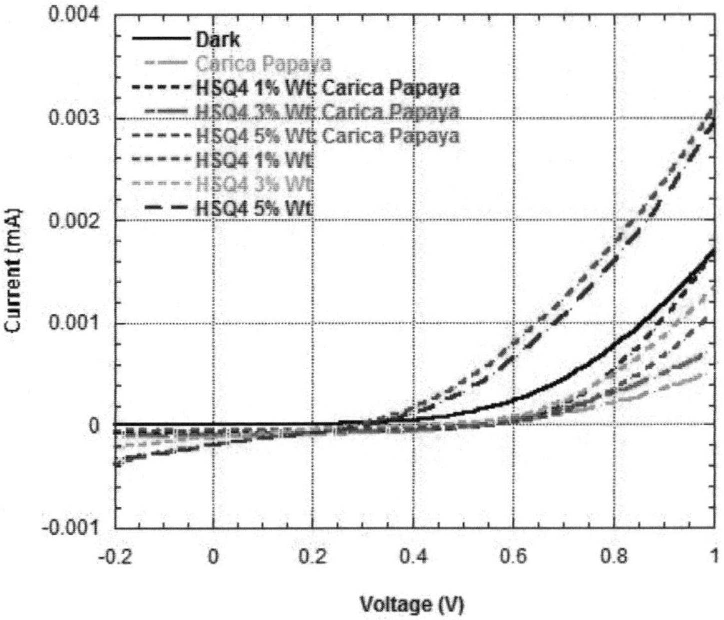

Figure 5. *I-V* characteristics from Dye of Carica papaya and Dye Synthesis HSQ_4

The *I-V* test resulted in the highest efficiency produced by cells with HSQ_4 working electrodes and TiO_2 from 0.93% to 1.14% and HSQ_4: Carica Papaya ie 0.14% to 0.35% showed an increase. So, it can be concluded that cell efficiency can increase greater by increasing the amount of concentration.

4. Conclusion

From the data generated it can be concluded that the performance of the platinum type DN-EP01 and TiO_2 type DN-EP03 is very good. Dye made from HSQ_4 powder which is easily soluble in ethanol can absorb and continue the visible light spectrum. Wavelengths are in the range of 400 nm - 800 nm. And the resulting energy gap is also in the range of 2.37 eV-2.43 eV, the smaller the gap of material energy, the higher the absorption of incoming light samples. The high level of sample uptake affects the level of electron mobility so that the efficiency of the resulting DSSC increases. Characterization of *I-V* DSSC Carica Papaya produced a V_{oc}value of 560 mV, I_{sc} 1.5 x 10^{-4} A, fill factor 0.47 and efficiency of 0.43%. After doping with HSQ_4 the concentration of 1 resulted in a V_{oc} value of 378 mV, I_{sc} 4.0 x 10^{-4} A, fill factor 0.49 and efficiency of 0.14%. An increase in efficiency in variation 3 is 0.24% and variation 5 is 0.35%, this shows the effect of HSQ_4 concentration on the coloring of papaya leaf extract can improve the performance of DSSC.

Acknowledgment

This study is partly supported by the Higher Education Project Grant with Contract Number 092/ SP2H/ LT/ DRPM/ 2019

References

[1] Manan, S. Solar energy is an efficient, reliable and environmentally friendly alternative energy source in Indonesia. Faculty of Engineering: Diponegoro University, Semarang. 2009

[2] Anh, Quan. Degradation of the Solar Cell Dye Sensitizer N719 Preliminary Building of Dye-Sensitized Solar Cell. Thesis. Denmark: Roskilde University. Vol. 2006
[3] M. Syahid, et al. Recent Advancement in Natural Dye Application: A Review. Journal of Cleaner Production. 2013 (1-22)
[4] C. P. Lee, R.Y. Lin, L. Lin, C. Li, T. Chu, S. Sun, J. T. Lin, K. Ho, Rsc Adv. 5 (2015) 23810
[5] Weiwei Zhang, Wenqin Li, Yongzhen Wu, Jingchuan Liu, Xiongrong Song, He Tian, and Wei-Hong Zhu (2016). *Novel Squaraine Cosensitization System of Panchromatic Light-Harvesting with Synergistic Effect for Highly Efficient Solar Cells*. ACS Sustainable Chem. Eng., Just Accepted Manuscript • DOI: 10.1021/acssuschemeng.6b00674. Pages 1-29
[6] Chuan Q, Youhei N, Shufang Z, Xudong Y, Ashraful I, Kun Z, Han C and Liyuan H. Novel Near-Infrared Squaraine Sensitizers for Stable and Efficient Dye-Sensitized Solar Cells. Advanced Functional Materials. 2014
[7] Ridwan, M.A., Noor, E., Rusli, M.S., Akhiruddin. (2018). Fabrication of Dye-Sensitized Solar Cell Using Chlorophylls Pigment from Sargassum. IOP Conf. Series: Earth and Environmental Science, 144(2018) 012039.
[8] Weihao, G. (2009). An Overview on P3HT PCBM, The Most Efficient Organic Solar Cell Material so Far. Solid State Physics II.
[9] Walter, M.G., Rudine, A.B., Wamser, C.C. (2010). Porphyrins and phthalocyanines in solar photovoltaic cells. Journal of Porphyrins and Phthalocyanines, 14, 759-792.
[10] Yogiraj V., Goyal P.K., Chauhan C.S., Goyal A., Vyas B. (2015). Carica papaya Linn: An Overview, International Journal of Herbal Medicine, 2(5), 01-08.
[11] Sumaryanti. (2011). Optical and Electrical Characterization of Spirulina SP Chlorophyll Solution as Dye-Sensitized Solar Cell (DSSC). Thesis. Sebelas Maret University.
[12] D. F. Rakhman et al, The Effect of Chlorophyll Concentration Variation on the Dye-Sensitized Solar Cell (DSSC) Output Power. Faculty of Electrical Engineering. Brawijaya University. Malang. 2014
[13] M N Arifin, A Supriyanto, A H Ramelan, F Nurosyid (2019). Influence of $Fe_2(SO_4)_3$ doping concentration into Chlorophyll (Carica Papaya L.) dye on the performance of Dye-Sensitized Solar Cells. IOP Conf. Series: Journal of Physics: Conf. Series 1153 012096 DOI: 10.1088/1742-6596/1153/1/012096
[14] J. H. Heo, D. H. Song, H. J. Han, S. Y. Kim, J. H. Kim, D. Kim, H. W. Shin, T. K. Ahn, C. Wolf, T.W. Lee, S. H. Im, Advances in Material. 27 (2015) 3424-3430
[15] L. K. Singh, T. Karlo, A. Pandey, Journal Renewable and Sustainable Energy, 5 (2013) 1-10
[16] L. K. Singh, T. Karlo, A. Pandey, Spectrochimica Acta Part A: Molecular and Biomolecular spectroscopy, 118 (2014) 938-943
[17] Mohammed, A.A., Ahmad, A. S. S and Azeez, W. A. (2015). Fabrication of Dye-Sensitized Solar Cell Based on Titanium Dioxide (TiO_2). Advances in Materials Physics and Chemistry, 5, 361-367
[18] J. Zhang, H-B, Li. S-L, Sun. Y. Geng, Y. Wu. Z-M. Su, Journal Material Science. 22 (2012) 568
[19] I. N. Obotowo et al. Organic Sensitizers for Dye-Sensitized Solar Cell (DSSC): Properties from computation, progress and future perspectives. Journal of Molecular Structure. 1122 (2016) 80-87
[20] Meng, S., Ren, J and Kaxiras, E. (2008) Natural Dyes Adsorbed on TiO_2 Nanowire for Photovoltaic Applications: Enhanced Light Absorption and Ultrafast Electron Injection. Nano Letters, 8, 3266-3272
[21] Z. Arifin, S. Soeparman, D. Widhiyanuriyawan, Suyitno, A.T. Setyaji, (2018). Improving Stability of Chlorophyll As Natural Dye for Dye-Sensitized Solar Cells, Jurnal Teknologi (Sciences & Engineering) 27-33.

IOP Conf. Series: Materials Science and Engineering **858** (2020) 012019 doi:10.1088/1757-899X/858/1/012019

Facile and rapid synthesis of tetrakis-2-amino-5-methylpyridinecopper(II) chloride pentahydrate

H Syaima[1], S B Rahardjo[1,2], V Suryanti[1,2]
[1] Chemistry Graduate Program, Graduate School, Universitas Sebelas Maret, Surakarta, 57126, Indonesia
[2] Department of Chemistry, Faculty of Mathematics and Natural Sciences, Universitas Sebelas Maret, Surakarta, 57126, Indonesia

E-mail: sentotbr@yahoo.com

Abstract. Pyridine and its derivatives are important compounds with tremendous applications. Pyridine derivatives, such as aminopyridine, have been widely investigated concerning their biological activity. It is reported that generally biological activities of the compound increased after complexation. Tetrakis-2-amino-5-methylpyridinecopper(II) chloride pentahydrate has been synthesized facilely in less than five minutes. It was formed by mixing $CuCl_2{\cdot}2H_2O$ and 2-amino-5-methylpyridine in 1:4 mole ratio in ethanol at room temperature. The complex was characterized to determine its proposed empirical formula and the structure. The forming of the complex was indicated by shifting of maximum wavelength (λ_{max}) of UV-Vis spectra towards smaller than $CuCl_2{\cdot}2H_2O$ from 850 nm to 675 nm. Atomic Absorption Spectroscopy (AAS) measurement showed the content of copper to be 9.63 % corresponding to the theoretical value of copper content in $Cu(2\text{-amino-5-methylpyridine})_4Cl_2(H_2O)_x$ (x= 4, 5, or 6). Thermogravimetric Analysis and Differential Scanning Calorimetry (TGA-DSC) showed a mass reduction of 13.02 % which was equivalent to the evaporation of five molecules of lattice water. The cation and anion charge ratio of the complex was 2:1. This four coordinated-complex was paramagnetic with an effective magnetic moment of 1.83 BM. The proposed empirical formula of the complex was $[Cu(2\text{-amino-5-methylpyridine})_4]Cl_2{\cdot}5H_2O$.

1. Introduction

Pyridine and its derivatives are interesting compounds among scientists. They are important with tremendous applications in the various fields [1-3]. Pyridine derivatives containing amino groups have been widely investigated concerning their biological activity. It is reported that the compounds posing exciting activity as antibacterial, antiviral, antifungal, and anti-inflammatory agents [4-7]. Further, it is reported that generally biological activities of the compound increased after complexation. Nitrogen atoms and its lone pair electrons in an aromatic environment make the pyridine group and its derivatives interesting compounds for coordination chemistry. Pyridine derivatives can act as a ligand and have been utilized in producing metal complexes with transition metal ions.

In several previous studies, pyridine derivatives containing amino groups were coordinating to different metal centers and different complexes were synthesized. Erdemir *et al.* (2019) have successfully synthesized Pd(II) complex with 2-aminopyridine through N-pyridine atom [8]. Kartal *et al.* (2018) prepared Zn(II) and Cd(II) complexes with 3-aminopyridine [9]. Ahmadi *et al.* (2011) have synthesized four novel cobalt complexes with 2-amino-4-methylpyridine, 2-amino-3-methylpyridine, and 2-amino-5-chloropyridine [10].

Copper(II) is stable and often used in many research [11]. Copper is also an important element for all living organisms such as humans, animals, microorganisms, and plants [12-13]. In this study, copper(II) complex with 2-amino-5-methylpyridine was synthesized to determine the empirical complex formula and properties formed.

2. Experimental

2.1. Materials

The chemicals and solvents were reagent grade and used without purification. The chemicals such as 2-amino-5-methylpyridine, nitric acid, $CuCl_2{\cdot}2H_2O$, $FeSO_4{\cdot}7H_2O$, $NiCl_2{\cdot}2H_2O$, $CuSO_4{\cdot}6H_2O$, and $AgNO_3$ were purchased from E. Merck.

2.2. Synthesis of the metal complex

The solution of $CuCl_2{\cdot}2H_2O$ (0.341 g; 2 mmol) in 10 mL ethanol was added dropwise into a 2-amino-5-methylpyridine solution (0.865 g; 8 mmol) in 15 mL ethanol with constant stirring. The brown solid precipitated. The solid was filtered and then washed with chloroform and ethanol. Then, the solid was dried in a vacuum desiccator over silica gel.

2.3. Instrumentation

The copper content in the complex was determined by Shimadzu AA-665 Atomic Absorption Spectrophotometer. Electronic absorption spectra were obtained with UV-Vis Double Beam Shimadzu PC 1601 spectrophotometer. The thermal behavior of the complex was recorded on Shimadzu 50 Differential Thermal Analyzer within temperature range 30-700 ^{o}C. Conductivity was measured on Jenway CE 4071 conductivity meter. The corrected molar magnetic susceptibilities have been measured at room temperature using Auto Sherwood Scientific 10169 Magnetic Susceptibility Balance. Infrared spectra for KBr discs were recorded on Prestige-21 Shimadzu spectrophotometers in the frequency range 4000–450 cm^{-1}.

3. Results and Discussion

3.1. Synthesis of the complex

Complex synthesis is carried out by mixing metals and ligands in a ratio of 1:4 in ethanol. When the metal solution was added dropwise into ligand solution, precipitate immediately formed followed by changing in color, from clear colored reactants into dark brown solid. The obtained yield was 68%. Directly formed deposits are possible due to fast complexing. According to spectrochemical series, C=N and NH_2 group in 2-amino-5-methylpyridine ligand pose high ligand field strength [14]. The complex formed has difference solubility with reactants in ethanol. Therefore, it immediately precipitated. The formation of the complex can be shown in electronic spectra. The maximum wavelength of Cu(II) solution shifted from 850 nm ($CuCl_2{\cdot}2H_2O$) to 675 nm (the complex) as shown in figure 1. The presence of blueshift in the spectra was due to the replacement of H_2O molecules in Cu(II) with 2-amino-5-methylpyridine ligands which have greater energy. These UV-Vis spectra are an early indication of complex formation. Then the complex needs further analyzed to determine the formula and characteristics of the complex.

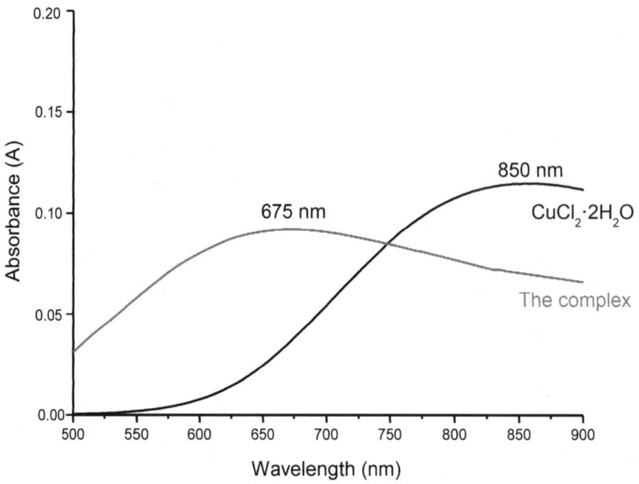

Figure 1. Electronic spectra of $CuCl_2{\cdot}2H_2O$ (a) and the complex (b) solution in methanol.

3.2. AAS result

AAS was used to estimate the possible empirical formula of the complex. The theoretical copper content in the complex is compared to the AAS results as shown in table 1. From the data, the possible formula of Cu(II)- 2-amino-5-methylpyridine is Cu(2-amino-5-methylpyridine)$_4$ $Cl_2(H_2O)_n$ (n = 4, 5 or 6).

Table 1. The copper content in some possible formulas of Cu(II)-2-amino-5-methylpyridine

	The formula of the complex	Theoretic calculation (%)	AAS result (%)
1.	Cu(2-amino-5-methylpyridine)$_4Cl_2(H_2O)_4$	9.94	
2.	Cu(2-amino-5-methylpyridine)$_4Cl_2(H_2O)_5$	9.67	9.63
3.	Cu(2-amino-5-methylpyridine)$_4Cl_2(H_2O)_6$	9.41	

3.3. Thermal analysis

The thermal behavior of the metal complex was characterized by TGA-DSC technique. The TG thermograms of the complex are represented in figure 2. The TGA curve of the complex shows one-step decomposition (mass loss: 13.02%) at 32-75°C. It corresponds to the evaporation of five hydrates or lattice water in the complex. It also strengthens by an endothermic process at 66 °C from DSC data. This process also happened in [Cu(Gapentine)(γ-aminobutyric acid)$(H_2O)_2$]Cl·H_2O at 27-87 °C which corresponding to the dehydration process [15]. The next step of decomposition appeared in the thermogram at the range 133-342 °C. It is probably assigned to the dissociation of the ligand. The third decomposition occurred within the range 352-700 °C appeared as a flat peak due to the last decomposition of the organic compound (ligand) leaving the CuO as a final residue [16].

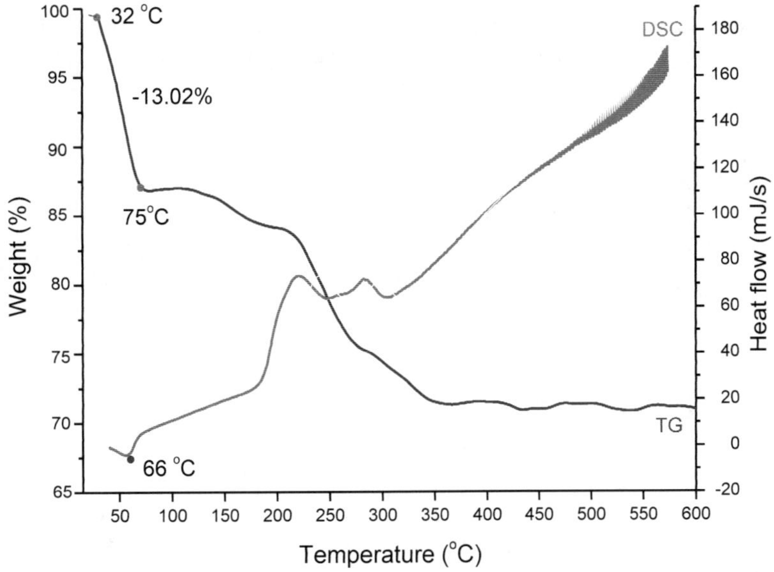

Figure 2. Thermogram of the complex

3.4. Qualitative chloride ion analysis

This analysis was used to determine whether the chloride ions were coordinated to the copper ion or acted counter ions. The testing was performed qualitatively by adding $AgNO_3$ solution into a complex solution. White solid of AgCl would precipitate according to the reaction [17]:

$$Cl^-_{(aq)} + AgNO_{3\,(aq)} \rightarrow AgCl_{(s)} + NO_3^-{}_{(aq)}$$

The results showed a white precipitate when the $AgNO_3$ solution was added dropwise to a solution of Cu(II)-2-amino-5-methylpyiridine. It explained that chloride acted as counter ions.

3.5. Electrical conductivity

Molar conductivity is used to determine the mole ratio of cation and anion and conductivity of the complex. This mole ratio was determined by comparing molar conductivity of the Cu(II) complex to molar conductivity of some known salts in methanol ± 1.10^{-3} M. From data in table 2, molar conductivity of the complex is in the range of 1:2 mole ratio of cation/anion. It means that chloride ions are not coordinated to Cu(II) and acting as counter ions. Therefore, the proposed formula of the complex is [Cu(2-amino-5-methylpyridine)$_4$]$Cl_2 \cdot 5H_2O$.

Table 2. Molar conductivity of Cu(II)-2-amino-5-methylpyiridine and known salts.

Solution	Λ_m ($S.cm^2.mol^{-1}$)	Mole ratio of cation/anion
$CuSO_4 \cdot 5H_2O$	2 ± 0.1	1 : 1
$FeSO_4 \cdot 7H_2O$	12 ± 0.1	1 : 1
$NiCl_2 \cdot 6H_2O$	170 ± 2.0	1 : 2
$CuCl_2 \cdot 2H_2O$	92 ± 1.5	1 : 2
Cu(II)-2-amino-5-methylpyridine	94 ± 0.2	1 : 2

3.6. IR spectra

FTIR spectra of the complex were compared with the free 2-amino-5-methylpyridine as shown in figure 3. The bands attributed to ν(N-H) vibration appear around 3444 and 3335 cm^{-1} for the ligand. These bands appear at the complex but shift into the absorption of 3444 and 3338 cm^{-1}.

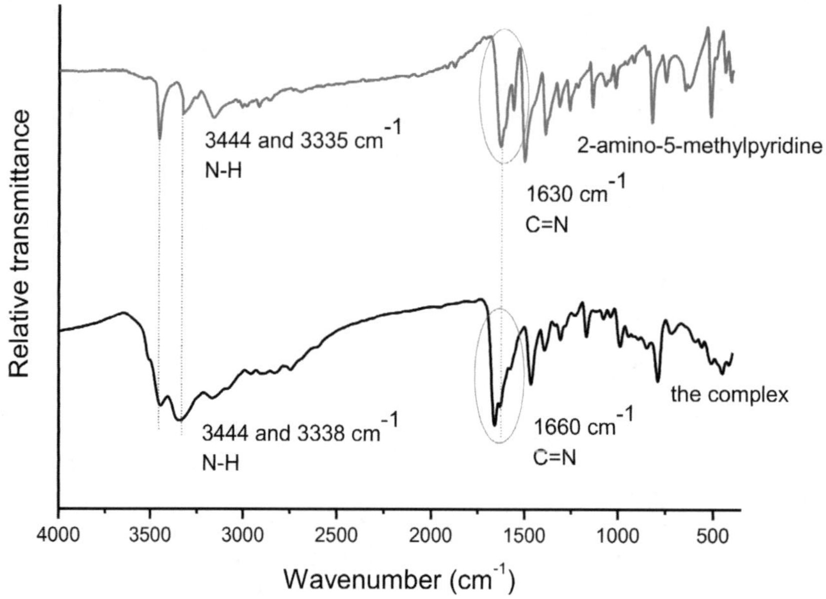

Figure 3. FTIR spectra of 2-amino-5-methylpyridine and the complex

It was also found that the bands due to the C=N group appeared at 1630 cm^{-1} in 2-amino-5-methylpyridine and 1660 cm^{-1} in the complex. It suggests that C=N suggesting its involvement in the coordination sphere. Shifting of coordinating group into higher wavenumbers also happen at other complexes [19-20]. The absorption of ν(N-H) of the ligand is almost not shifted. It shows that N-H group does not participate in chelation. Based on IR data, it was concluded that 2-amino-5-methylpyridine acting as the monodentate ligand.

3.7. Magnetic properties (μ_{eff}) and electronic spectra

The magnetic and electronic measurements can be used to estimate the geometry of the complex [16]. The value of the magnetic moment (1.83 B.M) of this complex confirmed the square planar geometry. Complex of Cu(II) with 7-((1H-benzo[d]imidazol-2-yl)diazenyl)-5-nitroquinolin-8-ol. has magnetic moment value 1.84 B.M which assigned to square planar geometry [16, 21-22]. The spectrum of Cu(II) complex displayed an absorption band at 675 nm assigned to d-d copper transition (sum of electronic copper transitions: $^2B_{1g}\rightarrow{}^2B_{2g}$, $^2B_{1g}\rightarrow{}^2E_g$, and $^2B_{1g}\rightarrow{}^2A_{1g}$ transition). This also assumes that the complex forming square planar geometry. It is reported that square planar copper complexes have a maximum wavelength of around 600 nm [21-24]. From the data, the proposed structure of [Cu(2-amino-5-methylpyridine)$_4$]$Cl_2\cdot 5H_2O$ is shown in figure 4. This μ_{eff} value is similar for paramagnetic Cu(II) complexes supporting the exclusion Cu-Cu [25].

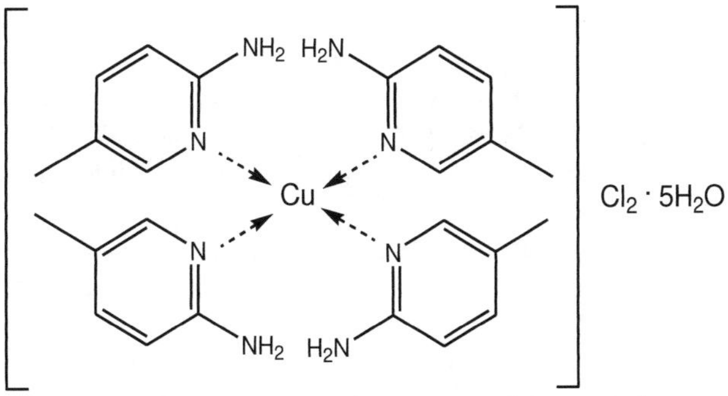

Figure 4. The proposed structure of the complex

4. Conclusion

The synthesis and the spectroscopic characterization of copper complex constructed by 2-amino-5-methylpyridine ligands have been described. The complex is estimated to be mononuclear square planar complex with empirical formula [Cu(2-amino-5-methylpyridine)$_4$]$Cl_2 \cdot 5H_2O$. The paramagnetic complex was obtained after coordination of N-pyridine from the ligand with the copper ion.

Acknowledgment

The authors gratefully express acknowledgments to Sebelas Maret University for financial support through Hibah PNBP Penelitian Fundamental with contract number: 5/6/UNS27.21/PP/2019.

References

[1] Raghavulu K, Sambaiah M, Gudipati R., Basavaiah K., Yennam S and Behera M 2019 *J. Saudi Chem. Soc.*
[2] Avcı D, Altürk S, Sönmez F, Tamer Ö, Başoğlu A, Atalay Y, Kurt B Z and Dege N 2018 *Tetrahedron* **74** 7198–7208.
[3] Sun B, Dong Y, Lei K, Wang J, Zhao L and Liu M 2019 *Bioorg.Med. Chem.* **27** 2427-2437.
[4] Altaf A A, Shahzad A, Gul Z, Rasool N, Badshah A, Lal B, Khan E 2015 *JDDMC* **1(1)** 1-11.
[5] Song Y-L, Tian C-P, Wu Y, Jiang L and Shen L-Q 2019 *Steroids* **143** 53–61
[6] Pal S 2018 *InTech* 76986
[7] Kadayat T M, Banskota S, Bist G, Gurung P, Magar T B T, Shrestha A, Kim J A and Lee E-S 2018 *Bioorg. Med. Chem. Lett.* **28** 2436–2441.
[8] Kartal Z, Şahin O and Yavuz A 2018 *J. Mol. Struct.* **1171** 578–586.
[9] Erdemir F, Celepci D B, Aktaş A, Gök Y, Kaya R, Taslimi P, Demir Y and Gulçin İ, 2019 *Bioorg. Chem.* **91** 103134.
[10] Ahmadi R A, Safari N, Khavasi H R and Amani S 2011 *J. Coord. Chem.* **64** 2056-2065.
[11] Sharma N, Prakasha R, Chaturvedib K 2012 *Scientific Review and Chemical Communication* **2(2)** 108-114.
[12] MacLean L, Karcz D, Jenkins H, McClean S, Devereux M, Howe O, Pereira M D, May N V, Enyedy É A and Creaven, B S 2019 *J. Inorg. Biochem.* **197** 110702.
[13] Kumar M, Kishor S, Kumar A, Bhagi A K and John R P 2020 *J. Mol. Struct.* **1199** 126996.
[14] Patra S G and Datta D 2018 *Comput. Theor. Chem.* **1130** 77–82.
[15] Mahmoud M A, Abbas A M, Zaitone S A, Ammar A M and Sallam S A 2019 *J. Mol. Struct.* **1180** 861–877.
[16] Gaber M, El-Wakiel N and Hemeda O M 2019 *J. Mol. Struct* **1180** 318–329.
[17] Ha H and Payer J 2011 *Elect. Acta* **56** 2781–2791.
[18] Kacar S, Unver H and Sahinturk V 2019 *Arab. J. Chem. In press*
[19] Smrečki N, Stilinović V, Jović O, Kukovec B M and Popović Z 2017 *Inorganica Chim. Acta* **462** 57–63.
[20] Salman Y, Barlas F B, Yavuz M, Kaya K, Timur S and Telli F Ç 2018 *Inorganica Chim. Acta*

483 98–105.
[21] Syaima H, Raharjo S B, Zein I M 2018 *IOP Conf. Ser.: Mater. Sci. Eng.* **349** 012025
[22] Syaima H, Raharjo S B, Suciningrum E 2018 *IOP Conf. Ser.: Mater. Sci. Eng.* **333** 012003
[23] Syaima H, Raharjo S B, Suciningrum 2017 *J. Phys.: Conf. Ser.* **909** 012080
[24] Rambabu A, Ganji N, Daravath S, Venkateswarlu K, Rangan and K Shivaraj 2020 *J. Mol. Struct.* **1199** 127006.
[25] Shanty A A, Raghu K G and Mohanan P V 2019 *J. Mol. Struct* **1197** 154–163.

IOP Conf. Series: Materials Science and Engineering **858** (2020) 012020 doi:10.1088/1757-899X/858/1/012020

PREPARATION COMPOSITE TiO_2 ANATASE WITH POLYDIMETILSILOXANE AND POLYTETRAFLUOROETHYLENE FOR SELF CLEANING ON GLASS SUBSTRATE

S Wahyuningsih[1*], A H Ramelan[1], A N Firdaus[1], R E Cahyono[1]
[1]Inorganic Material Research Group, Faculty of Mathematics and Natural Science, Sebelas Maret University, Surakarta, Indonesia

E-mail: sayekti@mipa.uns.ac.id

Abstract. Synthesis of TiO_2-PDMS and TiO_2-PTFE composites were carried out. Preparation of TiO_2 used ball milling technique. This study aims to determine the effect of adding PDMS and PTFE to the self-cleaning activity of TiO_2, which can be measured by determining the change in contact angle and photocatalytic character of TiO_2. The photocatalytic-based self-cleaning character was determined by Rhodamin B degradation.. The results showed the characteristic peaks of TiO_2 anatase in (2θ) = 25.390°; 37.781°; 48.07°; 54.10°; dan 62.902°. The existence of PDMS and PTFE is supported by the results of FTIR analysis that appear absorption at wavelengths around 795 cm^{-1} - 1265 cm^{-1} for PDMS, 1154 cm^{-1} and 1212 cm^{-1} for PTFE. In addition, it is also supported by AFM imaging results that show the roughness of the TiO_2-PDMS and TiO_2-PTFE composites. The Results of contact angles measurement from glass/TiO_2-PDMS is 133,47° and after being irradiated is 27,13°. The Results of contact angles measurement from glass/TiO_2-PTFE is 117,37° and after being irradiated is 34,70°. The highest degradation of Rhodamin B was 37.04% for glass/TiO_2-PDMS and 47.67% for glass/TiO_2-PTFE which is produced in the self-cleaning process for 20 minutes.

1. Introduction

Glass is a transparent material that is often used in buildings today. Glass has a variety of practical uses in various fields of industry, building construction, automotive, and others [1]. Glass can also adorn property and can be used to save energy. The increasing use of glass in buildings, especially in skyscrapers, creates another problem. The problem comes from how to clean it. If cleaning is done manually with human labor it can be risky. Various strategies have been used to modify the glass, one of which is self-cleaning technology [2].

Self-cleaning technology works by utilizing the photocatalytic effect produced from a chemical compound so that it can clean the glass substrate from impurities. Self-cleaning has two stages based on its nature in dealing with water. The surface of the glass can be hydrophobic or hydrophilic [3]. Materials that have self-cleaning properties are very useful for helping human work and have simple cleaning methods [4].

Titanium dioxide (TiO_2) can be applied in the fields of photocatalysts, solar cells, anti-fungal, antibacterial, antifungal, paint protective coatings, and wastewater cleaners. Application of TiO_2 in daily life can be used because TiO_2 is non-toxic, has high chemical stability, and high photocatalytic reactivity and is inert. To improve photo activity and photocatalytic efficiency of TiO_2 can be done by modifying the material. This modification can be done by doping the surface of TiO_2. The doping can be carried out using hydrophobic material. Polydimethylsiloxane (PDMS) and polytetrafluoroethylene (PTFE) are hydrophobic materials. In addition PDMS and PTFE also have other advantages that can absorb light in the ultraviolet and visible light, have inter-layer force with a substrate that is strong enough, low surface energy, and are a hydrophobic coating polymers that has good stability in acidic or basic environments. Increased photoactivity on TiO_2 will improve the self-cleaning process as indicated by the increased ability to respond to visible light. Composite self-cleaning activity was tested by degradation of rhodamine B dyes which were known from a decrease in dye concentration based on UV-Vis absorbance.

Conducted a study of photocatalytic activity, in which the coating was synthesized by the spray coating method using TiO_2 which was added with PDMS [5]. From the results of these studies obtained contact angles above 80° and after being irradiated with UV light the contact angles turned to hydrophilic. Kamegawa et al also conducted research on photocatalytic activity, where layers were synthesized using TiO_2 added with PTFE [6]. From the results of these studies obtained a contact angle of about 120° and after being illuminated with UV light the contact angle changes to hydrophilic.

2. Experimental

2.1. Preparation of TiO_2

TiO_2 preparation using the ball mill method refers to the method of Fadillah et al [7]. TiO_2 is ball milled with a ball ratio: TiO_2 = 20: 1 and the rotating speed is set at 1000 rpm for 4 hours. Ball-Milled TiO_2 were dissolved in 8 N NaOH then refluxed for 24 hours at 120 °C. The solution was neutralized with 2M HCl then filtered and dried at 60 °C for 12 hours and calcined at 400 °C for 2 hours. The dry precipitate was then characterized using XRD, FTIR and SEM.

2.2. Preparation of TiO_2-PDMS and TiO_2-PTFE composites

10 mg of TiO_2 ball milled was added to 70 mg, 75 mg, 80 mg, 85 mg, and 90 mg of PDMS or PTFE in 200 mL ethanol. Then TEOS was added with variations 0 mg, 5 mg, 10 mg, 15 mg, and 20 mg (Table 1). Characterization was carried out using FTIR and SEM.

Table 1. Comparison Ratio of TiO_2-PDMS and TiO_2-PTFE

TiO_2 (mg)	PDMS / PTFE (mg)	TEOS (mg)	Ethanol (mL)	Variations
10	90	0	200	1
10	85	5	200	2
10	80	10	200	3
10	75	15	200	4
10	70	20	200	5

2.3. Glass Substrate Preparation

Glass with a size of 7 cm x 3 cm x 0.3 cm is cleaned to remove organic or inorganic impurities that stick [8]. Glass substrate is inserted into a beaker containing acetone and then vibrated with an ultrasonic cleaner for 15 minutes. Then the washed glass is dried at 100 °C for 10 minutes. The washing process is continued with ethanol and distilled water alternately with the same time [9].

2.4. Deposition to Glass Substrate

Deposition to the glass substrate is done by the dip coating method. The synthesized TiO_2-PDMS and TiO_2-PTFE composites were respectively dissolved in 200 mL ethanol. Glass is inserted in the beaker containing the solution with a variety of compositions of TiO_2-PDMS and TiO_2-PTFE composites. The beaker that has been filled with glass and composites is sonicated with an ultrasonic cleaner for 30 minutes at a temperature of 25 °C and is repeated 4 times. Then the glass is heated at 110 °C for 30 minutes. The test is done by measuring the contact angle and with the Rhodamin B degradation test using a UV-Vis spectrophotometer.

2.5. Self Cleaning Activity test on Glass Substrate

2.5.1 Hydrophobicity of TiO_2-PDMS and TiO_2-PTFE Coated Glass

The hydrophobic properties of the coated glass was tested by measuring the contact angle. Contact angle measurement is done by dripping a dye (Rhodamin B) on the surface of the coated glass and then being photographed. The contact angle is measured using software image J.

2.5.2 Photodegradation of Rhodamin B

5 mL of 10 ppm rhodamine B dye was dripped on TiO_2-PDMS and TiO_2-PTFE coated glass. The glass was irradiated for 5, 10, 15, and 20 minutes later the degradation results were characterized using a UV-Vis spectrophotometer.

3. Results and Discussions

The preparation of TiO_2 material refers to research conducted by Fadillah et al [7] through a ball mill process with a speed of 1000 rpm for 4 hours. The results of TiO_2 ball milled were analyzed by X-ray diffraction using Cu Kα radiation (λ = 1.5406 Å). The peak obtained from the measurement results by XRD is matched with X-ray diffraction standards, namely the JCPDS (Joint Committee on Powder Diffraction Standards) [10].

The diffraction pattern of TiO_2 can be shown in Figure 1. The diffractogram shows that TiO_2 has a dominant peak at (2θ)= 25.390° (d_{101}=114,37 nm), (2θ)= 37.781° (d_{004}=76,33 nm), (2θ)= 48.07° (d_{200}=111,23 nm), (2θ)= 54.10° (d_{105}=122,19 nm), dan (2θ)= 62.902° (d_{211}= 90,41 nm). These results are areas of anatase phase TiO_2 characterization in accordance with JCPDS No. 21-1272.

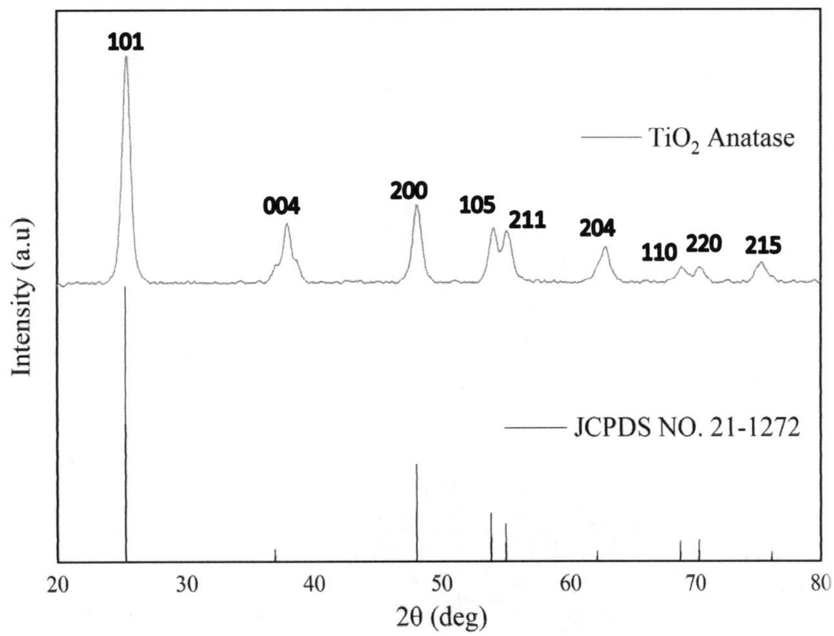

Figure 1. X-Ray Diffractogram of TiO_2.

Preparation of TiO_2-PDMS and TiO_2-PTFE were carried out by sonication method. PDMS and PTFE are used as dopants because PDMS and PTFE have hydrophobic and transparent surfaces. PDMS and PTFE are polymer materials with low surface energy, thus causing PDMS to have hydrophobic properties. The PDMS or PTFE material binds to the TiO_2 surface bridged by TEOS by means of which three ethyl groups from TEOS will break away so that the three O atoms can bind to TiO_2. TiO_2-PDMS composites were tested by infrared spectroscopy (FTIR) to determine the bonds formed. Based on the TiO_2-PDMS FTIR spectra shown in Figure 2a, the existence of TiO_2 can be seen in wave numbers 1154 cm^{-1} and 1214 cm^{-1} which indicate the presence of Ti-O bonds. Absorbance at wave numbers 1096 cm^{-1} and 1021 cm^{-1} indicates the presence of Si-O-Si bonds. Whereas the wave number 2964 cm^{-1} indicates the presence of CH_3. The presence of Si-CH_3 bonds originating from PDMS can be seen from the absorption at wavelengths of 800 cm^{-1} and 1262 cm^{-1}. Based on the TiO_2-PTFE FTIR spectra shown in Figure 2b, the existence of TiO_2 can be seen in wave numbers 1154 cm^{-1} and 1214 cm^{-1} which indicate the presence of Ti-O bonds. While the presence of PTFE can be seen at wave numbers 1154 cm^{-1} and 1212 cm^{-1}.

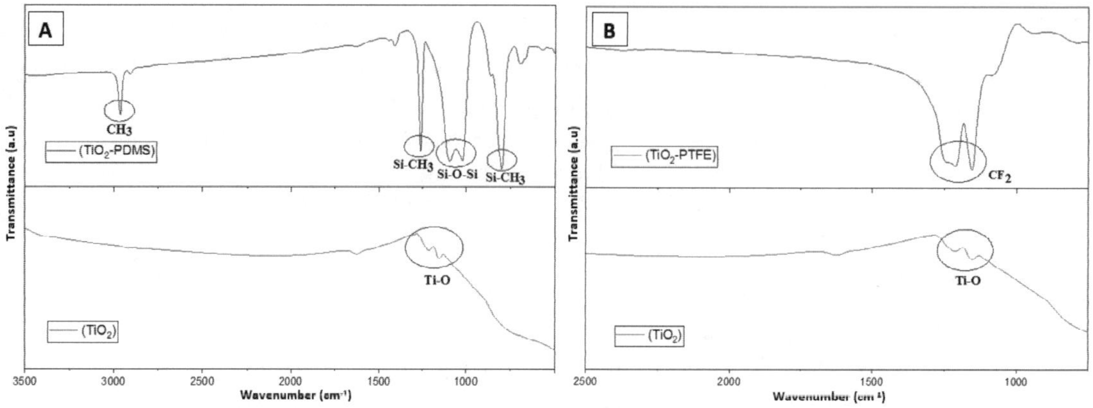

Figure 2. FTIR spectra of composites (A = TiO_2-PDMS, B = TiO_2-PTFE).

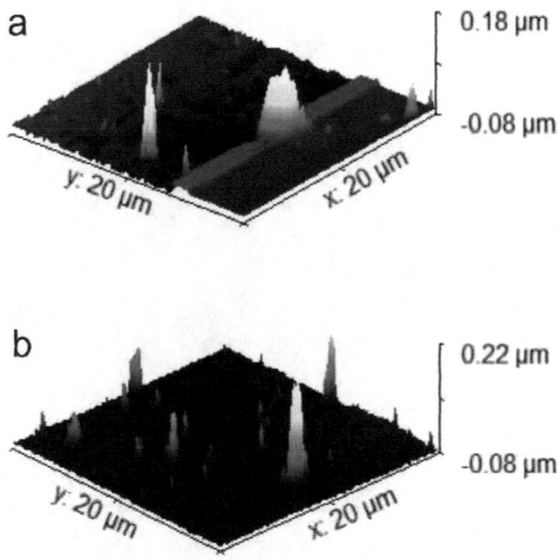

Figure 3. 3-dimensional AFM (20 x 20 μm^2) on glass substrate of (a) TiO_2-PTFE and (b) TiO_2-PDMS

Composite coating on glass substrate using dip coating method then analyzed using AFM. The results of the 3-dimensional analysis are shown in Fig. 3 and the roughness measurements are shown in Table 2. Analytical techniques using AFM following surface morphology have been proven to be very flexible and powerful tools for surface imaging at nanometers to submicrometer level and the relevation of surface characteristics of composite thin film.

IOP Conf. Series: Materials Science and Engineering **858** (2020) 012020 doi:10.1088/1757-899X/858/1/012020

Table 2. Rougness Analysis of TiO_2-PTFE and TiO_2-PDMS surface

Composite	RMS (nm)
TiO_2-PTFE	12.50
TiO_2-PDMS	11.74

Contact angle measurements are carried out using a liquid dropped on a glass surface coated with TiO_2-PDMS or TiO_2-PTFE. The liquid used is Rhodamin B. Photos of the droplets formed were analyzed using imageJ software to measure the contact angle. Based on these measurements obtained contact angles with various variations of the ratio of TiO_2: PDMS and TiO_2: PTFE before and after irradiation which can be seen in Table 3 below.

Table 3. Glass Contact Angles Before and After Irradiation

No	Composite Variation	Contact angle (before) (°)	Contact angle (after) (°)
1	Glass without coating	50,45	48,8
2	Glass/TiO_2	69,62	40,5
3	Glass/TiO_2-PDMS(1)	128,85	37,07
4	Glass/TiO_2-PDMS(2)	126,15	22,63
5	Glass/TiO_2-PDMS(3)	123,5	33,77
6	Glass/TiO_2-PDMS(4)	133,47	27,13
7	Glass/TiO_2-PDMS(5)	133,15	33,73
8	Glass/TiO_2-PTFE(1)	115,23	32,87
9	Glass/TiO_2-PTFE(2)	116,32	32,73
10	Glass/TiO_2PTFE(3)	117,27	35,5
11	Glass/TiO_2-PTFE(4)	117,37	34,7
12	Glass/TiO_2-PTFE(5)	108,77	34,2

The glass which was originally hydrophobic turned into hydrophilic after being exposed to ultraviolet light. The change in properties is caused by the occurrence of photocatalytic processes by thin layers of TiO_2-PDMS and TiO_2-PTFE. The hydrophilic nature can be achieved when stable hydroxyl forms gather on the surface. So that when there is water, the water is directly bound to the hydroxyl group. From this picture, it can be seen that the glass which is not coated and the glass which is only coated by TiO_2 experiences very small contact angle changes so that the glass does not yet have self-cleaning properties. While the glass coated by TiO_2-PDMS or the glass coated by TiO_2-PTFE has a significant angle change. TiO_2-PDMS(4) obtained the largest contact angle of 133.47°. The contact angle changes to 27,13° after exposure to light. While the TiO_2-PTFE(4) obtained the largest contact angle of 117.37°. The contact angle changes to 34,70° after exposure to light. This indicates that the glass coated by TiO_2-PTFE has high self-cleaning properties. This result is in line with research conducted by Tavares et al [5] and Kamegawa et al [6] where glass coated with TiO_2-PDMS has a contact angle of more than 80°, whereas glass coated by TiO_2-PTFE has a contact angle of 111°.

The glass coated with TiO_2-PDMS and the glass coated with TiO_2-PTFE were tested for self-cleaning ability by dripping 5 mL of rhodamin B 10 ppm on the glass and then irradiated using a 300 watt halogen lamp with a certain irradiation time. The self-cleaning activity test of glass coated with

TiO_2-PDMS and TiO_2-PTFE with varying irradiation time with various compositions of TiO_2: PDMS and TiO_2: PTFE can be seen in Figure 4a and 4b.

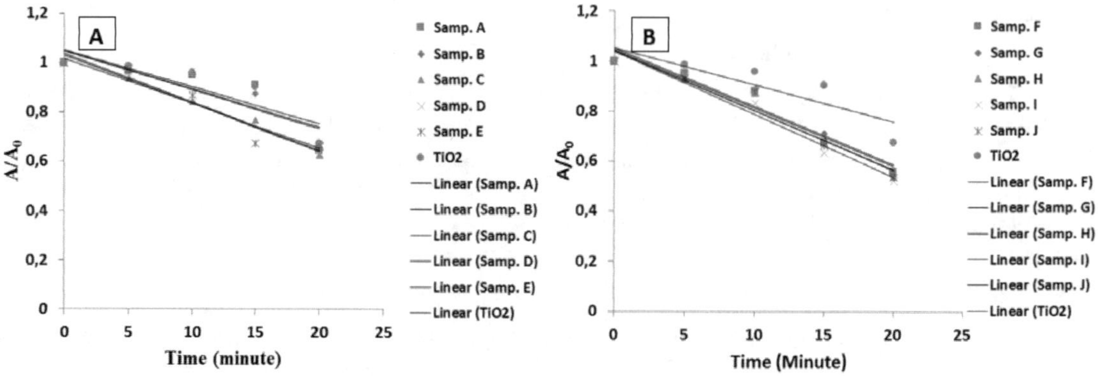

Figure 4. Rhodamin B Degradation Results on coated Glass (A = TiO_2-PDMS, B = TiO_2-PTFE).

The figure shows from this graph that it can be seen that glass coated by TiO_2-PDMS and TiO_2-PTFE has a higher decrease in absorbance than glass coated only by TiO_2. TiO_2-PDMS(3) has the greatest ability to degrade rhodamine B that is 37.04%, while the TiO_2-PTFE(4) has the greatest ability to degrade rhodamine B that is 47.67%. The increase in composite self-cleaning activities can be seen from the declining trend. Decreased absorbance of rhodamine B with increasing exposure time from 5 minutes to 20 minutes. This shows that the long exposure time will increase the self-cleaning activity on the composite. These results are in line with research conducted by Sunardi et al [11]. where the longer the glass is exposed to light, the smaller the absorbance value.

4. Conclusion

Coatings of TiO_2-PDMS and TiO_2-PTFE composites on glass substrates affect the hydrophobicity and hydrophilicity of the glass surface. Coating can increase the hydrophobicity of the glass surface. The glass coated by TiO_2-PDMS and TiO_2-PTFE has hydrophobic properties and after exposure to light it turns into hydrophilic. TiO_2-PDMS(4) the contact angle was 133.47°. The contact angle changes to 27,13° after exposure to light. In the TiO_2-PTFE(4) the contact angle was 117,37°. The contact angle changes to 34,70° after exposure to light. Addition of PDMS and PTFE to TiO_2 particles can increase the self-cleaning activity of TiO_2 on glass. The TiO_2-PDMS(3) was able to degrade rhodamine B by 37.04%, while the TiO_2-PTFE(4) was able to degrade rhodamine B by 47.67%.

Acknowledgment

The authors would like to acknowledge the Integrated Mathematics and Natural Science Laboratory of Sebelas Maret University for supporting and providing the facilities for research.

References

[1] Karasu B, Bereket O, Biryan E, and Sanoglu D 2017 El-Cezerî Journal of Science and Engineering **4** 2 209-233

[2] Cedillo E I G, Raffaele R, Monia M, Mauro M, Paulo P, and Cristina S 2014 Building and Environment **71** 7-14

[3] Li F, Li Q, and Kim H 2013 Applied Surface Science **226** 390-396

[4] Xu X, Michael R W, Yusuf m, and Stephen R H 2014 Composites Science and Technology **100** 220-227

[5] Tavares M T S, Santos A S F, Santos I M G, Silva M R S, Bomio M R D, Longo L, Paskocimas C A, and Motta F W 2014 Surface & Coatings Technology **239** 16-19
[6] Kamegawa T, Yuki S, and Hiromi Y 2012 Advanced Materials 1-4
[7] Fadillah G, Sayekti W, and Ari H R 2018 *2nd International Conference on Chemistry, Chemical Process and Engineering (IC3PE)* 1-7
[8] Yusrianto E, Noor B I, and Zahari I 2010 Solid State Science and Technology **18** 2 52-60
[9] Anuar and Dahyunir D 2016 Jurnal Fisika Unand **5** 3 228-232
[10] Hastuti E 2011 Jurnal Neutrino **4** 1 93-100
[11] Sunardi, Utami I, and Nora R S 2012 *Sains dan Terapan Kimia*. **6** 2 118-129.

Electrosynthesis of hydroxyapatite by multiple electrode arrangement

T Salsabila, W Widiyastuti and H Setyawan[1]
Sepuluh Nopember Institute of Technology, Sukolilo, Surabaya 60111, Indonesia

[1]E-mail: sheru@chem-eng.its.ac.id

Abstract. Hydroxyapatite ($Ca_{10}(PO_4)_6(OH)_2$) is a ceramic biomaterial which has high biocompatibility that makes it popular in the biomedical application as it can attach on human body without rejection by human immune system. Hydroxyapatite can be synthesized by various methods; one of them is electrochemical method. Synthesis of hydroxyapatite has been successfully conducted by electrochemical method using one pair of electrodes. However, the yield of hydroxyapatite produced by this pair of electrode was still low. To address those issue this research proposed synthesis of hydroxyapatite using multiple electrodes in different arrangements to improve the yield of the reaction, which were conventional and alternating arrangements. The potential value was fixed on 6 V to study the effect of electrodes arrangement on current density. Electrosynthesis of hydroxyapatite showed that multiple electrodes on alternating arrangement produced more hydroxyapatite mass and current density than the conventional arrangement. The alternating arrangement produced hydroxyapatite mass of 2.3483 g and current density of 19.32 mA/cm^2, while the conventional arrangement produced 2.3157 g and 15.05 mA/cm^2.

1. Introduction

Hydroxyapatite ($Ca_{10}(PO_4)_6(OH)_2$) or HA is one of the bioceramic materials types of calcium phosphate. HA is one of the minerals that compose human bones and teeth. [1] Due to high biocompatibility and bioactivity properties, HA is popular in its application in the biomedical field. [2] HA has been used in various biomedical applications such as bone and teeth filler, or bone coating to enhance bone hardening [1], tissue engineering [3], and wound healing [4].

HA can be produced using several methods, such as sol-gel [5], wet chemical precipitation [6], hydrothermal [7], emulsion [8], and electrochemical [9]. Electrochemical methods are starting to become a favorable choice for HA production. With electrochemical methods, particle size can be controlled easily through the current or power settings used in the system. [10] Nur and Setyawan have synthesized HA through the electrolysis method with one pair of platinum as the electrodes. [9] The electrolysis method from Nur and Setyawan (2016) is adopted in this research with the modification of carbon as the electrode material. HA production using one pair of carbon electrodes was still low, which was 0.5383 g. To address this issue, this research proposed a synthesis of hydroxyapatite powder using two pairs of electrodes with different electrodes arrangements, which was conventional and alternating arrangement. X-Ray Diffraction (XRD) and Scanning Electron Microscopy (SEM) analysis were used to observe the crystallinity and morphology of electrosynthesis product.

IOP Conf. Series: Materials Science and Engineering **858** (2020) 012021 doi:10.1088/1757-899X/858/1/012021

2. Experimental

2.1 Material

The electrolyte solutions for electrosynthesis of hydroxyapatite was composed of ethylenediaminetetraacetic acid or EDTA ($C_{10}H_{14}N_2Na_2O_8.2H_2O$, Merck), kalium dihydrogen phosphate $_4$(KH_2PO_4, Merck), and calcium chloride ($CaCl_2$, Merck) with a concentration ratio of EDTA/PO $^{3-}$/Ca^{2+} of 0.25/0.15/0.25 M. The electrodes material used was 2 pairs of 5×1.5 cm^2 carbon electrodes.

2.2 Experimental setup

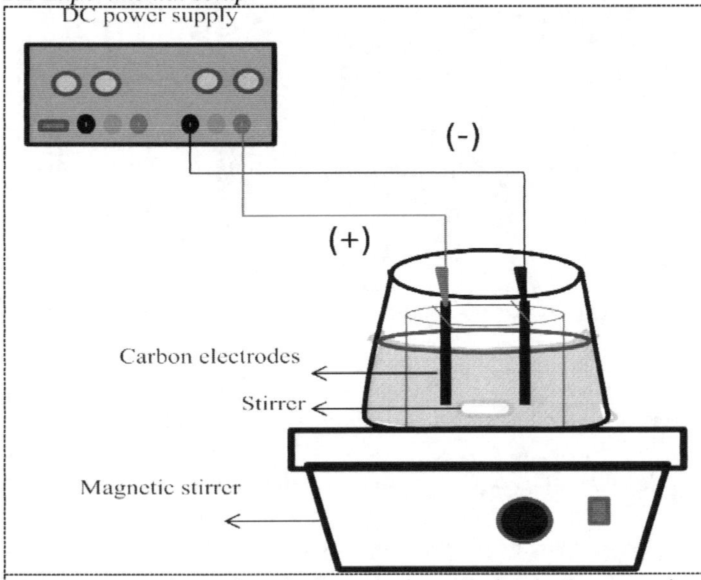

Figure 1. Schematic of electrolysis cell for hydroxyapatite synthesis.

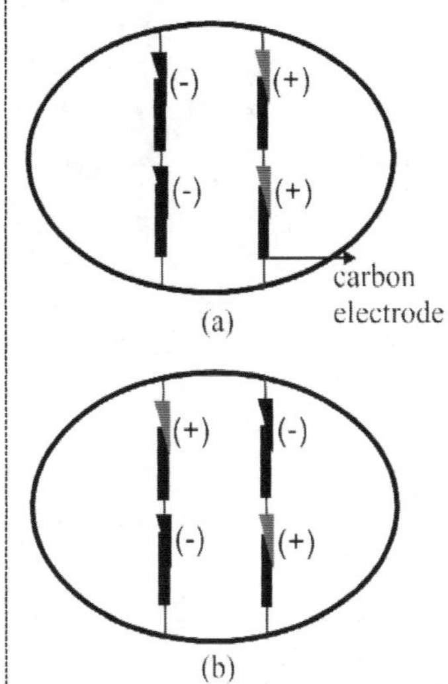

Figure 2. Schematic of two pairs of carbon electrodes in (a) conventional and (b) alternating arrangement.

3. Result and discussion

Figure 3 is the documentation of physical observation of hydroxyapatite electrosynthesis using 2 pairs of alternating electrodes arrangement for 6 hours with the time interval of 60 minutes. The first observation at t = 0 min showed bubbles formation at the anode and cathode. Electrosynthesis of hydroxyapatite was started by oxidation of H_2O at anode and reduction of H_2O at the cathode. [10] Gas bubbles at the anode indicated the formation of O_2 gas from the oxidation of H_2O, while gas bubbles at the cathode indicated the formation of H_2 gas from the reduction of H_2O. After time elapsed for 60 min, the electrochemical system started to show result of thin white solid deposited at the cathode. When t = 120 min, solution transparency started to changed to turbid. It was indicating that the electrolysis reaction started to occurr at the bulk solution. At t = 180 min, there was significant conversion of the solution. The bulk solution became white turbid and there was precipitate at the bottom of the reactor. With time elapsed during the electrolysis, there was thickening of the white solid deposited at the cathode, the bulk solution became more turbid, and more precipitate formed at the bottom of the reactor.

Figure 3. Observation of the solution alteration during electrolysis of hydroxyapatite in alternating arrangement.

The final product of hydroxyapatite electrosynthesis in conventional and alternating electrode arrangement was in the form of white solid deposited at cathode and white precipitate at the bottom of the reactor. The initial change at the cathode at t = 60 min shows that the main reaction of hydroxyapatite formation was on the cathode. This corresponds to the reaction mechanism proposed by Nur and Setyawan (2014) [10] that hydroxyapatite formation was initiated by OH^- ions from reduction of H_2O at cathode. Meanwhile, reaction product of precipitate formed at the bulk solution occurred because of part of highly concentrated OH^- ions generated from cathode transferred to low concentrated OH^- region. The pH of solution around cathode showed higher value than of the bulk solution. This condition occurred because of both the bulk solution also accepted H^+ ions, which were transferred from the anode, and H^+ ions from anode reacted with OH^- ions from cathode forming H_2O. Different ion concentration caused electrosynthesis of hydroxyapatite occurred immediately at the cathode, meanwhile at the bulk solution, electrosynthesis reaction could only be observed clearly at t = 180 min. The abundance supply of OH^- ions at the cathode region was qualifying for reaction occurred in a shorter time.

The results of hydroxyapatite synthesis through the electrolysis method were analyzed using X-Ray Diffraction (XRD) to observe the crystalline phase and Scanning Electron Microscopy (SEM) to observe the morphology of the particles formed. The analysis was carried out on powders attached to the cathode (C) and precipitate (P).

Figure 4 shows the XRD graph on the results of the electrosynthesis product by conventional and alternating electrodes arrangement. HA1C and HA1P referred to the product of electrolysis by conventional electrodes arrangement at cathode and precipitate, respectively. While HA2C and HA2P referred to the product of electrolysis by alternating electrodes arrangement. From the XRD graph, it was known that HA1C has 2θ diffraction angle peaks of 25.84°, 31.66°, 32.2°, 32.9° and HA1P of 25.8°, 31.54°, 32.04°. While HA2C has 2θ peaks of 25.77°, 31.64°, 32°, 32.9° and HA2P of 25.76°, 31.54°. All the diffraction peaks of the electrolysis product correspond to JCPDS data no 00-009- 0432 for hydroxyapatite compounds, which were 25.8° (002), 31.77° (211), 32.2° (112), 32.9° (300). Due to the sharpness of the diffraction peaks in figure 4, it shown that HA produced at the cathode region had better crystallinity than at the bulk solution (precipitate). This could be seen from the sharper diffraction peaks at HA1C and HA2C than HA1P and HA2P.

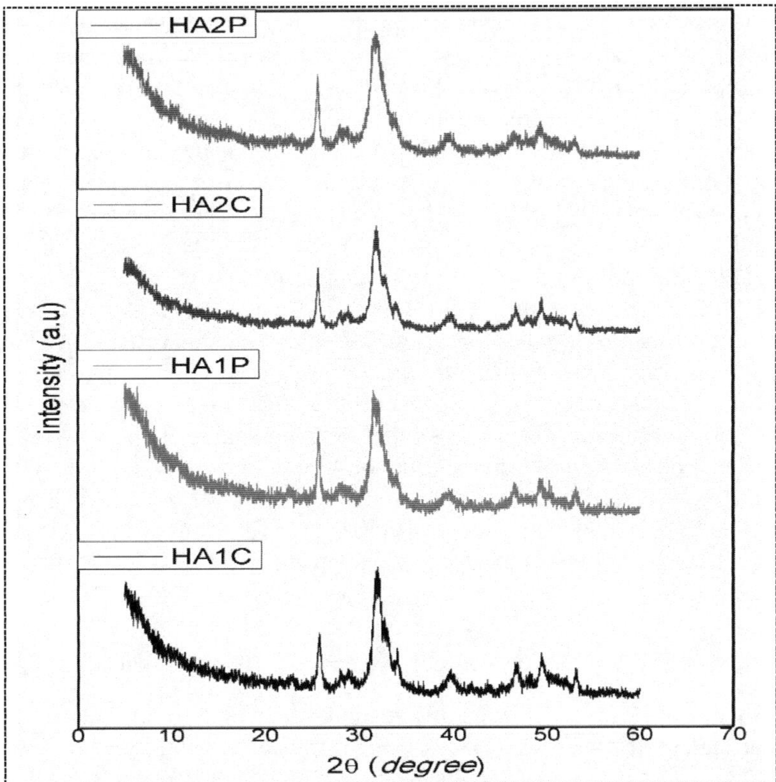

Figure 4. XRD graph of hydroxyapatite electrosynthesis product.

SEM analysis was performed on HA1 samples at the cathode and precipitate. Figure 5 shows the SEM result at the magnification of 10000x. HA formed at the cathode region had a spherical particle morphology, as well as smaller and uniform particle size compared to HA formed at the precipitate. HA from precipitate had irregular morphology, most of which were shaped like agglomerated spherical and some were shaped as long and sharp flakes.

Figure 5. Morphology of HA formed in conventional electrodes arrangement from (a) cathode region and (b) precipitate in at magnification of 10000x.

Through different electrode quantity and arrangement set for hydroxyapatite electrosynthesis at this research, it was known that the electrode quantity and arrangement affect the rate of HA production (Table 1). With the same fixed electrolyte volume, voltage and electrolysis time, HA production from 1 pair electrode system was much less than those of 2 pairs in conventional arrangement and alternating arrangement, which were 0.5383 g, 2.3157 g, and 2.3483 g, respectively. This trend of increasing mass production corresponded to the increasing of the average current density. The average current density of electrolysis system with 1 pair electrode, 2 pairs electrode on conventional arrangement and 2 pairs electrode on alternating arrangement were 13.29 mA/cm^2, 15.05 mA/cm^2, and 19.32 mA/cm^2, respectively. In an electrolysis process, the amount of electric charge that flows at the electrode to the solution affects the reaction yield. This relationship follows Faraday's first law. Potential value on this research was fixed on 6 V to study the effect of electrodes arrangement on current density. As the surface area of electrodes immersed was higher, there was more current flow from the power supply to the electrolyte solution. This could supply more electrical charge for electrolysis reaction, which consecutively affect the rate of HA production. While on the different arrangement of 2 pairs electrode, different value on the average current density indicated that the electrode arrangement could affect the electrical field on the electrolysis cell which then affect the production rate. The alternating arrangemet has higher value and random motion, which is horizontal and vertical, of electrical field than the conventional method. [11] Suggestion for further research is that the hypothesis of electrode arrangement effect on electrolysis of HA should be simulated to be confirmed.

Table 1. The effect of the electrodes arrangement on hydroxyapatite formation.

	Electrodes arrangement	Electrolysis time (minutes)	Voltage (V)	Current density (mA/cm^2)	Production mass (gr)
HA	1 pair	360	6	13.29	0.5383
HA1	2 pairs, conventional	360	6	15.05	2.3157
HA2	2 pairs, alternating	360	6	19.32	2.3483

4. Conclusion

Hydroxyapatite has been successfully synthesized through electrolysis with multiple electrodes on conventional and alternating arrangement. Both of the multiple electrodes arrangement can enhance the mass production of hydroxyapatite than from one pair electrodes system. The alternating electrodes arrangement gives average current density and hydroxyapatite mass production higher than the conventional electrodes arrangement. The multiple electrodes on the alternating arrangement produced hydroxyapatite mass of 2.3483 g and current density of 19.32 mA/cm^2, while one pair electrodes and multiple electrodes on the conventional arrangement produced hydroxyapatite mass and current density of 0.5383 g, 13.29 mA/cm^2, 2.3157 g and 15.05 mA/cm^2, respectively.

References

[1] Rigano L and Lionetti N 2016 Nanobiomaterials in galenic formulations and cosmetics *Nanobiomaterials in Galenic Formulations and Cosmetics* ed Grumezescu A M (Amsterdam : Elsevier) chapter 6 pp 121-148

[2] Hong Y, Fan H, Li B, Guo B, Liu M and Zhang X 2010 *Mater. Sci. Eng. R* **70** 225-242

[3] Liverani L, Abbruzzese F, Mozetic P, Basoli F, Rainer A and Trombetta M 2014 *Asia-Pac. J. Chem. Eng* **9 (3)** 407-414

[4] Okabayashi R, Nakamura M, Okabayashi T, Tanaka Y, Nagai A and Yamashita K 2009 *J. Biomed. Mater. Res. Part B Appl. Biomater.* **90 B (2)** 641–646

[5] Liu D-M, Troczynski T and Tseng W J 2001 *Biomaterials* **22 (13)** 1721–1730

[6] Michał W, Ewa D and Tomasz C 2015 *Colloid Polym Sci* **293** **(5)** 1561–1568

[7] Liu H S, Chin T S, Lai L S, Chiu S Y, Chung K H, Chang C S and Lui M T 1997 *Ceram. Int.* **23** **(1)** 19–25

[8] Bose S and Saha S 2003 *Chem. Mater.* **15** **(23)** 4464–4469

[9] Nur A and Setyawan H 2016 *J. Chem. Eng, Japan.* **49** **(2)** 144–151

[10] Nur A, Setyawan H, Widjaja A and Lenggoro I W 2014 *Bull. Chem. React. Eng. Catal.* **9** **(3)** 168–175

[11] Nurlilasari P, Widiyastuti W, Setyawan H, Faizal F, Wada M and Lenggoro I W 2019 *Chem. Eng. Sci.* **201** 112-120

Synthesis N doped TiO_2/PTFE by solvothermal method and their self-cleaning performance

F N Aini[1,2,3], S Hastuti[2], S Wahyuningsih[2,3]
[1] Master Program of Chemistry, Faculty of Mathematics and Natural Sciences, Universitas Sebelas Maret, Kentingan Jebres, Surakarta 57126, Indonesia
[2] Chemistry Department, Faculty of Mathematics and Natural Sciences, Universitas Sebelas Maret, Kentingan Jebres, Surakarta 57126, Indonesia
[3] Inorganic Material Research, Faculty of Mathematics and Natural Sciences, Universitas Sebelas Maret, Kentingan Jebres, Surakarta 57126, Indonesia

E-mail: sayekti@mipa.uns.ac.id

Abstract. Nitrogen doped TiO_2/PTFE with unique characteristic and photocatalytic self-cleaning activity under visible light were successfully produced using a modified solvothermal method with urea as the nitrogen source. This study combines photocatalyst of inorganic materials (TiO_2) with superhydrophobic polymers (PTFE) for self-cleaning glass coating. The X-Ray Diffraction show that the composite crystalline and new peak formed as indicated of PTFE. FTIR showed the presence of C-F groups on N doped TiO_2, which indicates that the composite material has been successfully formed. The morphology by scanning electron microscopy showed that agglomeration formed from N doped TiO_2/PTFE. The composite coatings showed a contact angle of 111.4°. The results confirm that the synthesized TiO_2 were successfully doped with nitrogen and composite with PTFE, allowing a higher photocatalytic activity. The composite N doped TiO_2/PTFE has completely coated on a glass substrate and enable to degrade methylene blue up to 79.9% for 1 hour irradiation.

1. Introduction
Recently, the use of glass that have transparent property used to in building, and as alternative save energy due to its benefits as room lighting any heating of building. The development of glass production process leads to a more widespread but there is a problem such as easy to be contaminated that will cause the reduction of lighting in a room and even can damage the surface materials. Various strategies have been developed to overcome these problems, one of them is through self-cleaning surface technology [1]. Titanium dioxide is one of extensive materials to be investigated, because has hydrophobic and hydrophilic. Titanium dioxide (TiO_2) is the best material for removal organic polutan so can act as photocatalyst [2].

There are three crystalline phase in TiO_2 they are: anatase, rutile (the most stable), and brookite, where rutile and anatase phases show better photocatalytic reactivity. The rutile phase seems to low photocatalytic performances than anatase because anatase has a larger of area so it has a high active side [2]. The disadvantage of TiO_2 is the band gap in the large pure anastase structure, which showed 3,2 eV that makes this material only can be activated under high-energy UV light [3]. Expansion of the TiO_2 thin films can be done by shifting the absorption of light from the UV light region through doping with non-metallic elements such as nitrogen [3]. Nitrogen can reduce the band gap and easily incorporated into the TiO_2 structure because its atomic size is proportional to oxygen, low ionization energy and high stability [4].

Enhancement the light absorption towards the visible range (longer wavelengths) and to improve the photocatalytic performances of TiO_2, many methodologies have been employed with: with organic dyes sensitization [5], deposition TiO_2 surface with metal ions [6], transition metals and non-metals (such as Nitrogen) doped [2]. Recent research reports that N is effective in narrowing the TiO_2 band gap, because N from 2p orbital can be intersection with 2p orbitals from O TiO_2. The electron in the valence band play an important role in the photocatalytic oxidation reaction, that because the holes in the valence band are used for the reaction [7]. Therefore N doping affects the catalytic activity and the ability of absorption of visible light.

On the other hand superhydrophilic photocatalytic materials has more disadvantages than superhydrophobic photocatalytic materials. The advantages of superhydrophobic materials application as self-cleaning surfaces, including the superior cleaning action of the lotus effect, in addition to providing other superhydrophobic properties such as water proofing [8]. However, there were a little examples in nature for superhydrophobic surfaces and minimum information for superhydrophobic photocatalytic surface act as photocatalysts [9]. Including fluorocarbon polymers such as polytetrafluoroethylene is one of organic materials with lower surface energy, were selected as hydrophobic agency [10].

Fluorocarbon polymer like polytetrafluoroethylene (PTFE), has been used in many field because it has high flexibility, heat and corrosion resistant [10]. Feng *et al* (2002) [11] report the use of alkyl fluoride as superhidrophobic material, alkyl fluoride like $-CF_2-$, $-CF_2H$, and $-CF_3$ effective enough to reduce surface free energy. Polytetrafluoroethylene (PTFE) has the lowest surface energy materials known, with excellent resistance to chemical reagents, and has a high temperature stability. The surface of a PTFE plate has a static water CA of 108°, and a SA >90°, indicating water adhesion to the PTFE surface [12].

The present research aimed to synthesize Nitrogen doped TiO_2 by hydrothermal. Composite N-TiO_2/PTFE were prepare by solvothermal and coating on glass to extend the self-cleaning photocatalytic activity of TiO_2 into the visible light region, thus obtaining a better photocatalytic efficiency. The Nitrogen doped TiO_2/PTFE coatings on glass were test for the degradation of methylene blue to identification the self-cleaning photocatalytic performance.

2. Experimental

2.1 Materials

Titanium (IV) isopropoxide ($Ti[OCH(CH_3)_2]_4$) or TTIP was bought from Merck. Ethanol for analysis (≥99.99%) was purchased from Merck. Isopropyl alcohol was bought from Merck. Methylene blue as a dye was bought from Merck. Urea as nitrogen source. Ultrapure water was used all through the experiments.

2.2 Synthesis N doped TiO_2/PTFE

Synthesis of TiO_2 was carried out by the hydrothermal method. TTIP is added to isopropyl alcohol and distilled water in a ratio of 1: 4. The mixture was added NH_4OH to pH 8 and keep stirer for 2 hours. Then added urea and hydrothermal at 200 °C for 10 hours. Then, precipitates were washing with ethanol and distilled water for 3 times. After drying at 60 °C in oven, the white N doped TiO_2 was calcined at 600 °C for 1 hour. Polytetrafluoroethylene (PTFE) micropowders was dissolved in ethanol and mixed with N doped TiO_2 with ratio of 1:1 and stirrer for few hours. After that, the mixture were poured into autoclave. The autoclave was stirred and keep at 200 °C for 10h. After that, the white precipitates was drying in oven at 80 °C.

2.3 Photocatalytic self-cleaning experiment

The activities of the materials self-cleaning N doped TiO_2/PTFE were evaluated by photocatalytic degradation of Methylene blue. Methylene blue was added on the glass where coated with N doped TiO_2/

PTFE and then irradiated using a 300 watt halogen lamp as a photon energy source in a closed reactor with various irradiation time of 0- 60 minutes with 5 minute increment.

2.4 Characterization

The X-ray diffraction (XRD) of each powdery TiO_2 was obtained with Cu Kα radiation (λ = 1.5418 Å) and caracterized 2θ from 5-80°. The characterization by Scanning Electron microscopy (SEM) to observe morphology of samples. Fourier Transform Infrared (FT-IR) Spectroscopy to identification IR spectra. Degradation of methylene blue were determined by UV Vis Spectrophotometer.

3. Results and Discussion

The result of XRD characterization shows by Figure 1. Nitrogen doping affect the crystallinity of TiO_2. The results shows the suitability with JCPDS standart number 21-1272. The change of relative intensity peaks on diffractogram showed the difference amount and value of crystal field that measured. That because the arrangement of N structure on TiO_2. Each dominant peak showed on 2θ = 25.4°$_{(101)}$, 38.2°$_{(004)}$, 48°$_{(200)}$, 53.8°$_{(105)}$, 54.9°$_{(211)}$ that identified as anatase phase. It is showed that every diffraction peack was an anastase phase and does not appear brookite or rutile phase. The presence of nitrogen not form a new difractogram, that because a low of dopan consentration [13] and the dopan only affect the crystal lattice. The characteristic peaks of anatase TiO_2 shift to higher angles that confirm the dopan nitrogen cause shift of angles. The difference of N and O binding characteristics influence the shift of TiO_2 angles. The addition of PTFE causes new diffractogram on 2θ=18.1 ° according to Surmenenv et al., 2017 [14].

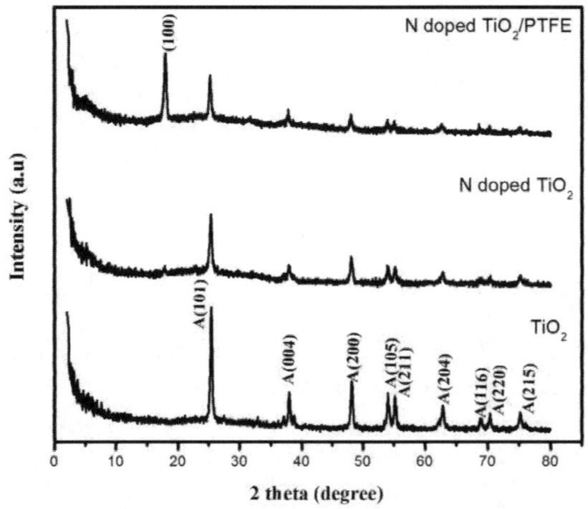

Figure 1. XRD Pattern of TiO_2, N doped TiO_2, and N doped TiO_2/PTFE.

FT-IR spectrum of TiO_2, nitrogen-doped TiO_2 samples with urea as nitrogen sources and N doped TiO_2/PTFE are shown in Figure 2. In 3434-1630 cm^{-1} peak, identified as water and hydroxyl. Vibration of Ti-O-Ti bond were identification about 500 cm^{-1}. The peak N-H bending of nitrogen atoms substitue into TiO_2 identification at at 1440 cm^{-1}. Polytetrafluoroethylene groups identified on 512 cm^{-1} peak, where on the peak occurred an overlapping between Ti and C-F.

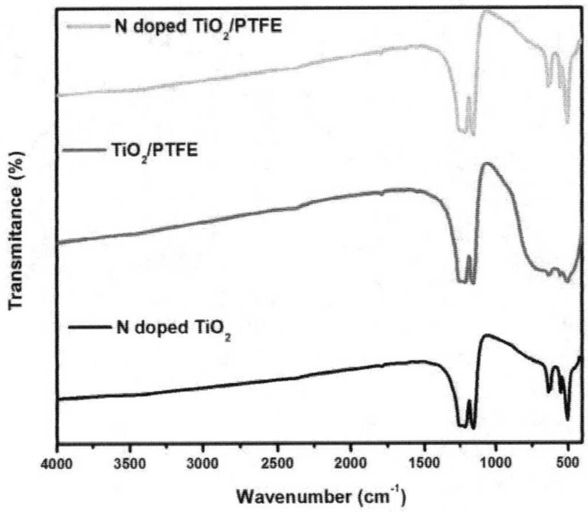

Figure 2. Spectrum of Nitrogen-doped TiO_2, TiO_2/PTFE and N doped TiO_2/PTFE characterized by FTIR.

The morphology of nitrogen-doped TiO_2 sample are shown in Figure 3 with 2500x magnification. It was found that the as-synthesized TiO_2 shown of bulky and several hundred micrometers in length. It can be seen Nitrogen addition affects the morphological characteristics of TiO_2. Nitrogen can inhibit the growth of TiO_2. The composite of Nitrogen doped TiO_2/PTFE is aglomerate and not structured, because TiO_2 blend with polymer.

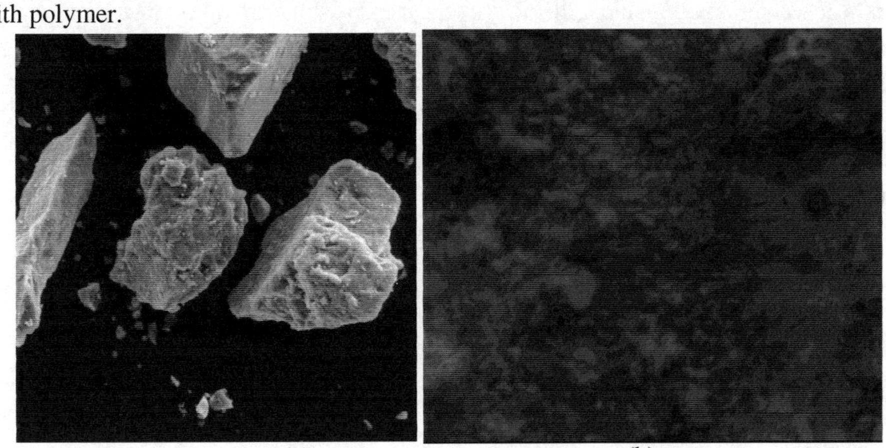

(a) (b)

Figure 3. (a) Morphology of Nitrogen doped TiO_2 with 2500x magnification (b) Nitrogen doped TiO_2/PTFE with 300x magnification.

As seen in Figure 4 and 5, N doped TiO_2/PTFE-coated glass substrate is hydrophobic (111.4°) and glass substrate uncoating N doped TiO_2/PTFE high wettability (72.7°). To understand the characteristic of hydrophobic surface, the glass layer has been coated with material and the glass without coating drops of various liquids like coffee, rhodamin b, and methylene blue. The results did not show any differences to the hydrophobic properties from N doped TiO_2/PTFE coated on glass.

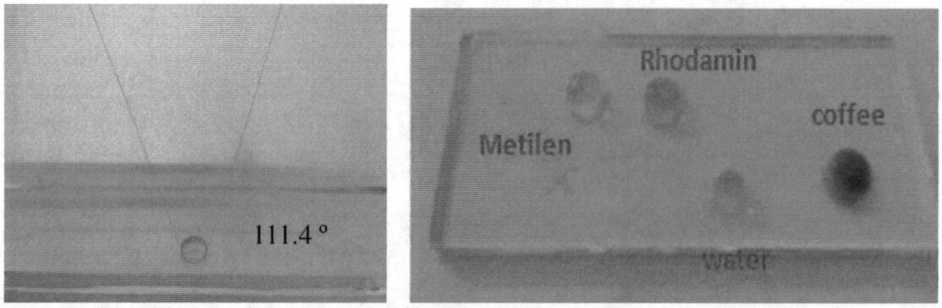

Figure 4. Repelling performance of different liquids on the substrat coating with N doped TiO_2/PTFE.

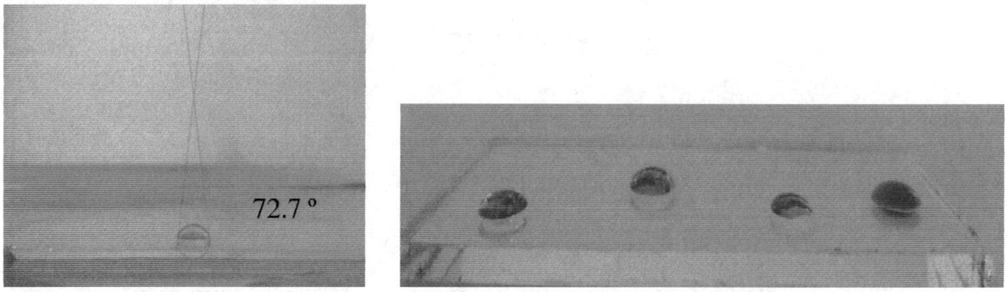

Figure 5. Repelling performance of different liquids on the substrat uncoating.

In real-life conditions, organic and inorganic particles will certainly contaminate surface coatings, hence the need for hidrophobic surfaces with self-cleaning properties. The advantages of hidrophopic surface is lower cost and effort consumption, because the surface can rapidly and effortlessly clean [10]. The photocatalytic self-cleaning properties of N doped TiO_2/PTFE coatings can demonstrate by ability of the film to degrade dye such as metylene blue.

The photocatalytic activity of modified materials was carried out with irradiation time variations of 0-60 minutes. Absorbance measurement is carried out at the maximum wavelength which can be used to obtain photodegradation. Methylene blue degradation results are shown in Figure 6 and Table 1. The methylene blue molecularly degraded because of the reaction with radical like O_2• and/ or OH• which formed during the photochemistry reaction. The process begin when the energy hits the modification material, electron will be excited from valence band to conduction band (e^-) and leaves positive charge, hole (h^+) in the valence band. The excited electron will trapped on the modified-TiO_2 surface and some of them recombined then results heat. While the trapped excited electron undergo reaction with O_2 yielding O_2•and also react with H_2O then produce OH• [15]. Both radicals are detoxication agent that contribute to the methylene blue degradation.

The best percentage degradation of methylene blue aqueous solution by nitrogen doped TiO_2/PTFE at 79.9% with visible light irradiation for 1h, this shows that N doped TiO_2 which is composed with PTFE hydrophobic polymer has a good self-cleaning photocatalyst ability.

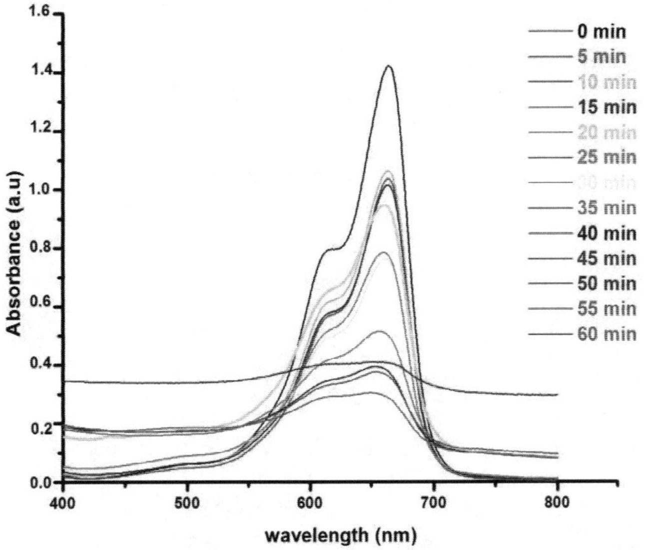

Figure 6. UV-Vis spectra of methylene blue under irradiation by N doped TiO_2/PTFE coating on glass substrate.

The existence of nitrogen is able to decrease the band gap energy so that it can suppress of recombination between electron-hole pairs produced. The irradiation time for At/A0 show that the longer irradiation (1h) increase photodegradation of methylene blue because the longer irradiation, more electrons are excited, hole (h+) that formed also increase so a radical amount hydroxyl which plays a role in the photodegradaton process increases. The irradiation time for N doped TiO_2/PTFE photocatalysts for methylene blue degradation is shown in Figure 7. The photocatalytic optimization is obtained from the percent degradation shown in Table 1.

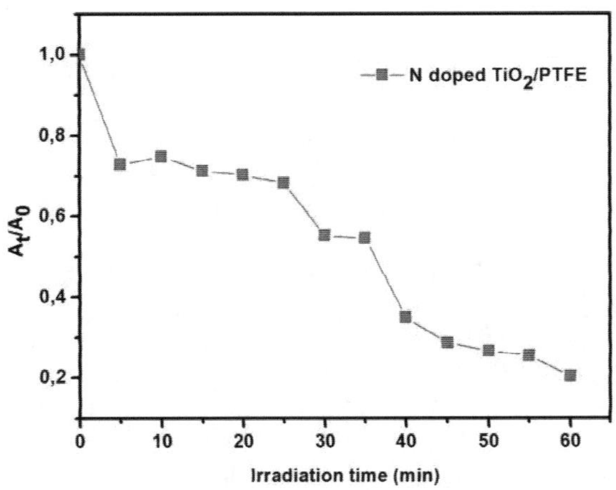

Figure 7. Photodegradation kinetic of methylene blue under visible light irradiation.

Table 1. Methylene blue degradation by N doped TiO_2/PTFE and N doped TiO_2 composites on glass substrate

composite variation	Percentage of Methylene blue degradation based on time variation (%)												
	0	**5**	**10**	**15**	**20**	**25**	**30 (minutes)**	**35**	**40**	**45**	**50**	**55**	**60**
N doped TiO_2/PTFE	0	27.1	25.3	28.7	31.7	45	45.7	53.3	65	71.4	73.6	74.6	79.9

4. Conclusion

Synthesis of Nitrogen-doped TiO_2 was successfully with hydrothermal method that combined with polymer PTFE by solvothermal method. After that, it was proved polytetrafluoroethylene groups identified on 512 cm^{-1} peak by FT-IR analysis. The nitrogen atom embedded in TiO_2 indicated at 1440 cm^{-1}. After nitrogen-doped and PTFE composite, the photocatalytic activity was greatly. The optimum percentage degradation of methylene blue by N doped TiO_2/PTFE at 79.9% with visible light irradiation for 1h. The addition of nitrogen increases photocatalytic activity and PTFE increases the contact angle of self-cleaning glass.

Acknowledgment
The writer proud and want to express their thank for supporting this research in financial to the Penelitian Mandatory UNS Program (PM-UNS).

References

[1] Gherardi F, Colombo A, D'Arienzo M, Di Credico B, Goidanich S, Morazzoni F, Simonutti R and Toniolo L 2015 *Microchem. J.* **126** 54–62
[2] Fisher M B, Keane D A, Fernández-Ibáñez P, Colreavy J, Hinder S J, McGuigan K G and Pillai S C 2013 *Appl. Catal. B Environ.* **130–131** 8–13
[3] Mahy J G, Cerfontaine V, Poelman D, Devred F, Gaigneaux E M, Heinrichs B and Lambert S D 2018 *Materials.* **4** 584.
[4] Pelaez M, Nolan N T, Pillai S C, Seery M K, Falaras P, Kontos A G and Entezari M H 2012 *Appl. Catal. B Environ.* **125** 331-349
[5] Guarisco C, Palmisano G, Calogero G, Ciriminna R, Di Marco G, Loddo V, Pagliaro M and Parrino F 2014 *Environ. Sci. Pollut. Res.* **21** 11135–11141
[6] Liu C, Yang D, Jiao Y, Tian Y, Wang Y and Jiang Z 2013 *ACS Appl. Mater. Interfaces.* **5** 3824–3832
[7] Hwang Y J, Yang S, and Lee H 2017 *App Catal B: Environ.* **204** 209-215
[8] Crick C R, Ismail S, Pratten J and Parkin I P 2011 *Thin Solid Films.* **11** 3722–3727
[9] Crick C R, Bear J C, Kafizas A and Parkin I P 2012 *Adv. Mater.* **26** 3505– 3508
[10] Wang Y, Huang Z, Gurney R S and Liu D 2019 *Colloids and Surf A: Physicochemical and Eng. Asp.* **561** 101-108
[11] Feng L, Li S, Li Y, Li H, Zhang L, Zhai J and Zhu D 2002 *Adv. Mater.* **24** 1857-1860
[12] Zhang J, Wu Q, Yu X, Zha P and Li H 2001 *Mater. Lett.* **48** 362–368
[13] Zhao W, Liu S, Zhang S, Wang R and Wang K 2019 *Catal. Today.*
[14] Surmenev R A, Surmeneva M A, Grubova I Y, Chernozem R V, Krause B, Baumbach T and Epple M 2017 *Appl. Surf. Sci.* **414** 335-344
[15] Banerjee S, Dionysiou D D, Pillai S C 2015 *App. Catal. B: Environ* **176** 396-428

IOP Conf. Series: Materials Science and Engineering **858** (2020) 012023 doi:10.1088/1757-899X/858/1/012023

Slow pyrolysis of coconut wood (*Cocos nucifera* L.) and bio-char compositions

Pranoto[1], D M Widjonarko[2], S Rafidah[3] and E Elmatiana[4]
Chemistry Department, Faculty of Mathematics and Natural Sciences, Sebelas Maret University, Surakarta

E-mail: [1]pak_pran@yahoo.com, [2]dianmwidjonarko@gmail.com, [3]sarahrafidah@gmail.com, [4]elmatianaelli@gmail.com

Abstract. Biomass in the form of coconut wood (*Cocos nucifera* L.) was pyrolysed in an externally heated batch reactor. In order to determine the lignin, cellulose and hemicellulose content was used a Van Soest method as reference method. The effects of final temperature and heating rate on the yields and compositions of the bio-char were investigated. Pyrolysis runs were performed using reactor temperatures were 375 °C, 475 °C and 575 °C with heating rates of 5 min^{-1}, 10 min^{-1} and 20 min^{-1}. The bio-char yields were significantly influenced by the process conditions. The result showed that the bio-char mass decreased with increasing the final temperature, but increased with the increase of heating rate. The activation energy increases with increasing final temperature and heating rate. While the highest activation energy was 38.89 kJ mol^{-1} and the obtained on bio-char product amounted to 6858.84 cal/g at the final temperature 575 °C and heating rate 5 min^{-1}. It was characterized by Fourier transform infrared spectroscopy. This study showed that FTIR spectra have the potential to be an important source of information for a quick evaluation of the chemical composition of coconut wood (*Cocos nucifera* L.).

1. Intoduction

As a result of the increase of energy demand and the decrease of fossil energy, there is a highinterest in development of alternative energy[6]. Biomass is one of the great potentials to be used as alternative energy[4]. Biomass includes all organic materials such as agricultural waste. One of agricultural waste is coconut wood (glugu). Coconut wood contain mixed components such as cellulose (45%), hemicellulose (35%) and lignin (15%), where the each component will be decomposed in the different rates[2]. Conversion of energy from biomass can be used a pyrolysis process. Pyrolysis is a decomposition process of material in a high temperature without oxygen presence or with limited oxygen and produces bio-oil, bio-char and gas[1]. The aim of the study was to investigate the effect of heating rates and final temperatures of pyrolysis process on coconut wood to determine their influence on bio-char physical and chemical properties. The bio-char physical properties of activation energy were determined by Arrhenius equation and calorific value were determined by ASTM D2015-00. Bio-char chemical characterization was determined by Fourier Transform Infrared (FTIR) and characterization of raw materials was determined by Van Soest Method.

2. Experimental

2.1 Materials

The materials used in this research were coconut wood waste which were collected from Kartasura, Sukoharjo regency. Samples were crushed by crusher equipment until passing 20 mesh and then dried until the water content is not more than 10%. The calorific value of the sample was tested by ASTM D2015 standard method.

2.2 Methods

The pyrolysis process was conducted by taking 50 grams sample and put it in to sample holder. Sample holder was placed in the pyrolysis reactor then all the apparatus were arranged as shown at figure 1. The heating rate variations of pyrolysis were 5 min^{-1}, 10 min^{-1} and 20 min^{-1} with final temperature variations were 375 °C, 475 °C and 575 °C and holding time of 20 min. The heating temperature was increased every minute using thermocontroller. Char yielded from pyrolysis process were tested for its calorific value using ASTM D2015 standard method.

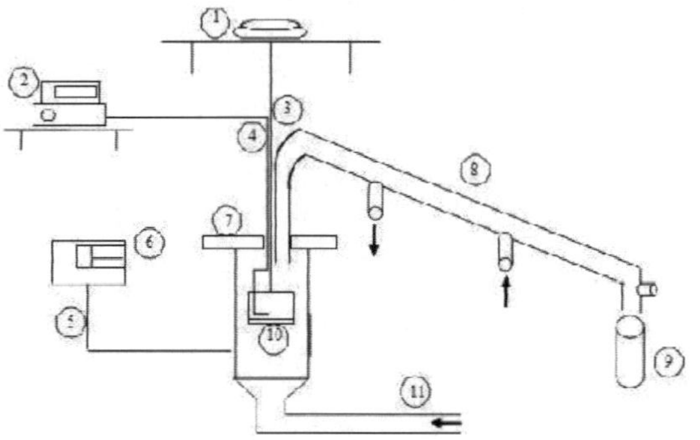

1. Balances
2. Thermocouple reader
3. Connecting wire
4. Thermocouple
5. Thermocouple
6. Thermocontroller
7. Furnace lid
8. Condenser
9. Tar container
10. Sample container
11. Nitrogen flow

Figure 1. Experimental apparatus schematic

2.3 Characterization

The raw material was tested for calorific value, cellulose, hemicellulose and lignin. The calorific value was determined using a bomb calorimeter with ASTM D2015-00 as standard method. Bio-char characterization was determined by calorific value and Fourier Transform Infrared Spectroscopy (FTIR).

2.4 Activation energy determination

Activation energy of pyrolysis process can be calculated by Arrhenius equation using first order reaction as the following equation .

$$k = A\exp(-\frac{E}{RT}) \qquad (1)$$

Where A is pre-exponential factor, E is activation energy, T is temperature then can be calculated with this following equation.

$$k_t = \frac{1\,dm}{(m-m_\infty)dt} \qquad (2)$$

where m_∞ is final mass, with this following equation:

$$\ln k = \ln A - \frac{E}{RT} \qquad (3)$$

by make a plot the relation between $\ln k$ and 1/T, gives a straight line with –E/R slope thus the activation energy can be defined.

3 Results and discussions

3.1 Characterization of raw materials

The Van Soest analysis showed that raw materials had a hemicellulose content, so the pyrolysis process was predicted can be occur in between 200-500 °C.

Table 1. Characterization of raw materials from Van Soest Method

Compositions	Content (%)
Cellulose	34.41
Hemicellulose	37.47
Lignin	17.83

3.2 Bio-char yields

Figure 2 shows the effect of heating rate and final temperature toward char yield in pyrolysis process. Higher heating rates caused the increasing of char yield which mean the weight-loss of the material was lower, while higher final temperatures caused the decreasing of char yield. This was because the higher final temperature will causes the operating time lasts longer, so carbon will be converted to liquid and gas.

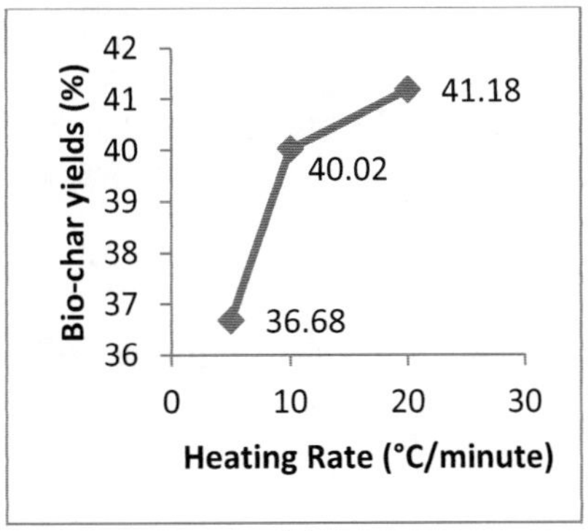

(a)

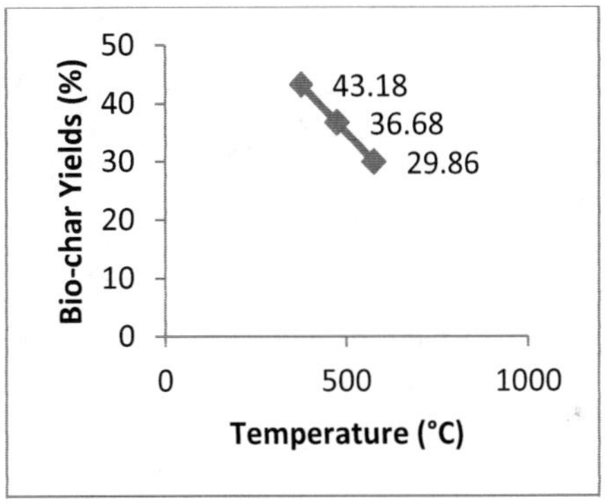

(b)

Figure 2. (a) Effect of heating rates on char yield; (b) Effect of final temperatures on char yield

3.3 Calorific value

Calorific value is the maximum amount of heat energy released by a fuel through a combustion reaction[5]. Calorific value of raw coconut wood is 4784.09 cal/g , while the calorific value of pyrolysis biochar were shown in table 2.

Table 2. Calorific value of biochar yielded from pyrolysis

Heating Rate (°C/minute)	Calorific Value (cal/g) Final Temperature (°C) 375	475	575
5	5646.61	6516.47	6858.84
10	5582.88	6451.74	6786.77
20	5482.85	6338.85	6694.49

The result of calorific value from pyrolysis of coconut wood with heating rate and final temperature variations are presented in table 2. The final temperature got higher thus the calorific value got bigger, because the increase of temperature caused the termination of lignocellulose bond which got much more so that pure carbon atom that was obtained got bigger[3] with the amount of carbon atom which was much therefore the calorific value got higher, but the pyrolysis process will reduce the caloric value of char with increasing heating rate. The heating rate of 5 °C/minute with a final temperature of 575 °C gives the highest calorific value of 6858.84 cal/g.

3.4 Activation energy

Activation energy was determined using Arrhenius equation with the first order of kinetics reaction which was often called global kinetics. The results of activation energy values in all samples are shown in the following table 3.

Table 3. The relation of activation energy and various heating rate

Activation Energy (kJ/mol)			
Final Temperature (°C)	Heating Rate (°C/minute)		
	5	10	20
375	33.88	26.88	29.07
475	34.19	29.54	35.44
575	21.88	25.40	38.89

Table 3 shows that the heating rates and final temperatures affects the activation energy of pyrolysis. Based on the results, pyrolysis process produces fluctuating activation energy values. The lowest activation energy was obtained at the final pyrolysis temperature of 575 °C with heating rate of 5 °C/minute. It shows that pyrolysis procces will be easier to occur because it only requires a small amount of energy to decompose.

3.5 FTIR characterization of bio-char

Figure 3 shows FTIR spectra of bio-char. The results of this study using FTIR showed that the bio char from pyrolysis process of heating rate variations didn't show any changes in the functional groups. Based on FTIR spectra of 375 °C showed that the stretching vibrations of –O-H groups with a strong and wide absorbance at 3314-3414 cm^{-1} represents dehydration of the cellulose and lignin components. The stretching vibrations of C-H_2 groups absorbance at 2925-2934 cm^{-1}. The carboxyl C=O stretching vibrations with an absorbance at 1707 to 1711 cm^{-1} indicates the presence of ketons and the vibration of C=C at 1455 cm^{-1} to 1459 cm^{-1} indicates the presence of an aromatic ring, while the vibrations at 812 to 874 cm^{-1} indicates the presence of aromatic C-H. FTIR spectra of 475 °C also showed that the stretching vibrations of –O-H groups at 3418-3443 cm^{-1}, the stretching vibrations of C-H_2 groups absorbance at 2923 cm^{-1}, the vibration of C=C at 1434 cm^{-1} to 1587 cm^{-1}, and the vibrations at 1229, 1222 and 1227 cm^{-1} represents of C-O stretching from alcohol. But, FTIR spectra of 575 °C only gives one vibration at 1500 cm^{-1} that indicates of C=C stretching.

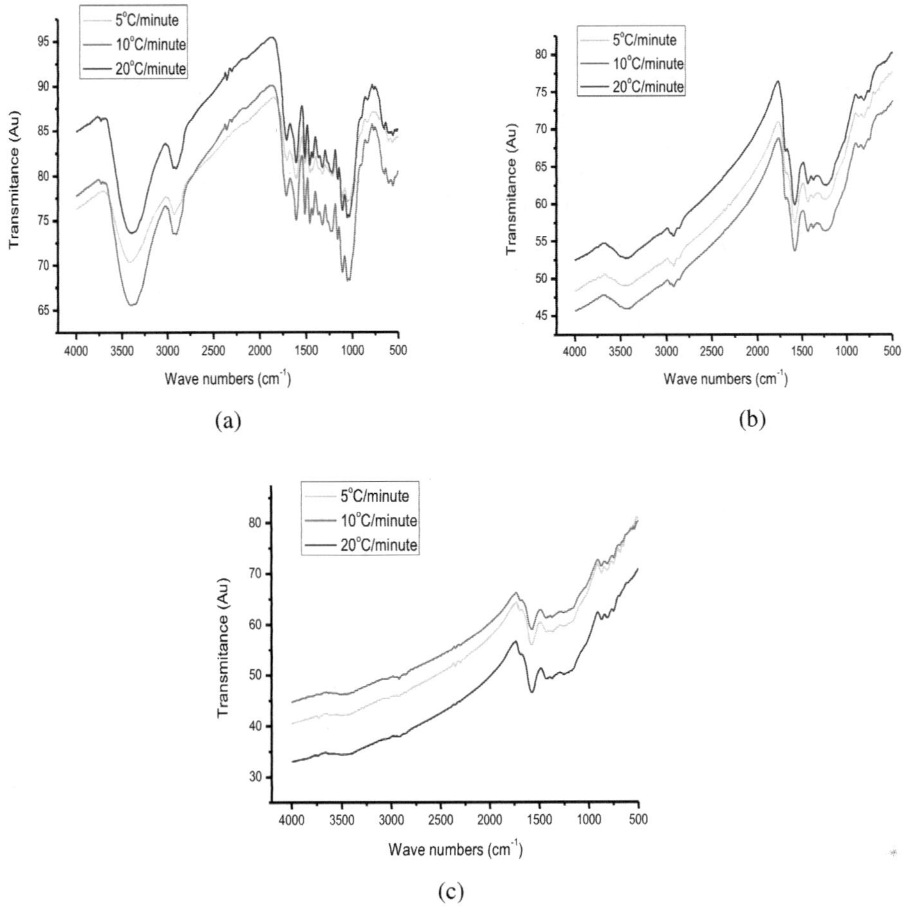

Figure 3. Effects of final temperature variations on FTIR spectra of coconut wood (*Cocosnucifera* L.) (a) 375 °C; (b) 475 °C; (c) 575 °C

4 Conclusions

In this study, slow pyrolysis experiments of coconut wood (*Cocos nucifera* L.) were performed in a batch reactor and the produced bio-chars were physically and chemically characterized. Heating rate and final temperature affects the activation energy and calorific value of a bio-chars. The effect of heating rate and final temperature produces fluctuating activation energy values. The lowest activation energy was obtained at the final pyrolysis temperature of 575 °C with heating rate of 5 °C/minute. It shows that pyrolysis procces will be easier to occur because it only requires a small amount of energy to decompose. Calorific value of the bio-chars was higher than its raw material. The increasing of final pyrolysis temperature will raised the calorific value of bio-chars. The highest calorific value of bio-chars was 6694.49 cal/g and was found at 575 °C with heating rate of 20 °C/min. FTIR spectra indicated that heating rate doesn't affect chemical compositions of bio-chars, but the final temperature does. Chemical compositions of bio-chars decreased as the final temperature increased.

References

[1] Cahyono, M.S. 2013. Pengaruh Jenis Bahan pada Proses Pirolisis Sampah Organik menjadi Bio-oil sebagai Energi Terbarukan. *Jurnal Sains dan Teknologi Lingkungan*, **5(2)**: 67-76.

[2] Hermiati, E., Mangunwidjaja, D., Sunarti, T.C., Suparno, O. and Prasetya, B. 2010. Pemanfaatan Biomassa Lignoselulosa Ampas Tebu untuk Produksi Etanol. *Jurnal Litbang Pertanian*, **29(4)**:121-130.
[3] Majedi, F., Wijayanti, W. and Hamidi, N. 2016. Perubahan Massa dan Nilai Kalor *Char* dengan Variasi heating Rate dan Suhu pada Pirolisis Serbuk Kayu Mahoni (Switenia Macrophylla). Rotor, **9(2)**: 59-64
[4] Patabang, D. 2012. Karakteristik Termal Briket Arang Sekam Padi dengan Variasi Bahan Perekat. *Jurnal Mekanikal*, **3(2)**:286-293.
[5] Santosa, S. and Soemarno. 2014. Peningkatan Nilai Kalor Produk pada Produk Proses Bio- Drying Sampah Organik. *Indonesian Green Technology Journal,* **3(1)**:29-38.
[6] Tampubolon, A. P. 2008. Kajian Kebijakan Energi Biomassa Kayu Bakar. *Jurnal Analisis Kebijakan Kehutanan*, **5(1)**: 29-37.

IOP Conf. Series: Materials Science and Engineering **858** (2020) 012024 doi:10.1088/1757-899X/858/1/012024

The effect of temperature on lignin purification from bioethanol production waste based on palm oil empty fruit bunch

W A Lestari[1*], M R Utami[1] , A Istiqomah[1], A H Setiawan[2]
[1]Chemistry Department, Faculty of Mathematic and Natural Sciences, Sebelas Maret University Surakarta, Indonesia.
[2]Research Centre for Chemistry, Indonesian Institute of Sciences, Kawasan PUSPIPTEK Serpong, Tangerang 15314, Indonesia

E-mail: windyayulestari7@gmail.com

Abstract. In this study, the optimal condition of lignin purification from bioethanol production waste based on palm oil empty fruit bunch has been conducted. Purification of Lignin through two steps are impurities precipitation in pH 7 and lignin purification in pH 2 simultaneously with variation temperature. The concentration of lignin determines by spectrophotometer UV-Vis and thermal analysis to determine the ratio of lignin content and residual content. It was found that in lignin purification step, the lowest lignin concentration is at temperature 30°C with the concentration of lignin is 8.388% and the highest residual content is at temperature 60°C with % residual combustion is 26.652%. The optimal temperature at pH 2 for lignin purification is 50 °C with a lignin concentration is 73.854% and the residual combustion is 27.956%.

1. Introduction

Indonesia is a country that has a fairly high agricultural wealth, especially in palm oil tree commodities. Palm oil tree commodity has been widely used in several industrial fields in Indonesia such as producing biofuel [1], composites [2], medicine [3] and the other utilities. Palm oil tree processing in some of its function gives a residual product in the form of Oil Palm Empty Fruit Bunch (OPEFB). In the previous growth of the palm oil industry, a large amount of palm biomass is generated every year, 23% of fresh fruit bunches become OPEFB [4]. OPEFB is the agricultural waste which has a high lignocellulose content [5]. Lignocellulose consists of lignin, cellulose and hemicellulose compounds [6]. The most abundant form of biomass is lignocellulose, with an annual production of around 170 billion metric tons [7]. The mainly components in lignocellulose materials are cellulose, hemicellulose and lignin. Among them, hemicellulose (20–30%) and cellulose (40–50%) are the polymers of C5 and C6 sugars [8]. By the main problem, we can enhance the application of palm oil waste i.e. OPEFB as the main ingredient of producing some ethanol. Lignocellulose compound can be derived into simple sugars which converted into ethanol through several processes and it is one of a potential renewable material for bioethanol conversion [9, 10]. But in bioethanol conversion there were some lack on the process, lignin contain in OPEFB can reduce the yield of the bioethanol. Therefore, we need to do some pre-treatment to relieve lignin compound.

The purpose of the pre-treatment process is to eliminate lignin content in bioethanol production. It is called delignification process that produce crude lignin [11]. Crude lignin is a waste that contains of lignin and a lot of impurities [5]. Lignin can be used commercially as a binder, adhesive, filler, surfactant, polymer product, dispersant and other chemical sources [12]. This crude lignin needs to be purified so that the product can be utilized [5]. Pre-treatment process can be done by lignin deconstruction between acid and alkali pre-treatment is apparent. Alkali-based pre-treatment utilizes various kinds of alkaline (e.g., sodium hydroxide, calcium hydroxide, ammonia, etc.), caused effects of lignin solubilisation or degradation [13]. Lignin purification has been carried out by the precipitation method with several of pH, by two steps at pH 7 then pH 2, and one stage at pH 2. Two-steps settling produces inorganic impurities then one-stage settling [6]. Some researcher has studied about the effect of variations in precipitation temperature on one-stage process that the percentage optimum lignin is obtained at 50°C and ash content increases followed by increasing the temperature [14]. Influence of temperature condition in two-steps precipitation process not conducted yet. Therefore, this research is focus on the effect of temperature variations on the purification of lignin by two-stage precipitation at pH 7 and pH 2. The concentration of lignin obtained using a UV-Vis spectrophotometer and thermal analysis.

2. Experimental

2.1. Materials

The materials in this research were commercially available from Merck without treatment and purification such as natrium hydroxide (NaOH), sulphate acid (H_2SO_4) 95-97% that used to control pH of solution, and 1,4-dioxane GR. Crude lignin obtained from bioethanol production waste in LIPI P2K Laboratorium, aquades.

2.2. Lignin Purification

The purification of lignin conducted at different temperatures and through two steps. First step is impurities precipitation. Crude lignin (10 gram) dissolved in 25 mL NaOH 5%. The solution stirred 15 minutes, 125 mL aquades added into solution. The solution stirred for the other 15 minutes, H2SO4 1% dropped into the solution until the solution reached pH 7. The solution stirred for 1 hour in variation temperature (30 °C, 40 °C, 50 °C, 60 °C). The solution precipitated for 16 hours at room temperature. The solution was centrifuged at 10.000 rpm in 10 minutes, filtrate and precipitate were separated. Precipitate dried at 50 °C for 16 hours and crushed (Compound 1). Weight all of compound 1 is measured.

The second step is lignin purification that conducted by dropped H2SO4 1% into the filtrate until the filtrate reached pH 2. The solution stirred for 1 hour in the same variation temperature (30 °C, 40 °C, 50 °C, 60 °C). The solution precipitated for 16 hours at room temperature. The solution separated by Buchner using Whatman filter paper 42. The precipitate dried at 50 °C for 16 hours and crashed (Compound 2). Weight all of compound 2 is measured.

2.3. Lignin Concentration Analysis

Calibration curve measured with variation concentration of standard lignin (5, 10, 15, 20, 50 ppm) by spectrophotometer UV-Vis at wavelength 280 nm. The equation of calibration curve used to calculate lignin concentration in samples (Equation 1).

$$y = 0.024x + 0.0145 \qquad (1)$$

Determining lignin concentration in compound 1 and compound 2 obtained by dissolved 5 mg compound in 10 mL dioxane:water (1:1), and measured by UV-Vis at wavelength 280 nm.

2.4. Thermal Analysis

Compound 1 and compound 2 was analysed using STA Linseis PT 1600 which heating from 20 °C to 600 °C and heating flow rate is 10 °C/minute. Concentration impurities can be obtained by % residual combustion.

IOP Conf. Series: Materials Science and Engineering **858** (2020) 012024 doi:10.1088/1757-899X/858/1/012024

3. Result and Discussion

3.1. Impurities Precipitation

Alkaline solution that commonly used as pretreatment solution is NaOH solution. During alkali-based pretreatment, the ether linkages between monolignols are partially broken [11]. Sodium hydroxide commonly used as catalysts for the base-catalyzed depolymerization of lignin. The reaction between alkaline metal hydroxides and lignin produce a mixture of simple aromatic compounds. Ether linkages (a- and ß-aryl ether), being the weakest bonds in the lignin structure, are predominantly broken down during the depolymerization process [12,11].

Figure 1a is a graph of lignin concentration with variation temperature. Based on those data, it can be concluded that the higher of temperature is the most lignin is taken at the precipitate. Precipitation step is step to separate between lignin and impurities by precipitate the impurities. Therefore, the lowest lignin concentration is taken in precipitation is the best condition to separate impurities and lignin in solution. It was found that the lowest lignin concentration is at precipitation temperature 30 °C.

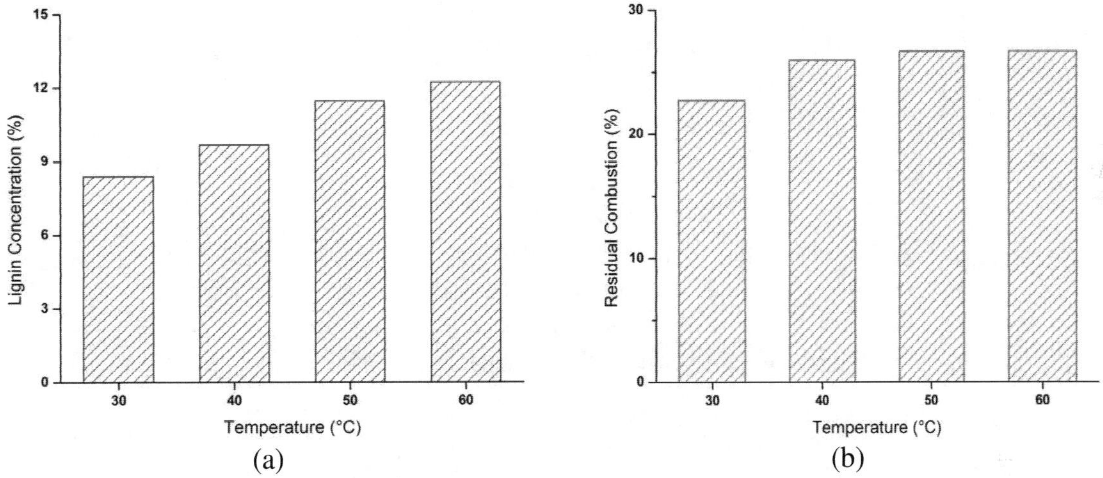

Figure 1. Lignin concentration (a) and residual combustion (b) in precipitation step.

Figure 1b is graph of % residual combustion in the precipitation step with variation temperature. Based on the graph, precipitation temperature at 40 °C, 50 °C, and 60 °C have not significantly different of % residual combustion that are 22.724%, 25.943%, 26.652% and 26.703% respectively. Purification temperature at 30 °C has the smallest % residual combustion that is 22.724%. Based on data, it can be concluded that purification temperature at 40 °C, 50 °C and 60 °C can precipitate more impurities.

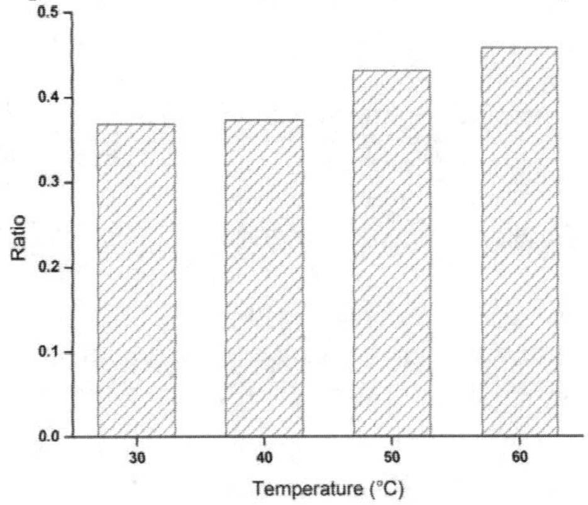

Figure 2. Ratio of lignin concentration of lignin and % residual combustion in precipitation step.

Ratio of lignin concentration and % residual combustion in precipitation step can be seen in figure 2. The lowest ratio is the most effective temperature because it has the lowest lignin concentration. This shows that the higher of temperature is the more lignin is taken at the precipitate. As well as from the impurity curve, it was found that the most impurity concentration at 30 °C. The most effective precipitation temperature is 30 °C.

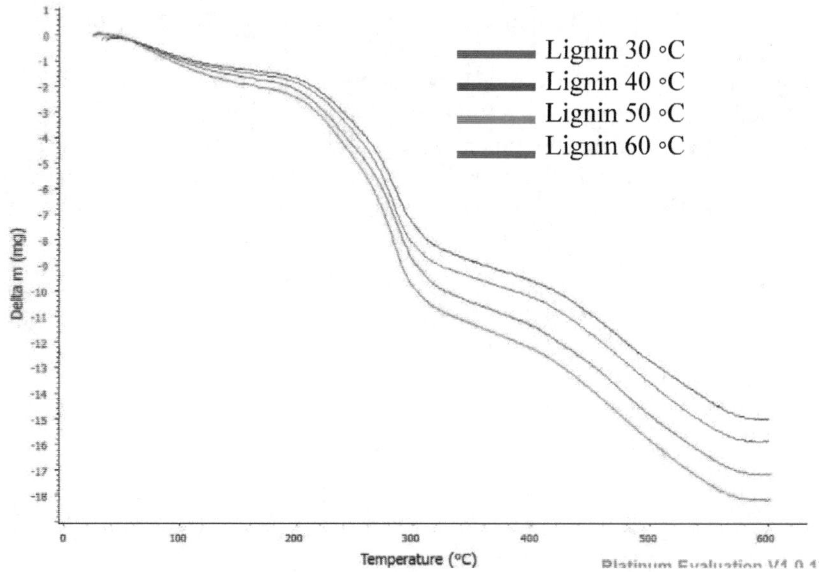

Figure 3. TGA spectra of precipitates in purification step.

Figure 3 is TGA spectra of precipitates in precipitation step at variation temperature. Degradation of lignin occurs around 350 °C-450 °C [12]. Lignin degraded at temperature around 200 °C-600 °C [13]. Precipitation temperature at 30 °C showed mass began to decrease from 200 °C-600 °C with % mass reduction is 22.724%. Precipitation temperature at 40 °C showed that mass reduction occur at temperature 200 °C-600 °C with % mass reduction is 25.943%. Precipitation temperature at 50 °C showed that the mass reduction occurs at temperatures 200 °C-600 °C with % mass reduction is 26.652%. Based on data, it was found that the mass began to decrease around 200 °C - 600 °C with % mass reduction is 26.703% at precipitation temperature 60 °C. The mass degradation reaction that occurs at any temperature variation is an endothermic reaction.

3.2. Lignin Purification

Lignin purification has been obtained by conditioning solution at pH 2 that produced lignin precipitates as dark brown solids. Figure 4a is graph of lignin concentration in the purification step with variation temperature. Based on the graph, lignin concentration at 30 °C and 50 °C are higher than lignin concentration at 40 °C and 60 °C. Lignin concentration at 30 °C and 50 °C are 63.229% and 73.854%, respectively. Lignin has a negative charge in alkaline conditions due to dissociation of phenolic and carboxyl groups in lignin. Repulsive forces between negative charge of lignin make lignin stable in solution. Decreasing of pH caused interaction between hydrogen ion and negative charge of lignin. The interaction cause neutral charges in molecular surface of lignin so that produces lignin precipitation [5]. Based on these data it can be concluded that high lignin concentration can obtain at temperature precipitation 30 °C and 50 °C.

IOP Conf. Series: Materials Science and Engineering **858** (2020) 012024 doi:10.1088/1757-899X/858/1/012024

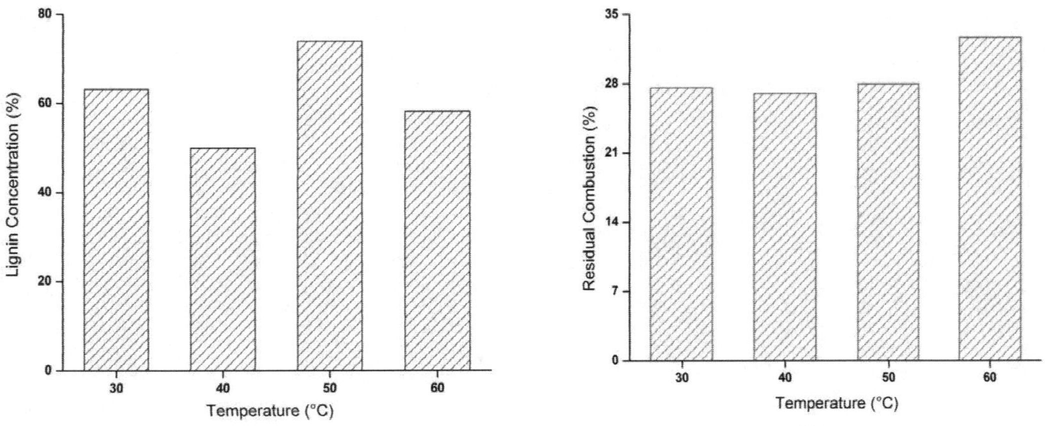

(a) (b)

Figure 4. Lignin concentration (a) and residual combustion (b) in purification step.

Residual combustion is impurities that precipitated with lignin in purification step. Figure 4b is graph of % residual combustion in the purification step with variation temperature. Based on the graph, purification temperature at 30 °C, 40 °C, and 50 °C have not significantly different with that are 27.602%, 27.012%, and 27.956% respectively. Purification temperature at 60 °C has the largest % residual combustion that is 32.655%. Based on data, it can be concluded that purification temperature at 30 °C, 40 °C, and 50 °C have low impurities.

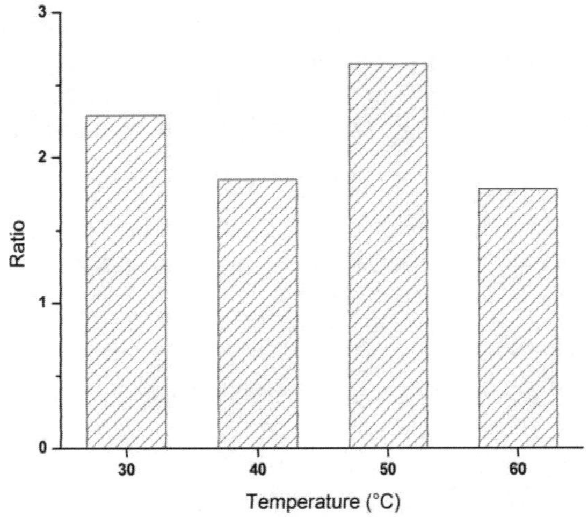

Figure 5. Ratio of lignin concentration of lignin and % residual combustion in purification step.

Ratio of lignin concentration and % residual combustion in purification step can be seen in figure 5. The highest ratio is the most effective temperature because it has the largest lignin concentration. The purification temperature at 50 °C has the highest ratio. It means that in the temperature condition 50 °C can be obtained the high concentration of lignin with the lowest impurities.

Figure 6 is TGA spectra of precipitates in purification step at variation temperature. Based on graph, each variation temperature has a relatively similar profile. It can be estimated that the contents of all samples are same. Lignin degraded between temperatures 350 °C - 450 °C, at temperature 100 °C - 200 °C is degradation of water and at temperatures 200 °C - 300 °C is hemicellulose is degradation [16]. Based on TGA spectra (Figure 6), it can be seen that mass degradation occurs at temperatures below 100 °C. The degradation is estimated as degradation of water molecules, because it is possible that the sample still contains water from the solvent. Mass degradation at temperatures around 350 °C - 600 °C is estimated of degradation of lignin polymers. The significant mass degradation occurs around 390 °C-400 °C that estimated of breaking the C-C bond.

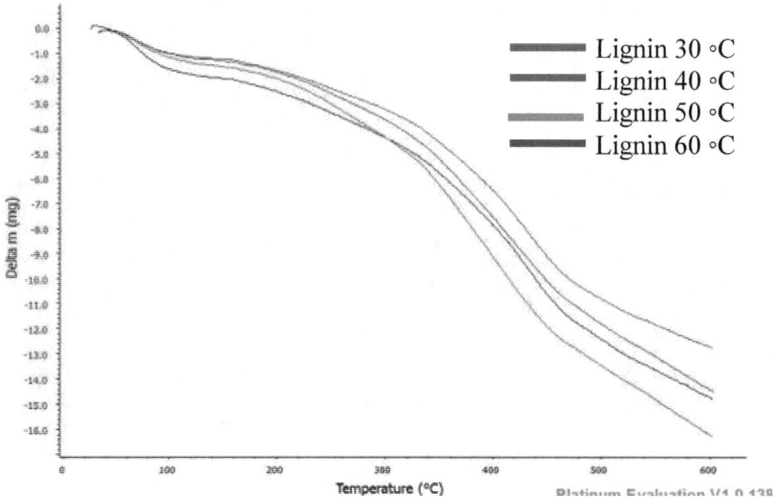

Figure 6. TGA spectra of precipitates in purification step.

4. Conclusion

The Purification of lignin as a side product from bioethanol processing can be done through two steps. Impurities precipitation step conducted by conditioning pH 7 and conditioning pH 2 for lignin uptake. Increasing of temperature is not directly increase lignin concentration that obtained. Temperature condition at 50 °C is the most effective condition to obtain high lignin concentration with low impurities.

Acknowledgement

Authors would like to acknowledge to all staff in laboratorium catalyst at Research Centre for Chemistry, Indonesian Institute of Sciences, Kawasan PUSPIPTEK Serpong, Tangerang for support through Kegiatan Magang Mahasiswa (KMM) 2015.

References

[1] Kristina, Evi R S and Novia 2012 *Chemical Chemistry Journal,* **18** 34-42
[2] Budiman H and Achmad H 2014 *Macromolecular Symposia* **353** 220–224
[3] Setiawan A H, Harry B, Evi T and Fauzan A 2014 *National Polymers Symposium*
[4] Sun Y and Cheng J 2002 *Bioresource Technology* **83** 1-11
[5] Tarmansyah 2007 *Pemanfaatan Serat Rami Untuk Pembuatan Selulosa* (Indonesia: Ministry of Agriculture's Balitbang Bulletin)
[6] Howard R L, Abotsi E L, Van R and Howard 2003 *African Journal of Bio-technology* **2** 602-619
[7] Ashutosh A, Masud R and Jeong H P 2018 *Fuel Processing Technology* **181** 115-132
[8] Simatupang, Harmaja, Andi N and Netti H 2012 *Mechanical Chemistry Journal of Sulawesi Utara University* **1** 20-24
[9] Shi J B, Yang QL and Lin 2013 *Polymer Degradation and Stability* **98** 550-556
[10] Franco C, Gianluca C and Andrea N 2014 *68th Conference of the Italian Thermal Machines Engineering Association* **45** 52-60
[11] Changzhi Li, Xiaochen Zhao, Aiqin Wang, George W. Huber, and Tao Zhang 2015 *American Chemical Society* **1** A-BN
[12] Carpenter, Westover T L, Czernik S and Jablonski W 2014 *Green Chem* **16** 384-406

[13] Boyu Du, ChaoLiu, Xing Wang, Ying Han, Yanzhu Guo, Haiming Li and Jinghui Zhou 2020 *Renewable Energy* **147** 1331-1339
[14] Wang H, Pu Y, Ragauskas A and Yang B 2019 *Bioresour Technol* **449-61** 271
[15] Jiadeng Zhu, Chaoyi Yan and Xin Zhang 2020A *Progress in Energy and Combustion Science* **76** 1-24
[16] Kim JY, Oh S, Hwang H, Kim UJ and Choi J W 2013 (Populus albaglandulosa *Polymer Degradation and Stabilit* **98** 1671-1678
[17] Jamin S and Togi S 2019 *Warta Perkaretan* **38** 1-6
[18] Harry B and Achmad H 2015 Macromol. Symp **353** 220–224

Antihypercholesterolemia activities of red ginger extract (*Zingiber officinale* Roxb. var rubrum) on wistar rats

S Jovita Nirvana[1], T Widiyani[1,2], A Budiharjo[1,2]

[1] Department of Bioscience, Graduate Program, Sebelas Maret University. Jl. Ir Sutami No. 36A, Kentingan, Surakarta,57126, Central Java, Indonesia.

[2] Faculty of Mathematics and Natural Sciences, University Sebelas Maret. Jl. Ir Sutami No. 36A, Kentingan, Surakarta,57126, Central Java, Indonesia.

Email: jovitashanghnesy@student.uns.ac.id

Abstract. The purpose of this study was to determine the phenolic content of red ginger, determine the effect on lipid profile and body weight. Total phenolic levels were measured by using UV/Vis spectrophotometry with the *Folin-Ciocalteau* reagent, at absorbance 760 nm wavelength. Hypercholesterolemia rat lipid profile was conducted *in vivo* by using a kit from *Dyasis*. Data was analyzed statistically with *Two Way ANOVA*. Rats were divided into 5 treatment groups with different of red ginger extract (0, 200, 350, 500 mg/kg bw, and simvastatin 7.2 mg/kg bw). Before the treatment of red ginger extract or simvastatin, rats were induced a high fat diet for 28 days. The treatment of red ginger extract and simvastatin were carried out for 2 weeks. The phenolic content of red ginger extract is 5194.15 ± 264.1 µg/g. The results of *in vivo* test showed that red ginger extract had a significant effect on lipid profile and body weight changes in hyperlipidemia rats at dose 200 mg/kg bw. Doses 200 mg/kg bw is not significantly different ($p \geq 0.05$) to positive control group (simvastatin 7.2 mg/kg bw). Total cholesterol levels decreases 47.9%, LDL 32.7% and triglycerides 64.3% and HDL increases 25.8%.

1. Introduction

Hypercholesterolemia is the main cause of atherosclerosis, coronary heart, stroke and causes many deaths around the world [1]. Hypercholesterolemia is abnormal lipoprortein metabolism characterized by increased levels of triglycerides, plasma total cholesterol, *low-density lipoprotein* (LDL), and *high-density lipoprotein* (HDL) [2]. In recent years, there have been many *in vivo* studies and clinical trials showing that phenolic compounds from plants have antioxidant activity that is beneficial in the prevention of hypercholesterolemia [3].

Phenolic compounds in this case reduce the risk of hypercholesterolemia by increasing the activity of antioxidant enzymes and reducing the formation of free radicals. Free radicals are molecules that have unpaired electrons. This compound is very reactive and high-energy, so it can attract electrons from other molecules [4]. Therefore, phenolic compounds are tasked with stabilizing free radicals by providing electrons to supplement the lack of electrons from free radicals, and inhibiting chain reactions. In addition, phenolic compounds with higher hydroxy groups will increase the ability to clean free radicals [5].

Red Ginger (Zingiber officinale Roxb. Var rubrum) contains flavonoids, and phenolics. They inhibit lipid peroxidation, increase antioxidant enzymes that regulate low density lipoprotein (LDL) receptors and 3-hydroxy-3-methylglutaryl coenzyme-A receptors (HMG-CoA), which affect cholesterol absorption in the liver [13] The purpose of this study was to determine the phenol content of red ginger extract, determine the effect of body weight and the effect of lipid profiles on the administration of red ginger extract.

2. Experimental

2.1 Materials

Materials used in this research includes rotary evaporator (RVO 400 SD Boeco Germany), waterbath (Grant), UV/Vis spectrophotometric (Perkin Elmer Lambda 25), vortex (Heidolph REAX top), measuring glass (Pyrex), aquades (General), and ethanol p.a (Merck).

2.2 The extract preparation (Zingiber officinale Roxb. var rubrum) uses the maceration method

Red ginger (*Zingiber officinale* Roxb. var rubrum) obtained from the Health Research and Development Agency Research and Development Center for Medicinal and Traditional Medicinal Plants (B2P2TOOT). 2 kg of red ginger powder soaked with 96% ethanol solvent in ratio 1: 5 (2 kg of powder: 10 L of 96% ethanol). Maceration was done in 3 x 24 hours. Every 24 hours the results of maceration were filtered using a glass funnel that has been given a flannel cloth or filter paper. The filtrate was collected in a separate container while the pulp was re-macerated with a new find. Remaseration was done in 3 times replication. Maceration results were evaporated using a rotary at 40 °C. Results of the evaporation then were heated on a waterbath at 50 °C.

2.2 Test the phenol levels of red ginger extract (Zingiber officinale Roxb. var rubrum).

1000 ppm gallic acid solution is made by dissolving 10 mg of gallic acid into 10 mL methanol as a stock solution. As much as 2.5 mL of stock solution was dissolved in methanol to 25 mL to make gallic acid solution with a concentration of 100 ppm, from this solution as much as 1 mL, 2 mL, 3 mL, 3 mL, and 5 mL into methanol up to 10 mL. 0.5 mL *Folin-Ciocalteu* reagent was added. Then added 1.5 mL of 20% sodium carbonate diluted with aquabidest until the volume reaches 10 mL. Then wait for more than two hours and move to the cuvette, still absorbing at a wavelength of 760 nm. Measurement of each ginger extract was done by adding 10 mg of ginger extract with 0.5 mL of *Folin-Ciocalteu* reagent and 7.5 mL aquabidest, incubated for 5 minutes. Subsequently 1.5 mL of 20% sodium carbonate was added to each solution, and diluted using aquabidest to the limit of the pumpkin marker 10 mL and incubated for more than 2 hours and transferred to the cuvette, still absorbing at 760 nm wavelength.

2.3 Grouping of test animals

In this study, test animals in the form of male white rats Galur Wistar with 25 animals, aged 11-12 weeks, weighing 150-200 grams. Obtained from Wistar rat breeding in Yogyakarta. Test animals were divided into 5 treatment groups. Test animals as research objects have received approval from the Ethics Committee of the University of Muhammadiyah Surakarta with document number EC No. 2388/A.1/KEPK-FKUMS/IX/2019.

2.4 Increased total cholesterol level

Test animals induce high fat diets that produce hypercholesterolemia rats. A high fat diet is administered orally with a composition of 1 mL of lard and 2.5 mL/200 grams of weight of quail egg yolk and drinking water containing Propylthiouracil (PTU) 2 mg/kg bw in all treatment groups. Rats were given a high fat diet for 28 days. Provision of propylthiouracil (PTU) aims to increase cholesterol levels by reducing the amount of thyroid hormone produced by the thyroid gland and blocking the conversion of thyroxine T3. The mechanism for reducing thyroid hormone cholesterol levels is, increasing LDL receptors on liver cells [6].

2.5 Extract treatment

Red ginger extract is given for 14 days (after animal hypercholesterolemia) given as much as 2 mL/200 grams bw morning and evening. The dosage of treatment is as follows:

Group 1: Red ginger 0 mg/kg bw.
Group 2: Simvastatin 7.2 mg/kg bw.
Group 3: Red ginger extract of 200 mg/kg bw.
Group 4: Red ginger extract of 350 mg/kg bw.
Group 5: Red ginger extract of 500 mg/kg bw.

The treatment dosage of ginger extract refers to the modified [7].

2.6 Weight weighing and serum preparation

Weighing the wistar rat body weight was done on days 0, 28, and 42, by weighing each wistar rat using a scale. Whereas the blood sample was taken on days 0, 28 and 42. Before the blood was taken the rats were fasted for 12 hours, but were still given drinking water. Blood samples were taken as much as 1.5 mL from the eye vein. The sample was then put into a vacutainer vacuum tube containing EDTA then centrifuged at a speed of 3000 rpm for 10 minutes. Blood is separated from serum and plasma. The serum was separated from eppendorf tube and then stored in a refrigerator at 20 °C before being analyzed.

2.7 Determination of plasma total cholesterol levels

Measurement of plasma total cholesterol by the CHOD-PAP enzymatic colorimetric and following the procedure from the DiaSys (*Diagnostic System International*) kit. Blood serum was taken as much as 10 μL, added cholesterol *reagents FS diasys* 1000 μL then vortex. The solution was incubated at 25 °C for 20 minutes. Furthermore, it was measured using a microlab 300 spectrophotometer with a wavelength of 500 nm.

$$\text{Total cholesterol mg/dl} = \frac{\Delta \text{ serum absorbance}}{\Delta \text{ standard absorbance}} \text{ x standart absorbance} \quad (1)$$

2.8 Determination of triglycerides

Measurement of plasma triglycerides by the enzymatic colorimetric *Glycerol-3-Phosphatase-Oxidase-Paminophenazone* (GPO-PAP). 10 μL of blood serum was added to the FS 1000 triglyceride reagent then vortexed, and incubated at 25 °C for 10 minutes. Furthermore, it was measured using a microlab 300 spectrophotometer with a wavelength of 500 nm.

$$\text{Trigliserides mg/dl} = \frac{\Delta \text{ serum absorbance}}{\text{serundart absorbance}} \text{x standart consentration} \quad (2)$$

2.9 High density lipoprotein (HDL)

Measurement of plasma HDL by the CHOD-PAP enzymatic and procedure from the DiaSys (*Diagnostic System International*) kit. 100 μL of serum sample and 250 μL of HDL precipitant reagent were inserted into the eppendorf tube. incubated at room temperature for 10 minutes. Subsequently centrifuged for 10 minutes at 4000 rpm, 100 μL supernatant was taken and 1000 μl cholesterol FS reagent was incubated at room temperature for 20 minutes and the absorbance was read at 500 nm.

$$\text{Cholesterol mg/dl} = \frac{\Delta \text{serum absorbance}}{\text{erum absorbance dl=}} \text{x standard consentration} \quad (3)$$

$$\text{Cholesterol HDL} = \text{cholesterol total} - \text{supernatant cholesterol} \quad (4)$$

2.10 Low density lipoprotein (LDL)

Measurement of plasma LDL by the CHOD-PAP enzymatic and the procedure from the DiaSys (*Diagnostic System International*) kit. 100 μl blood serum and 1000 μl precipitant LDL reagent were inserted into the epeendorf tube. Incubated at room temperature for 10 minutes, centrifuged at 2500 rpm for 15 minutes. 100 μL supernatant was added to 1000 μL cholesterol FS reagent, incubated for 10 minutes at 20-25 °C and the absorbance was read at 500 nm.

$$\text{Cholesterol mg/dl} = \frac{\Delta \text{ Cholesterol mg/}}{\text{Cholesterol mg/dl=}} \text{x standard consentration} \quad (5)$$

$$\text{Cholesterol LDL} = \text{total cholesterol} - \text{supernatant cholesterol} \quad (6)$$

2.11 Statistical data analysis

The quantitative data collected were analyzed by Two Way ANOVA statistical analysis. Significant differences were obtained at $p < 0.05$. The results are expressed as means and standard deviations (SD).

3. Results and Discussion

3.1 Result

Measurement of total phenol levels was carried out using gallic acid as a standard. Determination of total phenol levels of red ginger extract was carried out based on gallic acid standard curves. The use of gallic acid as a standard that is, because gallic acid is a derivative of hydrobenzoic which is a simple phenol acid which is pure and stable. Absorbance of five gallic acid concentrations was measured at a wavelength of 760 nm. The results are then used to form a regression curve and obtain a regression equation to measure phenolic levels of red ginger. Based on the regression curve, shows the gallic acid regression equation at concentrations of 10, 20, 30, 40, and 50 ppm is $y = 0.123272\ x + 0.049$, which is the standard solution of phenol compounds is linear between absorbance and concentration with the correlation coefficient (r) is 0.996. In the absorbance measurement shown by the correlation coefficient (r) of 0.996, this value (r) approaches the number 1 which shows that the regression equation is linear. Gallic acid standard curves are presented in Figure 1. Ginger extract solution was measured at a wavelength of 760 nm.

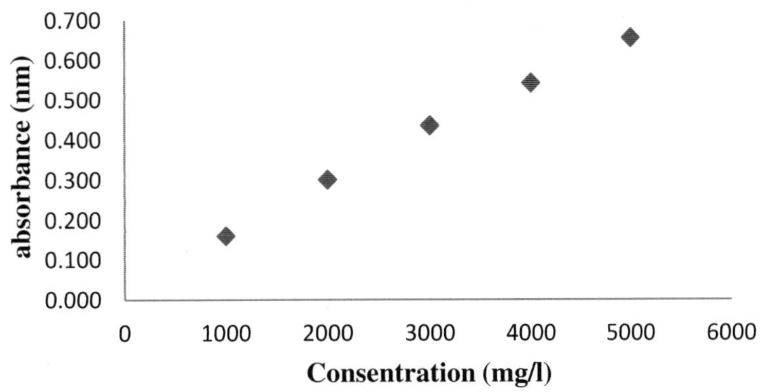

Figure 1. Total Curve Standard for Red Ginger Phenol (*Zingiber officinale Roxb.* var rubrum).

Red ginger extract has a significant effect on wistar rat body weight changes at an extract dose of 200 mg/kg BW which is equivalent to positive control (simvastatin 7.2 mg/kg bw) ($p \geq 0.05$). Giving extract of 200 mg/kg bw for two weeks showed a higher percentage of decline compared to the other groups, namely 22.5%. The effect of red ginger extract on body weight is presented in Table 1.

Table 1. Effect of ginger extract (*Zingiber officinale* Roxb. var rubrum) on body weight.

Body Weight (gram)					
Time	**Control**	**Simvastatin 7,2 mg/kg bw**	**Red ginger 200 mg/kg bw**	**Red ginger 350 mg/kg bw**	**Red ginger 500 mg/kg bw**
Day 0	187.6±9.55[a]	189.0±7.90[a]	183.0±9.56[a]	184.8±10.4[a]	185.6±8.20[a]
Day 28	229.2±8.16[ab]	218.6±13.9[a]	233.6±6.06[b]	224.4±6.45[ab]	229.2±1.48[ab]
Day 42	228.2±5.21[c]	170.8±2.28[a]	174.4±3.11[a]	174.6±7.76[a]	181.0±2.54[a]
Persentage (%)	0.4%	21.9%	22.5%	22.1%	21.0%

Note: The same letters behind the numbers in each column show no significance difference.

Red ginger extract had a significant effect on changes in the lipid profile of hypercholesterol rats at an extract dose of 200 mg/kg bw which was equivalent to positive control (simvastatin 7.2 mg/kg bw) ($p \geq 0.05$). The effect of red ginger extract on the lipid profile is presented in Table 2

Table 2. Effect of ginger extract (*Zingiber officinale* Roxb. var rubrum) on lipid profile at day 42.

Profil lipid (mg/dl)				
Treatment group	**Cholesterol**	**LDL**	**HDL**	**Trigliserida**
KN	73.1liser[b]	34.0liser[b]	27.3liser[a]	78.1liser[b]
KP	52.4liser[a]	22. 2iseri[a]	36.02iser[b]	31.82iser[a]
JA	48.82iser[a]	22.62iser[a]	35.52iser[b]	31.52iser[a]
JB	51.82iser[a]	25.82iser[a]	33.32iser[b]	38.92iser[a]
JC	64.52iser[ab]	22.92iser[a]	34.62iser[b]	34.52iser[a]

Note: The same letters behind the numbers in each column show no significance difference.

Red ginger extract has a significant effect on changes in hypercholesterolemia lipid profile at extract dose of 200 mg / kg body weight which is equivalent to positive control (simvastatin 7.2 mg/kg bw ($p \geq 0.05$). The percentage effects of red ginger extract on the lipid profile in Table 3

Table 3. Lipid profile change (%).

Lipid profile change (%)				
Treatment group	**Cholesterol**	**LDL**	**HDL**	**Trigliseride**
KN	-25.3	- 2.6	- 4.9	-15.8
KP	-42.2	-31.7	+28.1	-61.6
JA	-47.9	-32.7	+25.8	-64.3
JB	-43.4	-31.4	+27.1	-62.5
JC	-28.5	-30.4	+22.7	-57.5

Note: The (-) sign is decreasing, the (+) sign is increasing
KN: Negative control
KP: Simvastatin 7.2 mg/kg bw
JA: Red ginger 200 mg/kg bw
JB: Red ginger 350 mg/kg bw
JC: Red ginger 500 mg/kg bw

3.2 Discussion

The results of this study, the ethanol extract of *Zingiber officinale* Roxb. var rubrum has a higher total phenolic content of 5.19 mg/GAE/g compared to other Zingiberaceae plants such as *Curcuma aeroginosa* [8], *Kaemferia pulchra* [8], and *Boesenbergia pandurata* [9]. However, some plants such as *Curcuma longa* [10], *Zingiber zerumbet,* [11] and *Curcuma xanthorrhiza* [8] have higher total phenolic levels compared to *Zingiber officinale* Roxb. var rubrum.

Where as in this study to induce hypercholesterolemia using 1 mL of lard and 2.5 mL/200 grams bw of quail egg yolk and drinking water that already contains propylthiouracil (PTU) 2 mg / kg bw. This induction was able to increase body weight in all treatment groups on day 28. These results are consistent with a previous report by [12], which indicates that excess intake of fatty foods can greatly increase body weight. After 42 days the control group showed a weight loss of 0.4%. Whereas in the positive group giving simvastatin 7.2 mg/kg bw and red ginger extract 200, 350, and 500 mg/kg bw decreased by a percentage of 21.9%; 22.5%: 22.1% and 21.0%.

In the lipid profile test the control group showed a change in lipid profile, but the value of the percentage change in lipid profile was smaller compared with the treatment of simvastatin and red ginger extract with a percentage decrease in cholesterol levels of 23.3%, LDL 2.6%, triglycerides 15.8%, although a decrease in cholesterol, LDL and triglyceride levels but no increase in HDL. This shows that there is still an increase in the accumulation of fat in the liver resulting in an increase in the amount of acetyl CoA in the liver cells that produce cholesterol. Increased plasma cholesterol causes a decrease in liver LDL receptor activity resulting in decreased LDL absorption which causes plasma LDL to increase.

Whereas in the treatment group simvastatin 7.2 mg/kg bw (positive control) was able to provide changes in lipid profile levels of hypercholesterolemia rats which was able to reduce total cholesterol levels by a percentage of 42.2%, 31.7% LDL levels, and triglyceride levels 61.6% and able to increase HDL levels 28.1%. The mechanism of statins in inhibiting cholesterol formation by inhibiting the HMG-CoA reductase enzyme in the liver, a rate limiting enzyme in the cholesterol biosynthetic pathway. As a result, the conversion of HMG-CoA to mevalonate is inhibited and results in a reduction in cholesterol levels that are produced. This reduction activates a transcription factor called the sterol regulatory element binding protein (SREBP), which will be transported from the endoplasmic reticulum to golgi. LDL receptor transcription will be activated and will affect the increase in LDL receptor absorption there by reducing plasma LDL levels [13].

When compared with the treatment of red ginger extract treatment showed that red ginger extract significantly affected the changes in hypercholesterolemia lipid profile at extract doses of 200 mg/kg bw and 350 mg/kg bw which were equivalent to positive control of simvastatin dose 7.2 mg/kg bw ($p \geq 0.05$). Percentage reduction of 200mg / kg bw dose is plasma total cholesterol level of 47.9%, LDL 32.7% and triglyceride 64.3% and HDL increase of 25.8%. While the 350 mg/kg bw extract dose with a decrease in plasma total cholesterol levels was 43.4%, LDL 31.4%, and triglycerides 62.5% and increase in HDL 27.1%. These results indicate that the dose of 200 mg/kg bw showed the highest percentage reduction in plasma total cholesterol, LDL, and triglyceride levels compared to other treatment groups. But it has a lower HDL percentage increase compared to the 7.2 mg/kg bw simvastatin treatment. When compared with the treatment of red ginger extract dose of 350 mg /kg bw the percentage of HDL 200 mg/kg bw is smaller than the dose of 350 mg/kg bw. Percentage of red ginger extract on changes in lipid profile.

As is well known that red ginger extract contains phenolic compounds as antioxidant agents. Phenolic compounds have been shown to provide protection against metabolic disorders such as hypercholesterolemia. Hypercholesterolemia is the result of increased cholesterol levels and increased production of free radicals [14]. Phenolic compounds in this case reduce the risk of hypercholesterolemia by increasing the activity of antioxidant enzymes and reducing the formation of free radicals. In addition, antioxidant activity can accumulate superoxide anions and hydroxyl radicals, able to inhibit lipid peroxidation, and regulate low density lipoprotein (LDL) receptors [7].

The mechanism proposed in various studies so far is that ginger extract is able to inhibit the HMG-CoA enzyme in the liver (an enzyme that limits the rate of conversion of HMG-CoA to mevalonic acid). Enzyme inhibition leads to a reduction in cholesterol in the liver. However, the mechanism is better than both HMG-CoA reductase and LDL receptors, where HDL receptors act by increasing LDL absorption there by reducing plasma LDL levels [15],

4. Conclusion
Based on the results and discussion that has been submitted, it can be concluded that the red ginger extract has a total phenol level of 5194.15 ± 264.1 μg/g, and has a significant influence on changes in lipid profile and changes in body weight of hypercholesterolemia rats.

Acknowledgments
The writer would like to thank the University of Sebelas Maret especially the facilities provided by the integrated central laboratory so that the research can be carried out.

Reference
[1] Balamurugan G, and Shantha A 2010 *J. Pharm. Bioallied. Sci* **2**(4) 350-355
[2] Elly P, Wheni E R, and Cikra INHS 2018 *JDMS.* **17**(5) 55-59.
[3] Babak B A, Mahmoud R K, Mohammad M Z, and Mahmoud B 2015 *Der. Pharmacia. Lettre* **7**(12) 81-88
[4] Geetika P, Chirag S, Riyaz A H V, Pooja B, and Sibi *Res J. Med. Plant.* **9**(6) 300-306
[5] Mathew S, Abraham T E, and Zakaria Z A 2015 *JFST* **52**(9) 5790-5798
[6] Belal P, Rahim H, and Hassan M 2017 *VRF.* **8**(3) 185-193.
[7] Sunarti, Edy F, Urip H, Delzuzar, Tri W, and Lokot DL 2017 *Univ. med.* **36**(3) 228 - 235
[8] Alafiatayo A A, Ahmad S A, and Maziah M 2014 *Afr. J. Tradit. Complement. Altern. Med* **11**(3) 7-13
[9] Aweng E R, Karunakaran T and Zakia K 2018 *J. Pharm. Bio, and Chem. Sci* **9**(1) 962 - 971
[10] Array E J, Tonfack D F, Kingne K F, Kinge E E, and Womeni, H M 2019 *J. Food. R.* **3**(1) 86 - 90
[11] Vipada K, and Yingyong P 2012 *JDSA.* **7** 89-96
[12] Patrick, Ambrose T, Faith P, Samson H A, and Umaru 2015 *JBAH.* **5** (18) 123-127
[13] Meor F R M A S, Intan N S, Johnson S, Shariful H, and Subashini C T 2017 *IJEM.* **15**(2) E 43319
[14] Sushma D, and Randhir S 2017 *Phcog. J.* **9**(6) 807-814
[15] Salah M E S, and Reham A M 2016 *JPCBS.* **3**(4) 561-572

IOP Conf. Series: Materials Science and Engineering **858** (2020) 012026 doi:10.1088/1757-899X/858/1/012026

Optimizing antioxidant substances extraction produced by *Holothuria atra* using ultrasonic with response surface method

T Murniasih[1], Muthi'ah[2] and S B Widjanarko[2]
[1] Research Center for Oceanography, Indonesian Institute of Sciences,
Jl. Pasir Putih 1 Ancol Timur Jakarta 14430, Indonesia
[2] Department of Food Science and Technology, Faculty of Agricultural Technology-Brawijaya University, Jl. Veteran-Malang 65145, Indonesia

E-mail: tutikmurniasih71@gmail.com

Abstract. This study was to discover the effect of ultrasonic extraction based on the factors of extraction time and amplitude, to increase the yield recovery of H. atra extract and antioxidant activity (IC50). The process in optimizing ultrasonic extraction was performed using Response Surface Methodology (RSM) with Central Composite Design (CCD). H. atra extract produced by the 2 different extraction methods indicated a significant difference in yield (maceration 5.48%, ultrasonic 10.43%) and antioxidant activity (maceration 149.38 ppm, ultrasonic 84.85 ppm). The antibacterial activity against *B. subtilis, E. coli, and S. aureus*), also the toxicity LC50 at 457.77: 443.63 ppm were not significantly difference. It was found that ultrasonic extraction method needs shorter time with higher yield and more antioxidant activity. The sea cucumber extract that was produced from the ultrasonic method contains alkaloid, triterpenoid, and saponin, while steroid undetected, and from the maceration method contains no alkaloid and steroid.

1. Introduction

Concern to geographically, Indonesia is located between Pacific Ocean and the Indian Ocean, which means that this country is the best habitat for marine organisms, including sea cucumbers [1]. Indonesia is the world's highest imported of Holothuroidea [2]. The common species of Indonesian sea cucumber which are highly economic valued are golden sea cucumber *(Stichopus hermanii)*, black teatfish *(Holothuria nobilis),* white teatfish *(Holothuria fuscogilva)* and sandfish *(Holothuria scabra)*[2]. The most highly valued of sea cucumbers caused increasing of exploitation was being proposed to be listed in the endangered marine biota species and will become restricted for any kind of international commercial trade [3]. Another species of sea cucumber (such as black sea cucumber or *Holothuria atra*), is one of marine biota whose population is still considerably abundance, and this species is not highly-priced. Komala [4] says that 55% of the sea cucumber population in Kepulauan Seribu, Jakarta, is classified as *Holothuria atra.*

Bioactive substances content in *H atra* extract has been reported some researcher. The extract of *H. atra* contains phenolic compounds (chlorogenic acid, pyrogallol, rutoside, coumaric acid, catechin, and ascorbic acid) [6].As well as alkaloid, triterpenoid, steroid, saponin, anthraquinone, and glycoside which are considered to be the potential antioxidant [7]. It has also found that sea cucumber extract tends to act as antibacterial, antifungal [8] anticoagulant, antihypertension, antimalaria, antivirus [9] antithrombotic, overcome osteoarthritis [10] anti-inflammation, immune-stimulant [7], antitumor [11] and anticancer [12].

The common method used to extract sea cucumber was maceration. Extraction using maceration method generates the yield of about 6% [13] or 0.11% [14] in 3 x 24 hours. Using the maceration method will result in the low of extract yield in longer time processes. Thus, there should be any new method that could be increasing the extraction yield and cutting the time process. The ultrasonic method that has the cavitation phenomenon now being used as the alternative for reducing extraction time and multiplying the yield [15].

The main targets to be optimized were yield and the antioxidant activity, which is performed using Response Surface Methodology and Central Composite Design. This method is known as an effective statistic method for finding optimal condition [16]. As the independent variables of this experiment were the extraction time and the ultrasonic amplitude, which would be compared with the 'control' extraction yield generated from the maceration method. The other parameter such as chemical group screening and antibacterial activity would also be performed.

2. **Experimental**

2.1 Materials

The dried sea cucumber *(Holothuria atra)* were obtained from the Lampung in 2018. The fresh body wall of sea cucumber was separated with the viscera. The body wall was boiled for 1 hour and dried directly under the sun (above bamboo/*para-para*) for several days. The dried sea cucumber was continued for drying to reduce water content using the oven at 60 C during 24 hours. The sea cucumbers were powdered using a blender (Waring Commercial) and sifted using the mesh with 20 strainers. Powdering the sea cucumbers were easier to extract than slices once because the smaller size of materials it will have a broader surface. A broad material surface may enhance the efficiency of contact between the materials with the solvent [17], and the extraction of bioactive components can be optimized [18, 19, 20, 21].

2.2. Methods

2.2.1. Sea cucumber (Holothuria atra) extraction using ultrasonic and maceration method.

About 10 g of sea cucumbers powder was macerated with 96 % ethanol in 500 mL beaker glass, the ratio between sea cucumber powder and the solvent was 1:10. The samples were extracted using two methods, the first was maceration as a control method, processed for 3 x 24 hours in room temperature. The second was the ultrasonic method (treatment method) to perform comparison (Cole Palmer/cpx 130, 220 V, 130 Watt). Experiment design of ultrasonic method was performed using Response Surface Methodology with the model of Central Composite Design (**Table 1** and **2**) through statistic software so-called Design Expert. The filtrates generated from an extraction process using either maceration or ultrasonic methods were strained using Whatman no. 4 paper strainer and evaporated using rotary vacuum evaporator at the temperature of 40 degrees Celsius with the speed of 90 rpm to coagulate and turn it into coarse sea cucumber extract. The coarse extract would be analyzed and compared.

2.2.2.Experiment design using RSM and statistical analysis.

This experiment was performed using RSM with the model of Central Composite Design (CCD) on two independent factors, which were extraction time (XI) and extraction amplitude (X2) (**Table 1**).

Table 1. Independent variables of CCD experiment design.

Code	Factor	Variable Code				
		-α	-1	0	+1	+α
X_1	Time (minute)	23.78	30	45	60	66.21
X_2	Amplitude (decibel)	47.92	50	55	60	62.07

Each factor range was stipulated based on the preliminary research findings which the yield and antioxidant activity (IC_{50}) as the dependent variable. The completed design comprised 13 experimental points, including 5 center points, 4 factorial points, and 4 axial points. The CCD data were analyzed using multiple regressions to adjust it with the second-order polynomial model, as stated in the following equation:

$$Y = \beta_0 + \sum_{i=1}^{k} \beta_i X_i + \sum_{i=1}^{k} \beta_{ii} X_i^2 + \sum_{i=1,j=2}^{k-1,k} \beta_{i,j} X_i\, X_j + \varepsilon \qquad (1)$$

$\beta_0; \beta_i; \beta_{ii}$, and where the terms for regression coefficients in the intercept, linear, quadratic, and interaction; $X_i\,; X_{ii}\,; X_{i.j}$ were the independent variables; Y was the response and was the random error component.

Design Expert was previously used for obtaining the quadratic model of the polynomial coefficient. This model was more effective since easy to define R^2 value, Adjusted R^2 value, and adjusted with the significant statistic result with the F test.

Table 2. *H. atra* ultrasonic extraction with *Central Composite Design* (CCD).

Run	Factor code		Actual Factor		Response	
	Time (minute)	Amplitude (decibel)	Time (minute)	Amplitude (decibel)	Yield (%)	IC_{50} (ppm)
1	-1	-1	30	50	4.92	96
2	1	-1	60	50	8.8	125
3	-1	1	30	60	5.2	108
4	1	1	60	60	9.01	134
5	-1.41421	0	23.7868	55	4.5	88
6	1.414214	0	66.2132	55	10.4	129
7	0	-1.41421	45	47.92893	6.06	118
8	0	1.414214	45	62.07107	8.23	129
9	0	0	45	55	10.82	85
10	0	0	45	55	10.53	83
11	0	0	45	55	10.91	84
12	0	0	45	55	10.25	85
13	0	0	45	55	10.9	83
Prediction			45.52	54.92	10.7417	84.4356
Verification			45.52	54.92	10.43	84.85

Yield

The percentage of extraction yield was generated to the following formula:

$$\%\text{ of yield} = \frac{\textit{extraction yield}}{\textit{mass of material}} \qquad \textbf{(2)}$$

Antioxidant Activity (IC_{50})

Analysis of antioxidant activity (IC_{50}) was performed using 1,1-diphenyl-2-pyorihidrazil (DPPH) (concentration: 0.1 Mm or 39.4 mg DPHH in 1 L of ethanol). Approximately 100 µl of the sample was mixed with 3.9 mL DPPH. Incubation was performed in a dark with room temperature for 2 hours. The absorption was measured (λ = 517 nm) using UV-Vis spectrophotometer. Solvent control was prepared using 100 µl ethanol and 3.9 ml DPPH. The activity of free radical scavenger was counted with this formula:

$$\% \text{ of DPPH inhibition} = \frac{A^{\circ} - A\prime}{A^{\circ}} \times 100 \quad (3)$$

Note: A^0 = 'Control' Absorption; = Sample Absorption.

DPPH capability to prevent radicals was indicated by absorption reduction in the IC_{50} value measurement from the measurement in the inhibition percentage (%) as much as 50% [22, 23].

2.2.3. Characterization of H. atra extract.

The extract resulted from maceration and ultrasonic methods will be compared for secondary metabolite content (alkaloid, steroid, triterpenoid, and saponin), the yield, antioxidant activity, toxicity, and antibacterial. The chemical group content was done using Harbone method, (Brine Shrimp Lethality Test) using Mayer [24] method and antibacterial against *Bacillus subtilis*, *Escherichia coli*, *Vibrio eltor* and *Staphylococcus aureus* using agar diffusion method. The analysis was done in Oceanography LIPI laboratory, while the Chondroitin sulfate analysis was performed in PT. Saraswati Indo Genetech.

3. Results and Discussion

The value obtained from each response variable observed from each experiment is shown in **Table 2**. Dependent variable affected by independent factor was analyzed statistically, including fix summary model, analysis of variance, normal plots of residual, and graphic model of response.

3.1.Effect of extraction time and amplitude to the yield

Base to the yield response, which informed at **Table 2**, the highest yield of sea cucumber extract could be achieved with an extraction time of 45 minutes and amplitude of 55 decibels, and the yield was approximately 10.25-10.91%. The lowest yield of *H. atra* extracts achieved with extraction time 23 minutes and amplitude reach 55 decibels. These conditions were gained the extraction yield about 4.5%. The data were used in the statistical analysis through Design Expert for stipulating analysis model for various kinds of *H. atra* extraction yield responses.

The Lack of Fit Test generates model compatibility if the p-value is > 0.05 [25, 26]. If the R^1 is getting closer to 1, it indicates compatibility between observation value and prediction value [27, 28]. **Table 3** showed that the model can be counted on the experimental value, it can be used as the theoretical base for the yield response in the *H. atra* ultrasonic extraction process.

Table 3 indicated that there was a significant difference in yield response generates during *H. atra* ultrasonic extraction. The distinction between the percentages of the yield was a response to various treatments, including extraction time and amplitude. Based on the partial test for every variable, it can be seen that the actual distinction was in the extraction time (X_1), amplitude of ultrasonic extraction (X_2) and the quadrat of the two factors (X_1^2 and X_2^2), meanwhile, the interaction between both factors (X_1X_2) was not actually different because the p-value was > α (0.05).

Table 3. Regression coefficients and significance of the predicted second-order polynomial models.

Coefficient	Yield (%)	Antioxidant (IC_{50})
Intercept β_0	10.68**	84.00**
Linear		
β_1	2.00**	14.12**
β_2	0.44*	4.57**
Interaction		
$\beta_1 \beta_2$	-0.018	-0.75
Quadratic		
β_1^2	-1.69**	12.19**
β_2^2	-1.85**	19.69**
Mean	8.50	103.61
Standard deviation	0.44	1.13
R^2	0.98	0.99
Adjusted R^2	0.96	0.99
F-value (model)	< 0.0001	< 0.0001
F-value (lack of fit)	0.0932	0.3135

*$p < 0.05$; **$p < 0.01$

The length of time in the ultrasonic extraction process affected the yield. When the extraction time gradually increased, the yield continuously increased. When the time was passed to the optimum limit, the yield was reducing. The longer the extraction time, abundance yield will be reached. It can be explain because the chance of the solvent to contact with the material is higher due to the cavitation process [29, 30]. The extracted yield will optimum when the material is highly blended in the solvent until the solvent was completely saturated. Yield will be reduced when the extraction time is extended due to the degradation extraction compound [28]. Garcia [29] informed that the time needed by ultrasonic extraction process is shorter than the non-ultrasonic extraction process. The study performed by Zhao [28], regarding polysaccharides extraction, achieves the best time in the minute of 25th when the yield was abundantly generated.

The amplitude being observed by the researcher positively affected the extract yield. The yield was increased along with the enhancement of extraction time and amplitude. In the early extraction time with the amplitude of 50 decibels, the yield was continuously increased and the reduced when reaching a certain point of amplitude. Vinatoru [31] reported that the higher of amplitude, the higher chance for the cavitation bubbles (which are located in the membrane of the sample) to burst. This will generate a bigger micro facture. There is a relation between amplitude and energy (joule). The energy was directly proportional to the amplitude squared, it can be stated that the higher the amplitude, the higher the energy consumed [32]. The high amplitude which goes beyond the medium will generate pressure and shear force by the solvent molecule. This condition will generate local change in the density and modulus elasticity and increasing the mass movement interphase. It would be increasing the yield in less time [33].

3.2. Effect of extraction time and amplitude on the antioxidant activity

Antioxidant activity shown in the **Table 2** implied that the highest antioxidant activity could be achieved when the extraction time was 45 minutes and the amplitude was 55 decibels, with the IC_{50} of 83 ppm, and the lowest activity could be achieved when the extraction time was 60 minutes and the amplitude was 62.07 decibel, with the IC_{50} of 134 ppm. The data were used in the statistical analysis through Design Expert for stipulating analysis model for various kinds of *H. atra* extraction yield response.

Analysis of antioxidant (IC_{50}), based on the F (p-value) and the lack of fit in the **Table 3** described that ultrasonic extraction time and amplitude in the quadratic model significantly affected the antioxidant activity (IC_{50}). It was indicated by the F value of quadratic model < 5% ($p < 0.05$) as much as < 0.0001 which means the error chance was < 0.01%. F value of the lack of fit was in the amount of 0.3135 which indicates p > 5% ($p > 0.05$), so the model incompatibility is not significant. It can be concluded that the quadratic model was compatible. P-value and the lack of fit which was over to the limiting threshold indicated that the model was significantly implicated [25].

The antioxidant (IC_{50}) generated by the two methods was significantly different. This distinction happened as the response of treatment variation, including time and amplitude. it can be seen that the actual difference was in the extraction time (X_1), amplitude of ultrasonic extraction (X_2) and the quadrat of the two factors (X_1^2 and X_2^2), meanwhile, the interaction between both factors (X_1X_2) was not significantly different because the p-value was > α (0.05).

Time length of the ultrasonic extraction process significantly affected antioxidant activity (IC_{50}). In the early time of the process, the antioxidant was continuously increased because of the bioactive compounds. However, the time extension will reduce antioxidant activity. The ultrasonic method is suitable for to be applied to the extraction of a bioactive compound which cannot stand high heat, like the process of pomegranate extraction with the most optimal extraction time of 53 minutes [34]. Jiang [36] also stated that the early times of ultrasonic extraction generate a positive effect, but, if extended, then the quality of the composition will be reduced.

Ultrasonic extraction amplitude (X_2) was significantly affected the $IC_{50,}$ similar to square factors (X_1^2 and X_2^2) which significantly affect $IC_{50,}$ of *H. atra* extract. It was probably due to the cavitation phenomenon. Liu [15] stated that ultrasonic cavitation may break the cell membrane mechanically and increasing material transfer. This cavitation phenomenon brings physical effect and chemical effect on the extraction process if extraction time and amplitude increased. Jiang [35] stated that the higher of ultrasonic power or the higher length time of extraction process will increase the number of hydroxyl radicals, which could damage the chemical compound due to the oxidation brought by the ultrasonic wave. Balachandran [36] observed that antioxidant such as cyanidin-3-glucoside may be decreased to 20% in the 4 hours sonication. On the other hand, this radical can be used as food material function enhancer. The study performed by Balachandran [36] indicated that the sonication process on the phenol compound changes hydroxylation into ortho-, meta- and para-. So, it can be said that the establishment of OH▪, due to the sonication process, may enhance the antioxidant in an herbal compound such as flavonoid.

3.3. Optimization of time and amplitude

Based on the quadratic equation and the two or three-dimension contour graphic model (**Figure 1, 2 and 3**) were reflecting the relation between both independent variables. **Figure 1.** represents time and amplitude to yield response. Contour plot as shown in **Figure 1.** showed that extraction time affects the yield. Different color in contour plot graphic showed the yield value. Yellow to red color showed constantly increased yield, whereas blue color showed low yield. This contour described that the yield was increased when the time increases from 30-52.50 minutes, and decreased when the extraction time above 52.50 minutes. The difference in surface-level indicated different yield response value in each combination of formula components. The low area showed low response yield value, while a high area showed high yield response value. In the three-dimension graphic, it can be seen that response yield to the sonication time and amplitude started to stabilize in the range of 10.00% yield number up to 11%.

Figure 2. represents time and amplitude effect on antioxidant activity response (IC_{50}), It showed the extraction time which affects the antioxidant activity (IC_{50}). Different color in contour plot graphic showed different antioxidant activity. Green to blue color shoe high antioxidant activity (decreased IC_{50} number), otherwise, the red color showed low antioxidant activity (high IC_{50} value). Based on the graphic 3 dimensions, high antioxidant activity response starting from 30 minutes to 52.50 minutes and amplitude of 57.50 decibels.

IOP Conf. Series: Materials Science and Engineering **858** (2020) 012026 doi:10.1088/1757-899X/858/1/012026

Figure 3. represents extraction optimization result with time and amplitude treatment to the yield response and antioxidant. The optimum point in the extraction process (**Table 2**) wanted is in the form of goal, which is % yield maximally chosen and minimum antioxidant IC_{50}. The chosen formula solution is the formula with the period combination extraction of 45.52 minutes and 54.92-decibel amplitude. This formula was predicted to have 10.74% yield and 84.44% antioxidant. Desirability value of 0.972 means that the formula will produce a product having the characteristics corresponding to the optimization target of 97.2%. Desirability value reaching 1 showed good optimization function, in which the program can fulfill the purpose based on the criteria desired in the end product [25, 37]. The blue color in contour showed low desirability value while red color showed high desirability value. This contour showed the colors emerged are blue to yellow with maximum desirability value was 0.972.

Verification stage of optimization result was aimed to ensure that the optimum point solution obtained and Design Expert software can produce response value following the optimum response condition solution given by the software. Verification will be conducted by implementing the optimum solution in *H. atra* extraction process with ultrasonic during 45.52 minutes at 54.92 decibels, with three times repetition, and the resulted yield and antioxidant activity (IC_{50}) observation were performed. Verification and software prediction result will be analyzed with paired T-test using Minitab 17. P-value. The paired T-test above 5% showed that the prediction value and verification were not significant, it's mean that the resolution or prediction supports verification. Verification result of *H. atra* extraction with the ultrasonic method can be seen in **Table 2**.

The optimum extraction condition predicted by the program was 45.52 minutes at 54.92 decibels. The optimum response obtained by this prediction reached the yield 10.74717 % and antioxidant activity (IC_{50}) value was 84.4356 ppm. The verification result performed in the laboratory was 10.43% of yield and 83.85 ppm of antioxidant (IC_{50}). Verification result with Minitab 17, showed that T-test was not significantly different with the prediction resulted from Design Expert software. P-value resulted for yield was 0.30 and for antioxidant (IC_{50}) was 0.25. P-value > 0.05 can be interpreted that there is no significant difference between software prediction and the actual condition. It can be concluded that the model is accurate form extraction optimization with maximum yield and high antioxidant activity with low IC_{50} value. Wangtueai [38] said that if the differentiation between prediction value and research value is not more than 5% indicated that the model is correct.

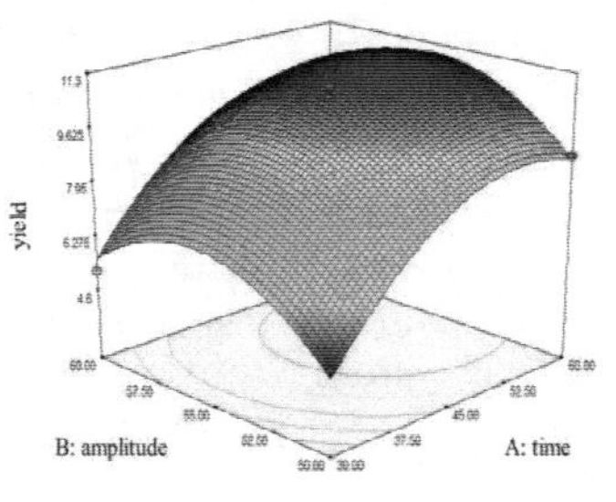

Figure 1. Graphic 3-dimensions of relation between sonication time and amplitude with yield response .

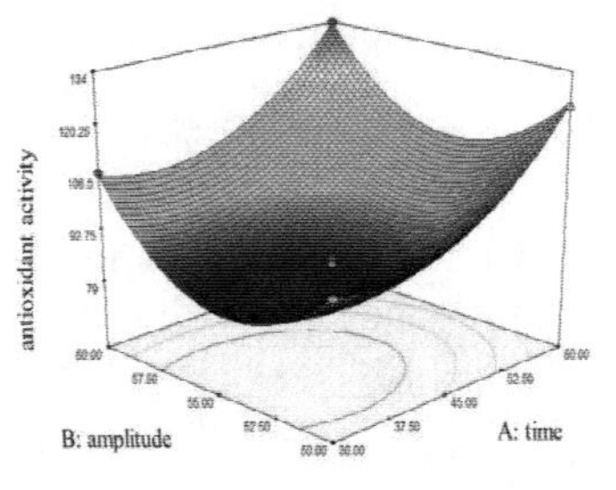

Figure 2. Graphic 3-dimensions of relation between sonication time and amplitude with antioxidant activity (IC_{50}).

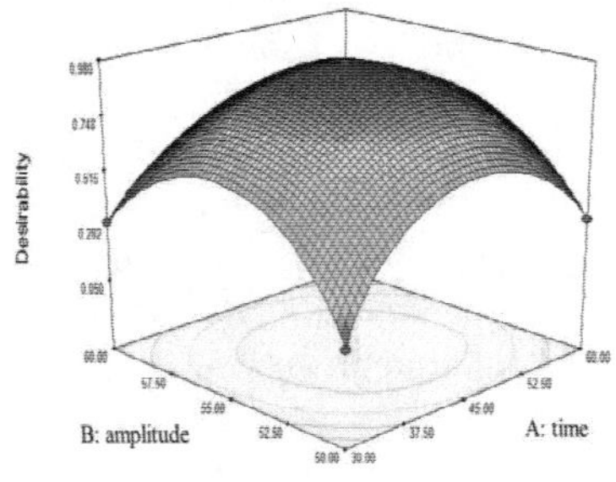

Figure 3. Graphic 3-dimension of relation between sonication time and amplitude with desirability value.

3.4. Characterization of H. atra Extract of Maceration and Ultrasonic Method

The filtrate obtained from extraction process with maceration or ultrasonic after filtering with Whatman no. 4, showed the difference color, filtrate result of ultrasonic method extraction is more dense, indicating that the antioxidant content is higher.

Table 4. Chemical group content in of *H. atra* extract by using maceration and ultrasonic method.

Phytochemical screening	The positive result according to literature	Result	
		Maceration	Ultrasonic
Alkaloid	Orange-red sediment is formed (Dragendorff reactor). White sediment is formed (Mayer reactor) brownish-red sediment is formed (Wagner reactor)	-	+
Steroid	The greenish-blue color is formed	-	-
Triterpenoid	Brownish or violet color is formed	+	+
Saponin	Stable foam is formed	+	+

Preliminary identification of secondary metabolite group with phytochemical analysis is an initial important step in research to know the chemical structural framework [39]. This analysis includes alkaloid, triterpenoid, steroid, and saponin test. Secondary metabolite compound analysis result of sea cucumber extract extracted with maceration and ultrasonic methods can be seen in **Table 4**. Secondary metabolite content of sea cucumber extract with 96% ethanol solvent-extracted with maceration and ultrasonic methods showed that on both methods there was no indication of steroid, but triterpenoid and saponin content were identified. While alkaloid in the extract using the ultrasonic method can be identified, and not detected in the maceration method. This result is in line with the research conducted by Putram [40] *H. atra* sea cucumber ethanol extract is positive contains triterpenoid, flavonoid, alkaloid, and saponin. Whereas there was a difference in the research result conducted by Rasyid [14], that *Holothuria sp.* extract didn't contain triterpenoid and alkaloid, but steroid was identified. It was possible because the difference in sea cucumber variety, also difference in solvent extraction will produce the difference in secondary metabolite.

3.5.The measurement result of yield (%)

The yield obtained from different extraction process can produce different yield. It can be seen in **Figure 4[a]**. The Yield generated from ultrasonic extraction method was 10.43% while from maceration was 5.48%. This data was similar to the T-test, maceration and ultrasonic methods that generated significantly different yield. By using ultrasonic would gain higher yield than maceration method. The benefit when the applied ultrasonic method in the extraction process was the higher energy transfer. It will increase the ability and reduce the time of response production level on the extraction process.[41].

3.6. The result of the antioxidant test

Antioxidant activity of holothurian extract using of maceration and ultrasonic method can be seen in **Figure 4[b]**. The result of the antioxidant activity test from the maceration method has IC50: 149.38 ppm, it is categorized into medium activity. Extraction using Ultrasonic method I generated higher antioxidant activity IC50:84.45 ppm and categorized into strong activity. Based on T-test of maceration and ultrasonic method have different antioxidant activity value and better that maceration, indicated by 0.003 p-values ($PV < 0.05$). It can be said that the ultrasonic extraction method can increase antioxidant activity compared to the maceration method.

The antioxidant content in *H. atra* extracts observed by Esmat [6] with analysis using high-performance liquid chromatography (HPLC), displaying phenolic components, such as chlorogenic, routine, coumaric acid, catechin and ascorbate acid. HPLC analysis is supported by GC-MS result showing the existence of 59 compounds belong to the flavonoid, phenolic, triterpenoid, saponin, and alkaloid category [7]. Murniasih research [42] exposed that *H. atra* extract separated with column chromatography on silica gel and elution with gradient system increases polarity. The active fraction of *H. atra* extract showed significantly restricted radical scavenging activity (35.3487%) in 1 mg/ml concentration. The result of Gas Chromatography-Mass Spectrophotometer analysis of *H. atra* produces 7 compounds, which is in fraction 2. Besides, fraction active 7 showed some compounds, which are cycloicosane (17), pentadecyl trichloroacetate (18), 14β-pregnane (21) and cis-4-siano-2-(2-hydroxycyclohexyl) piridin.

3.7.Result of antibacterial activity and toxicity test

The comparison of the antibacterial activity of *H atra* extract between maceration and ultrasonic method can be seen in **Table 5**. The data above shows that the sea cucumber capsule has antibacterial activity. It indicates that *H. atra* extract can inhibit the growth of *Bacillus subtilis, Escherichia coli, Staphylococcus aureus*, and *Vibrio El Tor.* Based on T-test, between maceration and ultrasonic methods didn't prove that there was a significant antibacterial activity, indicated with p-value > 0.05. Sea cucumber is known contained some bioactive antibacterial [43]. Abraham [8] reported the antimicrobial effect of *H. scabra* extract, *H. scabra* extract was reported active against *Aeromonas hidrofila, Escherichia coli, Enterococcus sp., Klebsiella pneumonia, Pseudomonas aeruginosa, Salmonella typhi, Staphylococcus aureus, Vibrio harveyi*, and fungi coming from fish, *Aspergillus.*

The toxicity test result was displayed in **Figure 4[c]**, there is no significant difference between maceration and ultrasonic method. The IC50 value of extract by using the maceration method was. 457.7 ppm, while ultrasonic was 443.6 ppm. Concern to this result both of them categorized moderately toxic. IC_{50} value will be more effective in preventing cancer cell growth, Dhinakaran [7] research stating that *H. atra* extracts shoes strong prevention activity to tumor growth in human's cervix cancer cell (HeLa) also as human's breast cancer cell (MCF-7). In Janakiram et al. [11] research, sea cucumber *(Curcumaria frondosa)* extract was stated has potential as anti-cancer for colon cancer patient since it contains monosulfated triterpenoid glycoside Frondoside A, the disulfated glycoside Frondoside B, the trisulfated glycoside Frondoside C, 12-methyltetradecanoic acid, eicosapentaenoic, and fucosylated chondroitin sulfate.

Table 5. Antibacterial activity assays of *H. atra* extract with maceration & ultrasonic method[a].

Sample	*B. subtilis* (cm)	*E. coli* (cm)	*S. aureus* (cm)	*V. El Tor*
Ampicillin (10 µg = 0.01 mg)	3.941± 0.17^a	2.303± 0.04^a	2.5925± 0.05^a	3.122± 0.12^a
H. atra maceration	1.39± 0.08^b	1.241± 0.08^b	1.211± 0.14^b	1.1123 ± 0.06^b
H. atra ultrasonic	1.48± 0.14^b	1.247± 0.08^b	1.383± 0.14^b	1.2823 ± 0.025^b

[a]Description: The value is in average ±deviation standard from 3 times repetition, Sample: 10 mg/ml

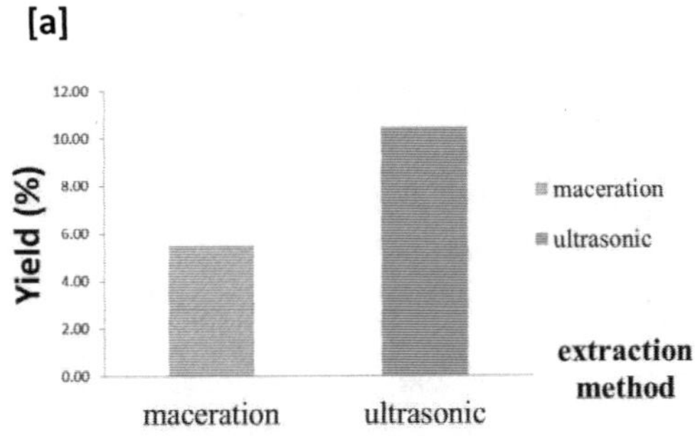

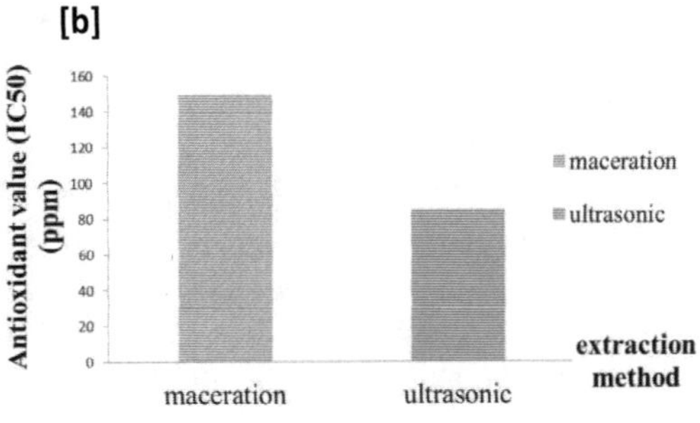

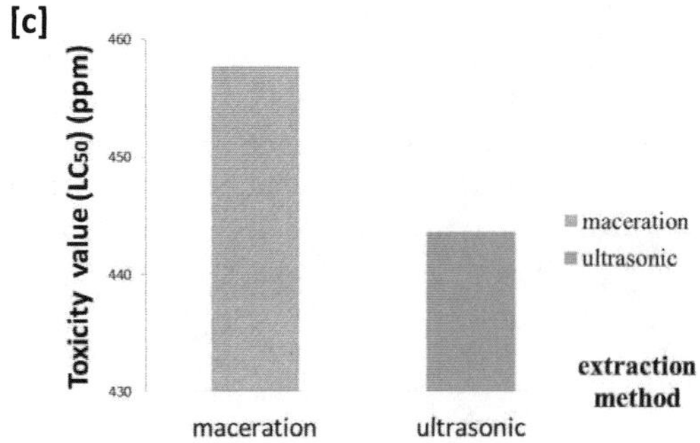

Figure 4. **[a]** Comparison graphic of *H. atra* yield with maceration and ultrasonic method. **[b]** Comparison of antioxidant activity result (IC_{50}) of *H. atra* with maceration and ultrasonic method. **[c]** *H. atra* toxicity comparison graphic of extraction with maceration and ultrasonic method.

4. Conclusion

Response Surface Method (RSM) could be used to optimize *H. atra* extraction using ultrasonic method. Two important parameter such as yield and antioxidant activity were affected by the extraction time and amplitude. The predicted and validated values are not significantly different on 95% trust level. The model in this research can be used to optimize the yield and antioxidant activity by time and amplitude variety. The optimum extraction process can be performed within 45.52 minutes of extraction time and 54.92-decibel amplitude. Extraction method using ultrasonic compared to maceration method showed differences in phytochemical contain and yield. The antioxidant activity were known to be higher in the ultrasonic method than maceration. While, it was not significantly different in bioactivity & toxicity (LC_{50}).

Acknowledgment

Authors very appreciate to Research Center for Oceanography LIPI for financial supporting of this research through DIPA-2019 program.

References

[1] Rasyid 2018 *Mengungkap Potensi Teripang dari Indonesia, Pusat Penelitian Oseanografi Lembaga Ilmu Pengetahuan Indonesia (LIPI)* (Yogyakarta: Deepublish Publisher)

[2] Choo P S 2008 Population status, fisheries, and trade of sea cucumbers in Asia. In *V. toral granda*, *A. lovatelli* and *M. vasconcellos* (eds). Sea cucumbers. A global review of fisheries and trade *FAO Fisheries and Aquaculture Technical Paper* **516** 81–118

[3] Darsono P 2007 Teripang (*Holothuroidea*): Kekayaan alam dalam keragaman biota laut *J. Oseana XXXII* **2** 1–10

[4] Komala R 2015 Diversity of sea cucumbers in seagrass and coral reefs ecosystem of Bira Besar Island, Kepulauan Seribu, Jakarta *Proc. Sem. Nas. Masy. Biodiv. Indon.* vol 1 (Depok: Masyarakat Biodiversitas Indonesia) pp 222–6

[5] Esmat A Y, Said M M, Soliman A A, El-Masry K S and Badiea E A 2013 Bioactive compounds, antioxidant potential, and hepatoprotective activity of sea cucumber (*Holothuria atra*) against thioacetamide intoxication in rats *Nutrition* **29** 258–267

[6] Dhinakaran 2014 Bioactive compounds from *Holothuria atra* of Indian ocean *Springer Plus* **3** 673

[7] Abraham T J, Nagarajan J and Shanmugam S 2002 Antimicrobial substances of potential biomedical importance from *Holothurian* species *Indian J. Mar. Sci.* **31** 161–4

[8] Mayer A, Rodríguez A D, Berlinck R G and Nobuhiro F 2011 Marine pharmacology in 2007–8: Marine compounds with antibacterial, anticoagulant, antifungal, anti-inflammatory, antimalarial, antiprotozoal, antituberculosis, and antiviral activities; affecting the immune and

nervous system, and other miscellaneous mechanisms of action *Comparative Biochemistry and Physiology* C **153** 191–222

[9] Bordbar B, Anwar F, Saari N 2011 High-value components and bioactives from sea cucumbers for functional foods: A review *Mar. Drugs* **9** 1761–1805

[10] Janakiram N, Mohammed A and Rao C V 2015 Sea cucumbers metabolites as potent anti-cancer agents *Mar. Drugs* **13** 2909–23

[11] Satari M H, Arma U, Ilyas S and Handayani D 2017 Anticancer potency of black sea cucumber (*Holothuria atra*) from Mentawai Islands, Indonesia *Padjadjaran J. of Dentistry* **29** 55–58

[12] Pangestuti R, Murniasih T, Putra M Y, Rasyid A, Wibowo J T, Ardiansyah A and Untari F 2016 Free radical scavenging activity of selected sea cucumber species from Mataram-Lombok, Indonesia *J. Teknologi (Sciences & Engineering)* **78** 4–2 179–185

[13] Rasyid 2017 Pengembangan formulasi produk makanan kesehatan dari teripang *Laporan Akhir Tahunan, Pusat Penelitian Oseanografi Lembaga Ilmu Pengetahuan Indonesia (LIPI)* Jakarta

[14] Liu Q M, Yang X M, Zhang L and Majetich G 2010 Optimization of ultrasonic-assisted extraction of chlorogenic acid from *Folium eucommiae* and evaluation of its antioxidant activity *J. Med. Plants Res.* **4** 2503–11

[15] Yang T, Zhang S, Wang R, Li D, Hu Y, Nie J, Zhao X, Wang Q, Chen Y, Zheng Y and Chen P 2016 Polysaccharides from *Rhizoma panacis Majoris* and its anti-oxidant activity *Int. J. Biol. Macromol.* **86** 756–763

[16] Heldman D R 2000 *Food Processing Engineering* (Westport Connecticut: The AVI Public Company)

[17] Astill C, Birch M R, Dacombe C, Humphrey P G and Martin P T 2001 Factors affecting the caffeine and polyphenol contents of black and green tea infusions *J. of Agric. and Food Chem.* **49** 5340–7

[18] Samaniego-Sánchez C, Inurreta-Salinas Y, Quesada Granados J J, Blanca-Herrera R, Villalón-Mir M, de la Serrana H L G and Martínez M C L 2011 The influence of domestic culinary processes on the trolox equivalent antioxidant capacity of green tea infusions *J. of Food Composition and Analysis* **24** 79–86

[19] Brown A 2014 *Understanding Food: Principles and Preparation* (Stamford: Nelson Education)

[20] Lee J E, Lee B J, Chung J O, Kim H N, Kim E H, Jung S, Lee H, Lee S J and Hong Y S 2015 Metabolomic unveiling of a diverse range of green tea (*Camellia sinensis*) metabolites dependent on geography *Food Chem.* **174** 452–9

[21] Badarinath A V, Malikarjuna K, Chetty C M S, Ramkanth S, Rajan T V S and Gnanaprakash K 2010 A reviews on in-vitro antioxidant methods: Comparisons, correlations, and considerations. *Int. J. of PharmTech. Research* **2** 1276–85

[22] Amponsah I K, Fleisher T C, Annan K, Dickson R A, Mensah A Y and Sarpong F M 2013 Anti-inflammatory, antioxidant and antimicrobial activity of stem bark extract and fractions of *Ficus exasperata Vahl. (Moraceae) J. of Pharmacognosy and Phytochemistry* **2** 38–44

[23] Meyer B N, Ferrigni N R, Putman J E, Jacbsen L B, Nicols D E and McLaughlin J L 1982 Brine shrimp: A convenient general bioassay for active plant constituents *J. Planta Medica* **45** 31-32

[24] Montgomery D C 2001 *Design and Analysis of Experiments, 5th Edition* (New York: John Wiley and Sons)

[25] Keshani S, Luqman Chuah A, Nourouzi M M, Russly A R and Jamilah B 2010 Optimization of concentration process on pomelo fruit juice using response surface methodology (RSM) *Int. Food Research J.* **17** 733–742

[26] Saniah K and Hasimah H A 2008 Development of *Morinda citrifolia* citrus-flavoured drink using response surface methodology (RSM) *J. Trop. Agric. and Fd. Sc.* **36** 87–97

[27] Zhao Q, Kennedy J F and Wang X 2011 Optimization of ultrasonic circulating extraction of polysaccharides from *Asparagus officinalis* using response surface methodology. *Int. J. of Biological Macromol.* **49** 181–7

[28] Garcia J L L and Castro M D L 2004 Ultrasound-assisted soxhlet extraction: An expeditive approach for solid sample treatment, application to the extraction of total fat from oleaginous seeds *J. of Chromatography* A **1034** 237–242

[29] Arasi G A, Rao G M and Bagyalakshmi J 2016 Optimization of microwave-assisted extraction of polysaccharide from *Psidium guajava L.* fruits *Int. J. Biol. Macromol.* **91** 227–232

[30] Vinatoru M, Toma M, Radu O, Filip P I, Lazurca D, Mason T J 1997 The use of ultrasound for extraction of bioactive principles from plant material *Ultrason Sonochem* **4** 135–9

[31] Zhang L and Wang M 2016 Optimization of deep eutectic solvent-based ultrasound-assisted extraction of polysaccharides from *Dioscorea opposita Thunb. Int. J. Biol. Macromol.* **95** 675–681

[32] Mason T J and Cintas P 2002 *Sonochemmistry Handbook of Green Chemistry and Technology* ed James Clark and Duncan Macquarrie (New Jersey : Blackwell Publisher)

[33] Zhu C P, Zhai X C, Li L Q, Wu X X and Li B 2015 Response surface optimization of ultrasound-assisted polysaccharides extraction from pomegranate peel *Food Chem.* **177** 139–146

[34] Jiang Y, Bai X, Lang S, Zhao Y, Liu C and Yui L 2019 Optimization of ultrasonic-microwave assisted alkali extraction of arabinoxylan from the corn bran using response surface methodology *Int. J. Biol. Macromol.* **128** 452–8

[35] Balachandran S, Kentish S E, Mawson R and Ashokkumar M 2006 Ultrasonic enhancement of the supercritical extraction from ginger *Ultrason Sonochem* **13** 471–9

[36] Das, Chittaranjan, Malgorzata, Henkel K and Schmeisser D 2016 Engineering of sub-nanometer six thickness in si photocathodes for optimized open circuit potential *ChemSusChem.* **9** 2332–6

[37] Wangtueai S and Noomhorm A 2009 Processing optimation and characterization of gelatin from Lizardfish (*Saurida spp*) scale. *LWT Food Science and Technology* **42** 825–834

[38] Verpoorte R and Alfermann A W 2000 Metabolic engineering of plant secondary metabolism (London : Kluwer Academic Publishers)

[39] Putram N M, Nursid M, Setyaningsih I and Tarman K 2017 Anticancer activity from active fraction of sea cucumber *JPHPI* **20** 53-62

[40] Reis S F, Coelho E, Coimbra M A, Ghannam N and Abu 2015 Improved efficiency of brewer's spent grain arabinoxylans by ultrasound-assisted extraction *Ultrason Sonochem* **24** 155–164

[41] Murniasih T, Putra M Y and Pangestuti R 2015 Antioxidant capacities of *Holothuria* sea cucumbers *Annales bogorienses* **19** 21-26

[42] Farouk A E, Faizal A H G and Ridzwan B H 2007 New bacterial species isolated from Malaysian sea cucumbers with optimized secreted antibacterial activity *American J. of Biochem. and Biotech.* **3** 60-65

Wound healing activity of skin incision and skin burn from spray gel formula contains combination of banana stem (*Musa acuminata* Colla) and *Aloe vera* ethanol extracts on mice

W Kundarto[1], K A Y Pangesti[1], F Shabrina[1], H Sasongko[1]
Department of Pharmacy, Faculty of Mathematics and Natural Sciences, Universitas Sebelas Maret. Jl. Ir. Sutami 36A Surakarta 57126, Central Java, Indonesia.

E-mail: wisnukundarto@staff.uns.ac.id

Abstract. Various studies reports the activity of wound healing from banana stem (*Musa species*) and also *Aloe vera* individually. This research purpose are to identify the wound healing activity of skin incision and skin burn animal model from spray gel formula contains combination of *Musa paradisiaca* Colla and *Aloe vera* L ethanol extracts on mice.This research were used randomized post test only control group design. Each of incision wound and burn wound models were used 28 mice and divided into 7 group. Group I was given with banana stem extract (BSE) 10%. Group II was given with *Aloe vera* extract (AVE) 10%. Group III was given with 7,5% BSE and 2,5% AVE. Group IV was given with 5% BSE and 5% AVE. Group V was given with 2,5% BSE and 7,5% AVE. Group VI and group VII (povidone iodine) as vehicle and positive control respectively. All group were sprayed twice a day from first day until seventh day after skin incision/burn induction, and length of wound /wound diameter were observed everyday. The result showed that combination of BSE and AVE had wound healing activity in both wound model which did not different significantly compared with individual extract.

1. Introduction

Wounds are a result of skin injuries that disrupt the other soft tissue. As a respon to injuries, wound healing is a complex process of tissue repair and remodeling [1]. Some research showed that *Musa paradisiaca* stem extract has wound healing activity in burn and incision wound [2,3]. *Aloe vera* also reported has wound healing activity in both wound [4,5]. Some research showed that combination of two or more extracts give a better wound healing activity [6,7]. This research purpose were to identify the wound healing activity of skin incision and skin burn from spray gel formula contains combination of *Musa paradisiaca* Colla stem and *Aloe vera* L ethanol extracts on mice.

2. Experimental

2.1. Material

Fresh *Musa paradisiaca* Colla stem and *Aloe vera* L were obtained from local regency (Sragen, Central Java Indonesia). All formula components were obtained from Pharmaceutical Laboratory Faculty of Mathematics and Natural Sciences Universitas Sebelas Maret Surakarta Central Java Indonesia. All mice were purchased from certified seller in Surakarta regency.

2.2. Preparation and extraction of Musa paradisiaca stem and Aloe vera.
Collected samples were determined in Laboratory Biology of Natural Sciences and Mathematic Faculty Universitas Sebelas Maret. Extraction were made by maseration methods [8].

2.3 Spray gel preparation.
Spray gel of *Musa paradisiaca* Colla and *Aloe vera* L ethanol extract and the blended extracts were prepared to the following formula (Table 1). Spray gels were prepared as previously described [9].

Table 1. Spray gel formula.

Component	Concentration (%)						
	Group I	**Group II**	**Group III**	**Group IV**	**Group V**	**Group VI**	**Group VII***
Banana stem extract	10	0	7,5	5	2,5	-	-
Aloe vera leaf extract	0	10	2,5	5	7,5	-	-
Carbopol 2%	10	10	10	10	10	10	-
Propylenglycol	2,7	2,7	2,7	2,7	2,7	2,7	-
Methyl Paraben	0,18	0,18	0,18	0,18	0,18	0,18	-
Propyl Paraben	0,02	0,02	0,02	0,02	0,02	0,02	-
Menthol	0,05	0,05	0,05	0,05	0,05	0,05	-
Oleum citri	4 drops	4 drops	4 drops	4 drops	4 drops	4 drops	-
Ethanol 70 %	ad 100	ad 100	ad 100	ad 100	ad 100	ad 100	-

*Group VII only contains povidone iodine as spray form

2.4 Experimental protocol.
A total of 56 male Swiss mices weighing between 20-25 g were purchased from local distributor in Surakarta, Central Java, Indonesia. The animals were adapted in caged with 12-h light/12-h dark cycles, controlled temperature, and free access to commercial pellet diet and water ad libitum for a week. All care was taken to minimize the suffering of the animals. The experimental protocol was approved by Health Research Ethics Committee Dr Moewardi General Hospital School of Medicine Universitas Sebelas Maret.

The animals were randomly divided into incision and burn wound groups with each group having 28 animals. Each group was further divided into seven subgroups composed of four animals per subgroup : 1) Banana stem extract (BSE) 10%-treated group , 2) *Aloe vera* extract (AVE 10%-treated group, 3) blended of BSE and AVE (7,5%:2,5%)-treated group, 4) blended of BSE and AVE (5%:5%)-treated group, 5) blended of BSE and AVE (2,5%:7,5%)-treated group, 6) vehicle group, and 7) Betadine ® (povidone iodine liquid) in spray form as positive group.

2.5 Wound healing activity from skin incision wound.
Animal model of skin incision wound was prepared as previously described [10] with modification. The back area of each animal was shaved approximately 2 cm and cleaned with ethanol 70%. The animals were anesthetized with inhalation of eter. Next, a 1,5 cm long and 0,2 cm depth, midline incision was made through the skin with sharp scalpel. After creating a wound, wound area was cleaned with saline water and sprayed twice a day according to each group for seven days. Healing of incision wounds was evaluated by measuring the length of wound and wound healing percentage for seven days.

Length of wound of each mice was measured three times using digital calipers. Wound healing was calculated using the formula : Wound healing = (length of wound at day 0 – length of wound at day n)/length of wound at day 0, and the result x 100% to get the wound healing percentage (%). Data of wound healing percentage from each group analyzed statistically. A p-value <0,05 was considered statistically significant.

2.6 Wound healing activity from skin incision wound.
Animal model of skin burn wound was prepared as previously described [10] with modification. The back area of each animal was shaved approximately 2 cm and was cleaned with ethanol 70%. The animals were anesthetized with inhalation of eter. Next, a partial thickness burn was made by putting a stainless steel hot plate with diameter 1 cm for 5 seconds on the shaved area. After creating a wound, wound area was cleaned with saline water and sprayed twice a day according to each group for seven days.

Healing of burn wounds was evaluated by measuring the diameter of wound and wound healing percentage for seven days. Diameter of wound of each mice was measured from four different way using digital calipers. Wound healing was calculated using the formula : Wound healing = (diameter of wound at day 0 – diameter of wound at day n)/diameter of wound at day 0, and the result x 100% to get the wound healing percentage (%). Data of wound healing percentage from each group were analyzed statistically. A p-value <0,05 was considered statistically significant.

3. Results and discussions

3.1. Skin incision wound.
Incision wound healing percentage of each group were showed in Table 2.

Table 2. Incision Wound Healing Percentage.

Day	Incision Wound Healing Percentage (%) (n=4)						
	Group* I	Group* II	Group* III	Group* IV	Group* V	Group VI	Group VII
1	6,978	11,206	7,261	10,685	10,818	0,559	6,695
2	14,266	23,553	14,567	21,265	22,924	2,058	14,602
3	24,113	35,651	24,151	32,262	33,887	5,368	24,073
4	33,553	47,043	34,026	42,425	44,954	9,550	33,147
5	43,206	60,013	45,040	53,944	58,555	13,867	41,684
6	51,871	73,065	55,742	64,649	69,996	18,206	50,152
7	61,369	85,889	65,974	75,438	81,395	22,143	58,572

* p<0,05 compared with negative control from day 1-7

3.2. Skin burn wound
Burn wound healing percentage of each group were showed in Table 3.

Table 3. Burn wound healing percentage .

Day	Burn Wound Healing Percentage (%)						
	Group* I	Group* II	Group* III	Group* IV	Group* V	Group VI	Group VII
1	8,93	13,42	9,23	9,49	8,60	2,61	4,42
2	16,06	23,52	18,49	20,21	20,36	7,18	7,99
3	20,37	29,15	20,69	25,57	26,42	10,58	14,75
4	24,72	34,88	25,00	30,42	31,00	13,39	19,52
5	29,60	40,22	29,92	34,50	34,98	15,30	23,99
6	33,07	44,47	33,97	38,15	38,99	16,84	28,00
7	36,61	50,21	37,20	41,40	43,15	18,88	31,58

* $p<0,05$ compared with negative control from day 1-7

Table 2 and Table 3 showed that both of *Musa paradisiaca* (banana) stem extract (BSE) and *Aloe vera* extract (AVE) in individual extract and also in blended extract has significant wound healing activity compared with vehicle group ($p<0,05$). Both of table also showed that wound healing activity from BSE and AVE formulas were not different significantly with individual extract. Nevertheless, Table 2 and Table 3 also showed that *Aloe vera* extract has higher wound healing percentage compared with *Musa paradisiaca* extract and blended extracts.

Based on literature review, *Aloe vera* chemical plant constituents composed of aloins A and B_2, aloe resin A, B_2, and C. The other important compounds include several sugars (glucose, mannose, cellulose) and various enzymes (oxidase, amylase, catalase), and also vitamins (B_1, B_2, B_6, C, E, folic acid) and minerals (calcium, sodium, magnesium, zinc, copper, and chrome) [11]. The healing property of *Aloe vera* is related to glucomannan. The glucomannan affects fibroblast growth factor and promotes the activity and proliferation of fibroblasts and in turn induces collagen production and secretion. The mucilage of *Aloe vera* increases amount of collagen and transversal connections among them in wound site. This action accelerates wound improvement [12]. The mucilago of *Aloe vera* also showed increasing production of fibroblastic growth factors (FGF) and transforming growth factor (TGF) β_1 in the wound area [13]. The mucilage also showed accelerates thrombosis, angiogenesis, and better arrangement of collagen on wound site [14]. *Aloe vera* mucilage also contain vitamins C and E as strong antioxidant and increasing collagen production [15]. The mucilage also possesses antioxidant enzymatic systems such as glutathione peroxidase and superoxide dismutase, which increases wound healing by neutralization the effect of free radicals produced on wound site and with their anti-inflammatory property [16].

Musa paradisiaca stem contain alkaloids, glycosides, saponins, tannins, flavonoids, and reducing sugars [17]. Tannins act as free radical scavenger. Flavonoids accelerates wound healing due to their astringent and antimicrobial activity. Flavonoids also possess potent antioxidant and free radical scavenging effect, and increasing the level of antioxidant enxymes in granuloma tissue. Saponins promote wound healing due to their antioxidant and antimicrobial activity [18].

Wounds usually generates superoxides and lipid peroxidation through neutrofil activation. Therefore any components that inhibits lipid peroxidation is believed to accelerate the viability of collagen fibrils by increasing the strength of collagen fibres, increasing the circulation, preventing the cell damage, and promoting DNA synthesis. Better collagenation seen in wound area under the influence of some herbal extracts may be due to improvement of antioxidant status [19]. Antioxidants can interfere with the oxidation process by reacting with free radicals, chelating catalytic metals, and by acting as oxygen scavenger. Free radicals and other reactive oxygen species (ROSs) are considered to be important causative factors in delaying the healing process [20].

Higher of ROSs concentration can promote severe tissue damage and impede the healing process by causing damage to cellular membranes, DNA, proteins, and lipids. Hence if an herbal extract having antioxidant and antimicrobial property, it could be a good therapeutic agent for increasing the wound healing process [19]. Oxidative stress also plays an important role in wound healing process. This oxidative stress may cause damage to the growing tissue at the repair site[21]. Hence the antioxidants present in the herbal extract could be expected to promote epithelization by controlling oxidative stress. Botanicals with antioxidant or free radical scavenging activity can play significant role in healing of wounds [22].

4. Conclusion

Both of spray gel contain banana stem extract (BSE), *Aloe vera* extract (AVE), and combination of blended extracts had wound healing activity in incision and burn wound model which did not different significantly compared with individual extract. This research has several limitations including duration of observation which only seven days, and povidon iodine which is used as positive control in both wounds given in spray liquid form. We also do not observed the skin tissue histology and phytochemical screening. Therefore further research need to explore this results.

Acknowledgment
This work was done with the help of Kartika Adi Yuni Pangesti and Fildza Shabrina.

References

[1] Thakur R, Jain N, Pathak R and Sandhu S 2011 *Evid. Based Complement. and Alternat. Med.* **2011** 1-17.
[2] Amutha K and Selvakumari U 2016 *Int. Wound J,* **13** 763-67.
[3] Weremfo A, Pappoe A N M and Adinortey M B 2011 *RJPPD* **6** 294-96.
[4] Maenthaisong R, Chaiyakunapruk N, Niruntraporn S and Kongkaew C 2006 *Burns* **33** 713-18.
[5}Yadav K H, Kumar J R and Basha I 2012 *Int. J. Pharma. Bio. Sci.* **3** 63-2.
[6] Aslam M S, Ahmad M S, Mamat A S, Ahmad M Z and Salam F 2016 *Evid. Based Complement. and Alternat. Med* **2016** 1-14.
[7] Elzayat E M, Auda S H, Alanazi F K and Al-Agamy M H 2018 *Saudi. Pharm. J.* **26** 733-38.
[8] Yusuf S, Agunu A and Diana M 2004 *J. Ethnopharmacol.* **93** 33-7.
[9] Iswandana R and Lidya K M S 2017 *PSR* **4** 121-31.
[10] Somboonwong J, Kankaisre M, Tantisira B and Tantisira M H 2012 *BMC Complement. Altern, Med.* **12** 1-7.
[11]Surjushe A, Vasani R and Saple D 2008 *Indian J. Dermatol.* **53** 163-66.
[12] Boudreau M D and Beland F A 2006 *J. Environ Sci Health* **24** 103-54.
[13] Atiba A, Ueno H and Uzuka Y 2011 *J. Vet.Med. Sci.* **73** 583-89.
[14] Oryan A, Naeini A T, Nikahval B and Gorjlan E 2010 *Vet Arh* **80** 509-22.
[15] Kashanian M, Lakeh M M, Ghasemi A and Noori S 2013 *J.Reprod. Med* **58** 34-8.
[16] Hajhashemi V, Ghannadi A and Heidari A H 2012 *Res.Pharm. Sci.* **7** 73-8.
[17] Onyenekwe P C, Okereke O E and Owolewa S O 2013 *Curr. Res. J.Biol. Sci.* **5** 26-9.
[18] Shenoy C, Patil M B, Kumar R and Patil S 2009 *Int. J. Pharm. Pharm. Sci.* **2** 167-75.
[19] Akkol E K, Koca U, Pesin I and Yilmazer D 2011 *Evid. Based Complement. and Alternat. Med.* **2011** 1-7.
[20] Farokhzad O C 2008 *Expert. Opin. Drug. Del* .**5** 927-29.
[21] Wannarat K, Tantisira M H and Tantisira B 2009 *Thai J Pharmacol* **31** 120-3.
[22] Kamath J V, Rana C and Chowdhury A R 2003 *Phytother. Res.* **17** 970-72.

IOP Conf. Series: Materials Science and Engineering **858** (2020) 012028 doi:10.1088/1757-899X/858/1/012028

The effect of heat treatment on antioxidant activity of ethyl acetate extracts Arbila beans from Kapan, East Nusa Tenggara

H Funome[1,3], A Pangastuti[2], R Setyaningsih[2]
[1]Department of Bioscience, Graduate Program, Universitas Sebelas Maret. Jl. Ir. Sutami No. 36A, Kentingan, Surakarta, 57126, Central Java, Indonesia
[2]Department of Biology, Faculty of Mathematics and Natural Sciences, Universitas Sebelas Maret, Jln. Ir. Sutami 36A, Surakarta 57126, Certral Java, Indonesia

Email: hisreidifunome@gmail.com

Abstract. Oxidative stress is a condition of imbalance between free radicals and antioxidants in the body. This condition is dangerous because free radicals easily react with biomolecules such as DNA and protein, causing damage. This oxidative stress can be prevented by intake of antioxidants in food or supplements. Arbila beans (*Phaseolus lunatus* L.) is a bean species that grows in East Timor and had been utilized by the local community from a long time ago. This research was aimed to analyze the antioxidant activity of Arbila bean ethyl acetate extract. The results of the research showed that antioxidants activity, which was expressed in IC_{50} value, was detected on the ethyl acetate extract of Arbila bean which is not cooked or cooked. The IC_{50} value of dry extract (LKE) and cooked extract (LRE) were 197.61 µg/mL and 207.93 µg/mL. Based on qualitative with HPLC, dry extract has 10 compounds and after given heat treatment is reduced to 6 compunds.

1. Introduction

Antioxidant is a chemical that is used to prevent oxidation by protecting cell components by neutralizing the damage effects of free radical [1]. Free radicals greatly affect regulation in the body, especially in cell, which causes imbalance in the control of signal transduction and gene expression [2]. One alternative of the problems caused by free radicals is natural antioxidant sourced from nature such as plants. Some plants are also known to have activity as safe antioxidants [3].

Arbila bean (*Phaseolus lunatus* Linn), found in Timor island, is a bean which grows wild and has a carbohydrate content of 201.70 mg/g and protein 68.40 mg/g [30] so as local people make it as an alternative food resource. Several studies have shown that Arbila beans have benefits as antidiabetic [4], antifungal and anti-proliferative [5], antimicrobial [6], cysteine proteinase inhibitors [7], hypercholesterolemia [8], antioxidants [9] and trypsin and chymotripsin inhibitors [10].

DPPH method is an easy antioxidant test that is used to analyze quantitative antioxidant activity. The mechanism of action of DPPH is that DPPH free radicals reacted with the appropriate reduction of the agent and produced new bonds, so that it can caused change the color of the solution. The color of the solution changes to indicate the electron has been taken by DPPH free radicals [11].

[1] Department of Bioscience, Graduate Program, Universitas Sebelas Maret. Jl. Ir. Sutami No. 36A, Kentingan, Surakarta, 57126, Central Java, Indonesia

Several studies have found that legume (Phaseolus) plants are known to have antioxidant benefits, such as *Phaseolus vulgaris* plants [12-17] and *Phaseolus radiatus* L.[18]. The purpose of this study is to evaluate the potential of ethyl acetate extract as a natural antioxidant.

Figure 1. *Phaseolus lunatus* Linn

2. Experimental

2.1. Meterials
Dry beans (*Phaseolus lunatus* L.) grow in Kapan, East Timor dan harvested in August 2018.

2.2. Sampel preparation
300 g Arbila beans cooked in 1.5 liter of distilled water (in 100°C, 3 hours) and dried in oven with temperature 55°C during 12 hours (cooked treatment). 300 g Arbila beans was dried in oven during 48 hours with temperature 40°C (dry treatment). Every treatment dry treatment and cooked treatment were powdered using blender and keep in box.

2.3. Extractions
Arbila beans extractions dry extract (LKE) and cooked extract (LRE) was using ethyl ecetate as eluent. LKE and LRE (100g) were macerated in ethyl acetate (1:5 w/v) separately during 48 hours by mixing three times in room temperature. The filtrate was filtered using filter paper and evaporated with rotary evaporator at temperature 40°C until thick. Extract LKE and LRE first test DPPH by comparison the color in crude extract that has been reacted with solution of DPPH. The indicator color of it was purple solid showed the absence of the reaction on scavenging of free radicals by crude extract. Indicator color yellow whitish showed indicates the presense of the reaction of scavenging of free radicals.

2.4. Determination of phenolic and flavonoid
Determination of total phenolic and total flavonoid content refers to Li [19] and Kiyani-Sam [20]. Determination of total phenolic made by taking 0.5 mL of extract LKE and LRE then dissolved in 7.5 mL of distilled water and then added with 0.5 mL of Folin-Ciocalteu reagent 10% (v/v). After 1 minute added 1.5 mL of sodium carbonate 20% (w/v). The solution was shaken and incubated in a dark room (25°C during 2 hours). The absorbance of the sampel was measured on a spectrophotometer UV-Vis Double Beam Shimadzu 1601 with a wavelength of 750 nm. Gallic acid (1-19 μg/mL) was used as a standard. The result was expressed together with Gallic acid (mgGAE/g dry weight). Determination of total flavonoids was determined based on the formation of flavonoid-aluminum. 1 mL samples of LKE and LRE are mixed with 3 mL of methanol and 0.2 mL of a solution of aluminum chloride 10% (w/v). Then added with 0.2 mL of 1M potassium acetate and 5.6 mL of distilled water then incubated for 30 min at room temperature. The sampels was measured the length absorbance at 415 nm using UV spectrophotometer. The standards used for the calibration curve is Rutin (2.5-20 μg/mL). Total flavonoid was the same with the Rutin (RE mg/g dry weight).

2.5. DPPH antioxidant test

Antioxidant test using DPPH as method to see radical scavenging ability from sample [21-22]. 200 mL of sample LKE and LRE were mixed to 3 mL of DPPH 0.1 mM and incubated for 30 minutes at room temperature in the dark. The samples were measured absorbance at 517 nm. All treatment were repeated three times. Radical scavenging activity was calculated by using formula = % Inhibition = (Abs DPPH-Abs sampel)/(Abs DPPH) x 100%. Radical scavenging activity by sample using DPPH reagent was indicated by change of the color from dark purple to yellow white which is measured the absorbance and compared with control that was methanol.

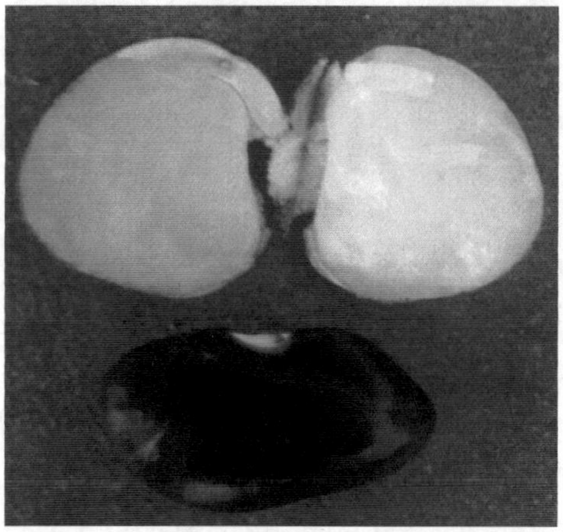

Figure 2. Seed core of *Phaseolus lunatus* Linn

3. Result and Discussion

3.1. Total Phenolic and total flavonoid

Total phenolic and flavonoid

Arbila bean seeds used are whole seeds consisting of seed coat and seed core. The content of phytochemical compounds is more contained in the seed coat. In general, beans with a seed coat that has a bright color show lower phenol levels compared to black beans [23]. In this study, Arbila beans have a deep black seed coat. The total phenolic content that is contained in the ethyl acetate extract with raw (LKE) and cooked (LRE) treatments ranged from 118.46 mg GAE/gr to 169.44 mg GAE/gr with gallic acid as standard (Table 1). The results of this study are the same as the results of research conducted by Silva [24] The content of phenolic compounds after cooking is reduced by 50.98 mg GAE/gr. The highest concentration is in the LKE sample. Although several studies such as Huber [25] prove that treatment cooked with maceration increases antioxidant activity with high concentrations of kaemferol and catechin. Anna [26] also proved that there was an increase in total phenolic after heat treatment. The results of research conducted by [15] stated that total phenolic experienced a significant increase when given heat treatment and given additional $NaHCO_3$. This contrasts with Granito [27] saying that processing legumes such as cooking might increase the degradation between the aromatic rings of phenolic compounds which can be structurally damaging and cause a phenolic decrease than the cooked beans. Huber [25] also added that providing cooked treatment can cause the loss of certain bioactive molecules. The results of research conducted by Huber [25] also prove that flavonoid levels are higher when cooked, but the results of this study decreased total flavonoids. Some studies show a decrease in total phenolic if given heat treatment [29-30]. Processing a food by providing heat can provide significant changes to the chemical compounds contained either increasing or decreasing in composition [29]. When cooking for longer can cause the diffusion of phenols from the seed coat to the cooking water even though it can provide activities similar to extracts without thermal treatment even though the phenol content is 80% lower [30]. This also happened in this study, namely the decrease in total phenolic when given heat treatment (cooked). Flavonoids are a group of compounds that give fruit and vegetables flavor and color. The flavonoid content in Arbila beans after boiling is reduced, because when given heat (cooked) some structures become degradation and have decomposition of thermolabile and are left behind in the solvent used to boil. Flavonoid compounds are included in one of the most compound groups in the phenol compound group. Warming can cause cell wall adhesion so that it can induce the release of phenolic compounds such as flavonoids that accumulate in the vacoule [15]. Flavonoids are mostly located in the seed coat, non-flavonoid phenolic compounds such as hydroxybenzoate and hidrosikinamat located in cotyledons. Hydrosisinamicic derivatives form the main phenolic component of peanuts. The longer the heating time with higher temperatures will cause the antioxidant compounds to weaken. Warming at a temperature of 100°C with a time of 3 hours can be a cause of a decrease in total flavonoids contained in LKE samples.

Table 1. Total phenolic content, total flavonoid content and antioxidant activity (IC_{50} value)

	Total phenolic (mg GAE/gr)	Total flavonoid (mg RE/gr)	DPPH Antioxidant activity (IC_{50} value)
LKE	169.44 ± 5.43	33.27 ± 3.10	197.61 µg/mL
LRE	118.46 ± 5.43	29.70 ± 3.10	207.93 µg/mL

Each value represents the mean value ± SD; GAE = Gallic Acid Equivalent; RE = Routine equivalent. LKE was dry extract; LRE was cooked extract

3.2. The DPPH antioxidant activity

The DPPH method is an antioxidant method that is often used. LKE and LRE samples showed antioxidant activity with the appearance of an antidote to DPPH free radicals. Methanol was blank to measured absorbance and abs DPPH that was measured is 0.945. The value of free radical inhibitory concentration was calculated by the IC_{50} value. Antioxidants detected by the DPPH method is classified according to Molyneaux [31] that LKE is classified as a weak antioxidant, and LRE is classified as a very weak antioxidant. Molyneaux classifies antioxidants in the level of strong and weak inhibit free radicals that are <50µg/mL (very strong), 50-100µg/mL (strong), 100-150µg/mL (moderate),150-200µg/mL (weak) and >200µg/mL (very weak). Some studies also show that heat processing can increase antioxidant capacity [34], and at temperatures above 120°C the antioxidant activity decreases significantly [30]. The appeal of BHT is used as a synthetic antioxidant. IC_{50} of BHT value is 166.99 µg/mL which is included in the group of weak antioxidants. On heating the sample at a temperature of 70°C-90°C causes the degradation of some stronger antioxidant compound that the difference in antioxidant activity is higher than without heat processing. Thermal treatment can affect permanent changes in nutrition depending on the level of the process and is able to change the total antioxidant capacity significantly higher than the total phenol raw seeds, DPPH free radical antioxidant activity [32]. The relationship between total phenolic and total flavonoids is directly proportional to the antioxidant activity of IC_{50} value. From this research, there is antioxidant bioactivity even though the range is weak or even very weak. This is due to the selection of the sample preparation that will be used should only be part of the seed coat only, not with the seed core. Akond [12] said that arbila beans had a carbohydrate content of 201.70 mg/g and protein 68.40 mg/g, so it could be concluded that the antioxidant activity was low because almost 70% of the extracted seeds were part of the seed core. Dewanto [32] also explained that most of the phenols were in the skin of the seeds and only a few were located in the core of the seeds. This is also proven by research conducted by Anton [33] that *Phaseolus vulgaris* L. beans with a brighter skin color that (red) has high antioxidant activity and also a high total phenolic as well.

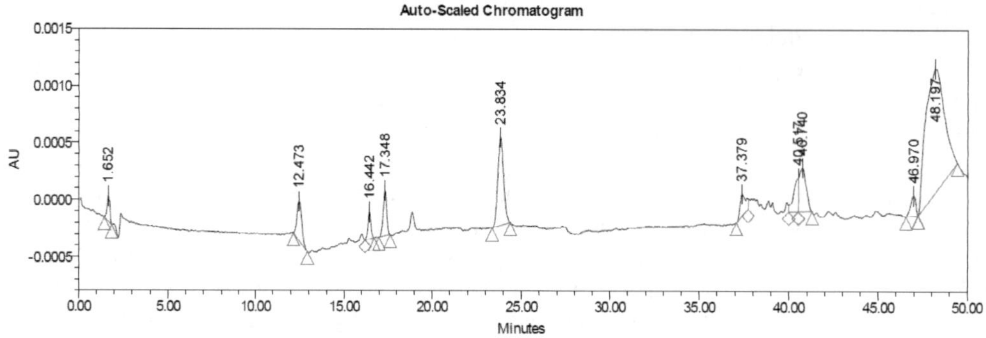

Figure 3. HPLC chromatogram of dry extract (LKE)

In the qualitative determination results by identifying the compounds present in the LKE extract and LRE were done by HPLC (High Performance Liquid Chromatography), in the LKE results there were 10 compounds identified, but after boiling (LRE) there were only 6 compounds indicated by the emergence of peaks at the time only certain retention. So it can be concluded that when given heat treatment is able to eliminate some of the compounds that previously existed (without treatment) or the missing compounds was thermolable compounds.

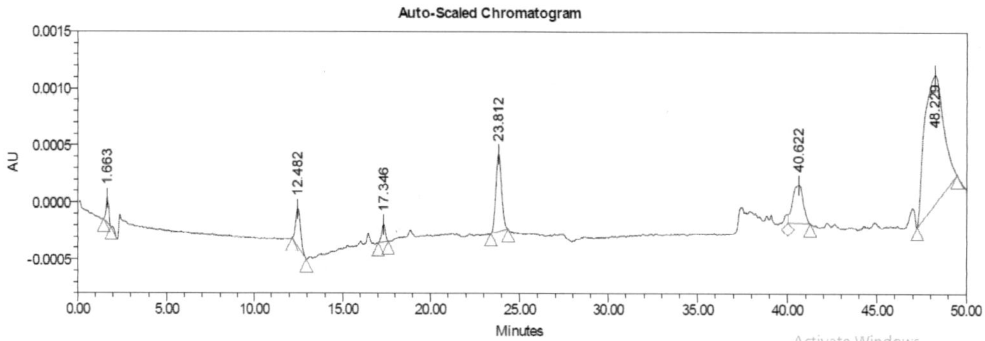

Figure 4. HPLC chromatogram of cooked extract (LRE)

4. Conclusion

This study concludes that the antioxidant activity of ethyl acetate extract especially LK (dry extract) without given heat treatment has the potential as a natural antioxidant even though with a group of weak antioxidant activity.

References

[1] Badarinath, A. V., RAo, M. K., Chetty, C.M.D., Ramkanth, S., Rajan, T.V.S., Gnanaprakash, K., 2010. A review on in-vitro antioxidant methods: comparisions, correlations and considerations. *International Journal of PharmTech Research*. CODEN (USA): IJPRIF ISSN : 0974-4304. Vol **2**(2):1276-1285

[2] Knight, Joseph A., 2000. Review: Free radicals, antioxidants, and the immune system. *Annals of Clinical & Laboratory Science*, vol. **30**(2)

[3] Krishnaiah, D., Sarbatly, R., Nithyanandam, R., 2011. A review of the antioxidant potential of medicinal plant species. *Food and Bioproducts Processing.* **8**(9):217–233

[4] Saleem, Z. M., Ahmed, S., & Hasan, M. M. 2016. *Phaseolus Lunatus* Linn : Botany , Medicinal Uses, Phytochemistry and Pharmacology., *World Journal Of Pharmacy And Pharmaceutical Sciences* **5**(11):87-93

[5] Ojo, R. J., Segilola, L. I., Ogundele, O. M., Akintayo, C. O., & Seriki, S. 2013. Biochemical evaluation of lima beans (*Phaseolus lunatus*) in alloxan induced diabetic Rats. *ARPN Journal of Agricultural and Biological Science,* **8**(4), 302–309.

[6] Wang, S., Rao, P., & Ye, X. 2009. Isolation and biochemical characterization of a novel leguminous defense peptide with antifungal and antiproliferative potency. *Applied Microbiology and Biotechnology*, **82**(1), 79–86.

[7] Wong, J. H., & Ng, T. B. 2005. Lunatusin, a trypsin-stable antimicrobial peptide from lima beans (*Phaseolus lunatus* L.). *Peptides,* **26**(11), 2086–2092.

[8] Lawrence, J. C., & Nielsen, S. S. 2001. Partial isolation and characterization of a cysteine proteinase inhibitor from Lima bean (*Phaseolus lunatus*). *Journal of Agricultural and Food Chemistry*, **49**(2), 1020–1025

[9] Zhao, Y., Du, S. K., Wang, H., & Cai, M. 2014. In vitro antioxidant activity of extracts from common legumes. *Food Chemistry*, **152**:462–466.

[10] Haynes, R., & Feeney, R. E. 1967. Fractionation and properties of trypsin and chymotrypsin inhibitors from lima beans. *Journal of Biological Chemistry*, **242**(22), 5378–5385

[11] Marathe, S.A., Rajalakshmi, V., Jamdar, S.N., Sharma, A., 2011. Comparative study on antioxidant activity of different varieties of commonly consumed legumes in India. *Food and Chemical Toxicology* **40**:2005-2012

[12] Akond, A.S.M. Golam, M., Khandaker, L., Berthold, J., Gates, L., Peters, K., Delong, H., Hossain, K., 2011. Anthocyanin, total polyphenols and antioxidant activity of common bean. *American Journal of Food Technology* Vol **6**(5): 385-394

[13] Cardador-Martiänez, A., Loarca-Pina, G., Dave Oomah, B., 2002. Antioxidant activity in common beans (Phaseolus vulgaris L.). *Journal of Agricultural and Food Chemistry*. **50**:6975-6980

[14] Anggraini, D.R., Widyawati,T., Sari, I.M., Wahyuni, S.H., Syarifah, S., Fratamasinaga, A.P., Antika, M.A., 2018. antioxidant assay of *Phaseolus vulgaris* L. beans ethanolic extract in vitro and in vivo. *Asian Journal of Pharmaceutical and Clinical Research*. Vol **11**, Special issue 1

[15] Akilliogle H.G. & Karakaya S., 2010. Changes in total phenols, total flavonoids, and antioxidant activities of common beans and pinto beans after soaking, cooked, and in vitro digestion process. *Food Science Biotechnology* Vol **19**(3): 633-639

[16] Guzmán-Uriarte,M.L., Sánchez-Magaña, L.M., Angulo-Meza, G. Y., Cuevas-Rodríguez, E.O., Gutiérrez-Dorado, R., Mora-Rochín, S., Milán-Carrillo, J., Valdez-Ortiz, A., Reyes-Moreno, C., 2013. Solid state bioconversion for producing common bean (*Phaseolus vulgaris* L.) functional flour with high antioxidant activity and antihypertensive potential. *Food and Nutrition Sciences,* **4**:480-490.

[17] Guajardo-Flores, D., Sergio O. Serna-Saldívar, Janet A. Gutiérrez-Uribe. 2013. Evaluation of the antioxidant and antiproliferative activities of extracted saponins and flavonols from germinated black beans (*Phaseolus vulgaris* L.). *Food Chemistry*,**141**: 1497–1503.

[18] Li, He., Cao, Dongdong., Yi, Jianyong., Cao, Jiankang., Jiang, Weibo., 2012. Identification of the flavonoids in mungbean (*Phaseolus radiatus* L.) soup and their antioxidant activities. *Food Chemistry* **135**:2942–2946

[19] Li, Xiaoli., Dong, M., Chen, X., Jiang, M., Lv, X., Yan, G., 2007. Antioxidant activity and phenolics of an endophytic Xylaria sp. from *Ginkgo biloba. Food Chemistry* **105**:548–554.

[20] Kiyani-Sam, M., Amjad, L., Ranjbar,M., 2015. Comparative study of antioxidant activity, phenol and flavonoid content in *Phaseolus vulgaris* and *Vigna sinensis* in different stages of seed germination. *Internastional Journal Of Biosciences* **6**(5):292-297

[21] Evangelho, J. A. do., Berrios, J. de J., Pinto, V. Z., Antunes, M.D., Vanier, N. L., Zavareze, E, da R., Antioxidant activity of black bean (*Phaseolus vulgaris* L.) protein hydrolysates. *Food Science and Technology*. **36**(Suppl.1): 23-27

[22] Dzomba, P., Togarepi, E., Mupa, M., 2013. Anthocyanin content and antioxidant activities of common bean species (*Phaseolus vulgaris* L.) grown in Mashonaland Central, Zimbabwe. African *Journal of Agricultural Research*. Vol **8**(25):3330-3333

[23] Barampama, L., & Simard, P. 1993. Oligosaccharides, antinutritional factors, and protein digestibility of dry beans as affected by processing. Journal of Food Science, **59**(5), 833–837.

[24] Silva, M.O., Brigide, P., Viva de Taledo, N.M., Canniatti-Brazaca, S.G., 2018. Phenolic compounds and antioxidant activity of two bean cultivars (*Phaseolus vulgaris* L.) submitted to cooked. *Brazilian Journal of Food Technology*, **21**

[25] Huber, K., Brigide, P., Bretas, E.B., Canniatti-Brazaca, S.G, 2014. Effect of thermal processing and maceration on the antioxidant activity of white beans. *Plos One* **9**(7)

[26] Anna, A,K., 2016. Kinetic changes of antioxidant capacity and physical quality of tempe during heating. *Thesis*. Bogor Agricultural University.

[27] Granito, M.; Brito, Y. Torres, A. 2007. Chemical composition, antioxidant capacity and functionality of raw and processed (*Phaseolus lunatus*). *Journal of the Science of Food and Agriculture*, **87**(15):2801-2809

[28] Chen, P.X., Dupuis, J.H., Marcone, M.F., Pauls, P.K., Liu, R., Liu, Q., Tang, Y., Zhang, B., Tsao, R. 2015. Physicochemical properties and in vitro digestibility of cooked regular- and non-darkening cranberry beans (*Phaseolus vulgaris* L.) and their effects on bioaccessibility, phenolic composition and antioxidant activity. *Journal of Agriculture and Food Chemistry*, **63**(48): 10448-10458

[29] Xu B.J., Chang S.K.C.. 2009. Isoflavones, Flavan-3-ols, Phenolic acids, total phenolic profiles, and antioxidant capacities of soy milk as affected by ultrahigh-temperature and traditional processing methods. *Journal of Agriculture and Food Chemistry* **57**:4706–4717

[30] Rocha-Guzman, N. E., Gonzales-Laredo, R.F., Ibarra-Perez, F.J., Nava-Berumen, C.A., Gallegos-Infante, J., 2007. Effect of pressure cooked on antioxidant activity of extracts from three common bean *(Phaseolus vulgaris* L.) cultivars. *Food Chemistry*, **100**:31-35

[31] Molyneux P., 2004. The use of the stable free radical diphenylpicrylhydrazyl (DPPH) for estimating antioxidant activity. *Songklanakarin Journal of Science and Technology* **26**:211-9.

[32] Dewanto V, Wu X, Adom KK, Liu RH. 2002. Thermal Processing Enhances the Nutritional Value of Tomatoes by Increasing Total Antioxidant Activity. *Journal of Agriculture and Food Chemistry*. **50**:3010-3014

[33] Anton, A.A., Fulcher, R.G., Amtfiled, S.D., 2009. Physical and nutritional impact of fortification of corn starch-based extruded sancks with common bean (*Phaseolus vulgaris* L.) flour: effect of bean addition and extrusion cooked. *Food Chemistry*. **113**:989-996

[34] Halvorsen B.L., Carlsen M.H., Phillips K.M., Bohn S.K., Holte K., Jacobs D.R., Blomhoff R. 2006. Content of redox-active compounds (ie, antioxidants) in foods consumed in the United States. *The American Journal of Clinical Nutrition*. **84**: 95–135.

[35] Obohl, H.A., Omofoma, C.O., 2008. The effect of heat treated lima beans (*Phaseolus Lunatus)* on plasma lipids in hypercholesterolemic rats. *Pakistan Journal Of Nutrition*. **7**(5): 636-639

Comparison of blasting cream anadara granosa with mytilus viridis in white rat skin(rattus novergicus)

Ariyanti Ariyanti[1] , Eni Masruriati[1] Sulistyani Mardaning Tyas[1], Khilya Aulia Nur Khasanah[1]
[1]Program Studi Sarjana Farmasi STIKES Kendal
riri99.cettaazzahra@gmail.com

Abstract. Collagen is the result of determination from chitosan blood shells (*Anadara granosa*) and green shells (*Mytilus viridis*). Previous research, extraction was carried out to obtain collagen. The yield of collagen produced is 2.03% with a concentration of acetic acid 0.75 M. Collagen can penetrate the main components of the dermis skin layer (the bottom of the epidermis) made by fibroblast cells. Based on the results of previous studies, the collagen produced will be made into a dosage form so that it can be better utilized. Based on the description above, researchers are interested in comparing blood clam shells and green mussel shells with a study entitled the comparison of physical test blasting creams of Anadara granosa with green shells Mytilus viridis against the skin of male white rats. Laboratory experimental research using a post test control design research design in the form of post only design which is divided into 5 treatment groups with each group consisting of 5 male white rats. The negative control group by giving a base cream and the test group by giving blasting cream Anadara granosan with Mytilus viridis concentrations of 10%, 15%, 20% and 25%. The parameter observed was skin moisture level. The results of statistical analysis using the ANOVA test.The results of this study are blasting cream Anadara granosa and Mytilus viridis can provide a moisturizing effect on rat skin. The most optimal concentration as a moisturizing effect on rat skin is a concentration of 25%

1. Introduction

Utilization of marine products in Kendal has been maximized, but the utilization of waste has not been maximized for daily income needs. Coastal communities only consume clam meat and do not utilize waste from shellfish. Shells from shells mostly only become waste and there is no utilization or as a form of craft [1]. In fact, meat and shellfish can be processed into beauty products and sources of vitamins for the body. Blood clam meat contains nutrients and minerals such as vitamin B1, iron, zinc, selenium as well as a source of collagen The content in the clam shell in the form of Ca + C 98.7%, Mg 0.0476%, Na 0.9192%, P 0.0183%, K 0.0398% and more than 71% of collagen. Collagen can be used in industrial fields such as facial, hair, body care products such as shampoo, conditioner, soap and other cosmetics. Collagen as a body treatment, especially on the skin is useful as a thickener and makes the skin look firmer, the process to get collagen must be through extraction [2]. Previous research, extraction from kitin was carried out to obtain collagen. The yield of collagen produced is 2.03% with a concentration of acetic acid 0.75 M. The immersion of collagen is influenced by the concentration of acetic acid. The higher the concentration of acetic acid, the greater the yield of collagen [3]. Collagen can penetrate the main components of the dermis skin layer (the bottom of the epidermis) made by fibroblast cells. Based on the results of previous studies, the collagen produced will be made into a dosage form so that it can be better utilized.

IOP Conf. Series: Materials Science and Engineering **858** (2020) 012029 doi:10.1088/1757-899X/858/1/012029

Based on the description above, researchers are interested in comparing blood clam shells and green mussel shells with a study entitled the comparison of physical test blasting cream of blood clam shells *(Anadara granosa)* with green shells *(Mytilus viridis)* against the skin of male white rats *(Rattus novergicus)*.

2. Experimental

2.1 Materials

Removal of non-collagen protein. The clam shell is cleaned. The sample was then weighed in a ratio of 1:20 (w / v) and then immersed in 0.1 M NaOH solution for 24 hours. Samples are filtered using filter paper, then washed with distilled water until the sample pH approaches or reaches pH 7.

2.2 Method

Collagen extraction was done by immersion in acetic acid according to modified Muyonga et al. (2004). The sample was weighed twice, with a ratio of sample weight and volume of solution 1:10 (w / v), then given the treatment code A. Treatment A was macerated in 0.75 M acetic acid macerated for 3 days. The result of extraction of treatment A was filtered and then salted out by adding NaCl 0.9 M. Then it was centrifuged for 10 minutes at a speed of 10,000 rpm to precipitate the wet collagen residue fibers. The separation results are then filtered and roasted at 60°C for one day to get dry collagen. The dried collagen is then weighed to calculate the amount of yield obtaine.

2.3 Yield measure collagen

The dried collagen obtained is prepared, then weighed and the results recorded. The yield calculation can be seen in equation (1).

$$\text{Yield } \% = \frac{\text{dry collagen weight}}{\text{powder weight}}\ x\ 100\% \qquad (1)$$

2.4 Making cream preparations

The formulation of blood clam shell collagen cream preparations made four preparations, each 30 g with a variety of the amount of collagen (active substance). Literature from the Indonesian Formulary [4] with the following formulation in table 1 and 2.

Table 1. 150 gram Cream Base Material Formulation

Material	Formula	Amount	Efficacy
Asam stearat	15%	$\frac{15}{100}x150 = 22{,}5$	Emulsifying
Vaselin album	25%	$\frac{25}{100}x150 = 37{,}5$	Base
Cera alba	15%	$\frac{15}{100}x150 = 22{,}5$	Emulsifying stabilizer
Trietanolamin	3%	$\frac{3}{100}x150 = 4{,}5$	Emulsifying agent
Propilenglikol	15%	$\frac{15}{100}x150 = 22{,}5$	Humectant
Nipagin	0,2%	$\frac{0{,}2}{100}x150 = 0{,}3$	Preservative
Aquadest	ad 150	150-(22,5+37,5 +22,5+4,5+22,5+0,3) = 40,2mL	Solvent

Table 2 Collagen taken per 30 gram formulation

Material	Formula	Amount	Base needed
Collagen	10%	$\frac{10}{100} x30 = 3$ g	27g
Collagen	15%	$\frac{15}{100} x30 = 4.5$ g	25.5 g
Collagen	20%	$\frac{20}{100} x30 = 6$ g	24g
Collagen	25%	$\frac{25}{100} x30 = 7,5$ g	22.5 g

2.5 Evaluate cream preparations

Cream evaluation is carried out at the beginning of manufacture and after one month of storage, cream preparations observed include:

1) Organoleptic
   The preparations that have been made are made observations on the appearance of the preparations including odor, color and consistency of the preparations. Organoleptic testing of color and consistency was repeated for each formula five times each.
2) Homogeneity
   Homogeneity test is done by means of a cream weighed ± 0.1 g then placed on a glass object and covered with other glass objects, then the cream preparation is given a load of 1 kg for 5 minutes, after which it is observed with a magnifying glass. Homogeneity testing was repeated for each formulation five times each.
3) Scattering power
   The scatter power test is done by placing ± 0.5 g of cream on a scale glass plate and on top of it covered with the same glass, then given a weight of 50 g and 100 g for 2 minutes, then measured the size of the spread diameter formed (cm). The spread test was repeated on each formula five times each.
4) Cream type testing
   Cream type examination is done by giving one drop of methylene blue solution to 0.1 g of cream, then the spread of methylene blue in preparations under a microscope is observed. If the color is spread evenly on the cream preparation, it means that the type of cream is oil in water (M / A), but if the color is only in spots it means the type of cream is water in oil (A / M).
5) Testing the pH of the cream
   The examination is carried out using a pH meter. The pH check is done by dipping the electrodes into 1 g of the cream preparation diluted with distilled water to 10 mL

2.6 Preparation of mouse skin

The skin of male white rats used are male rats aged approximately 2 months, weighing 200-400 g.Rat are shaved on the back using a shaver. Rat that have been shaved, measured diameter of 2.8 cm. Shaved skin is oriented first. Prepare mice that have been shaved then smeared cream base as a negative control orientation. Collagen cream is applied to the skin of rats that have been shaved as much as 50 mg in each sample until the skin is covered with a layer of collagen cream. After that it is observed between the skin with a cream base and the skin that has been smeared with collagen cream, in the time and minute how many changes in shape, color, odor and skin moisture note the results.

2.7 Skin evaluation

a) Organoleptic examination
   Organoleptic examination involves observing the shape, color, odor of the skin.
b) Skin moisture
   Moisture measurement that is using a skin moisture meter (skin meter) before being given a cream and after being given a blasting cream.

3. Result and Discussion

3.1 Collagen Yield Levels

The yield from the extraction of *Anadara granosa* and *Mytilus viridis* samples using a concentration of 0.75 M can be seen in table 3.

Table 3. Collagen yield levels

Sample	Initial weight	Dry collagen	Yield
Mytilus viridis	900,0005 g	11,3023 g	1,227%
Anadara granosa	900,0007 g	11,2003 g	1,257%

From table 3. shows the yield of collagen of *Mytilus viridis* 1.228% while *Anadara granosa* 1.257% means the concentration of acetic acid is best used 0.75 M based on previous research [3]. The concentration of acetic acid solution that can cause a decrease in protein levels because acetic acid will hydrolyze the peptide bonds more strongly so that there will be a loss of protein [5].

3.2 Anadara granosa and Mytilus viridis Collagen Cream Formulation

Collagen extract that has been produced from the extraction process is then used as ingredients for making cream. The resulting cream preparations are very good, non-sticky and homogeneous. Cream preparations in the form of thick solid white rather yellowish and easily applied to the intended object.

3.3 Test Results for Anadara granosa and Mytilus viridis Collagen Cream Evaluation

Cream evaluation test is an important element in determining the quality of preparations. Cream evaluation can be measured in blood clam shell collagen cream preparations including organoleptic, homogeneity, spreadability, pH test and type of cream. The results of characteristic testing can be seen in table 4.

Table 4. Test Results for Collagen *Mytilus viridis* and *Anadara granosa* Cream Evaluation

Formula *Anadara granosa*	Color	Smell	Dosage form	Homogenity	Spread power	pH	Cream type
F1	White	Typical	Solid	No coarse grains	3,44 cm	6,53	M/A
F2	Bone white	Typical	Solid	No coarse grains	3,53 cm	6,50	M/A
F3	Cream	Typical	Oily solid	No coarse grains	3,64 cm	6,47	M/A
F4	Yellowish	Typical	Oily solid	No coarse grains	3,95 cm	6,45	M/A
Base	White	Odorless	Solid	No coarse grains	3,45 cm	6,50	M/A
Formula *Mytilus viridis*	Color	Smell	Dosage form	Homogenity	Spread power	pH	Cream type
F1	White	Typical	Rather thick	Homogen	3,43 cm	6,50	M/A
F2	Bone white	Typical	Rather thick	Homogen	3,55 cm	6,50	M/A
F3	Cream	Typical	Rather soft	Homogen	3,66 cm	6,40	M/A
F4	Yellowish	Typical	Rather soft	Homogen	3,98 cm	6,50	M/A
Base	White	Typical	Thick	Homogen	3,53 cm	6,50	M/A

Information :

Basis = Basis Krim
F1 = Collagen concentration 10%
F2 = Collagen concentration 15%
F3 = Collagen concentration 20%
F4 = Collagen concentration 25%

Organoleptic testing of blasting cream evaluation results was carried out with visual observations including color, odor and dosage form. The results of organoleptic testing of collagen shells in blood and green shells in table 4. show that formula 1 is white, formula 2 is bone white, formula 3 is yellowish white, formula 4 is yellowish and formula 5 is white based. The *Mytilus viridis* and *Anadara granosa* cream formulation of form 1, 2 and 5 is in the solid form while formulas 3 and 4 are in the oily solid form while in the green tap shells formula 1 and 2 are rather thick, formulas 3 and 4 are somewhat mushy. There is a difference in the dosage form in the cream because the more active substances (collagen) that are used the more the oil content contained in the cream will be more moist. Collagen concentrations of 10% and 25% more moisturize on a cream base than previous studies with a collagen concentration of 5% [6].

The results of scatter power testing can be seen in table 4.2. The spread test is the ability to spread the cream to the skin which is used to determine when the cream is applied. A good cream should be spread evenly when used on the skin. The spreadability of the creams from the 5 formulations showed good results, namely 3.45 to 3.97 for preparations of *Anadara granosa* while 3.45 to 3.55 for preparations of green shells. The difference in spreadability occurs because if there is the addition of other substances in an oily cream, the spread will be longer or wider in diameter produced.
Homogeneity test is very necessary in making a cream. The advantage of homogeneity tests is to find out which creams are made homogeneous and when used can be evenly distributed on the surface of the skin without any coarse grains. The test results from 5 formulations of the 2 samples showed good results ie there were no coarse grains.

Determination of pH is tested to determine the pH of the preparation made. The results of testing the pH of the 5 formulations of the 2 samples indicate that the pH obtained ranged between 6.41-6.56. From the results of the cream pH test in accordance with the cream pH range ranged from 4.5 to 6.5 [7].

3.4 Moisture Cream Collagen Test Results Against Mouse Skin

The results of the rat skin test were measured in humidity before being smeared with blasting cream using a moisture meter. The humidity test results can be seen in Table 5.

Table 5. Mouse skin moisture test results

	Mean ± SD *Anadara granosa*									
	Base	± SD	F1	± SD	F2	± SD	F3	± SD	F4	± SD
Day1	10.3%	0.917	10.82%	0.208	11.46%	0.651	13.33%	0.361	15.94%	0.907
Day 2	10.3%	0.652	13.02%	0.436	3.41%	0.650	15.53%	0.529	17.66%	0.755
Day 3	10.3%	0.253	14.14%	0.513	14.46%	0.359	17.21%	0.529	19.44%	0.700
Day 4	10.04%	0.566	14.4%	0.500	15.46%	0.453	18.16%	0.503	20.16%	0.404
Day 5	10.03%	0.301	13.18%	0.100	14.43%	0.356	16.86%	0.153	18.9%	0.208
Day 6	10.2%	0.201	13.14%	0.265	13.43%	0.153	15.96%	0.503	18.3%	0.252
Day 7	10.2%	0.415	12.12%	1.082	12.46%	0.250	14.86%	0.153	17.44%	0.416

Mean ± SD *Mytilus viridis*										
Day 1	10.6%	0.907	11.13 %	0.200	11.8%	0.872	14.50%	0.656	16.3%	0.819
Day 2	10.3%	0.651	13.10 %	0.513	13.16%	0.884	15.83%	0.603	17.644 4%	0.404
Day 3	10.2%	0.252	14.13 %	0.503	14.36%	0.608	17.1%	0.579	19.46%	0.525
Day 4	10.05 %	0.569	14.26 %	0.529	15.90%	0.651	18.16%	0.416	20.16%	0.569
Day 5	10.00 %	0.300	13.20 %	0.436	14.66%	0.803	16.76%	0.531	18.94%	0.569
Day 6	10.04 %	0.200	13.26 %	0.306	13.3%	0.834	15.96%	0.500	18.46%	0.453
Day 7	10.02 %	0.416	12.20 %	0.385	12.30%	0.734	14.93%	0.872	17.46%	0.686

Keterangan : Base = cream base
±SD = Standar Deviasi
F1 = Collagen contentration 10%
F2 = Collagen contentration 15%
F3 = Collagen contentration 20%
F4 = Collagen contentration 25%

Moisture cream is a way to ensure that preparations made can provide a moisturizing effect that is safe to use. Blasting cream test in this study was carried out using the topical method directly on white male rat test animals. The use of male rats was chosen so that the study of active substances on the surface of the rat's skin did not affect the mouse's hormones. The number of male rats used for the collagen cream test was 25 3-month-old rats with a body weight of 160-200 grams. Selection of this mouse. The number of rats used for the collagen cream test as a moisturizer was between 10-30 male rats [8]. Male rats were divided into 5 groups, namely the negative control group with cream base and 4 test groups with collagen cream from concentrations of 10%, 15%, 20% and 25%. This research aims to learn more about the collagen used to moisturize mouse skin. The initial treatment of the rat's back is shaved first so that the hairs of the rat are not opened when applying blasting cream. The backs of the mice were shaved with a diameter of about 2.8 cm in each of the mice that would be used for trials. Furthermore, each rat measured its moisture before being smeared with collagen cream using a moisture test tool (skin mouisture meter).

Moisture results from each rat's skin before smeared with collagen cream were replicated during 5 tests, but the data used 3 tests were seen from the best results. The average results per day of initial skin moisture of rat on the basis of the cream obtained results are day 1 (10.5%), day 2 (10.3%), day 3 (10.1%), day 4 (10.03%), 5th day (10.01%), 6th day (10.0%), and 7th day (10.0%) for a sample of a blood clam while the results of a sample of a green mussel shell are a day 1st (10.6%), 2nd day (10.2%), 3rd day (10.1%), 4th day (10.04%), 5th day (10.01%), day 6th (10.0%), and the 7th day (10.0%). The process of reducing moisture on the skin of mice is due to the cream base reacting at the beginning of the application only because the treatment in this study was done only with one time applying the base to the skin and the treatment was seen from day 1 to day 7. When the initial application of skin rats would be high levels humidity compared to day 7 because the cream base is not absorbed into the skin of mice.

Testing the effect of clam shell shell collagen cream is done with the same treatment. Mice smeared with collagen cream as much as 50 grams of cream, applied to the back of the rats that have been shaved and then applying must be evenly distributed. Collagen cream testing was carried out for 7 days, then the data results were averaged from each formulation. The results of humidity testing data can be seen in table 5.

Humidity obtained from 4 concentrations shows the results that the collagen cream at the beginning of applying the skin of rats experiencing humidity from low then increases on Monday, then after that there is a decrease in humidity from Tuesday to Thursday. This happens because the process of absorption of active substances in the form of collagen at the beginning of the application appears to have increased from Saturday to Sunday but the peak of skin moisture increase occurs on Monday. The process of increasing humidity is due to a cream containing collagen which has perfect absorption for more or less on initial use for 4 days. The longer the collagen cream attached to the skin of the back of the rat, the collagen will not react because the application of collagen cream in this experiment is done only once and rub changes are seen from day 1 and day 7.

The benefits of collagen in a cream as an active ingredient that functions as a moisturizing effect when tested on the back of a rat's skin. Giving creams to the skin is a way to give medicine to the skin by applying which aims to maintain the skin from hydration or dryness, protect the surface of the skin and treat moisture to the skin. The higher the collagen applied to the skin of mice, it will produce a more supple skin. From the 7-day trial test of 4 concentrations of collagen, it was found that the skin with a cream base shows that the skin is still thin and pink, after being applied with a collagen cream, the concentration of 1% occurs, changes in skin color begin to change and the skin texture begins to spring. At concentrations of 10%, 15%,20% and 25% the change is more pronounced with the skin color getting thicker from the start and the texture of the skin more elastic according to the high concentration of collagen used in the cream.

Collagen cream testing was carried out on the back skin of mice. The choice of the back as a part of the rat used for the experiment is because the structure of the skin consists of edipermis, dermis and hypodermis. From there it can be concluded that the process of absorption of collagen active substances that occur in the skin from the skin's structural tissues occurs well because the method used is the method directly to the intended center, which is a cream against rat skin. Collagen cream begins to absorb into the skin tissue of mice, resulting in changes in skin moisture from skin types that are not so moist become more moist and skin texture is more supple. The process of absorption of collagen cream does not occur directly, but seen first for 1 day in a different time, from there it can be seen that the moisture of the skin of mice from 1 o'clock onwards changes occur. Changes that occur due to the active ingredient of collagen have an effect as a moisturizer on the skin of mice so that this study was successful (Setiadi, 2007).

3.5 Results of data analysis

Table 6. *Test of Normality*

	Formulation	Shapiro-Wilk	
		Df	Sig.
Humidity	Base cream	14	0.006
	Formulation 10%	14	0.200
	Formulation 15%	14	0.671
	Formulation 20%	14	0.834
	Formulation 25%	14	0.616

Table 7. *Test of Homogeneity of Variance*

Levene Statistic	df1	df2	Sig.
5,542	4	65	0.002

Hipotesis :
H0 = same variance
V1 = variance not same

Table 8 *Kruskall-Wallis Test*

	Humadity
Chi square	59,368
df	4
Asymp. Sig.	0,000

Hipotesis :

H0 = concentration does not affect humadity
V1 = concentration affects humadity

Table 9 Uji Mann Whitney

	Humadity
Mann-Whitney U	0,000
Wilcoxon W	100,000
Z	-2,510
Asymp.Sig(2-tailed)	0,000

Hipotesis :
H0 = concentration doesnt affect humadity
V1 = concentration affects humadity

In table 6, it can be seen that the significant value shows the result> 0.05 which means that each of the data is normally distributed. Then the variant homogeneity test was performed. This test is used to see one of the anova assumptions, namely the existence of the same variant of the analyzed sample. Homogeneity variance test results can be seen in table 7.

In table 7, the probability significance value of 0.001 or <0.05 means that H0 is rejected or the variance is different. The results of the transformation of the data showed a significant value <0.05. Therefore, one way ANOVA test could not be performed so a non parametic test was performed using kruskal-wallis to determine the average difference of each formulation. The results of kruskall-wallis calculations can be seen in table 8.

Based on the results of the kruskal-wallis calculation in table 8, the significant value indicates the number 0,000 or p <0.05; then H0 is rejected H1 is accepted or it can be concluded that concentration influences humidity.

Statistical analysis was continued with the Mann Whitney test to determine significant differences between humidity levels. The results can be seen in table 9.

The results of the mann whitney test are known that the Asymp value. Sig. (2-tailed) of 0,000 or p <0.05. It can be concluded that H1 is accepted or concentration influences humidity.

4. Conclusions and Recommendations

4.1 Conclusions

Based on the results of research that has been done, it can be concluded that blasting cream *Anadara granosa* and *Mytilus viridis* can provide a moisturizing effect on the skin of rat. The most optimal concentration as a moisturizing effect on rat skin is a concentration of 25%.

4.2 Recommendations

Suggestions that can be given for further research are further research needs to be done to find out what content is contained in the shells of blood clams and green shells. It is necessary to experiment using a rabbit test animal to determine the effect of humidity with a comparison of male rat test animals.

Acknowledgement

Thank to Kementerian Riset, Teknologi, dan Pendidikan Tinggi Republik Indonesia give a grant Penelitian Dosen Pemula 2019.

References

[1] A. A. Sinaga, S. Luliana, and A. Fahrurroji, “Losio Antioksidan Buah Naga Merah (Hylocereus polyrhizus Britton and Rose),” *Pharm. Sci. Res.*, vol. 2, no. 1, pp. **11–20**, 2015.
[2] Latifah, *Karakteristik Morfologi Kerang Darah*. Departemen Teknologi Hasil Perairan, Fakultas Perikanan dan Ilmu Kelautan, Institut Pertanian Bogor, Bogor., 2011.
[3] A. Ariyanti *et al.*, “comparison collagen content of the shell of a clam blood (anadara granosa) andshell of clam greens (mytilus viridis) in bandengan , kendal ,” vol. 1, no. 1, pp. **1–6**, 2017.

[4] D. RI, *Farmakope Indonesia. Edisi IV.* Departemen Kesehatan Republik Indonesia. Jakarta., 1978.

[5] N. A. Tridhar, *Perbandingan Produksi Kolagen dari Sisik dan Tulang Ikan Gurami (Osphronemus gouramy) Secara Kimia dan Enzimatis.* Fakultas Teknik Universitas Pasundan, Bandung, 2016.

[6] Waryani, "Pemanfaatan Kitosan Dari Cangkang Bekicot (Achatina Fulica) Sebagai Pengawet Ikan Kembung (Rastrelliger Sp) Dan Ikan Lele (Clarias Batrachus)," *Dep. Tek. Kim. Fak. Tek. Univ. Sumatera Utara*, 2015.

[7] D. Mahatmanti, Widhi, "Sintesis Kitosan dan Pemanfaatannya Sebagai Anti Mikrobia Ikan Segar," in *Laporan Pengabdian Kepada Masyarakat, Fakultas Matematika dan Ilmu Pengetahuan Alam, Universitas Negeri Semarang, Semarang*, 2011.

[8] Muyonga, *Characterisation of acids soluble collagen from skins of young and adult Nile perch (Lates niloticus).* Food Chemistry 85 : **81-89**, 2004.

Microencapsulation of riboflavin (vitamin B$_2$) using alginate and chitosan : effect of alginate and chitosan concentration upon encapsulation efficiency

YC. Danarto[1,2], Rochmadi[1], Budhijanto[1]
[1]Department of Chemical Engineering, Universitas Gadjah Mada, Indonesia
[2]Department of Chemical Engineering, Universitas Sebelas Maret, Indonesia

Corresponding autthor : yc.danarto@gmail.com

Abstract. Riboflavin (Vitamin B$_2$) plays an important role in the human tissue development, the production of red blood cells and helps release energy by breaking down proteins, fats and carbohydrates. Riboflavin cannot be produced the human body therefore must be supplied outside via dairy food. Riboflavin is very sensitive and unstable to environmental influence such as light and reduction agent. One of technology for maintaining the stability of riboflavin is microencapsulation which is composed of polymer matrix containing riboflavin. Na-alginate was chosen because it has properties such as biodegradable, biocompatibility, and non-toxic. Na-alginate also has a high loading capacity of riboflavin but Na-alginate is also easy to release riboflavin. Solving this issue, Na-alginate was coated with chitosan and the layer is then reinforced by crosslinking process with glutaraldehyde. The process of forming micro-sized particles was carried out by the emulsification ionic-gel method. This research aims to study the microencapsulation process of riboflavin with Na-alginate and chitosan. This research also studied the effect of Na-alginate and chiton concentrations upon encapsulation efficiency. The research showed that microencapsulation of riboflavin with alginate and chitosan can be done. The results showed that the Na-alginate and chitosan concentrations had an effect on the encapsulation efficiency. The increase in the concentration of alginate and chitosan will make the encapsulation efficiency higher until it finally reaches the optimum point. Encapsulation efficiency will drop past that point. The optimum point for Na-alginate and chitosan concentration are 3% and 2% respectively.

1. Intoduction

Riboflavin (Vitamin B$_2$) is water soluble vitamin that is needed for growth. It helps the body break down carbohydrates, proteins and fats to produce energy, and it allows oxygen to be used by the body. Riboflavin is 7,8-dimethyl-10-ribityl-isoalloxazine molecule and orange-yellow crystals. The chemical structure of riboflavin can be seen in figure 1. Riboflavin is flushed out daily and human body cannot provide riboflavin itself therefore it must be supplied from outside through dairy food such as eggs, milk, nuts, soybeans, green leaf, etc.

The limitation for supplying riboflavin to the human body is due to the highly sensitive to environmental influences such as light and reducing agents therefore encapsulation of riboflavin by polymers is needed.

IOP Conf. Series: Materials Science and Engineering **858** (2020) 012030 doi:10.1088/1757-899X/858/1/012030

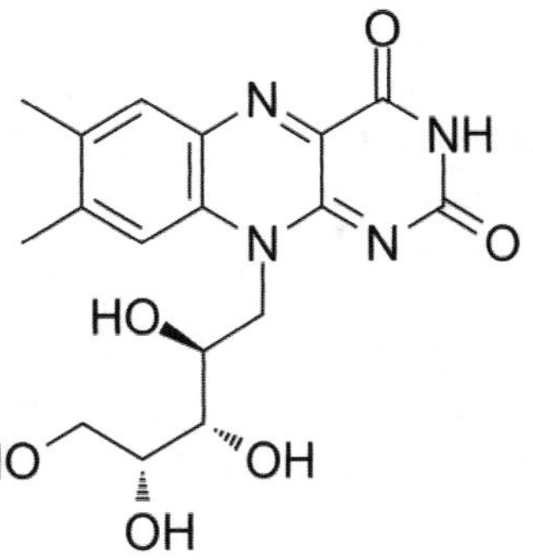

Figure 1. Chemical structure of riboflavin.

The polymers used for riboflavin encapsulation must be biodegradable, biocompatible and non-toxic. Alginate is the polymer that meet the criteria above. Sodium alginate is an anionic linear polysaccharide composed of alternating blocks of β-(1→4)-linked d-mannuronic acid (M) and α-(1→4)-linked l-guluronic (G) residues. The chemical structure of alginate can be seen in figure 2.

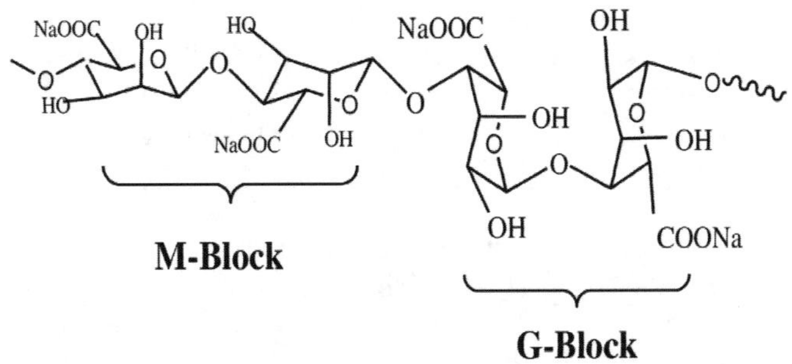

Figure 2. Chemical structure of Na-alginate with blocks arrangement M-G

The negatively charged carboxyl groups in alginates cause the straight alginate chains to repel each other and this results in a stable aqueous solution.

Alginate solutions can turn into sol-gel form when crosslinked with polyvalent ions such as Ca^{2+} and Zn^{2+} [1] but Ca^{2+} is more commonly used because it can bind to guluronic acid groups to form tissues with an egg-box structure [1]. Rastogi et.al.[2] explain that Na-alginate has a great ability for loading active compounds such as riboflavin but according to Liu and Krishnan [3], Na-alginate has the disadvantage of easily releasing active compounds therefore is not therefor is not suitable for control release active compounds.

Na-alginate needs to be maintained so it is not easy to release the active ingredient by coating it with a polymer such as chitosan. Chitosan is linear polysaccharide composed of randomly distributed β-(1→4)-linked D-glucosamine (deacetylated unit) and N-acetyl-D-glucosamine (acetylated unit). The chemical structure of chitosan can be seen in figure 3.

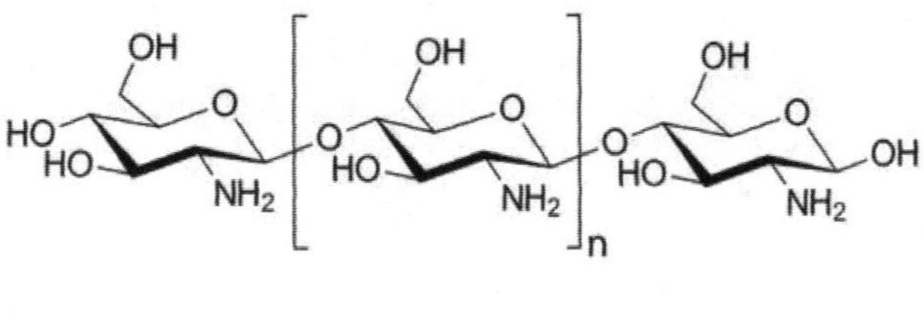

Figure 3. Chemical structure of chitosan

Similar to alginate, chitosan is biodegradable, biocompatible and anti-toxic. Amino groups in chitosan form strong electrostatic interaction with carboxylate groups in alginate generate chitosan/alginate complex which is able to withstand the release of active compounds. [4], [5]. Chitosan can be strengthened with glutaraldehyde by cross-linking process which is amino groups in chitosan react with aldehyde group in glutaraldehyde.

Microencapsulation is described as a process of enclosing micron-sized particles of solids or droplets of liquids or gasses in an inert shell, which in turn isolates and protects them from the external environment [6]. Some objectives of microencapsulation are protecting the sensitive substances from external environment, masking unpleasant odor or taste of substances, controlling the release of active substances, and safe handling the toxic material [7].

There are some research about microencapsulation of active substances with alginate and chitosan. These researches can be seen in table 1.

Table 1. Comparison with other studies about encapsulation active substances using alginate and chitosan

Active substances	Methods	References
riboflavin	dropwise alginate solution into the chitosan solution containing $CaCl_2$ using a hypodermic needle	Bajpai and Tankhiwale [8]
curcumine	ionotropic pre-gelation followed by policationic crosslinking	Das et.al. [9]
probiotic bact.	dropwise the alginate solution into the CaCl2 solution and then coated with chitosan solution	Trabelsi et.al. [10]
riboflavin	Dropwise alginate in oil phase emulsion into $CaCl_2$ solution and then coated with chitosan solution	This research

Encapsulation efficiency determines the number of active substances that can be contained in the microcapsules. The greater the encapsulation efficiency indicates that the microencapsulation process is also getting better because not much riboflavin is wasted during the process. Azevedo et.al. [11] studied the encapsulation efficiency of riboflavin in alginate-chitosan nanoparticles and it was about 55.9%. Jyothi et.al. [7] explained factors that influence the encapsulation efficiency such as :

a. Solubility of polymer in organic solvent
b. Solubility of organic solvent in water
c. Concentration of the polymer
d. Ratio of dispersed phase to continuous phase
e. Rate of sovent removal
f. Interaction between active substances and polymer
g. Solubility of active substances in continuous phase
h. Molecular weight of the polymer

The objective of this research was studying the microencapsulation riboflavin with Na-alginate and chitosan. This research also aims to study the effect of Na-alginate concentration dan chitosan concentration upon the encapsulation efficiency.

2. Experimental

2.1. Materials

Riboflavin was obtained from Sigma-Aldrich. Chitosan and alginate were obtained from local store. Chitosan solution was made by dissolving it in acetic acid solution 1%. Na-alginate solution was made by dissolving it in aqueous solution. Calcium chloride and glutaraldehyde were used as crosslinking agent for Na-alginate and chitosan. Paraffin oil was used as emulsification medium and span 80 was used as surfactant agent.

2.2. Methods

Riboflavin (about 160 mg) was mixed in 60 mL Na-alginate solution. The mixed solution was dispersed in 140 mL paraffin oil which contained 0,6 mL span 80 using magnetic stirrer for 20 minutes. The dispersed solution then was dripped slowly into 125 mL 1% $CaCl_2$ solution and stirred for 60 minutes. The microcapsules were filtered and washed. The microcapsules were then coated with

IOP Conf. Series: Materials Science and Engineering **858** (2020) 012030 doi:10.1088/1757-899X/858/1/012030

chitosan solution for 15 minutes and filtered. The coated microcapsules were then cross-linked with 2% glutaraldehyde solution for 60 minutes. Then microcapsules were washed and dried.

2.3. Encapsulation efficiency determination

The encapsulation efficiency was determined using this equation

$$EE\% = \frac{m_p}{m_i} . 100\% \quad (1)$$

which is

$EE\%$ = encapsulation efficiency

m_p = mass of riboflavin in microcapsules

m_i = initial mass of riboflavin

Initial mass of riboflavin (m_i) is the amount of riboflavin used in the microencapsulation process. Mass of riboflavin in microcapsules (m_p) is measured by extracting riboflavin in demineralized water and then determining its content in solution with a UV-Vis spectrometer.

3. Results and Discussions

3.1. Microcapsules

Microcapsules are formed by inserting riboflavin into Na-alginate solution and then microcapsules are formed by emulsion and crosslinking processes. These wet microcapsules can be seen in figure 4a. The microcapsules are then coated with chitosan, reinforced by crosslinking process using glutaraldehyde, and dried. This microcapsules can be seen in figure 4b. Compared to wet microcapsules, the dried microcapsules surface is more wrinkled. The average of microcapsules diameter range 380 – 610 μm.

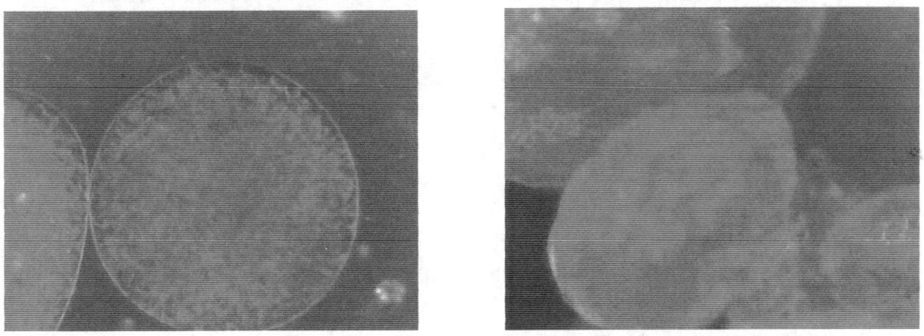

a. Wet microcapsules. b. dried microcapsules.

Figure 4. The microcapsules form.

3.2. The effect of Na-alginate concentration upon encapsulation efficiency

This research used Na-alginate solution in various concentration (2%, 2.5%, 3% and 3.5%). Microcapsules do not formed at alginate concentrations below 2% and mixing riboflavin with alginate solutions is imperfect if alginate concentrations are above 3.5%. The encapsulation efficiency for various Na-alginate concentration can be seen in table 2. The greater in Na-alginate concentration will increase the encapsulation efficiency until it reaches a peak at a certain point and then decreases.

The ability to load and hold riboflavin in the alginate hydrogel is due to the formation of an "egg-box" structure during the crosslinking process with Ca^{2+} ions. The greater concentration of Na-alginate will cause more riboflavin that can be loaded and held in the Na-alginate hydrogel. This is related to the number of "egg-box" structures that are formed. But if the concentration of Na-alginate is too large it will cause the formation of "egg-box" structure will be inhibited because it is difficult to react Ca^{2+} ions with Na-alginate and riboflavin contained in Na-alginate will easily diffuse out. The optimum point is 3% Na-alginate concentration.

Table 2. Encapsulation efficiency for various Na-alginate concentration.

Na-alginate concentration (%)	Encapsulation efficiency (%)
2	42.48
2.5	44.61
3	55.34
3.5	44.64

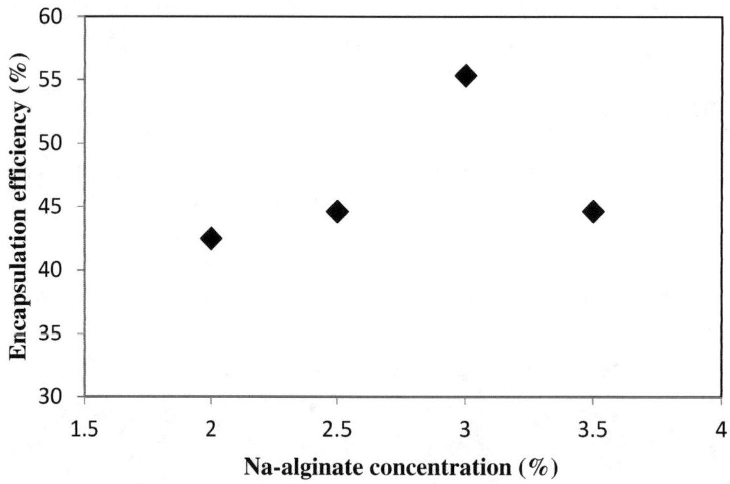

Figure 5. Encapsulation efficiency for various Na-alginate concentration.

3.3. The effect of chitosan concentration upon encapsulation efficiency

This research used chitosan solution in various concentration (1%, 1.5%, 2%, 2.5%, and 3%). Chitosan solution is difficult to form at chitosan concentration above 3%. The encapsulation efficiency for various chitosan concentration can be seen in table 3.

Table 3. Encapsulation efficiency for various chitosan concentration.

Chitosan concentration (%)	Encapsulation efficiency (%)
1.0	44.61
1.5	45.29
2.0	55.70
2.5	28.06
3.0	27.60

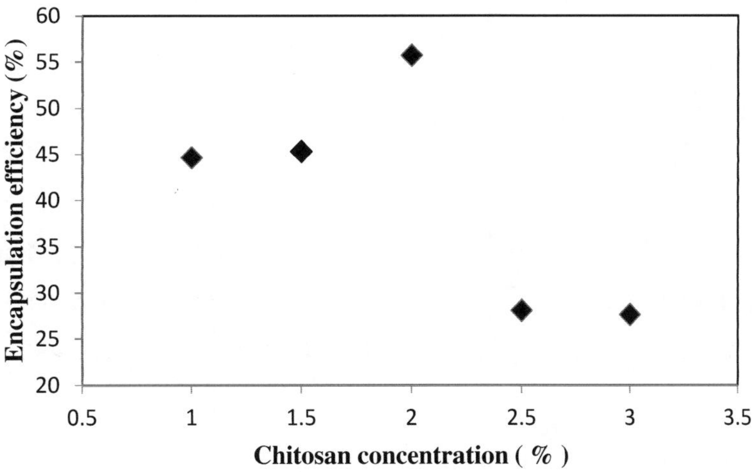

Figure 6. Encapsulation efficiency for various chitosan concentration.

The greater in chitosan concentration will increase the encapsulation efficiency until it reaches a peak at a certain point and then decreases. A large concentration of chitosan will make it easier to coat the alginate hydrogel and keep the riboflavin diffusion out. The concentration of chitosan that is too concentrated will have the opposite effect because it will push the diffusion of riboflavin out due to the large difference in the concentration of chitosan with alginate hydrogel. The optimum point is 2% chitosan concentration.

4. Conclusions

The research showed that microencapsulation of riboflavin with alginate and chitosan can be done. The results showed that the Na-alginate and chitosan concentrations had an effect on the encapsulation efficiency. The increase in the concentration of alginate and chitosan will make the encapsulation efficiency higher until it finally reaches the optimum point. Encapsulation efficiency will drop past that point. The optimum point for Na-alginate and chitosan concentration are 3% and 2% respectively

Acknowledgement

This research was supported by Doctoral Disertation Grant by Kemenristek-Dikti Indonesia.

References

[1] Grant GT, Morris ER and Rees DA 1973 *FEBS Letters* **32(1)** 195-8
[2] Rastogi R, Sultana Y, Aqil M, Ali A, Kumar S, Chuttani K and Mishra AK 2007 *Int. J. Pharmaceutics* **334** 71-7
[3] Liu P and Krishnan TR 1999 *J. Pharm. Pharmacol.* **51** 141-9
[4] Finotelli PV, Silva DD, Sola-Penna M, Rossi AM, Farina M, Andrade LR, Takeuchi AY and Rocha-Leao M H 2010 *Colloids and Surfaces B : Biointerfaces* **81** 206-11.
[5] Polk A, Amsden B, De Yao K, Peng T and Goosen MFA 1994 *Journal of Pharmaceutical Sciences* **83(2)** 178-85
[6] Ghosh SK 2006 *Functional Coating* (Willey Online Library) pp 1–28
[7] Jyothi NVN, Prasana, P M, Sakarkar SN, Prabha KS, Ramaiah PS and Srawan GY 2010 *Journal of Microencapsulation* **27(3)** 187-97
[8] Bajpai SK and Tankhiwale R 2006 *Reactive and Functional Polymers* **65** 1565-74
[9] Das RK, Kasoju N and Bora U 2010 *Nanomedicine : Nanotechnology, Biology, and Medicine* **6** 153-60.
[10] Trabelsi I, Bejar W, Ayadi D and Salah RB 2013 *Int. J. of Biological Macromolecules* **61** 36-42
[11] Azevedo MA, Bourbon AI, Vicente AA and Cerqueira MA 2014 *Int. J. of Biological Macromolecules* **71** 141-6

IOP Conf. Series: Materials Science and Engineering **858** (2020) 012031 doi:10.1088/1757-899X/858/1/012031

Inhibition Of *Pseudomonas aeruginosa* Virulence Factors Expression Regulated By Quorum Sensing System Using Ethyl Acetate Extract Of Temu Ireng (*Curcuma aeruginosa*)

A Pangastuti[1],SL Arum Sari[1], *ER Nugraheni[2], RT Astuti[3]
[1]Department of Biology, Faculty of Mathematic and Natural Sciences, Sebelas Maret University Surakarta, Indonesia
[2]Department of Pharmacy, Faculty of Mathematic and Natural Sciences, Sebelas Maret University Surakarta, Indonesia
[3]Department of Biology, Graduate School of Sebelas Maret University Surakarta, Indonesia.

E-mail: estu_rn@staff.uns.ac.id

Abstract. *Pseudomonas aeruginosa* is an opportunistic human pathogen that can cause serious problem in immunocompromised persons. Difficulty in treating *P. aeruginosa* infection is due to its abilities to develop multi drug resistence towards various antibiotic. *Quorum sensing* system is a potential target for *P. aeruginosa* infection because this system regulate the expression of virulence factors in this bacteria. In previous study, ethyl acetate extract of Temu Ireng rhizomes (*Curcuma aeruginosa*) was proved to inhibit *quorum sensing* in *Chromobacterium violaceum*. This study was aim to know the effect of *C. aeruginosa* extract towards expression of *P. aeruginosa* virulence factors which were regulated by *quorum sensing*, such as protease LasA activities, LasB and biofilm formation. Protease LasA activity assay was conducting using S*taphylococcus aureus* cell lysis activity. Protease LasB activity was measured by elastin congo red assay, and the formation of biofilm was monitored using PVC biofilm formation assay method. The result showed that ethyl acetate extract of *C. aeruginosa* rhizomes can decrease *P. aeruginosa* virulence controlled by *quorum sensing*, namely protease LasA, LasB, and biofilm formation by 66.92%, 37.8%, and 46% respectively.

1. Introduction

Quorum sensing is a system of communication between bacterial cells mediated by diffused molecules, which are released into the surrounding environment and can be sensed by the entire population so that they can act in harmony for certain purposes. This system regulates gene expression in response to cell density in the population. Quorum sensing is known to regulate gene expression associated with various physiological functions such as virulence, bioluminescence, sporulation, biofilm formation, pigment production, antibiotic production and interaction with higher organisms [1-3] .

Many pathogenic bacteria use the quorum sensing system to regulate the expression of virulence factors. this arrangement ensures that bacteria invade the host only when the amount has reached the quorum so that the host's defense system can be overcome. one of the pathogenic bacteria that uses the quorum sensing system for virulence is Pseudomonas aeruginosa. some virulence factors of P. aeruginosa which are under the regulation of the QS system are LasA and LasB Protease, pyocyanin pigment, and biofilm formation [4,5]. *P.aeruginosa* normally inhabit humans and act as saprophytes in the intestines and skin. However, *P. aeruginosa* infection is a serious problem in humans whose immune system decreases, especially hospital patients suffering from cancer, *cystic fibrosis* and burns [6,7]. *Pseudomonas aeruginosa* can cause infection in urinary tract, respiratory tract, surgical scars, Gastrointestinal tract, bones and joints, soft tissues, and various systemic infections.

Infectious diseases to date are still a problem for tropical countries. the re-emergence of infectious diseases that were previously extinct due to the development of pathogenic resistance to anti-microbial compounds. the search for new anti-microbial compounds continues to race with the development of resistance. one approach to overcome resistance is the search for new targets for antimicrobial compounds. QS system is one of the targets expected to be developed for the control of infectious diseases by bacteria that use this system for virulence control. the inhibition of the QS system is expected not to trigger the development of resistance, because the bacteria are not inhibited by growth but only the communication system is blocked, so that virulence factors are not expressed. Compounds which have the ability to interfere the QS system are called anti-pathogenic drugs [8]. some secondary metabolites from plants have been known to have the ability to block bacterial QS systems, such as cathecin, vanillin, zingerone, curcumin, etc [9].

Fitri (2010) had screened several plants for anti QS activity. The ethyl acetate extract of *curcuma aeruginosa* proved to have anti QS activity toward *C. violaceum* as a reporter strain . In this study we conducted further research to observe the effect of *C. aeruginosa* rhizome ethyl acetate extract to attenuate the virulence factors of *P. aeruginosa* which are under regulation of QS system, namely LasA protease, LasB elastase, and biofilm formation.

2.Experimental

2. 1. Extraction of Temu Ireng Rhizome

The Extraction of Temu ireng (*C. aeruginosa*) rhizome was conducted by maceration method using ethyl acetate as a solvent. Dried rhizome was pulverized into powder and then immersed in the solvent for 48 hours. The filtrate was separated by filtering using filter paper and separating funnel and then concentrated using a rotary evaporator at a temperature of 60 ^{0} C. The extract was dissolved in *DMSO* (*Dimethyl Sulfoxide*) with a concentration of 20 mg/ml w/v then sterilized by filtration using a 0.45 µm porous membrane. The solution is stored as stock in the freezer.

2.2. Anti QS assay on *Chromobacterium violaceum*

Qualitative assay of quorum sensing inhibition was carried out by the well diffusion method (Adonizio et al., 2006; Hussaini and Mahasneh, 2009). *Chromobacterium violaceum* is cultured overnight (16-18 hours) in LB broth medium then 100µL of the culture was spread on to LB agar medium. The wells with 6mm diameter were made on the LB agar. Extract was dissolved in DMSO and then put into the well. The concentrations of extracts that were put into the well were 200, 400, 600, 800 and 1000µg / mL. The plates were incubated at 30 ° C for 24 hours.

As a control, 20µL ethanol were included. As a comparison of antibiotic effect, tetracycline was added with a concentration of 10µg /mL into the well. Quorum sensing inhibition is indicated by the presence of a colorless cell growth zone ring around the well and quantify as the inhibition diameter. The diameter of the quorum sensing inhibition was measured from the outer edge of the disk to the end of the inhibition zone.

2.3. *Pseudomonas aeruginosa* Culture Maintenance and Preparation

Pseudomonas aeruginosa was grown on LB Agar media (LA) and incubated at 37 ° C. The culture was stored in a refrigerator as a stock culture. Just before used, bacteria was grown on LA using the streak plate method and incubated for 24 hours. A total of 1 bacterial colony were transferred to 20 mL LB media as inoculum and incubated at 37 ° C with shaking for 12 hours. Culture was then diluted 100 times in minimal AB media (for LasB protease activity testing) and LB media (for growth measurement, LasA protease activity and total protein) and the growth was measured using spectrophotometry at 600nm wavelength. When it reaches OD600 ~ 1.7, bacterial culture were grouped into three treatments: 1. culture without any supplementation 2. culture with supplementation of DMSO as a control and 3. culture with supplementation of extract. The final concentration of extract in the culture was 1 mg / ml. The activity of LasA and LasB as well as total protein was assayed using supernatant from bacterial culture at the end of the stationary phase. Meanwhile, The growth was measured every 2 hours for 24 hours.

2.4. Total Protein Assay
Total protein was determined by the Bradford method (1976) using *Bovine Serum Albumin* (BSA) as a protein standard. The total protein was then used to calculate LasA and LasB protease activity units.

2.5. Protease Activity Assay
LasA protease activity was determined by measuring the lytic activity of *P. aeruginosa* culture supernatant against *S. aureus* cells. A total of 60 ml of 12 hours aged *S. aureus* culture was boiled for 10 minutes then centrifuged for 10 minutes at 10,000xg. The pellets were resuspended in 10 mM phosphate buffer pH 4.5 to reach OD_{600} ~ 0.8. A total of 300 µl each of supernatant *P. aeruginosa* in LB medium (control and treatment) was added to 2,700 µl of *S.aureus* suspension . OD_{600} measurements were performed after 0, 10, 20, 30, 45 and 60 minutes of incubation. LasA protease activity is defined as a change in the OD_{600} value *S. aureus* per hour per mg of protein.

Elastolytic activity (LasB) was determined using the elastin congo red method. As much as 300µl of *P. aeruginosa* culture supernatan in AB minimal media is mixed with 2,700 µl elastin congo red buffer containing 20 mg of elastin congo red per liter and incubated with shaking for 3 hours at 37 °C. The mixture was centrifuged to separate the insoluble elastin congo red . Absorbance measurements were carried out at a wavelength of 495 nm. LasB protease activity was defined as OD_{495} per hour per µg protein.

2.6. Biofilm Formation Inhibition Assay
The effect of *C. aeruginosa* extract on biofilm formation was measured by the PVC biofilm formation assay method (O 'Toole and Kolter, 1998). The overnight culture of *P.aeruginosa* bacteria was resuspended in AB medium with 0.2% glucose and 0.5% casamino acid added with or without *C. aeruginosa* rhizome extract and diluted to reach OD_{600} value of 0.1.

The culture was transferred into polystyrene microtiter plate, 100 µL for each well and incubated for 10 hours at 30 ^{0} C. Culture medium was discarded, then the microtiter plate well was washed five times using distilled water to remove planktonic bacteria and biofilms that were not firmly bound. After being washed, 150 µL crystal violet included in each well and incubated 45 minutes. Crystal violet discarded and Microtiter plate was washed using sterile distilled water. Biofilm formed resuspended in 200µL 95% ethanol and then absorbance was measured using a spectrophotometry at 600 nm.

3. Result and Discussion
The ethyl acetate extract of *C. aeruginosa* rhizome had antiQS activity which was showed by the inhibition of violacein production in *Chromobacterium violaceum* (Figure 1)

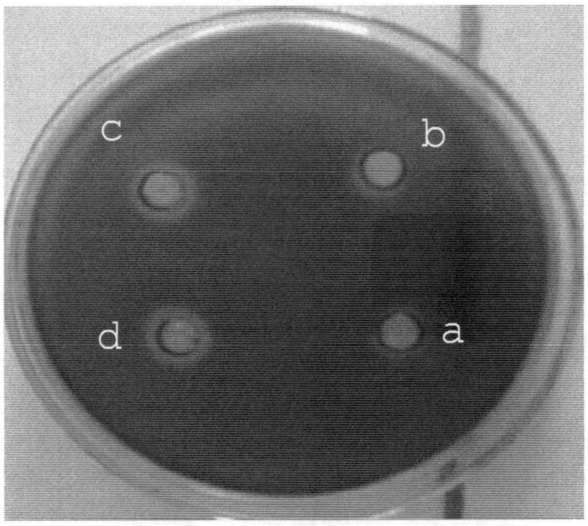

Figure 1. Anti QS activity of *C. aeruginosa* rhizome ethyl acetate extract on *Chromobacterium violaceum* violacein production : A.400µg/ml B.600µg/ml C.800µg/ml D.1000µg/ml

Ethanol extract of *C. aeruginosa* inhibited violacein production of *Ch. violaceum* as showed by the halo around the well which is indicating the growth of colourless cells. This QS inhibition zone was different from growth inhibition zone of tetracyclin since the bacteria still grew, the zone was not clear. The optimal concentration that inhibited QS was 1000 µg/ml which produced the largest inhibition zone. This concentration was then used to determine the attenuation of P. aeruginosa virulence factors.

The OD600 showed that the supplementation of DMSO and *C. aeruginosa* extract 1 mg/ml had no effect on the growth of *P. aeruginosa* bacteria (Figure 2). These results indicated that if the *P. aeruginosa* virulence factor expression decrease in subsequent tests, it was due to the inhibition of the *quorum sensing* system, not a decrease in the number of bacterial cells. Growth measurements were also carried out to determine the incubation time needed for bacterial culture to achieve optical density values of 1.7 which is the time for *C. aeruginosa* extract addition, to minimize the effects of extracts on bacterial growth at the beginning phase of growth. According to Adonizio (2008), supplementation of extracts at the beginning phase could affect the growth of *P. aeruginosa* so that the starting point for *quorum sensing* regulation between treatments did not occur at the same time.

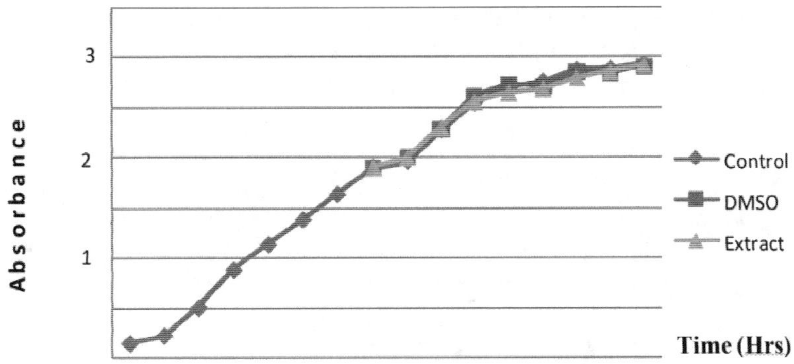

Figure 2. Growth curve of *P.aeruginosa* bacteria

The *C. aeruginosa* extract decreased the *P. aeruginosa* LasA activity. The biggest decrease occurred at 60 minutes incubation time (Figure 2). LasA protease activity in the the extract treatment decreased 66.92% comparing to the control. LasB protease activity was also decrease with the extract treatment (Figure 3). Not as high as in the LasA activity, approximately 37.8% decrease was observed in LasB activity of *P. aeruginosa* with the supplementation of extract, comparing to the control. The DMSO not significantly reduced the activity of LasA or LasB protease, suggesting that the decrease was due to the activity of the extract.

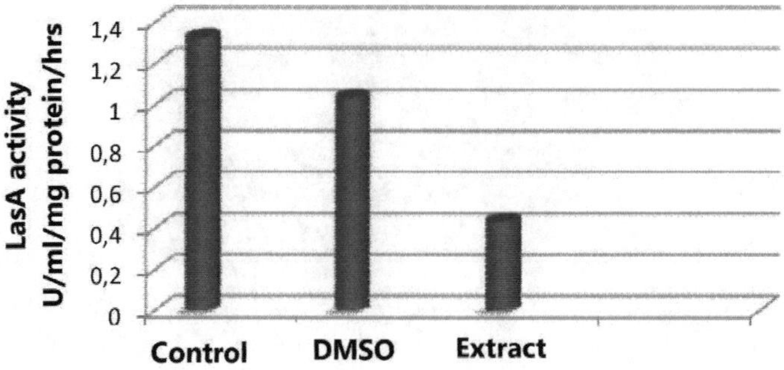

Figure 3. Las A activity of *P. aeruginosa*

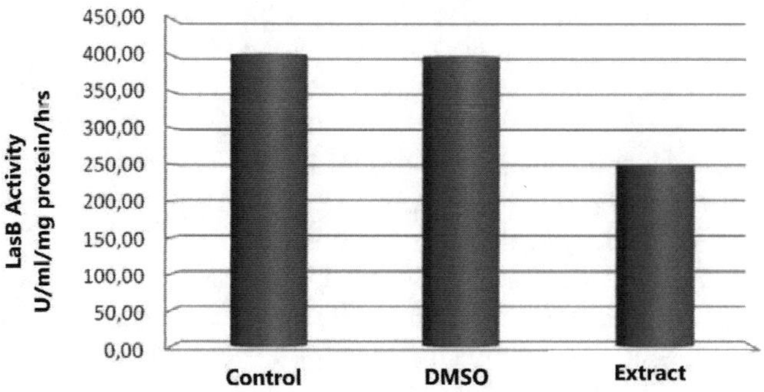

Figure 4. Las B activity of *P. aeruginosa*

Biofilm formation assay results showed differences in biofilm production in the control and extract treatment (Figure 4). Biofilm formation in the extract treatment was lower than the control treatment. *P. ae*ruginosa biofilm formation significantly decrease 46% in the presence of the extract.

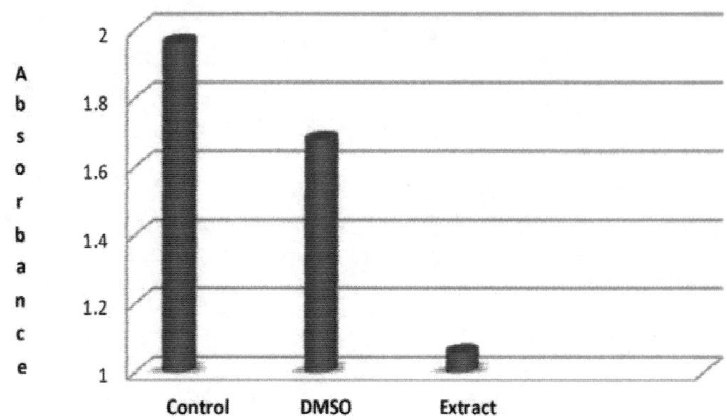

Figure 5. Biofilm formation of *P. aeruginosa*

Two proteolytic enzyme LasA and LasB mediated elastase activity in P. aeruginosa (Galloway 1991). These enzymes work synergistically in degradation of elastin, fibrin, interferon, complement and collagen. LasA is a serine protease which cuts elastin in certain Gly-Gly sequences and makes it more accesible to LasB. Protease LasA also has high *staphylolytic* activity (Kessler *et al* . 1993), which can provide selective benefits for *P. aeruginosa* in competition with *S. aureus* . Protease LasB (also called elastase), a zinc metalloprotease, is a dominant and most powerful protease released by *P. aeruginosa* . Protease LasB can degrade elastin, an extracellular matrix proteins, to help the invasion of this bacteria into the host (Galloway 1991). This protease is also be able to degrade and inactivate proteins from the immune system, including IgG and IgA (Heck et al. 1990), complement components (Hong & Ghebrehiwet 1992), and cytokines such as interferon gamma and tumor necrosis factor (Park et al. 1996). These activity play an important role in the virulence of these bacteria (Storey, 1998).

Pseudomonas aeruginosa possesses two QS systems: the LasR-LasI and the RhlR-RhlI, using N-(3-oxo-dodecanoyl)-L-homoserine lactone (OdDHL) and N-butyryl-L-homoserine lactone (BHL) as signal molecules. The LasR-LasI system controls the virulence factors LasA and LasB (Gambello, 1991; Storey *et al.*, 1998), however *rhlI/R* also controls activity to a lesser extent (Brint & Ohman, 1995; Pearson *et al.*, 1997). Several reports also showed the regulation of biofilm formation under the QS system. The effect of *quorum sensing* on biofilm formation can vary in different bacterial species. In *P. aeruginosa*, the *quorum sensing* system influences biofilm formation at the stage of maturation and dispersion (Davies, 1998; Heydorn, 2002). Besides, the biofilm effectiveness in the pathogenicity of bacteria is increasing as the cell density increase, cell to cell interaction in the biofilm acquiring the bacteria to be able to communicate each other, and this is conducted using QS signals.

Since many virulence factors expression are regulated under the QS system, a therapeutic compound could target this system to inhibit pathogenicity. The compound could be an analog to signal molecules, or targeting the receptor. One well known anti QS compound, the halogenated furanones, inhibit the virulence of P. aeruginosa by displacing the signaling molecule from its receptor (Manefield et al. 2002). In this study, compounds that inhibit *quorum sensing* and inhibition mechanism was not identified yet. Anti *quorum sensing* activity can also occur due to the influence of two or more chemical compounds and is not an activity of a single compound (Adonizio, 2008). Rukayadi and Hwang (2005) stated that *curcumin* compounds can inhibit *quorum sensing* in *C. violaceum*. Curcumin is also one of the active compound of *C. aeruginosa*. Corresponding with this study, *curcumin* could be an anti QS compounds in *C. aeruginosa*. Flavonoids specifically inhibit quorum sensing via antagonism of the autoinducer-binding receptors, LasR and RhlR (Paczkowski et al. 2017). In order to identify the compounds in *C. aeruginosa* that act as anti QS in *P. aeruginosa*, bioassay guided fractionation combined with chemical profile tests such as thin layer chromatography should be conducted.

4. Conclusion

The administration of *C. aeruginosa* rhizome ethyl acetate extract 1 mg / ml can reduce the expression of *P. aeruginosa* virulence factor in the form of LasA protease activity, LasB protease and biofilm formation, which is regulated under QS system, without affecting the growth.

References

[1] Rutherford S T and Bassler B L Bacterial Quorum Sensing: Its Role in Virulence and Possibilities for Its Control

[2] Kothari V, Sharma S and Padia D 2017 Recent research advances on Chromobacterium violaceum *Asian Pac. J. Trop. Med.* **10** 744–52

[3] Li Y-H, Tian X, Li Y-H and Tian X 2012 Quorum Sensing and Bacterial Social Interactions in Biofilms *Sensors* **12** 2519–38

[4] Williams P and Cámara M 2009 Quorum sensing and environmental adaptation in

Pseudomonas aeruginosa: a tale of regulatory networks and multifunctional signal molecules *Curr. Opin. Microbiol.* **12** 182–91
[5] De Kievit T R 2008 Minireview Quorum sensing in Pseudomonas aeruginosa biofilms
[6] Hoiby N 2002 New antimicrobials in the management of cystic fibrosis *J. Antimicrob. Chemother.* **49** 235–8
[7] Van Delden C and Iglewski B H 1998 Cell-to-cell signaling and Pseudomonas aeruginosa infections *Emerg. Infect. Dis.*
[8] Rasmussen T B and Givskov M 2006 Quorum-sensing inhibitors as anti-pathogenic drugs *Int. J. Med. Microbiol.* **296** 149–61
[9] Bouyahya A, Dakka N, Et-Touys A, Abrini J and Bakri Y 2017 Medicinal plant products targeting quorum sensing for combating bacterial infections *Asian Pac. J. Trop. Med.* **10** 729–43

Nanocomposite mesoporous silica-alginate hydrogels for extended delivery of antibiotics

Michael, S Caroline, J Lady, E A Prasetyanto[1]
[1] Dept. of Pharmacy, School of Medicine and Health Sciences, Atma Jaya Catholic University of Indonesia. Jl. Pluit Raya no 2, Jakarta 14440 Indonesia

E-mail: prasetyanto@atmajaya.ac.id

Abstract. Bacterial infection is one of major health problems in Indonesia. Improper use of antibiotics can further trigger antibiotics resistant bacteria which then worsen the situation. One of key challenges to avoid problem is to provide antibiotics delivery system that can release antibiotics in certain time window. In this work, we develop drug delivery system with tuneable release rate of antibiotics. The system will consist of porous silica nanoparticles of size of about 150 to 200 nm, that will be encapsulated in large micrometres size gel particles to allow the protection of the drug and avoid its early release. To understand the release behaviour, the system is tested under simulated gastric fluid.

1. Introduction

The use of antibiotics in a sub-MIC (Minimal Inhibitory Concentration) condition can cause a microbial resistance to antibiotics. This phenomenon can be found in the use of fluoroquinolone which can cause S. aureus bacteria to be able to transfer antibiotic resistant genes [1]. Treatment with conventional medicine needs to be taken many times to ensure that the plasma concentrations of antibiotics are sufficient, and the desired results can be achieved. Consumption of conventional antibiotics will increase the likelihood of a person having sub-MIC antibiotic concentrations that can trigger the mechanism of antibiotic resistance due to peak and valley effect [2].

One strategy to avoid sub-MIC (Minimal Inhibitory Concentration) condition in the blood plasm is to formulate antibiotics into extended release tablets [3, 4]. Other advantage of this extended release of antibiotics is to allow patients to consume antibiotics once a day [5]. This is important since patients tend to drop antibiotics medication once they feel better [6]. There are several extended releases of antibiotics in the market. The Biaxin® ER is a clarithromycin antibiotic that has extended release characteristics used to treat chronic bronchitis, pneumonia, and acute maxillary sinusitis. Another example of an extended release antibiotic that has been marketed is the Ceclor® CD tablet, a Cefaclor antibiotic. It is reported that both formulations were able to maintain antibiotics concentration higher than its MIC for longer period [7].

To control the release rate of the antibiotics, several materials have been used. Cellulosic polymers (such as hydroxypropylmethylcellulose, methylcellulose, hydroxypropylcellulose, hydroxyethylcellulose and sodium carboxymethylcellulose) are probably used most frequently to create swellable hydrophilic matrix [6]. In the case of Biaxin® ER and Ceclor® CD, they used substituted hydroxy cellulose (either as hydroxypropylmethyl cellulose or hydroxypropyl cellulose) and methacrylic acid copolymer [7]. In this work, we explore the use of nanocomposite mesoporous alginate as materials to control the release of the antibiotic from the matrices.

2. Experimental

All materials used in this experiment were analytical grade without further purification. FTIR and UV-Vis measurement were done on Shimadzu IRTracer-100 and Shimadzu UV-1800 UV Vis Spectrophotometer, respectively. Mesoporous silica was synthesized according to the reported literature [8]. Nanocomposite Mesoporous Silica-Alginate hydrogel beads were synthesized by dropping pre-mix composition Alginate/Mesoporous silica/Cefixime (200 mg of Alginate, 100 mg of Mesoporous Silica, and 100 mg of Cefixime) in total of 17.5 mL solution into solution of calcium chloride (3% w/v concentration). Simulated Gastric Fluid (without Pepsin) was prepared by mixing solution of 0.2% sodium chloride and hydrochloric acid in 1000 mL of deionized water with final pH of 1.2.

3. Results and Discussion

To produce the nanocomposite mesoporous silica-alginate, firstly we synthesized mesoporous silica nanoparticles. The mesoporous silica nanoparticles were synthesized using modified Stober technique under basic condition and CTABr was used as structured directing agent. The surfactant used during synthesis was then removed by calcination techniques. The morphology and the pore orientation of the particles were characterized by SEM (Figure 1a) and TEM (Figure 1b). From the both analysis techniques we can see that the homogeneous particles are spherical shaped and with a diameter of around 100 nm. The pores are clearly seen under TEM. The pores are arranged parallel with each other without any inter-connection in between. This means that all pores are accessible.

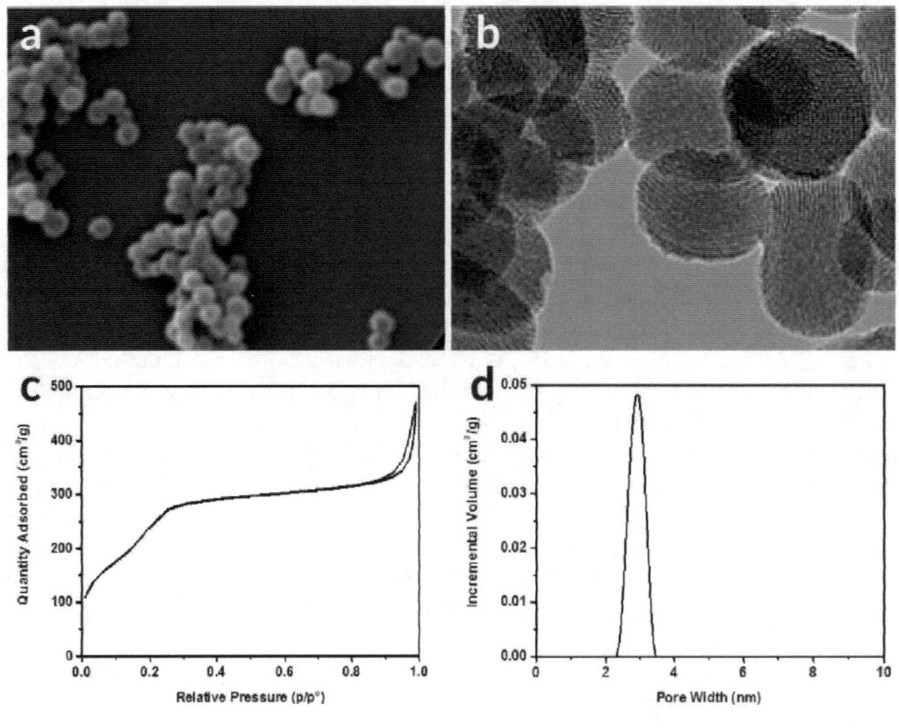

Figure 1. (a) SEM, (b) TEM image, (c and d) nitrogen adsorption isothermal curve and pore distribution of the MCM-41 type mesoporous silica particles.

From the nitrogen adsorption measurement, we can see that the adsorption and desorption gave us typical IV isothermal curve (Figure 1c-d). The capacity condensation of the nitrogen gas molecules in small mesopores during the adsorption process is overlapped with the region of the multiple-layer adsorption. It is suggested that we have small mesopores in MCM-41 type mesoporous silica. In the condensation process, no adsorption hysteresis was observed suggest that the pore diameter is constant without variation. By DFT model the pore distribution was calculated. The pores are mono-dispersed and the diameter of the pore is 2.95 nm. Since the diameter of the particles is only 100 nm, the aggregation of particles has formed large spaces (>50nm) in between the particles at high relative pressure in the right part of the isothermal curve. Traditionally BJH method has been applied for measuring the mesopore structure [9]. By BJH and DFT method, the obtained pore diameter is 2.1 nm and 2.95 nm respectively, the error of almost 30% is comparable with the earlier report [10].

The nanocomposite mesoporous silica alginate was synthesized by dropping pre-mix composition Alginate/Mesoporous silica/Cefixime (200 mg of Alginate, 100 mg of Mesoporous Silica, and 100 mg of Cefixime) in total of 17.5 mL solution into solution of calcium chloride (3% w/v concentration). The release behavior of antibiotics from the micrometer sized beads containing cefixime antibiotic were examined by mixing certain number of beads with pepsin-less simulated gastric fluid.

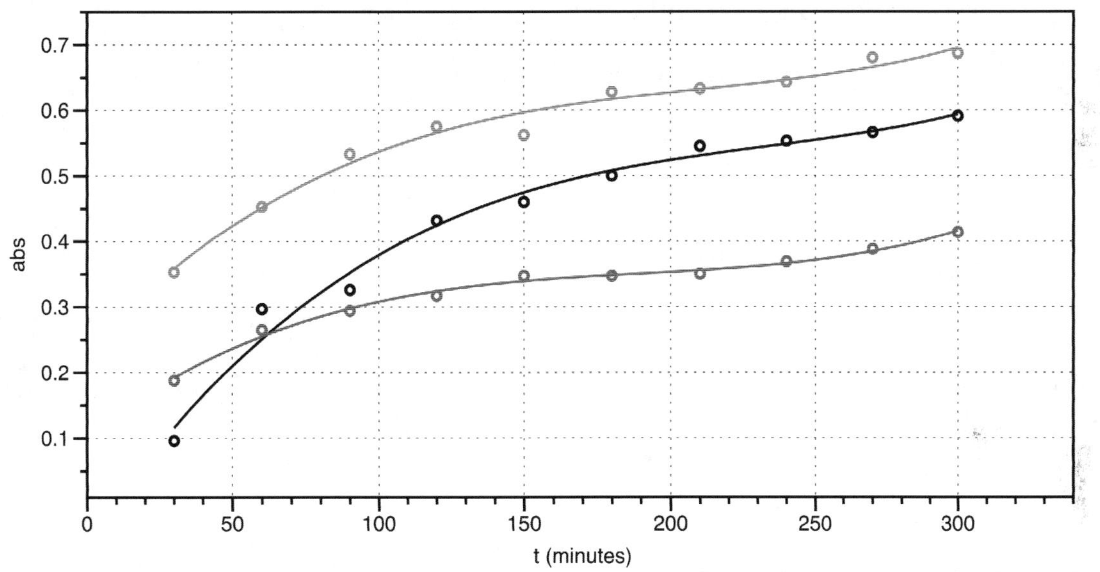

Figure 2. Release profile of cefixime Green: 40-Meso-alginate, Black: 40-alginate, Red: 20-Meso Alginate.

Figure 2 shows the release profile of cefixime under different condition. The green line depicted the release behaviour of release from pristine alginate beads having 40 mg of cefixime, green line depicted the release behaviour of release from pristine alginate beads having 20 mg of cefixime, and the red line depicted release behaviour from nanocomposite mesoporous silica-alginate having 40 mg of cefixime. Form the pictures we can see that nanocomposite beads gave lowest initial release, meaning that the mesoporous silica particles were able to keep cefixime from early release. A constant release can be observed and after 5 hours the nanocomposite did not reach the plateau. Prolonged experiment is needed to understand the release in long period of time.

4. Conclusion

The incorporation of mesoporous silica into nanocomposite mesoporous silica-alginate able to control the release of antibiotics in better way.

IOP Conf. Series: Materials Science and Engineering **858** (2020) 012032 doi:10.1088/1757-899X/858/1/012032

Acknowledgement
The authors would like to thank to Indonesian Ministry of Research, Technology and Higher Education for financial support.

References

[1] Andersson D I and Hughes D 2014 *Nat. Rev. Microbiol.* **12** 465
[2] Turnidge J D 1998 The pharmacodynamics of β-lactams *Clin. Infect. Dis.* **27** 10-22
[3] Bruschi M L 2015 *Strategies to Modify the Drug Release from Pharmaceutical Systems*: Elsevier Science)
[4] Kluin O S, van der Mei H C, Busscher H J and Neut D 2009 *Biomaterials* **30** 4738-42
[5] Jackson J, Fernandes A W and Nelson W 2006 *Clin. Ther.* **28** 1462-71
[6] Gao P, Nie X, Zou M, Shi Y and Cheng G 2011 *The Journal of antibiotics* **64** 625-34
[7] Gotfried M H 2003 *Expert Rev. Anti Infect. Ther.* **1** 9-20
[8] Donato L, Atoini Y, Prasetyanto E A, Chen P, Rosticher C, Bizzarri C, Rissanen K and De Cola L 2018 *Helv. Chim. Acta* **101** e1700273
[9] Barrett E P, Joyner L G and Halenda P P 1951 *J. Am. Chem. Soc.* **73** 373-80
[10] Kruk M, Jaroniec M and Sayari A 1997 *J. Phys. Chem. B* **101** 583-9

IOP Conf. Series: Materials Science and Engineering **858** (2020) 012033 doi:10.1088/1757-899X/858/1/012033

Development of Hybrid Gelatine/Alginate/PVA Hydrogels for Extended Delivery of Antibiotics

A. Mujono, J. Evelyn, and E.A. Prasetyanto[1]
Dept. of Pharmacy, School of Medicine and Health Sciences, Atma Jaya Catholic University of Indonesia.
Jl. Pluit Raya no 2, Jakarta 14440 Indonesia

E-mail: prasetyanto@atmajaya.ac.id

Abstract. Low compliance during consumption of antibiotics will result higher risk to antibiotic resistance. By modifying the release rate of the antibiotic, the patient will be able to consume antibiotics in less frequency while maintaining a stable drug concentration in the plasma, thus lowering the risk of antibiotic resistance. In this work, we report the use of hybrid hydrogels based on the Gelatine/Alginate/PVA as drug release controller of Cefixime, one of the most used antibiotics in Indonesia. The presence of Gelatine, PVA and Cefixime in the hydrogel system was analysed by FTIR spectroscopy. The release of Cefixime from the system is monitored under simulated gastric fluid. The results show that the release rate of the cefixime can be controlled by varying formulation of the hydrogels system. The presence of PVA in the formulation increase the release rate of the Cefixime.

1. Introduction

Typhoid fever is an infectious disease caused by *Salmonella enterica* serovar *Typhi*. Typhoid fever is common in populations that have poor sanitation and have difficulty accessing clean water. The management of typhoid fever is by administering antibiotics. One of the antibiotics that can be used is the class III cephalosporin class of antibiotics. Cephalosporins are one of the antibiotics belonging to the β-lactam group. Like other β-lactam antibiotics, cephalosporins have β-lactam rings and work by inhibiting the synthesis of bacterial cell wall peptidoglycan [1]. Cephalosporins consist of several classes, namely the first to third generation, antipseudomonal cephalosporins, and anti-MRSA cephalosporins (Methicillin-resistant *Staphylococcus aureus*). Each cephalosporin group has different antimicrobial activity. The first class cephalosporins have good activity against gram-positive bacteria and good activity against gram-negative bacteria. The second class cephalosporins have better activity in gram negative bacteria but not as good as the third generation cephalosporins. Third generation cephalosporin activity against gram-positive cocci is less active but has very good antistreptococcal activity. In addition, third generation cephalosporins also have better activity against Enterobacteriaceae compared to previous generation cephalosporins. Antipseudomonal cephalosporins have better activity against gram negative bacteria compared with third generation cephalosporins especially against *Pseudomonas aeruginosa*. The last class, namely anti-MRSA cephalosporin is an antibiotic that is able to fight the MRSA bacteria by sticking to the modified PBP (Penicillin Binding Protein) surface. Cefixime has t1/2 plasma for 3-4 hours and is excreted in urine and eliminated in bile. The standard dose for adults is 400 mg/d for 5-7 days and is extended for cases of *S. pyogenes* infection. Standard doses for children under 6 months and older and weighing less than 45 kg are given based on body weight (8 mg/kg/d). Cefixime is given orally to fight urinary tract infections caused by *E. coli* and *P. mirabilis*, otitis media caused by *H. influenzae* and *S. pyogenes*, and pharyngitis caused by *S. pyogenes*. Cefixime has a high effectiveness for curing typhoid fever [2, 3]. There are 2 main classifications of the Drug Delivery System, namely immediate release and modified release [4, 5]. Modified releases can be divided into extended release, delayed release, targeted release, dan orally disintegrating tablets (ODT) [6, 7]. Antibiotics are generally produced in the form of immediate release drug (IR). IR is a drug whose speed or absorption is not manipulated. IR is formulated to release the drug directly after consumption. Slow release drug is a variation of extended release drug. Extended release drugs can control the release of drugs in a certain time, speed, duration, and location. The use of slow release drugs will help reduce the minimum frequency of drugs, which in turn will reduce the number of non-compliance with taking antibiotics [8, 9].

In this work, we would like to encapsulate third generation cephalosporins such as cefixime, which are usually used to treat gastrointestinal infection in a nano-hybrid hydrogel system for oral drug delivery [10]. The system will consist of PVA/Alginate, in which the drug will be loaded, that will be encapsulated in a micrometres size gel particle to allow the protection of the drug and avoid its early release. The gel microparticles is a biocompatible soft material permeable to small molecules and through which the drug can be easily released. The regulator of slow release in slow release drugs can be made with various materials. In this study, gelatine was used as a regulator of slow release. Gelatine is a natural polymer. Natural polymers have the advantage of being biocompatible, biodegradable and do not cause systemic toxicity when administered. In addition, gelatine is also easily available at inexpensive prices.

2. Materials and Methods

All materials used in this experiment were analytical grade without further purification. UV-Vis and FTIR measurement were done on Shimadzu UV-1800 UV Vis Spectrophotometer and Shimadzu IRTracer-100 Spectrophotometer, respectively. Pepsin-less Simulated gastric fluid was prepared by dissolving 2.0 g of sodium chloride and 7.0 ml of hydrochloric acid in 1000mL of deionized water with final pH of 1.2. Alginate hydrogel beads and hybrid alginate/gelatine/PVA hydrogel beads were synthesized by pipette dropping pre-mix composition alginate/gelatine/PVA/Cefixime (200 mg of Alginate, and 100mg of Cefixime) in total of 17,5 mL solution into solution of calcium chloride (3% w/v concentration).

3. Results and Discussion

The millimetre size Gelatine/Alginate/PVA hydrogels beads were obtained by drop casting a mixture alginate/gelatine/PVA/Cefixime into solution of calcium chloride. After stirring for 25 minutes in the calcium chloride solution, a semi hard hydrogels beads with opaque colour were formed. The longer the beads stay in the calcium chloride solution harder beads will be obtained due to higher degree of crosslinking. After drying of the beads in the desiccator, the beads were able to swell again with quite high swelling ratio (up to 300% in volume).

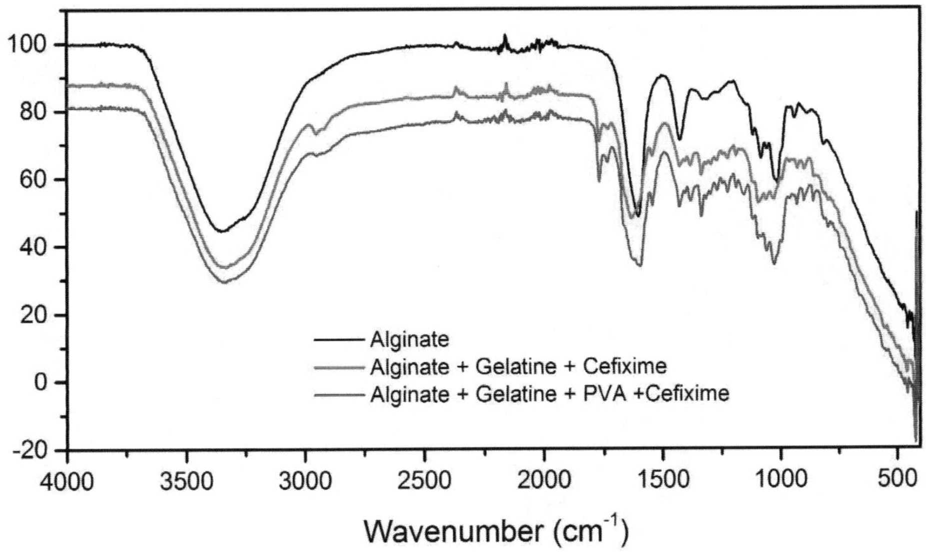

Figure 1. FTIR spectra of dried alginate, hybrid alginate/gelatine loaded with cefixime and hybrid alginate/gelatine/PVA loaded with cefixime.

Figure 1 shows the FITR spectra of the alginate, hybrid alginate/gelatine loaded with cefixime and hybrid alginate/gelatine/PVA loaded with cefixime (all in dried bead condition). The dried samples were measured by attenuated total reflection (ATR) technique with the range of from 4000 to 480 cm^{-1}. From the spectra shown in Figure 1, it is clearly showed the presence of Cefixime and Gelatine in the hybrid hydrogels system. The presence of cefixime was recognized by CH stretching band at around 2800 cm^{-1}. While the incorporation of gelatine gave additional bands 1750 cm^{-1} and 1500 cm^{-1}. The measurement was done in triplicate with different samples and the spectra were consistent meaning the hybrid system are relatively homogeneous. Further addition of PVA did not alter the spectra of the samples.

The release profile of cefixime from the hybrid hydrogels we examined under condition mimicking human gastric. For this purpose, the hydrogels containing cefixime were stirred in the pepsin-less simulated gastric fluid. From the UV-Vis spectroscopy analysis, cefixime has absorption maximum at 285 nm. Figure 2 shows UV-Vis spectra of simulated gastric solution after in contact with cefixime contained hybrid hydrogels for certain minutes.

The samples were obtained by vigorously mixing Hybrid Gelatine/Alginate and Hybrid Gelatine/Alginate/PVA Hydrogels containing 40 mg (Figure 2a and c) and 80 mg (Figure 2b and d) in 100mL of pepsin-less simulated gastric fluid. Samples were taken every 30 minutes for 5 hours and immediately analysed by UV-Vis spectroscopy. The concentration of cefixime in the simulated gastric fluid increased by increasing of contact with hydrogel beads. In comparison with hybrid system without PVA, the Hybrid Gelatine/Alginate/PVA Hydrogels release cefixime faster. The release of cefixime is not yet done at 5 hours of contact time.

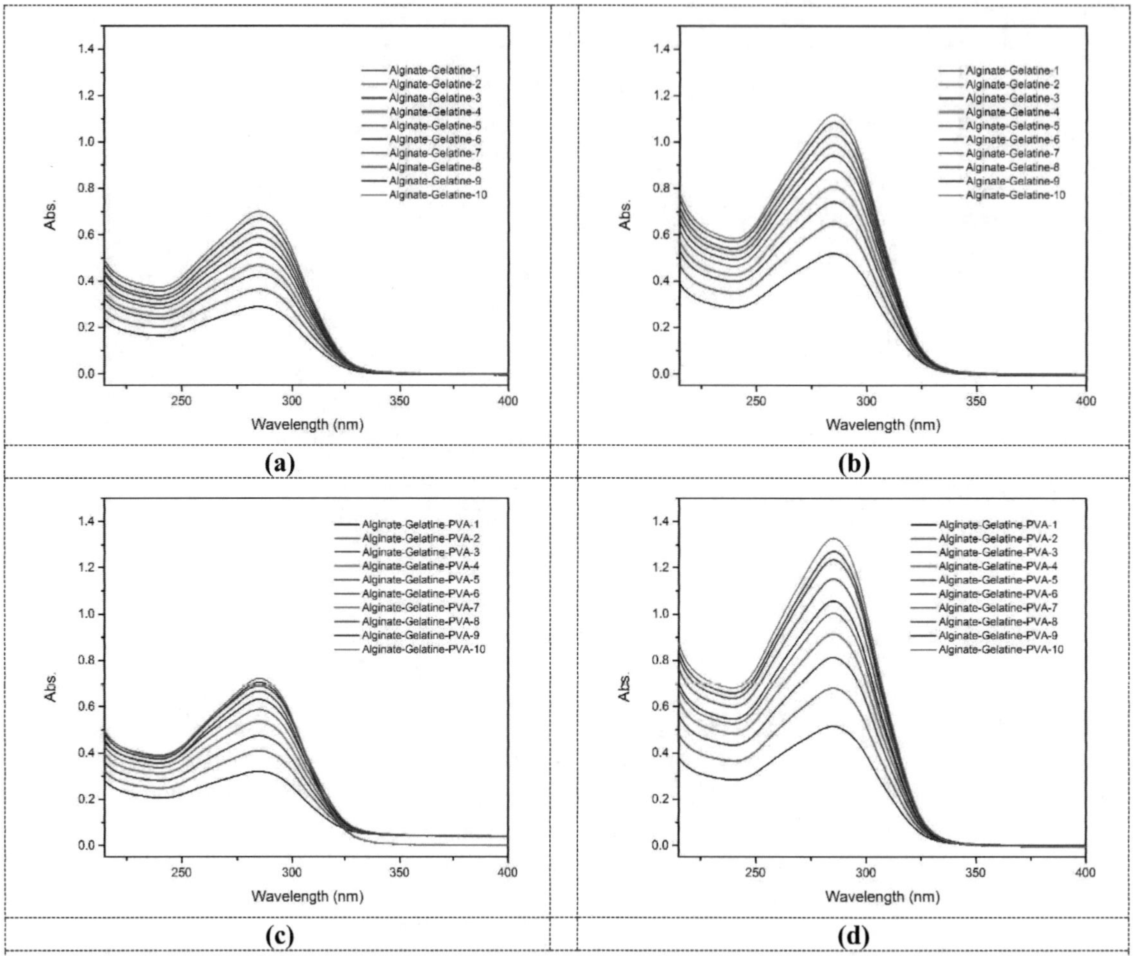

Figure 2. Release profile of cefixime from the Hybrid Gelatine/Alginate/PVA Hydrogels over time in the pepsin-less simulated gastric fluid.

4. Conclusions
We successfully incorporated cefixime into the Hybrid Gelatine/Alginate/PVA Hydrogels and the results show that the release rate of the cefixime can be controlled by varying formulation of the hydrogels system. The presence of PVA in the formulation increase the release rate of the Cefixime.

5. Acknowledgement
The authors would like to thank to Indonesian Ministry of Research, Technology and Higher Education for financial support.

6. Reference
[1] Brunton L L, Knollmann B C and Hilal-Dandan R 2018 Goodman & Gilman's : The Pharmacological Basis of Therapeutics

[2] Joshi R D, Khadka S, Joshi D M, Shrestha B, Dangal G, Acharya K P, Shrestha S and Dongol Y 2018 Antimicrobial Sensitivity Trend in Blood Culture Positive Enteric Fever *J Nepal Health Res Counc* **16** 228-32

[3] Nawaz A, Aslam A and Haq G 2018 Potent Antibiotic to Treat Typhoid Fever in Patients of Pediatric Age-group *Asian Journal of Research in Medical and Pharmaceutical Sciences* **3** 1-5

[4] Kogan S, Zeng Q, Ash N and Greenes R A 2001 Problems and challenges in patient information retrieval: a descriptive study *Proc AMIA Symp* 329-33
[5] Turnidge J J C i d 1998 The pharmacodynamics of β-lactams **27** 10-22
[6] Venkatesh G M, Stevens P J and Lai J-W 2012 Development of orally disintegrating tablets comprising controlled-release multiparticulate beads *Drug Dev. Ind. Pharm.* **38** 1428-40
[7] Bruschi M L 2015 *Strategies to Modify the Drug Release from Pharmaceutical Systems*: Elsevier Science)
[8] Fair R J and Tor Y 2014 Antibiotics and bacterial resistance in the 21st century *Perspect Medicin Chem* **6** 25-64
[9] Gao P, Nie X, Zou M, Shi Y and Cheng G J T J o a 2011 Recent advances in materials for extended-release antibiotic delivery system **64** 625
[10] Kaurthe J 2013 Increasing antimicrobial resistance and narrowing therapeutics in typhoidal salmonellae *J Clin Diagn Res* **7** 576-9

Cathafiline from *Cassytha filiformis* and BR-Xanthone A from *Garcinia mangostana* as potential bone morphogenetic protein receptor type I (BMPR-I) inhibitor for iron-refractory iron deficiency anemia *in silico*

Y H Suselo[1,2], A Alhaqq[3] and D Indarto[1,2]

[1] Department of Physiology, Faculty of Medicine, Universitas Sebelas Maret
[2] Biomedical Laboratory, Faculty of Medicine, Universitas Sebelas Maret
[3] Undergraduate Medical Student, Faculty of Medicine, Universitas Sebelas Maret

E-mail: yulianaheri@staff.uns.ac.id

Abstract. Oral iron supplementation has become a standard therapy for anemia around the world but some anemic patients are not responsive to it, as well known as Iron Refractory Iron Deficiency Anemia (IRIDA). One of the causes of IRIDA is BMPR-I activation. LDN-193189 is a synthetic small molecule that is developed for inhibition of BMPR-I. In the human liver, this receptor consists of 4 isoforms of Activin Like Kinase (ALK) 1, 2, 3, and 6. However, the synthetic molecule is not specifically bind to BMPR-I. This study aimed to identify Indonesian phytochemicals that can inhibit BMPR-I as IRIDA therapy with molecular docking approach. This bioinformatics study used 517 phytochemicals, which were registered in HerbalDB, had molecular structure and met the criteria for Lipinski's rule of five. Three dimensional structure of LDN-193189 as a standard compound was found in complex with BMPR-I subtype ALK2 and obtained from Protein Data Bank (ID: 3Q4U). The AutoDock Vina 1.1.2. software was used to perform molecular docking between LDN-193189-ALK2 and phytochemicals-ALK-2. Binding complexes of ALK2 and LDN-193189/phytochemicals were visualized using PyMol 1.3 and Chimera 1.12 programs. The potential candidate of BMPR-I inhibitor was analysed based on docking score, binding site and conformation of phytochemicals toward ALK2. It revealed that BR-Xanthone A had lower docking score than LDN-193189 (-11.40 kcal/mol vs 11.30 kcal/mol). Phytochemicals which have hydrogen bonds to ATP binding site of ALK2 and similar conformation with LDN-193189 were Cathafiline and BR-Xanthone A. Both of those phytochemicals met the criteria of Lipinski's rule of five. Cathafiline and BR-Xanthone A were potential as inhibitor BMPR-I for treatment of IRIDA.

1. Introduction

Iron deficiency anemia is an imbalance of iron intake, iron storage, and iron loss in the body that cause insufficiency of iron reserves to produce erythrocytes [1]. The prevalence of anemia in Indonesia is still relatively high at 28.1% in children aged 12-59 months and 37.1% in pregnant women. About 50% of those cases are iron deficiency anemia [2,3].

First-line therapy for iron deficiency anemia is oral iron supplementation. However, there are some patients who do not respond to oral administration. Level of serum hepcidin in unresponsive patients is higher than responsive patients of oral supplementation [4]. Low oral responsiveness can be caused by chronic inflammation or polymorphism in genes involved in iron regulation such as Divalent Metal Transporter (DMT) -1 or TMPRSS6 (Transmembrane Protease, Serine 6) [5]. Polymorphism in TMPRSS6 gene causes hepcidin production increase, therefore iron export from lumen into the circulation decrease. This condition is called Iron Refractory Iron Deficiency Anemia (IRIDA), which patients will be unresponsive of oral iron therapy [6]

Other therapies targeting the hepcidin axis have been developed at this time [7]. Hepcidin is the main regulator of iron which can prevent the release of iron from the intestinal lumen into the circulation [8]. Hepcidin axis consist of BMP (Bone Morphogenetic Protein)/ SMAD (Sma and Mothers Against Decapentaplegic) pathway and IL-6 (Interleukin-6)/ STAT (Signal Transduction and Activator of Transcription) pathway. One molecular therapy that targets the hepcidin axis is LDN-193189, a small molecule inhibitor of BMPR-I (Bone Morphogenetic Protein Type I), specifically the ALK (Activin Like Kinase) 2 and ALK3 subtypes [7]. LDN-193189 works by ATP-mimetic method, which is to compete with ATP to occupy the ATP binding site of ALK2 [9,10]. ALK2 has some ATP binding site in the ATP binding pocket which are Val214, Gly215, Lys216, Gly217, Arg218, Tyr219, Gly220, Val222, Ala233, Lys235, Thr283, His284, His286, Ser290, Lys340, Asr34 Leu343, and Asp354 residue [11]. High concentration of LDN-193189 can inhibit other kinases that also require ATP in their activities therefore it has low specificity towards ALK2 and ALK3 [12]. Thus, LDN-193189 is not developed anymore. Even so, BMPR-I inhibitors are proven to be effective therapy for IRIDA, so alternative therapy other than LDN-193189 should be further developed [13].

Indonesia has 9,600 plant species that have pharmacological activities [14]. So far, no research has been conducted to identify phytochemical compounds of Indonesian herbal plants that have BMPR-I inhibitor activity to suppress the production of hepcidin in IRIDA. The method that is widely used in drug development today is the molecular docking method, which is a computational method for predicting the bond between ligands and macromolecules efficiently [14]. So this study aimed to identify Indonesian phytochemical compounds that have the potential as BMPR-I inhibitors of IRIDA by molecular docking approach.

2. Experimental

The samples of this bioinformatics study were obtained from the HerbalDB database and their three-dimensional structures were obtained from PubChem. The standard compound used in this research was LDN-193189. 3D structure LDN-193189 binding to ALK2 subtype of BMPR-I was obtained from Protein Data Bank with code: 3Q4U. AutoDock Vina 1.1.2 was used to analyze the binding energy of molecular docking between phytochemicals toward ALK2. The docking results was visualized using PyMol 1.3 and Chimera 1.12.

Before running the molecular docking, the structure of ALK2 was prepared first. ALK2 structure was separated from its ligand (LDN-193189) using AutoDock Tools version 1.5.6. Water molecules were then removed from ALK2 and polar hydrogen molecules were added. The initial visualization of ALK2 and LDN-193189 binding was performed using Chimera 1.12 to get the information of LDN-193189 binding site on ALK2 and to determine the grid box that would be used in docking process then. The established grid box was at X center = +22.65 size=22 Å, Y center = -19.05 size=30 Å, and Z center = +6,56 size=24 Å. Secondly, validation of LDN-193189 toward ALK2 was performed using AutoDock Vina 1.1.2. The validation aimed to get the standard molecular docking score between ALK2 and LDN-193189 for phytochemicals screening. The validation was performed at least three times to get valid score. Then phytochemicals were molecularly docked with ALK2 to get the binding energy scores. The analysis was performed by comparing docking scores, binding locations, binding types, and conformation of phytochemical toward ALK2.

3. Results and Discussion

Initial visualization was performed firstly to get the information of LDN-193189's binding site on ALK2. The result of initial visualization was LDN-193189 interacted with ALK2 on several non-polar amino acids such as Val214, Val222, Ala233, Leu263, Tyr285, Leu343, and Ala353. In addition, there were also interactions between LDN-193189 with several amino acids such as Lys235, His286, Thr283, Glu287, Gly289, Asp293, Lys340, and Asn341 to strengthen LDN-193189's position towards ALK2. There was a hydrogen bond between the N atom on LDN-193189 and the H atom on His286 with a distance of 2,212Å. LDN-193189 was also predicted having water-mediated hydrogen bond at Glu248. However, this study used Chimera 1.12 software to perform visualization therefore water-mediated hydrogen bond could not be visualized. Those interactions helped us to determine the grid box needed in docking process then.

Secondly, interaction of LDN-193189 and ALK2 was validated to get the standard binding energy score. The obtained binding energy score from the interaction between LDN-193189 with ALK2 is -11,30 kcal/mol (Table 1).

Table 1. Validation results of LDN-193189 interaction toward ALK2

Ligand	Docking score (kcal/mol)					Docking score mean (kcal/mol)	Deviation standard
	I	II	III	IV	V		
LDN-193189	-11.30	-11.30	-11.30	-11.30	-11.30	-11.30	0

Validation results showed that LDN-193189 interaction toward ALK2 was found at Val214, Val222, Ala233, Lys235, Leu263, Thr283, Tyr285, His286, Glu287, Gly289, Lys340, Asn341, Leu343, and Ala353 as shown in figure 1. Water-mediated hydrogen bond could not be visualized because water molecules had been removed during the preparation. The absence of water-mediated hydrogen bond caused a slight shift in the position of LDN-193189 when compared to the prediction in initial visualization as shown in figure 2, so the hydrogen bond with His286 could not be seen. However, there were still interactions between LDN-193189 and His286, which was the interaction of N atom on His286 with H atom on LDN-193189 with a distance of 2.497Å and the interaction of N-N atoms with a distance of 3.49Å.

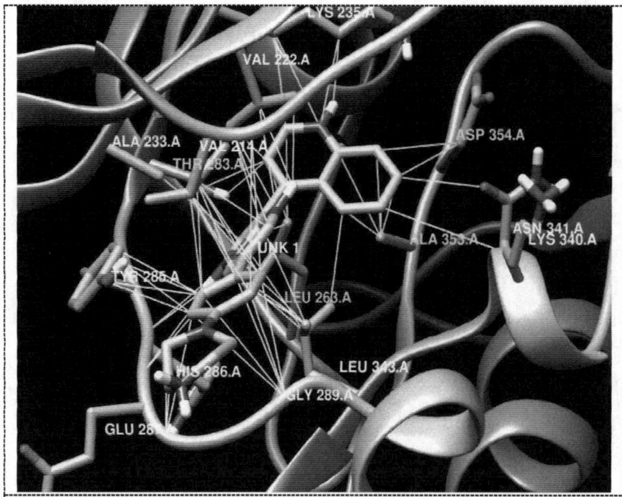

Figure 1. The binding of LDN-193189 toward ALK2. Visualization was using Chimera 1.12.

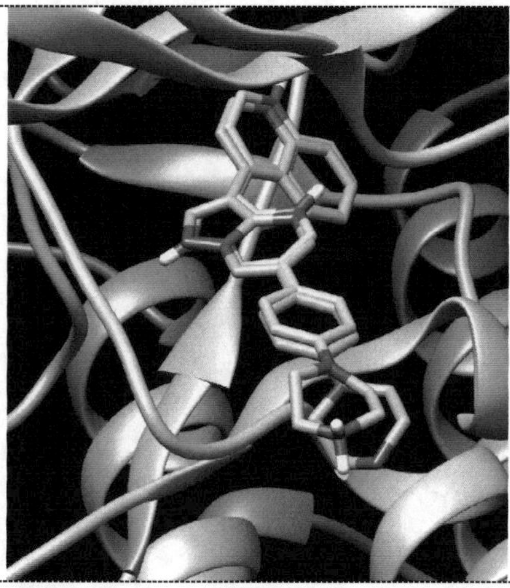

Figure 2. Visualization of the validation result (light blue) compared with the prediction (purple). Visualization was using Chimera 1.12.

Molecular docking was performed on 517 Indonesian phytochemicals in the HerbalDB database which had PubChem ID to identify their potential as BMPR-I inhibitor. Molecular docking was performed three times to get valid docking scores. It revealed that there was only one compound having docking score lower than the standard, which was BR-Xanthone A, with a docking score of -11.40 kcal/mol. It means that BR-Xanthone A binds ALK2 easier than the standard do. BR-Xanthone A had hydrogen bonds with Thr283 and Asp293 as shown in figure 3. In addition to hydrogen bonds, BR-Xanthone A also had hydrophobic interactions with non-polar amino acids such as Val214, Val222, Ala233, Leu263, Tyr285, Leu343, and Ala353. There were also interactions of BR-Xanthone A with other amino acids, such as His286, Glu287, Gly289, Asp293, and Lys340 (Table 2). BR-Xanthone A also had similar conformation with the standard as shown in figure 4 although it had different binding site of hydrogen bonds. Figure 5 shows all interactions of BR-Xanthone A toward ALK2.

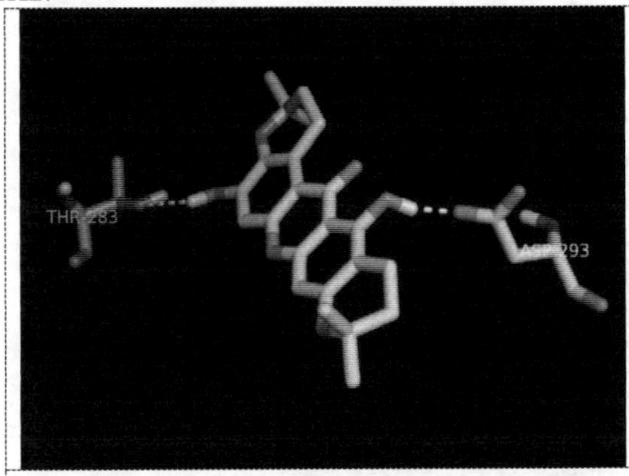

Figure 3. The hydrogen bonds of BR-Xanthone A toward ALK2. Visualization was using PyMol 1.3.

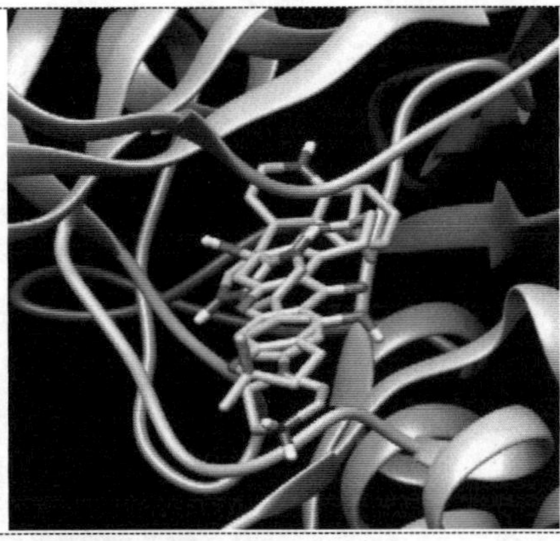

Figure 4. Visualization of the BR-Xanthone A (light blue) compared with the standard (purple). Visualization was using Chimera 1.12.

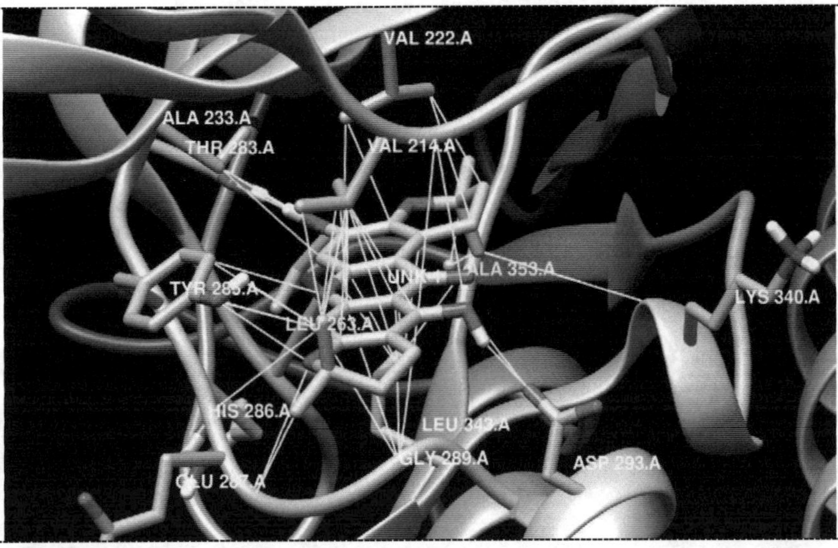

Figure 5. Interactions of BR-Xanthone A toward ALK2. Visualization was using Chimera 1.12.

IOP Conf. Series: Materials Science and Engineering **858** (2020) 012034 doi:10.1088/1757-899X/858/1/012034

This study assessed the phytochemical's potential not only by the docking score, but also the binding types, the number of ATP binding sites occupied, as well as the conformation or phytochemical position toward ALK2 compared with the standard. Therefore, this study also analysed seven other phytochemicals with the lowest docking scores to compare their potential with BR-Xanthone A and standard. Another phytochemical having higher potential among others was Cathafiline. It had docking score -10.60 kcal/mol and similar binding sites compared with the standard. It also had hydrogen bonds on ATP binding sites more than the standard had. Cathafiline had hydrogen bonds on Lys235, His284, and His286, hydrophobic interactions with Val214, Val222, Ala233, Leu263, Leu281, Tyr285, Leu343, and Ala353, and several other bonds (Table 2). All interactions of cathafiline toward ALK2 can be seen in figure 6, while its hydrogen bonds are shown in figure 7. Cathafiline also had similar conformation compared with LDN-1933189 as shown in figure 8.

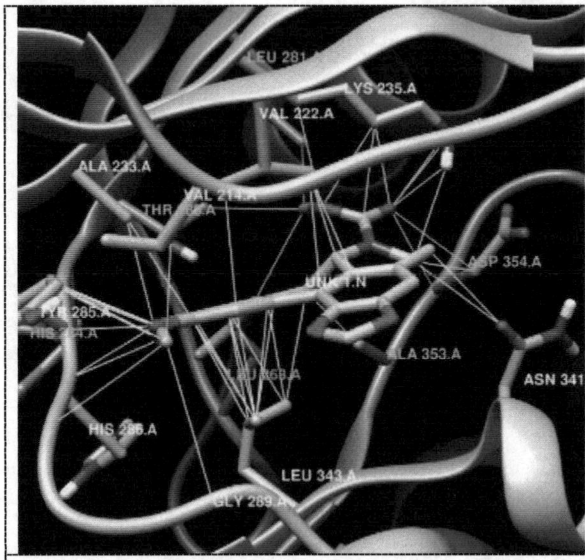

Figure 6. Interactions of Cathafiline toward ALK2. Visualization was using Chimera 1.12.

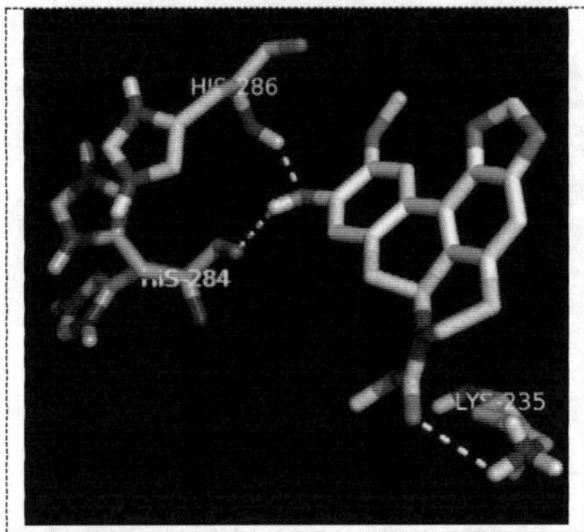

Figure 7. The hydrogen bonds of Cathafiline toward ALK2. Visualization was using PyMol 1.3.

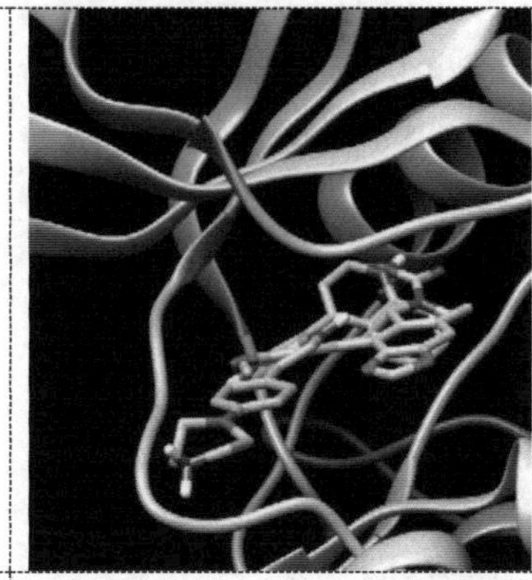

Figure 8. Visualization of the Cathafiline (light blue) compared with the standard (purple). Visualization was using Chimera 1.12.

Table 2. Interactions of phytochemicals toward ALK2 compared with LDN-193189[a]

Ligand	Docking score (kcal/mol)	Hydrogen bond location	Other interactions	ATP binding sites occupied	Conformation
LDN-193189	**-11.30**	**His286**	**Val214, Val222, Ala233, Lys235, Leu263, Thr283, Tyr285, His286, Glu287, Gly289, Lys340, Asn341, Leu343, Ala353, Asp354**	Val214, Val222, Ala233, Lys235, Thr283, His286, Lys340, Asn341, Leu343, Asp354	-
BR-Xanthone A	-11.40	**Thr283**, Asp293	**Val214**, **Val222**, **Ala233**, **Leu263**, **Tyr285**, **Leu343**, **Ala353**, **His286**, **Glu287**, **Gly289**, Asp293, **Lys340**	Val214, Val222, Ala233, Leu343, His286, Lys340, Thr283	Similar
Cathafiline	-10.60	**Lys235**, His284, **His286**	**Val214**, **Val222**, **Ala233**, **Lys235**, **Leu263**, Leu281, **Thr283**, His284, **Tyr285**, **His286**, **Gly289**, **Asn341**, **Leu343**, **Ala353**, **Asp354**	Val214, Val222, Ala233, Lys235, Thr283, His284, His286, Asn341, Leu343, Asp354	Similar

[a] Bold residues were the same binding sites as standard's

Lipinski's rule of five is a theory used to determine the bioavailability of a drug compound. If a compound meets these criteria, the compound can be developed as a drug, because it will be well absorbed by the body. There are four things that must be fulfilled by a compound to be absorbed by the body, which are: the donor hydrogen bond < 5, the acceptor hydrogen bond < 10, molecular weight < 500 Daltons, and the logarithmic coefficient of the octanol-water ratio in the compound (logP) < 5 [15]. BR-Xanthone A and Cathafiline have met the criteria of Lipinski. BR-Xanthone A has a molecular weight of 396,439 Da, lower than LDN-193189, indicating that the absorption rate is better than LDN-193189. Cathafiline has a molecular weight of 369,373 Da, lower than the standard. It also has lipophilicity (LogP) less than the standard, therefore Cathafiline is more hydrophilic and has lower toxicity (Table 3).

Table 3. Lipinski's criteria of LDN-193189 and phytochemicals

Active compound	Molecular Formula	Lipinksi's Rule of Five			
		Molecular weight (<500) (Da)	H-Bond Donor (<5)	H-Bond Acceptor (<10)	Lipophilicity (<5)
LDN-193189	$C_{25}H_{22}N_6$	406,493	1	5	3,3
BR-Xanthone A	$C_{23}H_{24}O_6$	396,439	2	6	4,8
Cathafiline	$C_{20}H_{19}NO_6$	369,373	1	6	2,8

BR-Xanthones A is a compound found in the skin of *Garcinia mangostana* (mangosteen) [16]. This compound belongs to the family of xanthonoid type, pyranoxanthone class. This compound is considered to have antioxidant activity [17]. Cathafiline is a compound of *Cassytha filiformis* [18]. This plant has been used as a therapy for gonorrhea, cancer, kidney disease, as a diuretic, and other diseases. Parts of plants used to obtain pharmacological activities are leaves. This plant has pharmacological activity as an antioxidant, antitrypanosomal, antiplatelet, and vasorelaxant [19].

We realized that this study has some limitations. AutoDock Vina used in this study needs removal of water molecule when doing the receptor preparation. Therefore, water-mediated hydrogen bonds cannot be performed. In addition, Chimera 1.12 dan PyMol 1.3 used in this study cannot identify other bonds than hydrogen bonds. This study is just a prior study, therefore next development study is needed.

4. Conclusion

Cathafiline and BR-Xanthone A were potential as BMPR-I inhibitor for treatment of IRIDA *in silico* with the docking score of -10.60 kcal/mol and -11.40 kcal/mol respectively. Both compound had higher amount of hydrogen bonds than the standard and similar conformation as the standard. Those compound can be found in *Cassytha filiformis* and *Garcinia mangostana* respectively.

References

[1] Miller J L 2013 *Cold Spring Harb Perspect Med* **3** a011866
[2] Departemen Kesehatan 2013 *Riset Kesehatan Dasar (Riskesdas) 2013* (Jakarta: Badan Litbangkes)
[3] WHO 2015 *The global prevalence of anaemia in 2011* (Geneva: WHO)
[4] Bregman D B, Morris D, Koch T A, He A and Goodnough L T 2013 *Am J Hematol* **88** 97-101
[5] Nairz M, Theurl I, Wolf D and Weiss G 2016 *Wien Med Wochenschr* **166** 411-23
[6] Falco L D, Sanchez M, Silvestri L, Kannengiesser C, Muckenthaler M U, Lolascon A, Gouya L, Camaschella C and Beaumont Ce 2013 *Haematologica* **98** 845-53
[7] Fung E and Nemeth E 2013 *Haematologica* **98** 1667-76
[8] Ganz T and Nemeth E 2015 *Nat Rev Immunol* **15** 500-10
[9] Sanvitale C E *et al.* 2013 *PLoS One* **8** e62721
[10] Imai K and Takaoka A 2006 *Nat Rev Cancer* **6** 714-27
[11] Marchler-Bauer A *et al.* 2017 *Nucleic Acids Res* **45** D200-3
[12] Vogt J, Traynor R and Sapkota G P 2011 *Cell Signal* **23** 1831-42
[13] Blanchette N L, Hanz D H, Torti F M and Torti S V 2016 *Expert Rev Hematol* **9** 169-86
[14] Yanuar A, Mun'im A, Lagho A B A, Syahdi R R, Rahmat M and Suhartanto H 2011 *Int J Com Sci* **8** 180–3
[15] Bielska E, Lucas X, Czerwoniec A, Kasprzak J M, Kaminska K H and Bujnicki J M 2011 *BioTechnologia* **92** 249-64
[16] Shinbo Y *et al.* 2006 *Biotechnology in Agriculture and Forestry* **57** 165-6
[17] Zhao Y, Liu J, Lu D, Li P and Zhang L 2010 *Nat. Prod. Res.* **24** 1664-70
[18] Chang F, Chao Y, Teng C and Wu Y 1998 *J. Nat. Prod.* **61** 863-6
[19] Mythili S, Gajalakshmi S, Sathiavelu A and Sridharan TB 2011 *Asian J Plant Sci Res* **1** 77-83

Effect of TiO_2 Addition on Optical Properties of TeO_2-ZnO-PbO Glasses

Ahmad Marzuki*, **Devara Ega Fausta**
Department of Physics, Sebelas Maret University, Surakarta 57126, Indonesia

*E-mail: amarzuki@mipa.uns.ac.id

Abstract. Glasses with composition (in mol%): $60TeO_2$-(30-x)ZnO-10PbO-xTiO_2 (TZPb:Ti) (where x = 0, 1, 2, 3 and 4) have been fabricated and characterised in order to study their physical and optical properties. Density and refractive index of glasses were measured at room temperature by applying Archimedes's principle and Brewster's angle method, respectively. Absorption spectra within range of UV-VIS-NIR were measured at room temperature using UV-VIS Spectrophotometer while those at IR range using FTIR. We have shown that density decreases with the increase of TiO_2 while refractive index and optical energy band gap behave conversely.

Keywords: Tellurite glasses, TiO_2 doped glasses, IR absorption spectra, UV-VIS absorption spectra, refractive index.

1. Introduction

In the last few decades, tellurite glasses has received more attention among other glass families [1] especially due to its potential applications such as optical amplifier, fiber sensors, and lasing materials [2, 3, 4], i.e., utilizing their unique properties such as low melting point, relatively stable against crystallization, large transparency window, high refractive index and large rare earth ion solubility [5,6,7,8,9]. Tellurium oxide, however, is an intermediete glass former, glass cannot be formed by itself. In order to form a glass, addition of alkali oxides or alkaline earth oxide and heavy metal oxides is required [10].

Effect of addition of heavy metal oxide on the glass structure, thermal stability and optic have been carried [11,12]. TiO_2 is one of metal oxides that have been widely used in many applications [13]. In glass technology, TiO_2 addition were studied from different point of view, such as structural change, crystallization tendency and optical properties [14, 15]. In this paper, TiO_2 were added to TeO_2-ZnO-PbO glass aimed at investigating its effect on molar volume, glass density, refractive index, absorption spectra within UV-VIS-NIR and IR range.

2. Experiment

Tellurite glass used as a base have compositions: $60TeO_2$–(30-x)ZnO–10PbO–xTiO_2 (x = 0, 1, 2, 3, 4 mol%). Powders of all components (TeO_2, ZnO, PbO, and TiO_2) were mixed in agate mortar. The mixture was then poured into platinum crucible and melted in electrical furnace at 900°C for 35 min. Stirring was carried out to obtain a bubble-free melt. Casting was carried out by pouring the melt into a preheated square brass mold and subsequently annealed at temperature of 250°C for 6 hours and then cooled very slowly to room temperature (cooling rate 1 min/sec). The glasses were then optically polished for the purpose of optical characterizations.

Glass density *(ρ_g)* of all glasses were measured by applying Archimedes principle using pycnometer. The immersion liquid for this measurement was distilled water *(ρ_d* = 1,000 g/cm^3 at room temperature). Measurements were carried out based on the following equation:

$$\rho_g = \frac{\rho_d(w_{ga}-w_{la})}{[(w_{ga}-w_{la})-(w_{gt}-w_{lt})]} \quad (1)$$

where w_{ga} is mass of empty glass, w_{gt} is mass of distilled water filled in the glass, whilst w_{la} and w_{lt} are weights in open air and destilled water. Molar volumes were then calculated based on density data using equation:

$$V_m = \frac{\sum_i x_i M_i}{\rho_g} \quad (2)$$

where x_i is mole fraction for each glass component *i* which has molecular weight (*Mi*).

Refractive index for all samples were obtained from measurement Brewster's angle method using setup as reported in [16]. Absorption spectra within UV-Vis-NIR range (from 200-900 nm) were obtained from UV-Vis Spectrometer Perkin Elmer Lamda 25 while for that within IR area (400-4000 cm^{-1}) were obtained from Shimadzu Prectige IR 21 Spectroscopy. All the above experiments were carried at room temperature.

3. Results and Discussion

3.1. Density and refractive index

Table 1 shows the density and molar volume which change as the TiO_2 concentration change. It can be seen from Figure 1 that the density of glass decreases with increasing TiO_2 concentration. This result can be easily understood as Zn has higher atomic mass (65.387 amu) compared to that of Ti (47.867 amu). Increasing molar volume as more TiO_2 was added might be easily linked to the atomic radius of Zn (139 pm) compared to that of Ti (147 pm). In addition, this increase might also be linked to the increase of the ratio between nonbridging oxygen (NBO) to bridging oxygen (BO).

Table 1. Density and Molar Volume of TZPbTi Glasses

Composition Code	Density (g/cm^3)	Molar Volume (cm^3/mol)
TZPbTi 0	6,1921	23,0120
TZPbTi 1	6,0877	23,4044
TZPbTi 2	5,9338	24,0087
TZPbTi 3	5,9270	24,0337
TZPbTi 4	5,6993	24,9911

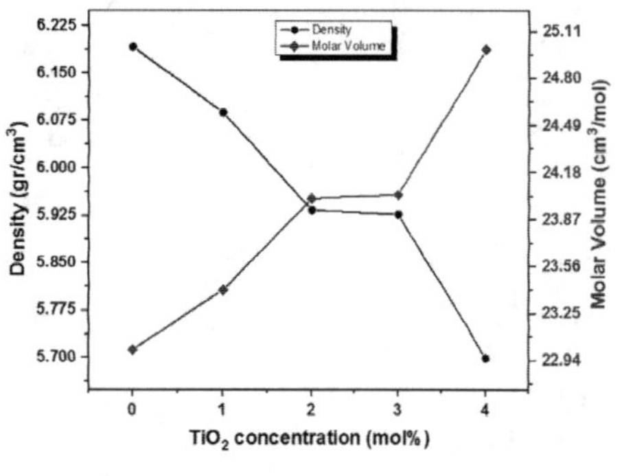

Figure 1. Relations between Density and Molar Volume with TiO_2 concentration.

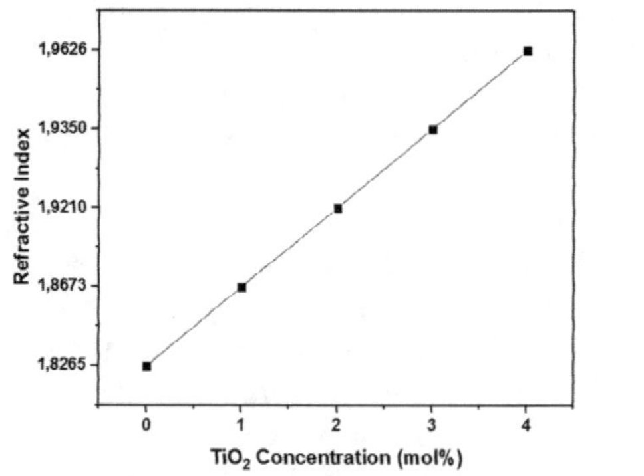

Figure 2. Relations between Refractive Index with TiO_2 concentration

Refractive index change as function of TiO_2 addition is tabulated in Table 2 and visualized in Figure 2. It is seen that refractive index increases with the increase of TiO_2 concentration. Linked to the increasing molar volume as discussed above, this result might be related to the increasing the number of new non-bridging oxygen (NBO). It was reported that the increase of NBO/BO ratio can raise the glass polarizability and result in increasing refractive index [17]. From view point of ionic polarizability, Ti ions have greater ionic polarizability (Ti^{2+}= 0.7012871, Ti^{3+} = 0.6242542 or Ti^{4+} = 0.3912914) than Zn ions (Zn^{2+}= 0.1752656, Zn^{4+} = 1.1759126). Consequently, glass polarizability increases with the increase of Ti ions contents and therefore increase glass refractive index.

Table 2. Refractive Index (n) and Optical Band Gap Energy value (E_g) of TZPb:Ti Glasses

Composition Code	n	E_g (eV)
TZPbTi 0	1,8265	2,63
TZPbTi 1	1,8673	2,62
TZPbTi 2	1,9210	2,60
TZPbTi 3	1,9350	2,55
TZPbTi 4	1,9626	2,52

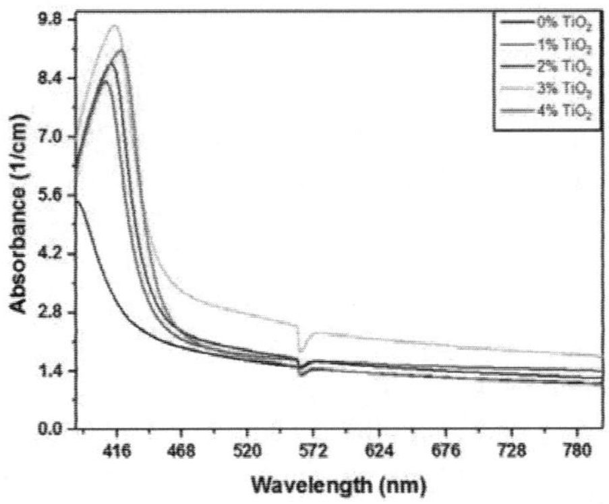

Figure 3. Absorbance of TZPbTi Glasses

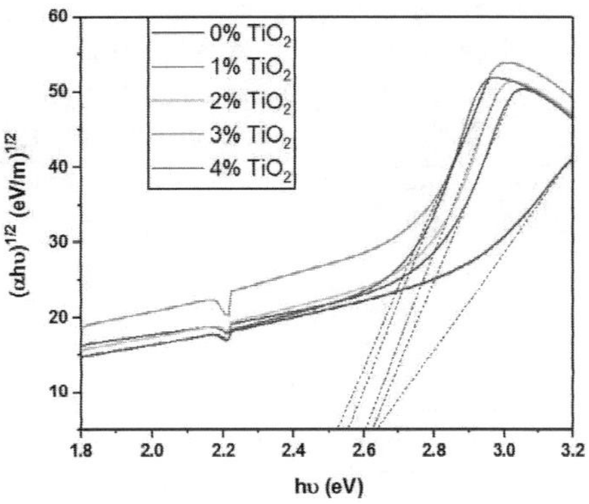

Figure 4. Band gap energy determination of TZPb:Ti glasses

3.2. UV-VIS and IR Absorption Spectra

Absorption spectra of all samples are shown in Figure 3. It can be seen that absorption spectra of samples increases with increasing the TiO_2 content. Absorption peak at around 400 nm is shown in Table 3, the same result was also reported by Rao, et al. 2018 [18] which is resulted for electronic transition in Ti^{3+} ions assigned to $^2B_{2g}\rightarrow{}^2B_{1g}$ transition.

Table 3. Refractive Index and Optical Band Gap Energy value of TZPb:Ti Glasses

Composition Code	Wavelength (nm))	Absorptivity coefficient (1/cm)
TZPbTi 0	384	5,42
TZPbTi 1	407	8,32
TZPbTi 2	412	8,75
TZPbTi 3	414	9,68
TZPbTi 4	420	9,07

Figure 4. shows the band gap energy determination by applying Tauch's plot method. Firstly, graph relating $(\alpha h\upsilon)^{1/2}$ and photon energy (hυ) was constructed. Defining band energy as the energy boundary determining the range of energy transmitted or absorbed by the materials of interest, the energy band gab can be obtained by taking extrapolation of the linear part of the curve indicating the boundary between transmission and absorption area. It is shown that the band gap energy decreases with the increase of TiO_2 content.

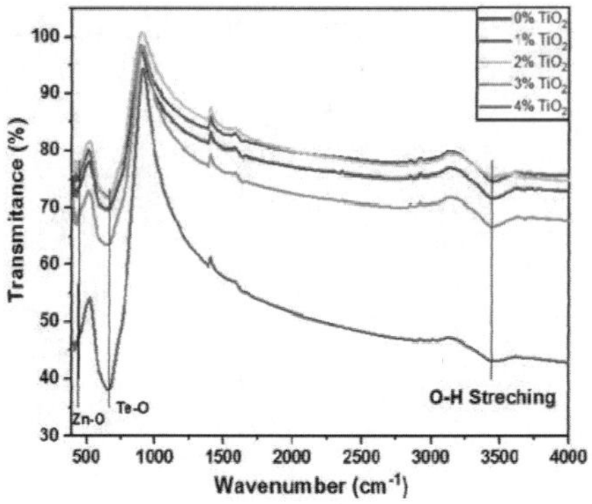

Figure 5. Transmitance at IR region of TZPbTi Glasses

Figure 5 shows the IR transmitance spectra of all samples within the range of 400-4000 cm^{-1}. The O-H stretching group was identified in the range of wavenumbers 3400 cm^{-1} [19]. The Te-O group is in the range of wavenumbers 650 cm^{-1} [20]. Zn-O group is in the range of wave number 436 cm^{-1} [21]. The addition of TiO_2 results in the absorption peak at wavenumber of around 673 cm^{-1} which corresponds to asymmetric vibrations of TiO_6 [22] is stronger for the glass with higher content of TiO_2.

4. Conclusions

Glasses with composition (in mol%): $60TeO_2$-(30-x)ZnO-10PbO-$xTiO_2$ (TZPb:Ti) (where x = 0, 1, 2, 3 and 4) have been studied and characterized in term of density, refractive index, UV-VIS and IR absorption spectra. Addition of TiO_2 ions causes decrease in density and optical band gap energy. Refractive index of glasses increases due to the increase NBO/NO ratio. Asymmetric vibrations of TiO_6 is found stronger as the content of TiO_2 increases.

References

[1] Sidek H A A, Elazoumi S H, El-Mallawany R, Zaid M H M, Matori K H and Halimah M K 2019 *J. Non-Cryst. Solids* **523**, 119640 https://doi.org/10.1016/j.jnoncrysol.2019.119640.

[2] Jauhariyah M N R, Cari and Marzuki A 2016 *Materials Science Forum,* 864, 37-41. https://doi.or/10.4028/www.scientific.net/MSF.864.37.

[3] Nazrin S.N, Halimah, M K, Muhammad F D, Yin J S, Hasnimulyato L, Faznny M F and Zaitizilah, I 2018 *J. Non-Cryst. Solids* **490**, 35-43. https://doi.org/10.1016/j.jnoncrysol.2018.03.017).

[4] Su X, Zhou M, Cheng P and Zhou Z 2018 *J. Alloy. Compd* **739**, 149-159. https://doi.org/10.1016/j.jallcom.2017.12.223.

[5] Baki, S O and Halimah M K 2019 *Solid State Phenom.* **290** 9-15. https://doi.org/10.4028/www.scientific.net/SSP.290.9.

[6] Marzuki A, Fausta D E 2018 IOP Conf Ser Mater Sci Eng **333** (1) 012015. https://doi.org/10.1088/1757-899X/333/1/012015.

[7] Marzuki A, Yunianto M, Riyatun, Purwanto H 2018 KEM **772** 85-89. https://doi.org/10.4028/www.scientific.net/KEM.772.85..

[8] Kaur N, Khanna A, Krishna P S R and Gonzalez F 2015 *AIP Conf Proc* **1665** 070029 https://doi.org/10.1063/1.4917893.

[9] Zhu X, Pirson A C, Milanese D, Lousteau J, Boetti N G, Pugliese D and Peyghambarian N 2018 *Advances in Glass Science and Technlogy* 124-150. https//dx.doi.org/10.5772/intechopen.74664.

[10] Naresh P, Kavitha B, Inamdar H K, Sreenivasu D, Narsimlu N, Srinivas Ch, Sathe V and Kumar K S 2019 *J. Non-Cryst. Solids* **514** 35-45. https://doi.org/10.1016/j.jnoncrysol.2019.03.042.

[11] Griebenow K, Bragatto C B, Kamitsos E I and Wondraczek L 2018 *J. Non-Cryst.* **481** 447-456. https://doi.org/10.1016/j.jnoncrysol.2017.11.041.

[12] Marzuki A, Fausta D E 2019 IOP Conf Ser. Mater. Sci. Eng. **578**:012043 https://doi.org/10.1088/1757-899X/578/1/012043

[13] Haider A J, Jameel Z N and Al-Hussaini I H M 2018 Energy Procedia **157** 17-29. https://doi.org/10.1016/j.egypro.2018.11.159.

[14] Brauer D S, Karpukhina N, Law R V and Hill R G 2010 J. Non-Cryst. Solids **356** (44–49) 2626-2633. https://doi.org/10.1016/j.jnoncrysol.2010.03.022.

[15] Leal J J, Narro-García R, Flores-De los Ríos J P, Gutierrez-Mendez N, Ramos-Sánchez V H, González-Castillo J R and Rodríguez E 2019 *J. Lumin.* **208** 342-349. https://doi.org/10.1016/j.jlumin.2019.01.004.

[16] Marzuki A, Girana L B, Gianti M S and Purnomo F A 2019 *JPCS* **1204** 012098 https://doi.org/10.1088/1742-6596/1204/1/012098.

[17] Azlan M N, Halimah M K, Shafinas S Z and Daud W M 2015 *Mater. Express* **5** (3) 211-218. https://doi.org/10.1166/mex.2015.1236.

[18] Rao Ch.S, Seshulatha K, Gurupria Y S, Sagar P, Kumar M V, Vali S S, Ramanjaneyulu D and Kotireddy S 2018 *Mater. Today* **5** (13) Part 1 26217-26222. https://doi.org/10.1016/j.matpr.2018.08.070.

[19] Perumal S, Sambandam C G, Prabu K M and Anathakumar S 2014 *IJRET* **3** 651-657. https://doi.org/10.15623/IJRET.2014.0304116.

[20] Li Y, Zhang Q, Song J, Gao X, Tang X and Lu A 2018 *J. Non-Cryst. Solids* **483** 43-49. https://doi.org/10.1016/j.jnoncrysol.2017.12.053.

[21] Elkhoshkhany N, Khatab M and Kabari M 2018 Mater. Chem. Phys. **215** 489-498. https://doi.org/10.1016/j.matchemphys.2018.05.007.

[22] Chen Q, Zhang M, Wang H, Wang Q, Ma Q and Li J 2017 *J. Non-Cryst. Solids* **468** 58-66. https://doi.org/10.1016/j.jnoncrysol.2017.04.020.

IOP Conf. Series: Materials Science and Engineering **858** (2020) 012036 doi:10.1088/1757-899X/858/1/012036

Temperature-dependent photoluminescence of H_2TPP and ZnTPP thin films on Si substrates

Nurhayati[1], V Suendo[1,2*], A A Nugroho[3], A Alni[4]
[1] Inorganic and Physical Chemistry Research Division, Faculty Mathematics and Natural Sciences, Institut Teknologi Bandung, Jl. Ganesha 10, Bandung 40132, Indonesia.
[2] Research Center for Nanosciences and Nanotechnology, Institut Teknologi Bandung, Jl. Ganesha 10, Bandung 40132, Indonesia.
[3] Physics of Magnetism and Photonics Division, Faculty Mathematics and Natural Sciences, Institut Teknologi Bandung, Jl. Ganesha 10, Bandung 40132, Indonesia.
[4] Organic Chemistry Research Division, Faculty Mathematics and Natural Sciences, Institut Teknologi Bandung, Jl. Ganesha 10, Bandung 40132, Indonesia.

E-mail: vsuendo@chem.itb.ac.id

Abstract. Temperature-dependent photoluminescence (PL) properties of tetraphenyl porphyrin (H_2TPP) and Zinc-tetraphenyl porphyrin (ZnTPP) thin films on the silicon substrates has been investigated. The photoluminescence (PL) properties are observed within the temperature range of 163–543 °K. The 405 nm diode laser beam is used to excite the molecule during the photoluminescence measurement. The integrated PL intensity exhibits abnormal behaviour, the intensity decreases with temperature in the temperature range of 163-273 °K, and then followed by the increase in the further temperature range of 273-543 °K. This anomaly reflects the competition between the radiative recombination process and the non-radiative recombination processes, i.e. carrier capture, thermalization, and carrier relaxation processes. The peak energy of 0-0 and 0-1 transitions show a red-shift with temperature, while their full width at half maximum increases.

1. Introduction

Porphyrins are macrocycle organic compounds of particular interest with tunable electronic properties. We can tune the electronic properties of porphyrins either by the insertion of metallic atoms into the porphyrins ring or by the substitution of peripheral substituents that are attached to the ring. Besides, this tuning process also leads to the modification of photophysical, electrochemical, and optical properties of porphyrin. The possibility of tailoring their optical properties and the chemical stability make porphyrins as promising candidates for active material in optoelectronic devices such as organic light-emitting diodes (OLEDs) [1], organic solar cells [2], sensors [3,4], and optical limiters [5]. In the optoelectronic device applications, the carrier (electron-hole pair) recombination processes are fundamental concepts. Carrier recombination is the opposite process of carrier generation. Carrier recombination can be distinguished depends on the type of energy released during recombination, whether it is a radiative or non-radiative recombination. A radiative recombination process releases a photon when an electron in the excited state recombines with a hole in the ground state. During a non-radiative recombination process, the electronic energy is released instead of light. Here, the electronic energy is commonly released as the vibrational energy of atomic/molecular lattice, i.e., phonons or macroscopically observed as heat. The transformation of the electronic excitation energy into heat will make low internal quantum efficiency. Consequently, non-radiative recombination is an undesirable process in optoelectronic applications. However, the understanding of the non-radiative process is essential information to understand and to improve the performance of luminescent materials. Also, the radiative recombination could be enhanced if the non-radiative routes can be prevented. Temperature-dependent PL spectra are excellent to explain the emission origin [6] and can provide useful information to understand the non-radiative process [7]. This method is one of the most powerful optical and effective characterization techniques widely used to study exciton-phonon coupling and non-radiative processes [8-10]. In this work, we report a study on the temperature-dependent PL spectra of porphyrin derivatives thin films using the steady-state PL spectroscopy. The integrated PL intensity, peak energy, full width at half maximum (FWHM) as a function of temperature are studied systematically. Our new observation reveals the abnormal behaviour of the integrated PL intensity of the H_2TPP and ZnTPP thin films on Si substrates.

2. Experimental

H_2TPP and ZnTPP (>95% purity) were purchased from Dr. Silviu Pharmachem Pvt.Ltd and used after the crystallization process by the slow evaporation technique from the chloroform solution. After the evaporation process, the porphyrin crystal was diluted in chloroform. The porphyrin solution in concentration 10^{-3} M was dropped on Si substrates. Photoluminescence spectra were recorded with an ultraviolet-visible mini-spectrometer (Ocean optics USB4000). A 120× magnification microscope objective lens was connected to the spectrometer through an optical fiber with an SMA 905 connectors. The microscope objective lens was set up with Semrock EdgeBasic, BLP01-405R-25 (a 405 nm long-wave pass edge filter). The samples were excited by a 405 nm diode laser beam. The laser was equipped with a focusing lens setup and Semrock MaxLine, LL01-405-12.5 (a 405 nm laser clean-up filter). The spectra were recorded as a function of temperature between 163-543 K. Liquid nitrogen was used as the coolant in the temperature-controlled measurement (Linkam FTIR600). The acquired PL spectra in the wavelength unit (nm) were converted quantitatively into the energy unit (eV) by the Jacobian Conversion [11] and fit using the Gaussian function.

3. Results and Discussion

Figure 1(a) and 1(c) shows the temperature-dependent of H_2TPP and ZnTPP thin films on Si substrates, respectively. The integrated PL intensity is plotted versus the inverse temperature in Figure 1(b) and 1(d) for H_2TPP and ZnTPP, respectively. The integrated PL intensity decreases with temperature in the range of 163-273 ^{o}K. This is due to the thermal escape of charge carriers from the excited state to the non-radiative recombination centres. However, as the temperature further increases over 273 ^{o}K, the anomalous enhancement on the PL intensity starts and rising up to 543 ^{o}K. We suggest that by the increase of temperature, some of the charge carriers that previously trapped in the non-radiative centres at lower temperature may escape by thermalization to the radiative recombination centres, and then gradually release photons at higher temperature. This phenomenon leads to an increase in PL intensity for temperature higher than 273° K.

In this research, we focus on the emission of H_2TPP (free-base porphyrin) and ZnTPP (metalloporphyrin). Figu

re 2(a) shows the emission band of ZnTPP is blue-shifted compared to the H_2TPP. This blue shift is due to the strong metal-ligand interaction. The PL spectra of H_2TPP and ZnTPP consist of two peaks produced by the transition from the excited state to the two vibrational states in the ground state, namely 0-1 and 0-0 transitions. The diagram of the radiative recombination of porphyrin for 0-1 and 0-0 transitions, as shown in Figure 2(b). Temperature-dependent of the integrated PL intensity, PL peak energy, and FWHM of 0-1 and 0-0 transitions are able to reveal the reason for the anomalous behaviour of the total integrated PL intensity in Figure 1(b) and 1(d).

IOP Conf. Series: Materials Science and Engineering **858** (2020) 012036 doi:10.1088/1757-899X/858/1/012036

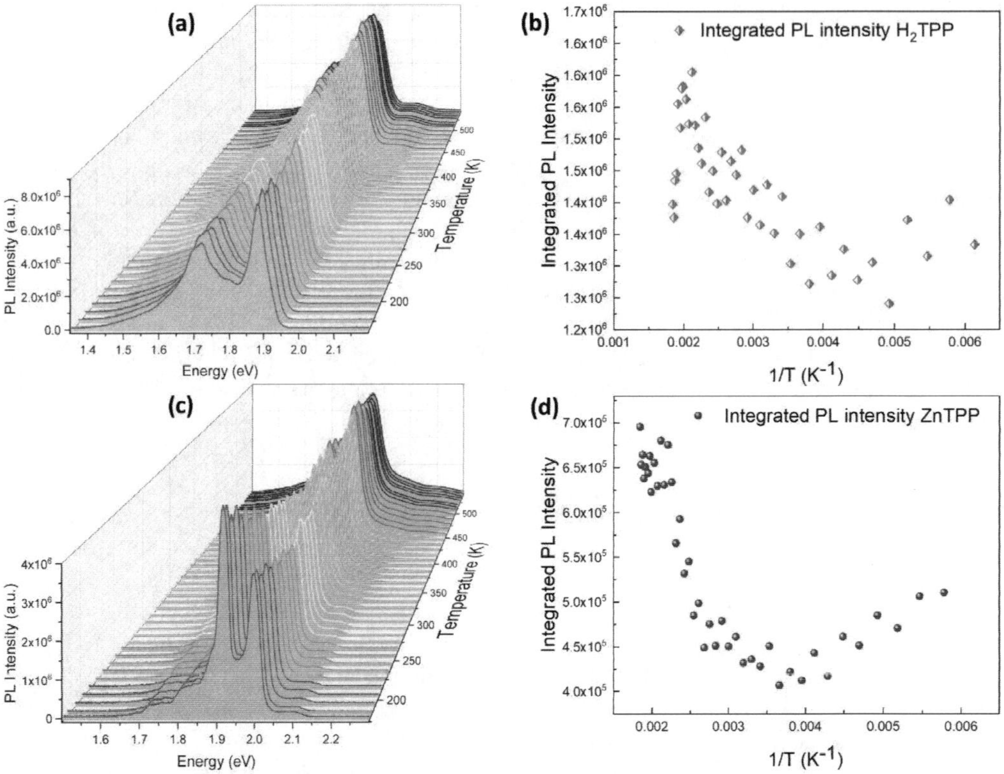

Figure 1. (a) Photoluminescence measurements of H_2TPP thin film on Si substrates as a function of temperature. (b) Integrated PL intensity of H_2TPP is plotted versus the inverse temperature. (c) Photoluminescence measurements of ZnTPP thin film on Si substrates as a function of temperature. (b) Integrated PL intensity of ZnTPP is plotted versus the inverse temperature.

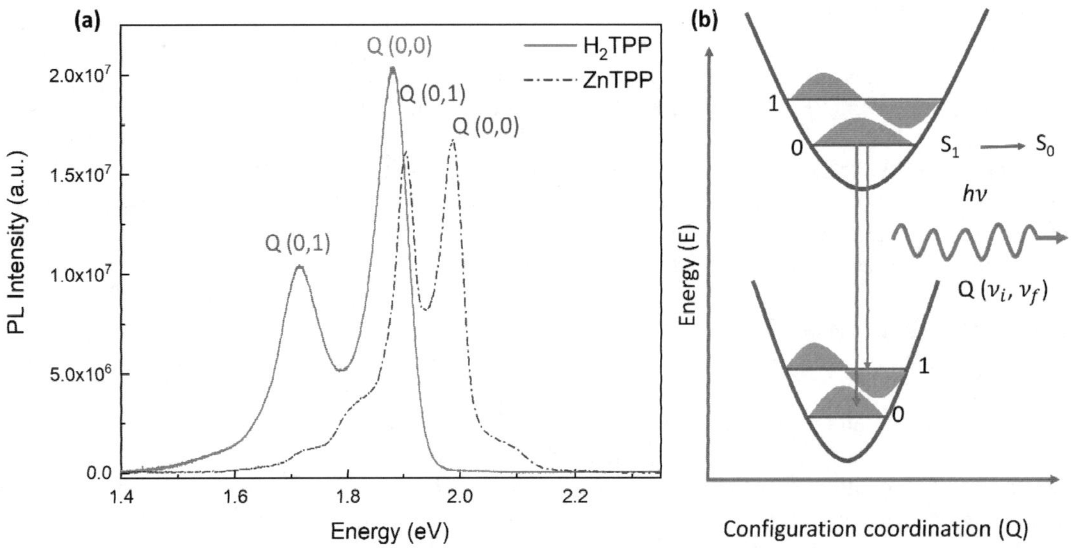

Figure 2. (a) 0-1 and 0-0 transitions in the emission spectra of H_2TPP and ZnTPP thin films on Si substrates. (b) The diagram of the radiative recombination of porphyrin for 0-1 and 0-0 transitions.

Figure 3(a) and 3(c) shows the integrated PL intensity of H_2TPP 0-1 and 0-0 transitions, respectively. The integrated PL intensity of the 0-1 transition of H_2TPP exhibits similar behaviour with the total integrated PL intensity in Figure 1(b) and 1(d). It first decreases from 163 to around 300 °K and then increases with increasing temperature. However, the integrated PL intensity of 0-0 transition increases with increasing temperature from 163 to around 300 °K and then saturated with increasing temperature. Besides, the temperature-dependent peak energy and FWHM for the 0-1 and 0-0 transition of H_2TPP on Si substrates, as shown in Figure 3(b) and 3(d). The temperature-dependent PL peak energy of 0-1 and 0-0 transitions show the well-known S-shape shift behaviour. The PL peak energy is red-shifted in the range of 163-300 °K, then blue-shifted between 300 and 325 °K, and finally red-shifted again between for higher temperature up to 543 °K. This behaviour is associated with the localized carriers or non-radiative centres [12]. At the range of temperature 163 to 300 °K, the peak energy decreases due to the charge carriers that gradually trapped into the non-radiative centres, i.e., the surface states. The peak energy increases with temperature until 325 °K. This is the evidence that the carriers are activated into the radiative centres by the thermal energy. Finally, the PL peak energy decreases due to the shrinkage of bandgap with temperature. This charge carrier thermal activation and redistribution characteristic correlates with the changes of FWHM as a function of the temperature [13]. In the 0-0 transition of H_2TPP, FWHM shows the inverted S-shape behaviour that blue-shifted between 163 and 300 °K, then red-shifted in the range of 300-325 °K, and finally blue-shifted again at higher temperature up to 543 °K. The U-shaped behaviour of the 0-1 transition FWHM with minimum around 300 °K clearly reveals a redistribution of the charge carrier in the non-radiative centres [14,15].

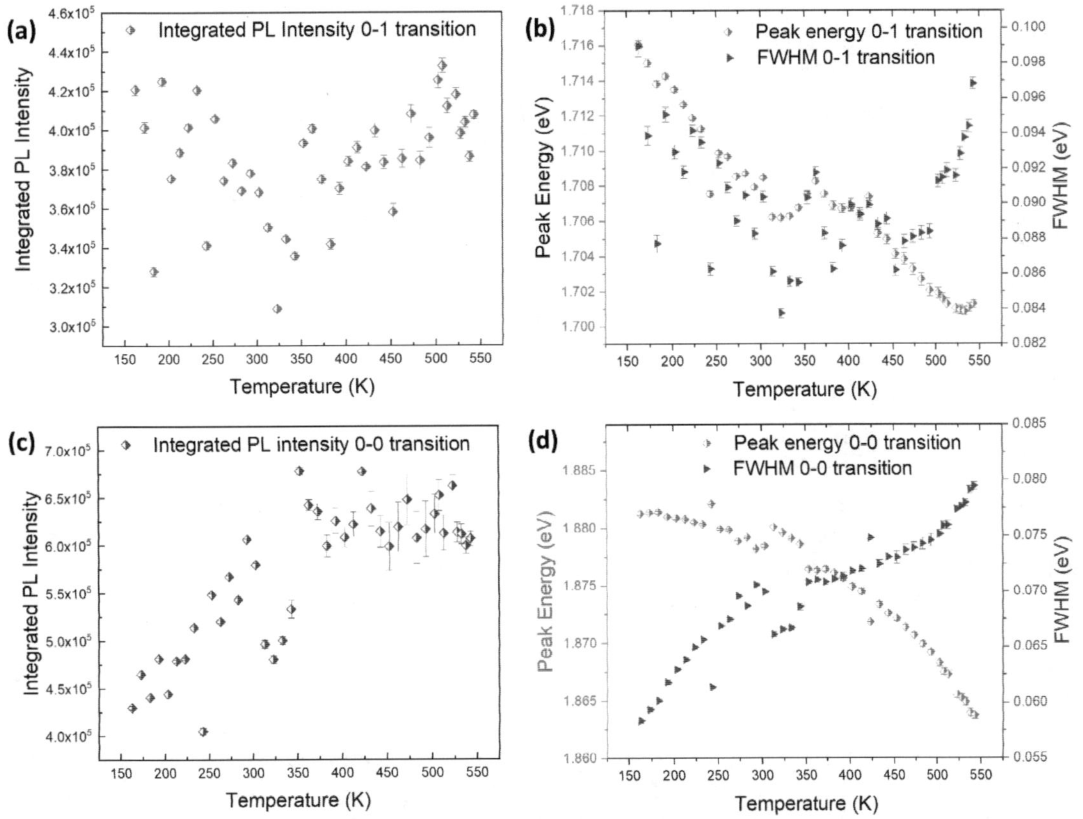

Figure 3. (a) Integrated PL intensity of the H_2TPP 0-1 transition is plotted versus the inverse temperature. (b) Temperature-dependent peak energy and FWHM for the H_2TPP 0-1 transition. (c) Integrated PL intensity of the H_2TPP 0-0 transition is plotted versus the inverse temperature. (d) Temperature-dependent peak energy and FWHM for the H_2TPP 0-0 transition.

Figure 4(a) and 4(c) shows the integrated PL intensity of ZnTPP 0-1 and 0-0 transitions, respectively. These integrated PL intensities show similar behaviour with the total integrated PL intensity in Figure 1(b) and 1(d). The integrated PL intensity decreases from 163 to around 300 °K and then increases with temperature. Temperature-dependent peak energy and FWHM for the 0-1 and 0-0 transition of ZnTPP on Si substrates are shown in Figure 4(b) and 4(d). The S-shape shift behaviour of the PL peak energy and the inverted S-shape shift behaviour of FWHM can be qualitatively interpreted as a competitive process between radiative and non-radiative recombination. The successive red/blue/red-shifts of PL peak energy is accompanied by the blue/red/blue-shifts of FWHM that clearly exhibits the thermally carrier redistribution.

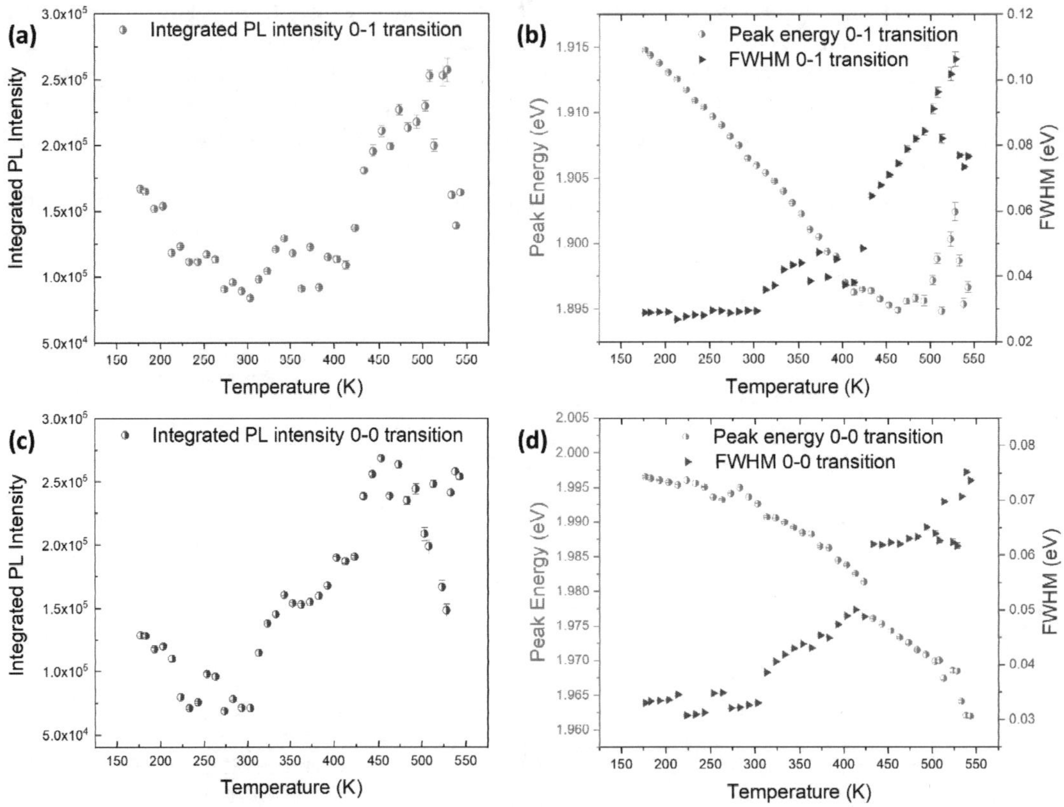

Figure 4. (a) Integrated PL intensity of the ZnTPP 0-1 transition is plotted versus the inverse temperature. (b) Temperature-dependent peak energy and FWHM for the ZnTPP 0-1 transition. (c) Integrated PL intensity of the ZnTPP 0-0 transition is plotted versus the inverse temperature. (d) Temperature-dependent peak energy and FWHM for the ZnTPP 0-0 transition.

4. Conclusions

We briefly reported the temperature-dependent PL properties of H_2TPP and ZnTPP thin films on the silicon substrates. The analysis of the changes on the integrated PL intensity, PL peak energy, and FWHM with temperature indicates that thermally activated charge carrier redistribution is present. The anomalous on the evolution of integrated PL intensity with temperature reveals the activation of trapped charge carriers at high temperature. Here, some trapped charge carriers in the non-radiative centres are excited back to the radiative recombination centres by thermalization process. This phenomenon gives an important impact to improve the performance of organic optoelectronic devices operated at high temperature.

Acknowledgments
The Indonesia Endowment Fund for Education (LPDP), Ministry of Finance, Republic of Indonesia financially supported this work. This research was partially supported by the Asahi Glass Foundation Overseas Research Grant 2017. V. Suendo acknowledges the Ministry of Research, Technology and Higher Education (Menristekdikti) of Indonesia for financial support through the Penelitian Terapan Unggulan Perguruan Tinggi (PDUPT) 2016 and 2019 research grants.

References
[1] Lima N M A, Avila H J C, Marchiori C F N, Sampaio S G, Mota J P F, Ribeiro V G P, Clemente C S, Mele G, Cremona M and Mazzetto S E 2019 *Materials* **12** 1063
[2] Mahmood A, Hu J Y, Xiao B, Tang A, Wang X, Zhou E 2018 *J. Mater. Chem A* **36** 16769-16797
[3] Sanij M D, Bahrampour A, Bahrampour A R 2019 *Photonic Sensors* **9** 1 60-68
[4] Natale C D, Salimbeni D, Paolesse R, Macagnano A, D'amico A 2000 *Sensors and Actuators B* **65** 220-226
[5] Biswal B P, Valligatla S, Wang M, Banerjee T, Saad N A, Mariserla M B M K, Chandrasekhar N, Becker D, Addicoat M, Senkovska I, Berger R, Rao D N, Kaskel S, Feng X 2019 *Angew. Chem* **58** 21 6896-6900
[6] Liao M, Feng Z, Yang S, Chai C, Liu Z, Yang J and Wang Z 2002 *Solid State Commun* **121** 287
[7] Boruc Z, Fetlinski B, Kaczkan M, Turczynski S, Pawlak D, Malinowski M 2012 *J. Alloys Comp* **532** 92
[8] Zhao Y, Riemersma C, Pietra F, Koole R, Donega C M, Meijerink A 2012 *ACS Nano* **6** 9058-9067
[9] Gaponenko M S, Lutich A A, Tolstik N A, Onushchenko A A, Malyarevich A M, Petrov E P, Yumashev K V 2010 *Phys. Rev.B* **82** 125320
[10] Jing P, Zheng J, Ikezawa M, Liu X, Lv S, Kong X, Zhao J, Masumoto Y 2009 *J. Phys. Chem. C* **113** 13545-13550
[11] Mooney J, Kambhampati P 2013 *J. Phys. Chem. Lett* **2** 3316-3318
[12] Mazzucato S, Potter R J, Erol A, Balkan N, Chalker P R, Joyce T B, Bullough T J, Marie X, Carrere H, Bedel E, Lacoste G, Arnoult A, Fontaine C 2003 *Physica E* **17** 242-244
[13] Yuan Q, Liang B, Zhou C, Wang Y, Guo Y, Wang S, Fu G, Mazur Y I, Ware M E, Salamo G J 2018 *Nanoscale Research Letters* **13** 387
[14] Mano T, Notzel R, Gong Q, lippen T V, Hamhuis G J, Eijkemans T J, Wolter J H 2005 *Jpn. J. Appl. Phys* **44** 6829-6832
[15] Li Y, Jiang Y, Die J, Wang C, Yan S, Wu H, Ma Z, Wang L, Jia H, Wang W, Chen H 2018 *Chin. Phys. B* **27** 9 097104

Starch-based biofoams reinforced with microcrystalline cellulose from banana stem: Hydrophobicity and Biodegradability.

S Fatrozi[1], L Purwanti[1], S K Sari[1], M N Ariesta[1,], T Kusumaningsih and S D Marliyana[1]

[1]Chemistry Department, Faculty of Mathematics and Natural Sciences, Sebelas Maret University, Jl. Ir Sutami 36A Surakarta, Indonesia

E-mail: msoerya@staff.uns.ac.id

Abstract. Microcrystalline cellulose was isolated from the banana stem, an abundant but underutilized bioresource, by using three main steps, specifically bleaching, alkaline treatment and acid hydrolysis. The isolated MC was used to reinforce starch-based biofoams with different loadings of 0%, 2%, 4%, and 6%. The result of contact angle measurement using the LB-ADSA method achieved remarkable increases of contact angle value as the MC loading increased from 0% to 6%. In addition, compared to starch-based biofoams, starch/MC composite biofoams showed improved biodegradability. The starch/MC composite biofoams displayed good hydrophobicity properties that can potentially represent a sustainable and green alternative for packaging and insulation materials.

1. Introduction

Indonesian banana trees are abundant but are underutilized and just to be waste because only the fruit is taken. Therefore, the integrated utilization of banana tree residues is a necessity. Banana stems have a substance of fibers that are suitable for textiles, composite materials, paper, and adsorbent materials [1]. Its have a good fiber contains several compounds such as lignin, alpha-cellulose, hemicellulose, and water [2]. This supported by [3] where the banana tree has a specific gravity of 0.29 g/cm and fiber size of 4.20 - 5.46 mm and lignin of 33.51%. In the previous study [4], reported natural fiber, essential oils, and also use enzymes can be improved the characteristic of biofoam as a filler. Then, according to Averous [5], the addition of cellulose fibers up to 15% can increase water resistance while increasing its tensile strength. The addition of fiber also affects increasing the hydrophobic nature of Biofoam due to the ability of the fiber to absorb water that is smaller than starch [6].

Given their characteristics, cellulose fibers can reinforce biofoam materials [7]. Biofoam based on starch, in reality, are porous materials which in wet conditions had poor mechanical strength. Such weaknesses can be addressed through the incorporation of cellulosic fiber, particularly when converted as microcrystalline cellulose, which can improve water resistance. Moreover, microcrystalline cellulose in combination with starch is supposed to be fully biodegradable, enhancing its potential packaging ability. In the study, the synthesis new class of biofoam consisting of starch-based biofoam

[1] msoerya@staff.uns.ac.id

with microcrystalline cellulose from the banana stem, which was tested for their hydrophobicity and biodegradability.

2. Materials and Methods

2.1 Materials

Microcrystalline cellulose (MC) was obtained from banana stem (*Musa paradisiaca* L) waste. The chemicals were purchased from Merck and Bratachem, polyvinyl alcohol (PVA) (Merck) Technical grade cassava starch (Bratachem), and Mg Stearate (Bratachem).

2.2 Preparation of biofoam

Various biofoam formulations (Table 1) were prepared using a conventional molding machine for compression. Starch, PVA, Magnesium stearate, distilled water, a varied amount of MC were mixed until a homogenous mixture was obtained. The mixture was poured into a 170 mm × 110 mm x 2 mm mold, which was placed in the compression molding machine. The mold temperature was set at 150°C. The process was carried out at a load pressure of 100 bar for 4 mins of holding time. The variables in this paper was MC content (0%; 2%; 4% and 6% MC) for 30 g of Starch content (in weight) each variation contain 100 mL of water.

Table 1. Compositions of the starting materials per 30 g of dry cassava starch

Formulas	Code	Composition (%)			
		Starch	MC	PVA	Mg stearate
Starch/PVA/Mg Stearate/MC 0%	M0	85	0	10	5
Starch/PVA/Mg Stearate/MC 2%	M1	82	2	10	5
Starch/PVA/Mg Stearate/MC 4%	M2	80	4	10	5
Starch/PVA/Mg Stearate/MC 6%	M3	78	6	10	5

2.3 Water-Contact Angle

The water-contact angle value was measured using ImageJ ver-1.52a (Wayne Rasband, National Institute of Health, USA) software with Low Bond-Axisymetri Drop Shape Analysis (LB-ADSA) method.

2.4 Water Sorption

The Starch/PVA/Mg Stearate/MC biofoam water sorption was measured according to ABNT NBR NM-ISO 535. Four composites were cut into 2x2 cm specimens and conditioned at room temperature by using a chamber containing water. After 1minute immersion time the average value of biofoam water sorption was calculated:

$$\text{Water absorption (\%)} = \frac{W_a - W_0}{W_0} \text{ X } 100\% \quad (1)$$

where W_0 and W_a respectively, are the specimen mass before and after water absorption. The tests are performed for each specimen by triplicate.

2.5 Differential Scanning Calorimetry

Differential Scanning Calorimetry analysis of Starch/PVA/Mg Stearate/MC biofoam was performed using Linsesis type STA PT 1600 at a heating rate of 10 C/min under nitrogen atmosphere.

2.6 Biodegradability test

In biodegradability analysis, the biofoam composite ware was cut into rectangular samples of 30-40 mm. The biodegradability of the samples buried in compost soil was determined by calculating weight

IOP Conf. Series: Materials Science and Engineering **858** (2020) 012037 doi:10.1088/1757-899X/858/1/012037

loss. Every sample was buried in the soil of compost and each specimen was dug out of the soil of compost after being buried for 30 days. The proportion of weight loss W (%) was determined using the formula:

$$\text{Weight loss (\%)} = \frac{W_{initial} - W_{final}}{W_{initial}} \times 100\% \qquad (2)$$

Where $W_{initial}$ and W_{final} are the weights of specimen before and after buried in compost soil.

3. Results and Discussion

3. 1 Water-Contact Angle

The hydrophobicity of biofoam can be obtained by measuring the contact angle (**θ**) of the water droplets on the biofoam surface. The results obtained show the addition of MC will have a significant influence on the contact angle (Table 2). The difference in contact angle between biofoam with 0% MC and biofoam with 2% MC is around 15°, indicates that the addition of MC enhances the hydrophobic nature of biofoam. Meanwhile, as shown in Table 2, no significant differences between biofoam that contain MC.

Tabel 2. Water-contact angle of biofoam

Composites	Contact angle (°)
Biofoam, 0% MC	74,27
Biofoam, 2% MC	89.19
Biofoam, 4% MC	90,70
Biofoam, 6% MC	94,97

The difference contact angle may be related to surface roughness and / or availability OH on biofoam surfaces [8]. In this case, the difference between the contact angles may be related with OH availability. Starch and MC have strong hydrogen bond interactions so that it will reduce water interactions. An increase in the hydrophobic nature of biofoam will occur with increasing MC content due to the hydrophobic nature of MC [9].

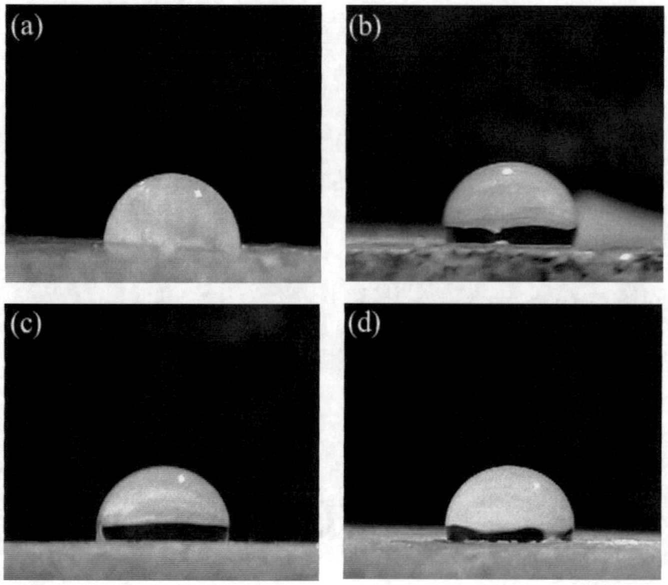

Figure 1. Water-contact angle image of (a) Biofoam with 0% MC, (b) Biofoam with 2% MC, (c) Biofoam with 4% MC, and (d) Biofoam with 6% MC

IOP Conf. Series: Materials Science and Engineering **858** (2020) 012037 doi:10.1088/1757-899X/858/1/012037

3. 2 Water Sorption

In the equilibrium moisture environment, water sorption per unit mass (percent) of the composite biofoams were examined. For increased MC content, lower water sorption occurs. The biofoam/0% MC absorbed approximately ~11% of water, while the absorption was decreased almost linearly with the addition of MC to a minimum of 7%. The value obtained in this study is lower than previous studies which range from 23-35%[10]. The decrease in water absorption can be caused by the interaction of strong hydrogen bonds between starch and MC, in addition to the presence of crystalline PVA which can also affect[11]. Optimum water absorption ability on Biofoam/6% MC content.

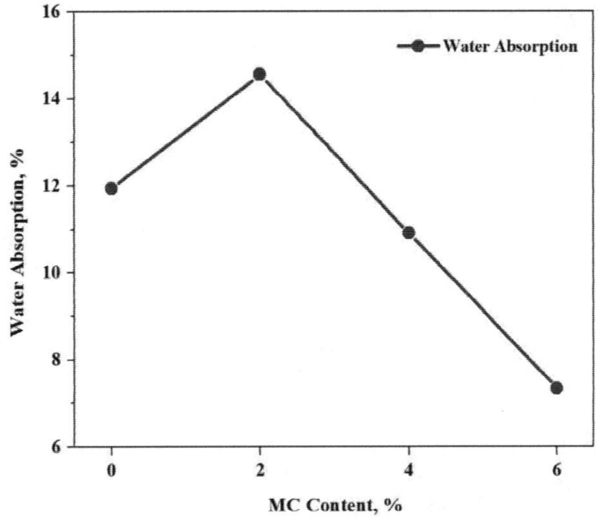

Figure 2. Water absorption graph of biofoam

3. 3 Differential Scanning Calorimetry

Biofoam thermal stability is characterized by DSC [12]. The interaction between MC polymer chains and starch is influenced by moisture concentration. Decreasing the water content in the composite will increase the interaction between the MC polymer chain and starch which will cause the structure to become more rigid [13]. The heating process up to 150°C is accompanied by applying pressure of 100 bar causing the water content in the composite is lost so that a rigid biofoam is obtained. Figure 3 shows the DSC curve of biofoam, the glass transition temperature of biofoam was found to increase around 270°C to 287°C. The glass transition temperature is highly dependent on the mobility of the polymer chain [14].

It has been reported that strong interactions between polymer chains will cause an increase in Tg because the mobility of the polymer chains will decrease [14]. Picture. 3 shows that in the presence of MC, glass transitions occur at higher temperatures due to the interaction of strong hydrogen bonds thereby reducing the mobility of the cellulose chain and the starch chain. The high Tg indicates that the material has good thermal stability properties and can include several aspects for packaging materials such as resistance to degradation by high storage temperatures [15].

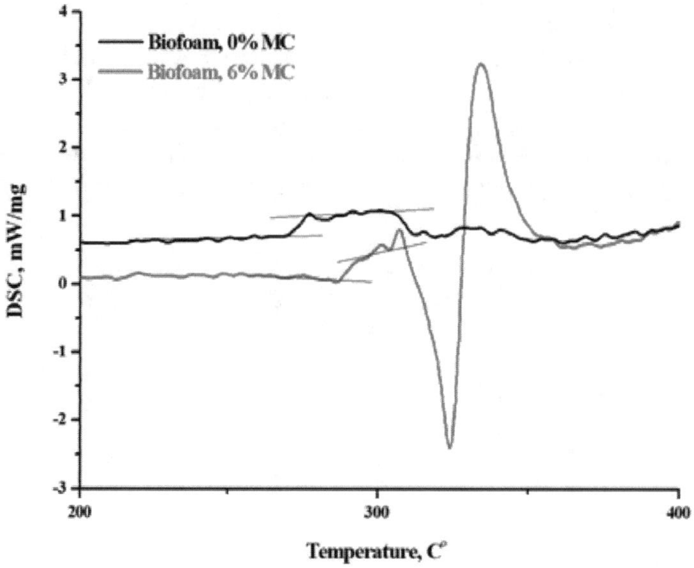

Figure 3. DSC graph of biofoam with 0% MC and biofoam with 6% MC

3. 4 Biodegradable Test

The weight loss of biofoam is illustrated in Figure. 4. The composite of Biofoam/6% MC has the highest weight loss followed by that of Biofoam/4% MC, Biofoam/2% MC and Biofoam/0% MC, respectively. The reasons for this are the uniformity of the distribution of MC in the composite and the interfacial compatibility of MC and composite. The composite reinforced with MC has many inner defects and poor compatibility, leaving the composite vulnerable to the outer microorganism leading to the lowest weight loss.

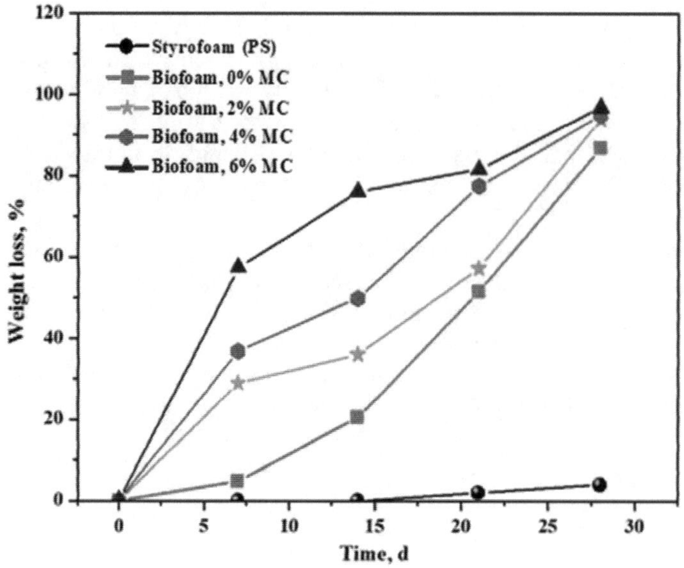

Figure 4. Biodegradable graph of styrofoam (PS) and biofoam

The linear polymers in chemical degradation reactions cause a decrease in molecular weight or length chain reducting. Starch has acetal bonds which are very easy to degrade. The amount of mass reduction is caused by the composition of biofoam which consists of natural materials that are easily digested by microbes. The main factor of a polymer that can be degraded naturally was a natural polymer containing a hydroxyl group (-OH), this group was easily degraded microbially [16].

4. Conclusion

Microcrystallibe Cellulose (MC) obtained from banana stem were prepared and used to reinforce staech biofoam. The incorporation of MC into the starch biofoam improved hydrophobicity and Biodegradability properties. Starch and MC interact strongly through hydrogen bonds which causes an increase in the water contact angle and reduces water absorption, the results obtained are close to the performance of PS foam. In addition, high Tg values and better biodegradability properties than PS foam indicate that this material has potential as a packaging material.

Acknowledgment

We gratefully acknowledge financial support from the Higher Education Directorate, Ministry of Education, the Republic of Indonesia for this research (PKM budget).

References

[1] Pereira A L S, Nascimento D M., Souza M S M, Casseles A R, Morais J P S, Paula R C M, Rosa M F, and Feitosa J 2014 *BioResources* **4** 1-6

[2] Goswami T, Kalita D and Rao P G 2008 *Indian J. Chem. Technol.* **1** 1-7

[3] Syafrudin 2004. *Pengaruh Konsentrasi Larutan dan Waktu Pemasakan Terhadap Rendemen dan Sifat Fisis Pulp Batang -Pisang Kepok (Musa spp) Pascapanen* (Yogyakarta : Universitas Gadjah Mada)

[4] Ketkaew S, Kasemsiri P, Pongsa U and Hiziroglu S 2016 *Eng. Mater. and Technol.* **718** 1-5

[5] Avérous L C, Fringant L and Moro 2001 *Polymer* **42** 6571-78.

[6] Benezet J C, Davidovic A S, Bergeret A, Ferry L and Crespy A 2011 Mechanical and physical properties of expanded starch, reinforced by natural fibres *Ind. Crops Prod.* **1** 435-440.

[7] Lopez-Gil A, Silva-Bellucci F, Velasco D, Ardanuy M, Rodriguez-Perez MA. 2015. Cellular structure and mechanical properties of starch-based foamed blocks reinforced with natural fibers and produced by microwave heating *Ind. Crops Prod.*194–205

[8] Guz L, Famá L, Candal R, dan Goyanes S 2017 *Carbohydr. Polym.* 1611–19

[9] Medronho B, Romano A, Miguel M G, Stigsson L dan Lindman B 2012 *Cellulose* **3** 581–587

[10] Etikaningrum 2016 Pengembangan berbagai modifikasi serat tandan kosong sawit pada pembuatan biofoam *Jurnal Penelitian Pascapanen Pertanian* **3** 1-10

[11] Cinelli P, Chiellini E, Lawton J W and Imam S H 2006 *Polym Degrad Stabil* **91** 1147-55

[12] Roig F, Dantras E, Dandurand J, Lacabanne C 2011 *J. Phys. D: Appl. Phys.* **44** 3727–29

[13] Tunc S dan Duman, O 2010 *Appl. Clay* **48** 414–424

[14] Zheng K, Chen L dan Cui Y Li P 2004 *Polym. Eng. Sci.* **44** 1077–82

[15] Cairul Iqbal M, Gumah Abadi A, Ahmad N, Katas H dan Azdina J J 2012 *Sains Malays* **41** 561–68

[16] Nissa, R. C., Fikriyyah, A. K., Abdullah, A. H. D and Pudjiraharti, S. 2019 *IOP Conf. Series: Earth and Environ. Sci.* **277** 1-5

IOP Conf. Series: Materials Science and Engineering **858** (2020) 012038 doi:10.1088/1757-899X/858/1/012038

The analytical and numerical solutions of two dimensional heat transfer equation in a multilayered composite cylinder

S Alaa[1*], **Irwansyah**[2], **D W Kurniawidi**[1] and **S Rahayu**[1]
[1] Physics Department, Universitas Mataram, Jl. Majapahit No. 62 Mataram, Indonesia
[2] Mathematic Department, Universitas Mataram, Jl. Majapahit No. 62 Mataram, Indonesia

E-mail: siti.alaa@unram.ac.id

Abstract. Heat transfer is one of the most observed phenomena in the fields of aerospace, industry, nuclear, power generation, automotive, etc. Currently, materials for heat transfer are made of composites of various types so that they have different properties from their constituent materials. To understand the phenomenon of heat transfer in order to save costs and risks that can arise, it can be done by simulation. In this paper, we describe analytical and numerical solutions for two dimensional (2D) heat transfer equation in a multilayered composite cylinder. In particular, we use eigen function method to find the analytical solution and finite difference method to generate the numerical solution. The simulation of heat transfer for different time frame carried out using Python 3.7 with Numpy library.

1. Introduction

The heat transfer phenomenon appears immediately after energy is produced, exchanged, or changed. Researchers are trying to understand the phenomenon of heat transfer to improve equipment's design or on explaining the cause of errors in some devices that utilize heat transfers such as in the chemical, nuclear and power generation industries. From this understanding, the researchers are able to make some modifications on certain components in order to meet safety or cost requirements and improve the work's efficiency [1, 2].

The experimental approach has been widely used in the past, but costs have become quite expensive when dealing with very high temperatures, pressure, speed or potentially dangerous products (chemical or nuclear industry) [3, 4]. In addition, an experimental approach in place often lacks the flexibility needed when sensitivity studies are desired. On the other hand, with strong computing facilities available on the market at affordable costs, analytical calculations and simulations with a numerical approach appear to be a very promising and accurate way to predict thermal phenomena and their effects in all situations [5].

In industry, multilayered composite materials are often used [6]. The advantages of these ingredients are combining physical, mechanical, and thermal properties of different substances. The idea of the cylindrical layer is also widely used in investigating the thermal properties of composite materials. In this case, it is assumed that the particles forming a composite matrix are cylinder in shape. The analytical and numerical methods used to investigate the heat conduction of multilayered transients are analogous to those used to investigate single-layer transient heat conduction. Among the analytical methods indicated are the variable separation method, finite difference method, Laplace-transform method, integral method to transform, and expansion of eigen functions and Green function methods [7].

In this study heat transfer simulations were carried out on multilayered composites using the finite difference method. This simulation can be used in the industrial sector so that it can save costs and produce new composites that have better characteristics/properties.

2. Experimental

Composite models are made in cylindrical shape with a constant heat source in the center of the cylinder as shown in Figure 1.

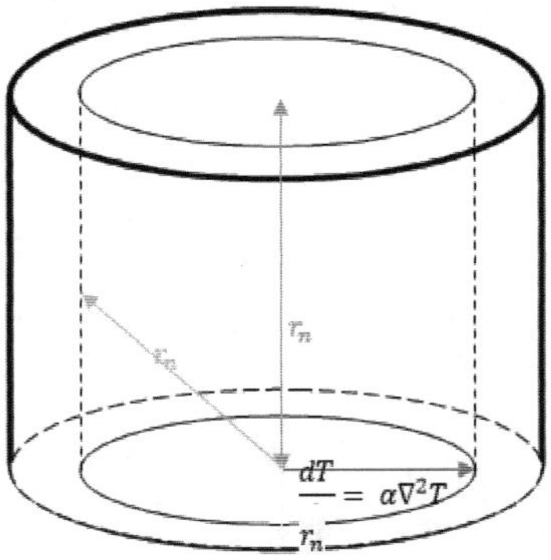

Figure 1. Multilayered composite design model

Consider a multilayer cylindrical composite with coordinates

$$r_0 \le r \le r_n$$
$$0 \le \theta \le \varphi$$
$$0 \le z \le L$$

We assume all the layers are isotropic and the thermal contact is perfect. When $t = 0$ the temperature in the $-i$th layer is $f_i(r,\theta,z)$. When $t > 0$ there are boundaries heat source $g_i(r,\theta,z,t)$. The heat equation in cylindrical coordinates is as follows [8].

$$\frac{dT}{dt} = \alpha\nabla^2 T$$
$$= \alpha\left(\frac{1}{r}\frac{\partial}{\partial r}\left(r\frac{\partial T}{\partial r}\right)\right) + \frac{1}{r^2}\frac{\partial^2 T}{\partial \theta^2} + \frac{\partial^2 T}{\partial z^2}$$

In multilayered case, we have

$$\frac{1}{\alpha_i}\frac{\partial T_i}{\partial t} = \frac{\partial T_i}{\partial r^2} + \frac{1}{r}\frac{\partial T_i}{\partial r} + \frac{1}{r^2}\frac{\partial^2 T_i}{\partial \theta^2} + \frac{\partial^2 T_i}{\partial z^2} + \frac{g_i(r,\theta,z,t)}{k_i} \quad (1)$$

Where k_i is a dependent constant related to heat properties of corresponding material and $T_i = T_i(r,\theta,z,t)$ is the temperature on the i layer. We also assume the following boundary for each variable.

$$r_0 \le r \le r_n$$
$$r_{i-1} \le r \le r_i$$
$$0 \le \theta \le \varphi$$

$$\varphi \leq 2\pi$$
$$0 \leq z \leq L$$
$$t \geq 0$$

With boundary condition

$$A_{in}\frac{\partial T_i}{\partial r}(r_0,\theta,z,t) + B_{in}T_i(r_0,\theta,z,t) = C_{in}$$
$$A_{out}\frac{\partial T_n}{\partial r}(r_n,\theta,z,t) + B_{out}T_n(r_0,\theta,z,t) = C_{out}$$

If $\theta = 0$, we have $T_i = T_i(r,0,z,t) = 0$ or $\frac{\partial T_i}{\partial \theta}(r,0,z,t) = 0$. If $\theta = \varphi$, then $T_i = T_i(r,\varphi,z,t) = 0$ or $\frac{\partial T_i}{\partial \theta}(r,\varphi,z,t) = 0$.

3. Results and Discussion

3.1. Analytical Solution

In this part, we derive an analytical solution in two dimensional heat transfer system of equations (1). If z is fixed, then equation (1) would become

$$\frac{1}{\alpha_i}\frac{\partial T_i}{\partial t} = \frac{\partial^2 T_i}{\partial r^2} + \frac{1}{r}\frac{\partial T_i}{\partial r} + \frac{1}{r^2}\frac{\partial^2 T_i}{\partial \theta^2} + \frac{g_i(r,\theta,z,t)}{k_i}$$

where

$$r_{i-1} \leq r \leq r_i$$
$$0 \leq \theta \leq \varphi$$
$$\varphi \leq 2\pi$$
$$t \geq 0$$

In the 1st layer ($i = 1$) we have

$$A_{in}\frac{\partial T_1}{\partial r}(r_0,\theta,t) + B_{in}T_1(r_0,\theta,z,t) = C_{in}.$$

And in the outer layer ($i = n$), we have

$$A_{out}\frac{\partial T_n}{\partial r}(r_n,\theta,t) + B_{out}T_n(r_0,\theta,t) = C_{out}.$$

Moreover, for $\theta = 0$, we have $T_i(r,\theta,t) = 0$ or $\frac{\partial T_i}{\partial \theta}(r,0,t) = 0$. And, for $\theta = \varphi$, we have $T_i(r,\varphi,t) = 0$ or $\frac{\partial T_i}{\partial \theta}(r,\varphi,t) = 0$.

In the ith layer, for the inner interface of two layers, we have

$$T_i(r_{i-1},\theta,t) = T_{i-1}(r_{i-1},\theta,t)$$

and

$$k_i\frac{\partial T_i}{\partial r}(r_{i-1},\theta,t) = k_{i-1}\frac{\partial T_{i-1}}{\partial r}(r_{i-1},\theta,t)$$

Also, for the outer interface of two layers, we have

$$T_i(r_i,\theta,t) = T_{i+1}(r_i,\theta,t)$$

IOP Conf. Series: Materials Science and Engineering **858** (2020) 012038 doi:10.1088/1757-899X/858/1/012038

and

$$k_i \frac{\partial T_i}{\partial r}(r_i, \theta, t) = k_{i+1} \frac{\partial T_{i+1}}{\partial r}(r_{i+1}, \theta, t)$$

To find the analytical solution, we need to find the related eigen function φ of the equations. The eigen function φ satisfies the following equation.

$$\nabla^2 \varphi = -\lambda^2 \varphi \tag{2}$$

In the cylindrical coordinates with fixed z, we have

$$\frac{\partial^2 \varphi_i}{\partial r^2} + \frac{1}{r}\frac{\partial \varphi_i}{\partial r} + \frac{1}{r^2}\frac{\partial^2 \varphi_i}{\partial \theta^2} = -\lambda^2 \varphi_i \tag{3}$$

We assume that

$$\varphi_i(r, \theta) = R_i(r)\theta_i(\theta) \tag{4}$$

Therefore, we have

$$\frac{\partial \varphi_i}{\partial r} = \theta_i \frac{dR_i}{dr}$$
$$\frac{\partial^2 \varphi_i}{\partial r^2} = \theta_i \frac{d^2 R_i}{dr^2}$$
$$\frac{\partial^2 \varphi_i}{\partial \theta^2} = R_i \frac{d^2 \theta_i}{d\theta^2}$$

So, (3) would become

$$\theta_i R_i + \frac{1}{\mathrm{r}} \mathrm{R_i}'\theta_i + \frac{1}{\mathrm{r}^2} \mathrm{R_i}\theta_i = -\lambda^2 R_i \theta_i,$$
$$\frac{R_i''}{R_i} + \frac{1}{r}\frac{R_i'}{R_i} + \frac{1}{r^2}\frac{\theta_i''}{\theta_i} = -\lambda^2 \tag{5}$$
$$r^2 \frac{R_i''}{R_i} + r\frac{R_i'}{R_i} + \frac{\theta_i''}{\theta_i} = -\lambda^2 r^2 \tag{6}$$

where

$$r^2 \frac{R_i''}{R_i} + r\frac{R_i'}{R_i} + \lambda^2 r^2 = -\frac{\theta_i''}{\theta_i} = v^2$$
$$r^2 R_i'' + rR_i' + R_i(\lambda^2 r^2 - v^2) = 0 \tag{7}$$
$$\Theta_i'' + v^2 \Theta_i = 0 \tag{8}$$

Now, let

$$R_i(r) = \sum_{m=0}^{\infty} a_m r^m.$$

We have,

$$R_i^{\prime(r)} = \sum_{m=1}^{\infty} m a_m r^{m-1}$$

and

$$R_i^{\prime\prime(r)} = \sum_{m=2}^{\infty} m(m-1) a_m r^{m-2}$$

Therefore, equation (7) can be written as

IOP Conf. Series: Materials Science and Engineering **858** (2020) 012038 doi:10.1088/1757-899X/858/1/012038

$$r^2 \sum_{m=2}^{\infty} m(m-1)a_m r^{m-2} + r \sum_{m=1}^{\infty} m a_m r^{m-1} + (\lambda^2 r^2 - v^2) \sum_{m=0}^{\infty} a_m r^m = 0$$

The solution for (7) is as follows.

$$R_{ij}(r) = C_1 J_{vij}(\lambda_{ij} r) + C_2 Y_{vj}(\lambda_{ij} r) \quad (9)$$

Where J_{vij} is the Bessel function of the 1st kind and Y_{vj} is the Bessel function of 2nd kind. Meanwhile, the solution for equation (8) is as follows.

$$\begin{aligned} \Theta_{ij}(\theta) &= C_3 \sin(v_{ij}\theta) + C_4 \cos(v_{ij}\theta) \\ &= C_3 \sin(v_j\theta) + C_4 \cos(v_j\theta) \end{aligned}$$

Therefore, we have

$$T_i(r, \theta, t) = \sum_{k=1}^{\infty} \sum_{j=1}^{\infty} T_{ikj}(t) R_{ij}(r)\, \Theta_k(\theta)$$

The heat source can be written as

$$g_i(r, \theta, t) = \sum_{k=1}^{\infty} \sum_{j=1}^{\infty} g_{ikj}(t) R_{ij}(r)\, \Theta_k(\theta),$$

where

$$g_{ikj}(t) = \frac{\int_0^{\varphi} \int_{r_{i-1}}^{r_i} g_i(r,\theta,t) R_{ij}(r) \Theta_k(\theta) r dr d\theta}{\int_0^{\varphi} \int_{r_{i-1}}^{r_i} r {R_{ij}}^2(r) {\Theta_k}^2(\theta) r dr d\theta}$$

The coefficient T_{ikj} can be written as

$$T_{ikj} = \frac{\alpha_i}{k_i} e^{-\Gamma_{ikj}(t)} \int_0^t g_{ikj}(t)\, e^{-\Gamma_{ikj}} dt + a_i e^{-\Gamma_{ikj}(t)}$$

where

$$a_i = T_{ikj}(0)$$

$$a_i = \frac{\int_0^{\varphi} \int_{r_{i-1}}^{r_i} f_i(r,\theta,z) r R_{ikj}(r) \Theta_k(\theta) Z_i(z) dr d\theta\, dz}{\int_0^{\varphi} \int_{r_{i-1}}^{r_i} r {R_{ikj}}^2(r) {\Theta_k}^2(\theta) {Z_i}^2(z) r dr d\theta}$$

Hence, the final solution is

$$\Gamma_{ikj} = -\alpha_i \left(\frac{R_{ikj}''(r)}{R_{ikj}(r)} + \frac{1}{r} \frac{R_{ikj}'(r)}{R_{ikj}(r)} + \frac{1}{r^2} \frac{\theta_j''(\theta)}{\theta_j(\theta)} + \frac{Z_i''(z)}{Z_i(z)} \right)$$

3.2. Numerical solution

In this part, we derive numerical solution of equation (1) using finite difference method. Heat equation in the i-th layer is as follows.

$$\frac{\partial^2 T_i}{\partial r^2}+\frac{1}{r}\frac{\partial T_i}{\partial r}+\frac{1}{r^2}\frac{\partial^2 T_i}{\partial \theta^2}+\frac{\partial^2 T_i}{\partial z^2}+\frac{g_i(r,\theta,z,t)}{K_i}=\frac{1}{\alpha_i}\frac{\partial T_i}{\partial t}$$

Using Taylor expansion, we have

$$\begin{aligned} T_i(x,t+\Delta t) &= T_i(x,t)+\frac{\partial T_i}{\partial t}\Delta t+\frac{\partial^2 T_i}{\partial t^2}\frac{\Delta t^2}{2!}+\frac{\partial^3 T_i}{\partial t^3}\frac{\Delta t^3}{3!}+\cdots \\ &= T_i(x,t)+\frac{\partial T_i}{\partial t}\Delta t+\frac{\partial^2 T_i}{\partial t^2}\frac{\Delta t^2}{2}+\frac{\partial^3 T_i}{\partial t^3}\frac{\Delta t^3}{6}+\cdots+\frac{\partial^n T_i}{\partial t^n}\frac{\Delta t^n}{n!} \end{aligned}$$

So, we have

$$\frac{\partial T_i}{\partial t}\Delta t = T_i(x,\theta,z,t+\Delta t)-T_i(r,\theta,z,t)$$

or

$$\frac{\partial T_i}{\partial t}=\frac{T_i(x,\theta,z,t+\Delta t)-T_i(r,\theta,z,t)}{\Delta t}$$

Also, we use Taylor series to get the following equations,

$$\begin{aligned} T_i(r+\Delta r,\theta,z,t) &= T_i(r,\theta,z,t)+\frac{\partial T_i}{\partial r}\Delta r+\frac{\partial^2 T_i}{\partial r^2}\frac{\Delta r^2}{2!}+\frac{\partial^3 T_i}{\partial r^3}\frac{\Delta r^3}{3!}+\cdots \\ &= T_i(r,\theta,z,t)+\frac{\partial T_i}{\partial r}\Delta r+\frac{\partial^2 T_i}{\partial r^2}\frac{\Delta r^2}{2}+\frac{\partial^3 T_i}{\partial r^3}\frac{\Delta r^3}{6}+\vartheta(\Delta r^4) \end{aligned} \quad (10)$$

Also, we have

$$\frac{\partial T_i}{\partial t}\Delta r = T_i(r+\Delta r,\theta,z,t)-T_i(r,\theta,z,t)$$

or

$$\frac{\partial T_i}{\partial r}=\frac{T_i(r+\Delta r,\theta,z,t)-T_i(r,\theta,z,t)}{\Delta r}$$

Then,

$$\begin{aligned} T_i(r-\Delta r,\theta,z,t) &= T_i(r,\theta,z,t)-\frac{\partial T_i}{\partial r}\Delta r+\frac{\partial^2 T_i}{\partial r^2}\frac{\Delta r^2}{2!}-\frac{\partial^3 T_i}{\partial r^3}\frac{\Delta r^3}{3!}+\cdots \\ &= T_i(r,\theta,z,t)-\frac{\partial T_i}{\partial r}\Delta r+\frac{\partial^2 T_i}{\partial r^2}\frac{\Delta r^2}{2}-\frac{\partial^3 T_i}{\partial r^3}\frac{\Delta r^3}{6}+\vartheta(\Delta r^4) \end{aligned} \quad (11)$$

By adding (10) and (11), we have

$$\frac{\partial^2 T_i}{\partial r^2}+\vartheta(\Delta r^4)=\frac{T_i(r+\Delta r,\theta,z,t)-2T_i(r,\theta,z,t)+T_i(r-\Delta r,\theta,z,t)}{\Delta r^2}$$

or approximately,

$$\frac{\partial^2 T_i}{\partial r^2}\approx\frac{T_i(r+\Delta r,\theta,z,t)-2T_i(r,\theta,z,t)+T_i(r-\Delta r,\theta,z,t)}{\Delta r^2}$$

By similar way, we have the following equation in θ and z,

$$\frac{\partial^2 T_i}{\partial \theta^2}=\frac{T_i(r,\theta+\Delta\theta,z,t)-2T_i(r,\theta,z,t)+T_i(r,\theta-\Delta\theta,z,t)}{\Delta\theta^2}$$

and

$$\frac{\partial^2 T_i}{\partial z^2} = \frac{T_i(r,\theta,z+\Delta z,t) - 2T_i(r,\theta,z,t) + T_i(r,\theta,z-\Delta z,t)}{\Delta z^2}$$

Note that, when z is fixed, $\frac{\partial T_i}{\partial z} = 0$ and $\frac{\partial^2 T_i}{\partial^2 z} = 0$. Let $l, m,$ and n are discrete indices for variables $r, t,$ and θ, respectively. Then, we have the following equations.

$$\frac{\partial T_i}{\partial t} = \frac{T^i(x,n,m+1) - T^i(l,n,m)}{\Delta t}$$
$$\frac{\partial T_i}{\partial r} = \frac{T^i(l+1,n,m) - T^i(l,n,m)}{\Delta r}$$
$$\frac{\partial^2 T_i}{\partial r^2} = \frac{T^i(l+1,n,m) - 2T^i(l,n,m) + T^i(l-1,n,m)}{\Delta r^2}$$
$$\frac{\partial^2 T_i}{\partial \theta^2} = \frac{T^i(l,n+1,m) - 2T^i(l,n,m) + T^i(l,n+1,m)}{\Delta \theta^2}$$

Therefore, the final numerical solution is

$$T^i_{l,n,m+1} = T^i_{l,n,m} + \frac{\alpha_i \Delta t}{\Delta r^2}(T^i_{l+1,n,m} - 2T^i_{l,n,m} + T^i_{l-1,n,m}) + \frac{\alpha_i \Delta t}{r_l \Delta r}(T^i_{l+1,n,m} - T^i_{l,n,m})$$
$$+ \frac{\alpha_i \Delta t}{r_l^2 \Delta \theta^2}(T^i_{l,n+1,m} - 2T^i_{l,n,m} + T^i_{l,n-1,m}) + \frac{\alpha_i \Delta t}{l_i} g_i(r_n, \theta_m, z, t_l) \quad (12)$$

This numerical solution (12) then being simulated using Python 3.7. and Numpy Library with additional assumption that the heat source time independent [9, 10]. The results depicted in Figure 2 and 3. The Figure 2 shows the temperature profile for the first layer. As we can see, the temperature increases with time. This is because the heat source is constant and the heat delivered through the cylinder and meet the temperature profile in boundary condition from the second layer. When the heat delivered to the second layer, the temperature decreases as shown in Figure 3, and the curve's negative gradient depends up on the thermal diffusivity (α) of the medium.

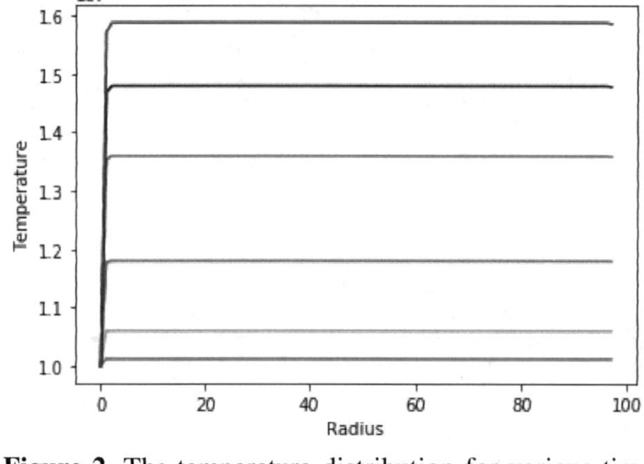

Figure 2. The temperature distribution for various time indices in the first layer with $\alpha = 1,1.10^{-3}$

IOP Conf. Series: Materials Science and Engineering **858** (2020) 012038 doi:10.1088/1757-899X/858/1/012038

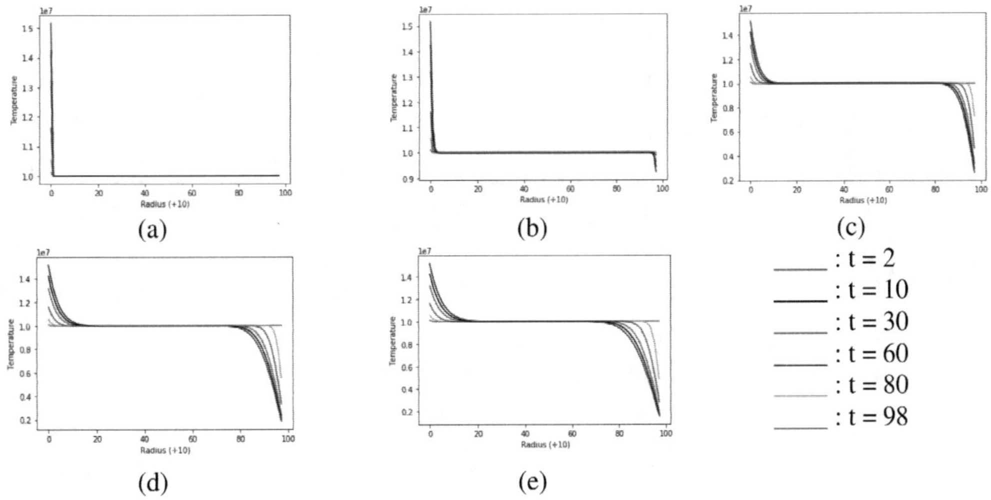

Figure 3. The temperature distribution in the second layer for various α, (a) $\alpha = 1,1.10^{-4}$, (b) $\alpha = 1,1.10^{-2}$, (c) $\alpha = 1,1.10^{-0,5}$, (d) $\alpha = 1,1.10^{-0,2}$, (e) $\alpha = 1,1.10^{-0,05}$

4. Conclusion

The heat transfer profile in a two dimensional multilayered composite cylinder was determined using analytical and numerical method. The temperature distribution profile in different time and different thermal diffusivity for two layers was calculated using Python 3.7.

Acknowledgment

This research was supported by PNBP Universitas Mataram, 2019 Grant.

References

[1] Alaa' S, Khairurrijal, and Abdullah M 2013 Materials Science Forum **737** 209-214

[2] Mehendale S S and Jacobi A M 2000 Applied Mechanics Reviews **53**(7) 175

[3] Dalkilic A S and Wongwises S 2010 Energy Conversion and Management, **51**(12), 2535–2547

[4] Lacarrière B, Trombe A and Monchoux F 2006 Energy and Buildings **38**(3) 232–237

[5] Peniguel C 1998 International journal of heat and fluid flow **19**(2) 102-114

[6] Buckley T M 2002 Patent No. US 6,464,672 B1

[7] Dalira N and Nourazar S 2014 Journal of Engineering Physics and Thermophysics, **87**(1)

[8] Mazumder S 2016 Elsevier USA

[9] Mckinney W 2013 O'Reilly Media USA

[10] Zoder M, Balke J, Hofmann M and Tsatsaronis G 2018 Energies **11**(10) 2609

IOP Conf. Series: Materials Science and Engineering **858** (2020) 012039 doi:10.1088/1757-899X/858/1/012039

The Effectiveness of Zeolite/Claystone/Activated Charcoal Composite in Reducing Levels of Mercury (Hg) in the Waste Resulting from the Activities of Unlicensed Gold Mining (PETI) in Sintang

Dwi Rizaldi Hatmoko[1], Ari Handono Ramelan[1], Pranoto[1]

[1] Program Studi Magister Ilmu Lingkungan Pascasarjana UNS, Jl. IrSutami No.36 A, Surakarta, Indonesia

Email: dwirizaldi_hatmoko@student.uns.ac.id

Abstract. Research on zeolite/clay/activated charcoal composites as Hg heavy metal adsorbents has been carried out. Zeolite and clay were previously activated using NaOH while activated charcoal was activated using HCl. The adsorbent characterization was carried out by Fourier Transform Infrared Analyzer (FTIR), X-Ray Diffraction (XRD), and mercury levels using Atomic Absorption Spectroscopy (AAS). Hg metal adsorption experiments were carried out by varying the composition of zeolite / claystone / activated charcoal. The location of the study was in Gold Mining without Permit (PETI) in the Kapuas Sub-watershed, Ulak Jaya Village, Sintang. The results showed that zeolite / claystone / activated charcoal succeeded in reducing mercury levels in mining waste.

1. Introduction

The gold mining industry is one of the sectors that the Indonesian government has relied on to bring in foreign exchange. In addition to bringing in foreign exchange, the mining industry also opens up employment opportunities and for districts and cities is a source of local own-source revenue (PAD) [1]. Gold mining in Indonesia in general still uses mercury (Hg) in the panning process as a binder of the gold element. This mining technique harms the environment in the form of the release of heavy metal mercury (Hg) into the atmosphere that can endanger the ecosystem and cause serious harm to health [2]. Mercury is known to be a heavy metal which has very high toxicity. The use of mercury in an industry often results in environmental pollution, primarily through wastewater [3]. Most areas in West Kalimantan have high gold content. The gold contained in the land attracts the interest of small scale miners / traditional miners. One of the locations of illegal gold mining (PETI) is in the Village of Ulak Jaya Sintang, Sintang District, Sintang District, West Kalimantan. This gold mining is done traditionally, where the processing process does not use high technology and only uses elementary equipment.

Unlicensed Gold Mining Habits (PETI) in Ulak Jaya Sintang Village, Sintang District, Sintang District, West Kalimantan, which dispose of waste directly into the Kapuas River Basin (DAS) to become black and smelly. Tugaswati in her writing explained that the metal concentration of Hg in several fish species in the Kapuas River and the Landak River reached 50 μg/g; 100 times higher than the frequency permitted by the World Health Organization (WHO) for seafood products, 0.5 mg/kg [4]. Seeing how this pollution will be very dangerous if left unchecked, it is necessary to take anticipatory steps by utilizing zeolite, claystone and activated charcoal to reduce mercury content in gold mining wastewater. This step is considered effective also can reduce costs in practice prevention. In principle, the process is the absorption or adsorption of natural zeolite and metal ions in the gold mining liquid waste. According to Sutopo it is known that zeolites can reduce iron (Fe) levels from 1.466 mg/L to 0.120 mg/L, manganese (Mn) from 2.220 mg/L to 0.060 mg/L, copper (Cu) from 0.430 mg/L to 0.030 mg/L, and zinc (Zn) from 0.180 mg/L to 0.110 mg/L [5].

1.1. Zeolite as an absorbent (absorbent) of mercury (Hg)

Experiments using zeolites as absorbent (absorbent) of mercury (Hg) contained in the water this is possible because zeolites have the nature of absorbing and have hollow structures. According to Government Regulation No. 82 of 2001 glasses of water containing mercury as raw material for drinking water (class I) a maximum of 0.0010 mg/L, if the number exceeds this number will interfere with the health of its consumption.

The zeolite structure can be seen in Figure 1. Based on the structure of porous solids that have cavities and great thermal stability, zeolite-like material can be used as a carrier of catalyst metals or adsorbents.

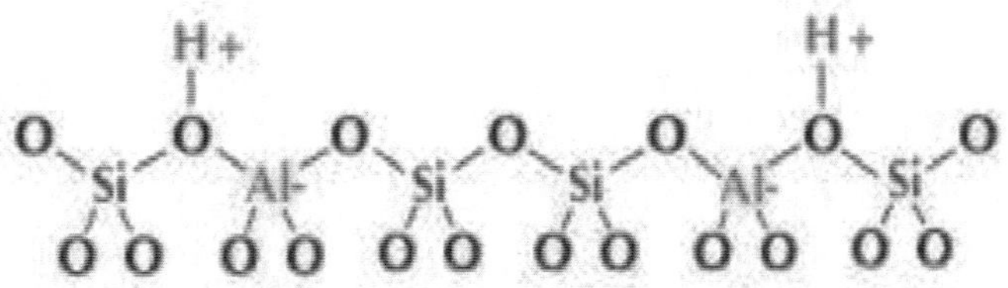

Figure 1. Composition and Structure of Zeolites

1.2. Claystone

Clay or claystone is a pozzolanic material by-product from the rest of the coal. Claystone is formed because of the non-combustible minerals contained in coal, such as Silica (Si), Alumina (Al) and Iron (Fe). The mineral content in claystone varies depending on the type of coal used. Because of the high content of silica and alumina in it, claystone can be used as a source of silica and alumina in the synthesis of zeolite material. Clay minerals or claystone can be used as a catalyst, adsorbent and as a resin for ion exchange, for the purpose of exchanging ions, clay must first be activated with the aim of increasing the absorption of clay [6]. Claystone or claystone has great potential as an adsorbent because of its abundant, large surface area with active groups such as silanol and aluminol in its skeletal structure and easily dispersed in water [7].

Rahmawati research states that the adsorption data of fourteen organic molecules on Cox claystone through experiments for polar compounds, namely acetate, adipate, benzoate, phthalates and citrate in decarbonated and clay fractions (<2 mm) [8]. Adsorption from a single carboxylic function is low, as evidenced by acetate and adipate. Adsorption of small polycarboxylic acids is significant as evidenced by phthalate, succinate, oxalate, gluconate, EDTA and citrate. Nusa's research proved that claystone can reduce the pH content of 6.83, BOD 16 mg/L, TSS 0.761 mg/L and Oil Fat 4.21 mg/L. Although in the ANOVA test, there was no significant effect between stirring time and different concentrations on pH and BOD. But discharged household wastewater is still above the standard quality of the Decree of the Minister of Environment No.112 of 2003 for pH 6-9, BOD of 100 mg/L, TSS of 100 mg/L and Fat oil of 10 mg/L [9].

1.3. Activated Charcoal
Mercury (Hg) absorbent using activated H3PO4 activated charcoal that activated charcoal is very good in the absorption process. This is proven by the absorption of Ni (II) metal which is absorbed by 95.08% while the Pb (II) metal is absorbed by 52.77%. Activated charcoal has a composition composed of elements of carbon, calcium, sulfur, phosphorus, lignite and cellulose. High carbon and cellulose content (36.00%) and lignite (20.90%) contained active charcoal so that it can be used in the adsorption process. The content contained in rice husk charcoal is not much different from the element content contained in activated carbon. The ability of activated charcoal as absorbent material is not the same from one another, because absorption is not necessarily good for the other absorption process. Differences in pore particle size and activation levels can affect the optimization of the use of activated charcoal [10].

This research uses experimental methods in the laboratory to obtain results data. This research broadly includes four stages, namely the preparation and characterization of adsorbents, adsorbent activation, adsorbent performance testing, and determination of adsorption isotherms. Furthermore, the results of the data are tested in the field with the aim to get a match between the laboratory results and the conditions in the field.

The data in this study consisted of two types, namely, quantitative data and qualitative data. Quantitative data are in the form of laboratory results on the effectiveness of zeolite, claystone and activated charcoal composites in the processing of mercury (Hg) waste in unlicensed Gold Mining (PETI) activities. While the qualitative data is questionnaire data about the social impact of the use of zeolite, claystone and active charcoal composites in the processing of mercury (Hg) waste in the Unlicensed Gold Mining activities (PETI) in the Kapuas Sub-Watershed, Sak Jaya Village, Sintang.

2. Results
2.1. Zeolite Analysis and Activation of gold mining waste
FTIR analysis was performed to determine zeolite functional groups and the effect of the zeolite activation process was carried out with 1M HCl. The use of this concentration was in accordance with the results reported by Sastiono [11], which activated mordenite and clinoptilolite zeolite types and obtained CEC results from the zeolites increased, but the use of HCl of more than 1 N has decreased the CEC value. According to Tri zeolites will experience dealumination that will continue to occur if the greater the concentration of acid used [12].

The results of IR zeolite spectra, before and after activation are shown in Figure 2. and summarized in Table 1:

IOP Conf. Series: Materials Science and Engineering **858** (2020) 012039 doi:10.1088/1757-899X/858/1/012039

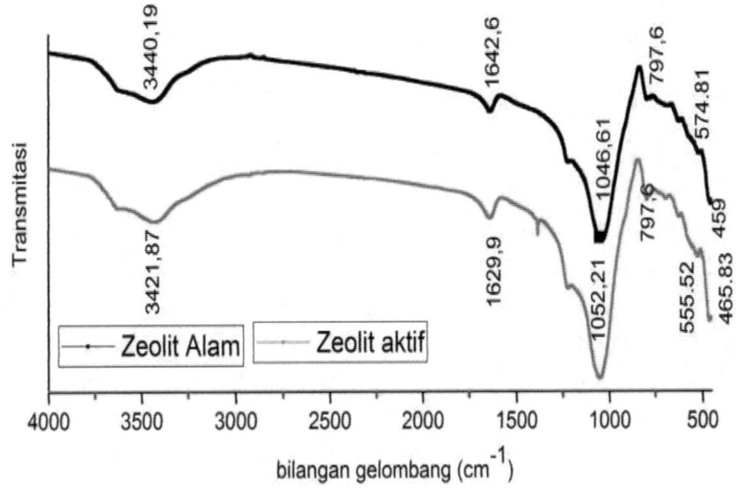

Figure 2. FTIR Spectra of Natural and Activated Zeolite Samples

From Figure 2 the absorption of natural zeolite FTIR identified at wave number 3440.19 cm-1 is the peak of the hydrated -OH group, this ftir spectrum is in accordance with Research Heraldy also shows that OH absorption occurs at successive wave numbers 3442.7; 3435.0 and 3424 cm-1 [13]. The vibration of buckling OH from H_2O adsorbed according to Suroto occurred at wave number 1637.5 also occurred in natural zeolites at wave number 1642.6 cm-1. In the study of Heraldy Al-O or Si-O stretching vibrations appeared at wavenumbers 1052.3 and 1043.3 cm^{-1} whereas in studies from Suroto Al-O or Si-O stretching vibrations occurred at wave number 1055.0 cm^{-1}. In this study, Al-O or Si-O stretching vibrations from natural zeolites appeared at wave number 1046.61 cm^{-1}. The vibration of Si-O buckling natural zeolite according to research by Heraldy appeared at wavenumber 794.6 cm^{-1} while research from Suroto vibration buckling Si-O appeared at wavenumbers 777.4 cm^{-1} and 796.5 cm^{-1}. In this study, natural zeolites have the vibrations of buckling Si-O at wave number 797.6 cm^{-1}, which is not much different than that of the reference. Figure 2 shows that active zeolite spectra are almost the same as natural zeolite spectra. This shows that the role of activation does not change the structure of natural zeolites.

Table 1. Functional Group of Natural and Activated Zeolite Samples

Functional Group	Wave number (cm^{-1})		
	Literature	Natural Zeolite	Active Zeolite
-OH hidrated	3442,7(1) 3435(3) 3442,7(2) 3424(4)	3440,19	3421,87
vibration buckling OH from H_2O	1637,5(2)	1642,46	1629,92
stretching vibrations Al-O or Si-O	1055,0(2) 1043,3(1) 1052,3(1)	1046,61	1052,21
vibration buckling Si-O	796,5(2) 777,36(2) 794,6(1)	797,6	797,6
vibration buckling T-O	471,6(1)	441	465,83

	445,5[2]	459	
Al-OH *bending* absorption	690-705[5]	695,37	695,37
vibration buckling Si-O atau Al-O	579 485[6]	574,81	-

Note: (¹) Heraldy [13], (²) Suroto [14].

2.2 Analysis and Activation of claystone on gold mining waste

Activation of claystone is made from clay samples washed with distilled water several times and filtered until the clay is completely clean from impurities, then dried in an oven at 120 °C for 2 hours, after which it is crushed and sieved with 80 mesh size. Clay samples that have been prepared are then taken ±100 g and immersed in an Erlenmeyer containing NaOH solution with a concentration variation of 1.5 M, then heated for 8 hours. After it was washed two times with distilled water, one time with a composite of methanol and distilled water (1: 1) and two times with methanol, then dried at room temperature.

The following graph shows before and after activation of claystone on gold mining waste.

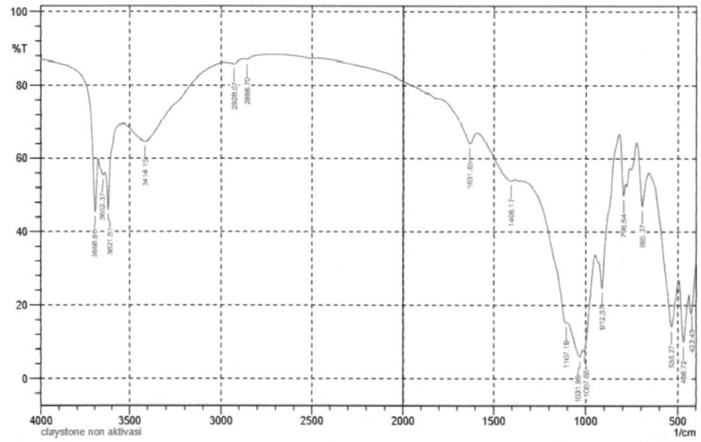

Figure 3. Claystone chart before activation

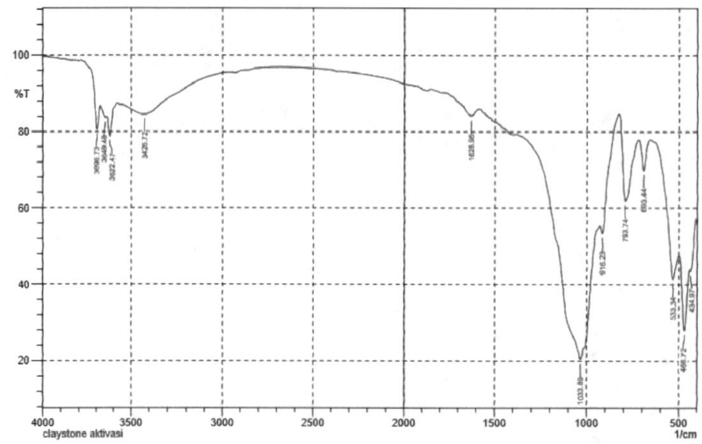

Figure 4. Claystone chart after activation

From the two graphs above shows a very significant change after gold mining waste is activated by claystone. The concentration of mercury (Hg) drops, which means the levels are reduced, and it shows that claystone is effective for trapping harmful substances in waste.

2.3. Activation and activation of activated charcoal from gold mining waste
The chemical activation of activated charcoal is carried out by the addition of a 4 M HCl solution which aims to dissolve impurities on the surface of the charcoal so that the pores of the charcoal will open and its surface area increases. The effect of activation on activated charcoal can be observed in the FTIR spectra shown in Figure 6. The results of the FTIR spectra in Figure 6 show the appearance of new peaks in activated charcoal after activation. A new peak appears at wave number 1037 cm^{-1} which is the absorption of C-O. The emergence of new peaks is caused by the dissolution of impurities during the activation process. Based on the data in Table 2 shows that there are some differences between activated charcoal before and after activation, such as the shifting of wavenumbers from 3427 cm^{-1} to 3421 cm^{-1} which shows the absorption of the -OH group experiencing peak widening.

The absorption peaks of C=O, C=C, and C-O after activation look sharper, indicating an increase in absorption intensity. Increased intensity is possible because of impurity on the activated charcoal after it dissolves by the activation.

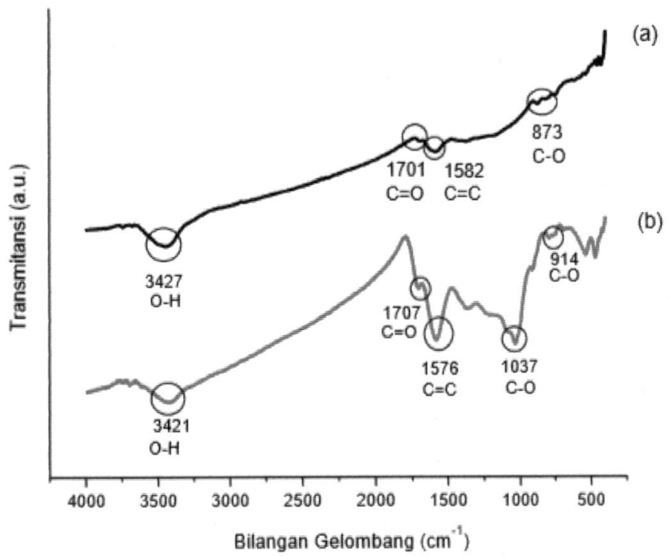

Figure 5. FTIR spectra of activated charcoal before activation (a) and after activation (b)

Table 2. Functional Group

No	Functional Group	Wave number (cm^{-1})		
		Literature	Before activated	After activated
1.	O-H	3425(1)	3427	3421
2.	C=O	1705(1)	1701	1705
3.	C=C	1589(1)	1582	1576
4.	C-O	1300-800(2)	873	914; 1037

Note (1) Kurniati *et al* (2011),(2) Silverstain *et al* (2000)

The results of FTIR analysis showed that there were O-H, C=O, C=C and C-O uptake in the activated char spectra, which showed the presence of active groups. Analysis of the mineral content of activated charcoal was done by XRD. The results of XRD analysis of activated charcoal before and after activation are shown in Figure 6. The activated charcoal diffractogram was identified in the minerals cristobalite, fayalite and standard manganoan contained in the ICSD (Inorganic Crystal Structure Database) shown in Appendix 4. Peak diffraction of activated charcoal before activation is shown in 2θ = 24.22° which is the mineral cristobalite while the peak of diffraction of activated charcoal after activation is shown at 2θ = 21.91° (cristobalite), 26.723 (fayalite), 27.86 (manganoan) and 35.84 (manganoan + fayalite). The diffractogram comparison after activation shows the appearance of a new peak. The appearance of a new peak indicates that the impurity has dissolved due to the activation process. Soluble impurity will open the pores on the surface of activated charcoal and cause an increase in mineral content.

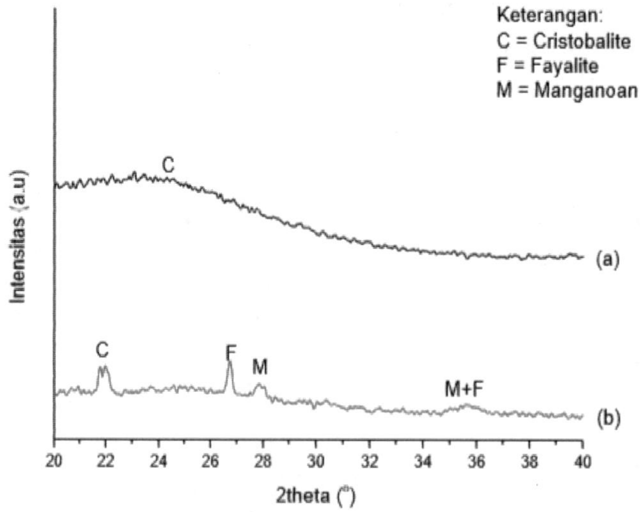

Figure 6. Diffractogram of activated charcoal before activation (a) and after activation (b)

2.4. Effectiveness of Zeolite, Claystone and Activated Charcoal Composites in suppressing Mercury

Per the results of the Baristan laboratory tests, when using the leading composites from Zeolite, Claystone and Activated Charcoal can reduce levels of Mercury (Hg) in wastewater from gold mining (Table 3).

Table 3. Composite test results for waste

Number of analysis	Code of Sample	Result / Mercury / Hg Total (mg/L)
P. 3223 -19	First waste	64,0
P. 3224 -19	$HgNO_2$ (prepar)	10,7
P. 3225 -19	Final waste	< 0,0002
P. 3226 -19	Zeolit (pure) / P	0,570
P. 3227 -19	Claystone(pure) / P	61,7
P. 3228 -19	Charcoal(pure) / P	40,0
P. 3229 -19	Zeolit50:25:25 / P	3,349
P. 3230 -19	Clay 50:25:25 / P	3,409
P. 3231 -19	Charcoal50:25:25 / P	< 0,0002
P. 3232 -19	Zeolit (pure) / L	< 0,0002
P. 3233 -19	Clay (pure) / L	29,6
P. 3234 -19	Charcoal (pure) / L	35,7
P. 3235 -19	Zeolit50:25:25 / L	0,914
P. 3236 -19	Clay 50:25:25 / L	< 0,0002
P. 3237 -19	Charcoal50:25:25 / L	< 0,0002

Based on the data in the above test results found the value of Mercury contained in the initial waste of 64.0 then through laboratory tests combined with Zeolite, Claystone and Activated Charcoal that has not been composite or pure each of them can only reduce the mercury content in the waste by Zeolite Pure 0.570, Claystone 61.7, and Activated Charcoal 40.0. These results indicate that the reduction of mercury levels in the waste does not comply with the mercury standard that is equal to 0.005 Hg by the Decree of the Minister of Environment No.202 of 2004 concerning Wastewater Quality Standards for Businesses and Mining Activities of Gold Ore and or Copper (Table 4).

Table 4. Wastewater Quality Standards for Gold and Copper Ore Mining Activities

Parameter	Measure	Maximum level	Analysis Method
pH	-	6 – 9	SNI 06-6989-11-2004
TSS	mg/L	200	SNI 06-6989-3-2004
Cu	mg/L	2	SNI 06-6989-6-2004
Cd	mg/L	0,1	SNI 06-6989-18-2004
Zn	mg/L	5	SNI 06-6989-7-2004
Pb	mg/L	1	SNI 06-6989-8-2004
As	mg/L	0,5	SNI 06-2913-1992
Ni	mg/L	0,5	SNI 06-6989-22-2004
Cr	mg/L	1	SNI 06-6989-22-2004
Hg	mg/L	0,005	SNI 06-2462-1991

But after being tested using Zeolite, Claystone and Activated Charcoal which has been composite, the results are quite significant where the mercury levels are reduced and are below the quality standard. Based on the results of laboratory tests, the levels of Mercury (Hg) in gold mining waste decreases to 0.914 with a Zeolite composite with a ratio of 50:25:25. While using Claystone composites with the same comparison value, the mercury content decreases to 0,0002 and using Activated Charcoal with the same ratio can also reduce mercury levels up to 0,0002. This finding is by the provisions of the standard quality standard for mercury content, meaning that the use of Zeolite, Claystone and Activated Charcoal composites to reduce mercury levels in the gold mining waste at Sintang is very useful.

3. Conclusion

Based on the discussion above, it is confirmed that zeolite, claystone and activated charcoal are capable and effective in reducing mercury (Hg) levels in waste products from unlicensed gold mining (PETI) in the Kapuas Sub-watershed, Ulak Jaya Sintang Village. With these findings so that it can be utilized as well as possible to anticipate the increasingly severe natural damage due to gold mining waste. The goal is that the preservation of nature can be utilized more optimally by residents around the mine.

Reference

[1] Yudhistira 2008 *Kajian Pencemaran Merkuri (Hg) Terhadap air sungai Menyuke Dan Gangguan kesehatan Pada Penambangan Sebagai Akibat Penambang Emas Tanpa Izin (Peti) di kecamatan Menyuke Kabupaten landak Kalimantan Barat* (http://Journal tambang emas rakyat) diakses 08 September 2018 pukul 21.00.)

[2] Chinthia R H 2012 Jurnal Kesehatan Masyarakat **1**(2) 973 – 982

[3] Dash H R and Das S 2012 Biodegrad. **75** 207-213

[4] Tugaswati, Athena F B, and Lubis A 2003 *Studi pencemaran merkuri dan dampaknya terhadap kesehatan masyarakat di daerah Mundu Kabupaten Indramayu* (Indramayu: Badan Litbang Kesehatan Puslit Ekologi Kesehatan)

[5] Sutopo F X R 1991 *Pengkajian Karakteristik Zeolit Cikalong Tasikmalaya dan Pemanfaatannya dalam Pengolahan Air, Laporan teknik Pengolahan* (Bandung: PPTM.)

[6] Catherina, Bijang M, Sekewael S J, and Koritelu J A 2014 *Indonesia Jurnal Chemichal Res* **2014**(1) 93–98

[7] Pranoto, Masykur A, and Nugroho Y A 2018 *IOP Conf. Ser.: Mater. Sci. Eng.* **333** 2018 012057

[8] RahmawatiNA 2010*Pengaruh Limbah Cair Industri Sarung Tangan Pada Air Irigasi dan Pengaruhnya terhadap Produksi Padi dan Kualitas Lingkungan* (Tesis Yogyakarta: Program Ilmu Lingkungan Sekolah Pasca Sarjana. Universitas Gadjah Mada Yogyakarta)

[9] F Zhang S Y Xie D W Hu J F Shao and B Gatmiri 2012 Journal Applied Clay Science **69**(2012) 79–86)

[10] Pari G Hendra D and Pasaribu R A 2004 Jurnal Industri Hasil Perkebunan **12**(224) 33-46

[11] Sastiono A 1993 *Perilaku Mineral Zeolit Dan Pengaruhnya Terhadap Perkembangan Tanah* (Disertasi Bogor: Program Pascasarjana, Institut Pertanian Bogor)

[12] Trilianty Lestaria 2010*Faktor-faktor Yang Berhubungan Dengan Keracunan Merkuri (Hg) Pada Penambang Emas tanpa Ijin (PETI) di Kecamatan Kurun Kabupaten Gunung Mas* (Tesis Semarang: Program Pascasarjana Ilmu Lingkungan UNDIP)

[13] Heraldy E Hisyam S W dan Sulistiyono 2000 Indonesian J. Chem **3** (2)

[14] T Suroto 2004 *Kajian Pengaruh Konsentrasi Asam Klorida Terhadap Distribusi Ukuran Pori zeolite Alam dan Uji Kemampuan Sebagai Adsorben Untuk Kemurnian Minyak Daun Cengkeh* (Skripsi. Yogyakarta: Universitas Gajah Mada)

IOP Conf. Series: Materials Science and Engineering **858** (2020) 012040 doi:10.1088/1757-899X/858/1/012040

The assessment of claystone, quartz and coconut shell charcoal for adsorbing heavy metals ions in acid mine drainage

F Mukarrom[1,2], Pranoto, R Karsidi, E Gravitiani, F Astuti, W Maharditya
[1]Doctorate Program of Enviromental Science, Postgraduate, Sebelas Maret University
[2]Mining Engineering Department, Institut Teknologi Nasional Yogyakarta, Yogyakarta

E-mail :faisolmk@student.uns.ac.id

Abstract. This research aims to determine the assessment of claystone, quartz and coconut shell charcoal for adsorbing heavy metals ions in acid mine drainage. Claystone is activated by using NaOH 3M, quartz sand is activated by using Na_2CO_3, while coconut shell charcoal is activated with HCl 4M. The adsorbent characterization was done by Fourier Transform Infrared Analyzer (FTIR) and X-Ray Diffraction (XRD). The results of the characterization of claystone FTIR showed the presence of active groups of silanol (Si-OH) and aluminol (Al-OH), XRD claystone characterization showed the presence of montmorillonite, kaolinite, illite and quartz minerals. While the results of FTIR characterization of quartz sand showed the presence of active groups of silanol (Si-OH) and siloxan (Si-O-Si), XRD characterization of quartz sand showed the presence of quartz mineral content. Furthermore, coconut shell charcoal showed the presence of hydroxyl (OH) and carboxyl active groups. (COOH) which is known from FTIR analysis as well as the mineral content of cristobalite, fayalite and manganoan which are known from XRD analysis. From the results of the FTIR and XRD analysis, it can be concluded that the three adsorbents have the potential as adsorbents. So that they can be used as an acid mine water absorber.

1. Introduction

Mining industry has several distinctive characteristics. Each phase of the activity requires land clearing (deforestation), land and and minerals excavations in large volumes that can make changes the landscape. Mining activities cause significant changes of the landscape will lead to the degradation of environmental function and quality [1], which include the potential of erosion, sedimentation, decreased air quality, the forming of acid mine drainage, and enlowering land productivity. The elements of heavy metals in acid mine drainage has serious effects on human health. Toxicologically, this heavy metal can be divided into two types. The first type is an essential heavy metal, where its presence in a certain amount is needed by living organisms, but in such exceed amount in human's body's cause *toxic* effects. Some of these heavy metals include Zn, Cu, Fe, Co, Mn and so on. The second type is non-essential heavy metals in the human's body. These heavy metals are not yet known for its benefits such as Hg, Cd, Pb, Cr and others. Active treatment method of acid mine draiange is able to precipitate dissolved metals and form *sludge* or mud deposit [2].

One of the passive treatment methods which is considerable is the adsorption method. The availability and the affordabillty of the adsorbent are very important. The abundant availability and low perational

IOP Conf. Series: Materials Science and Engineering **858** (2020) 012040 doi:10.1088/1757-899X/858/1/012040

costs are also as important factors in implementing the method. This research will use claystone, coconut shell charcoal, quartz sand and zeolite as the adsorbtion media. Claystone, could be perform as adsorbent since it is composed from aluminum granules and fine silica . This mineral known has a strong adsorption (sorption) on certain heavy metals dissolved in water [3]. Quartz could be used as an adsorbent because it has a size of sand which is dominated by silica mineral (SiO_2) hexagonal texture, in its formation crystallized. Utilization of coconut shell one of which is used as charcoal for fuel and can also be used as an adsorbent. Coconut shell is the best material used as activated carbon, because activated carbon produced from coconut shell has carboxyl (COOH) and hydroxyl (OH) functional groups, more micropores, smaller ash content, high solubility in high water and high reactivity [4].

2. Experimental

2.1. Materials

Claystone, quartz, coconut shell, NaOH p.a (Merck), HCl p.a (Merck), Na_2CO_3 p.a (Merck).

2.2. Preparation of Adsorbent

Claystone is cleaned from impurities and dried by aerating it in the open air until it is dry, then claystone is crushed until smooth. Claystone is then sifted with a 150 mesh sieve. Powder that escapes 150 mesh is immersed in distilled water and filtered, then dried at 105 ºC for 4 hours. The quartz sand obtained is cleaned of impurities, sifted and finely ground. After smooth, proceed with a sieve size of 100 mesh. Then wash with 0.1 M HNO_3 while stirring with a magnetic stirrer for 1 hour. Washing is done 3 times and the residue obtained is dried [5]. The coconut shell obtained was then cut into small pieces, washed 3 times, dried in the sun to dry, purified 400 °C for 1 hour, grinded and sieved using a 150 mesh size sieve [6]. Adsorbents were characterized using FTIR and XRD.

2.3. Adsorbent Activation

The claystoneas much 50 gram is mixed with 250 mL of NaOH 3 M solution. The mixture is stirred at 70 °C for 5 hours then cooled, filtered and washed with distilled water to a neutral pH. Furthermore, activated claystone is dried in an oven at 105 °C for 4 hours [7]. 165 g of quartz sand and 135 g of sodium bicarbonate are mixed and the mixing process is carried out evenly. Then put in a crucial and then heated to a temperature of 1100 °C for 2 hours. Then the melt is cooled and crushed into powder [8]. The coconut shell charcoal as much 50 gr soaked with 100 mL 4 M HCl solution for 24 hours, then filtered and washed with distilled water. The activated carbon produced is then dried in an oven at 110 °C for 3 hours, then cooled in a desiccator [9]. Activated adsorbents were characterized by FTIR and XRD.

3. Result and discussion

3.1. FTIR Characterization

3.1.1. Claystone

The changes of claystone before and after activation such as shifting of wave numbers from 1031 cm^{-1} to 1033 cm^{-1} in Si-O-Si / Al-O-Al groups and wave numbers from 912 cm^{-1} to 916 cm^{-1} in Al groups -OH which shows an increase in intensity because the impurities have been lost during the activation process so that the functional groups become more active. The absorption peak of Al-OH-Al, -OH, H-O-H has decreased intensity.

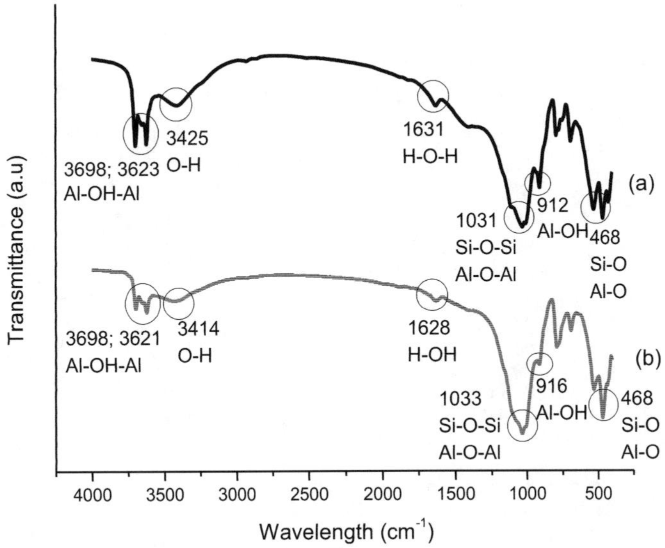

Figure 1. FTIR of claystone before (a) and after activation (b)

Table 1. Comparison of FTIR claystone adsorption before and after activation

No	Functional Group	Wavelength (cm^{-1})		
		Reference	Before activation	After activation
1	Al-OH-Al	3616[10]	3698; 3623	3698; 3621
2	-OH	3500-3400[11]	3425	3414
3	H-O-H	1639[12]	1631	1628
4	Si-O-Si / Al-O-Al	1039[10]	1031	1033
5	Al-OH	920-690[13]	912	916
6	Si-O / Al-O	470,6[14]	468	468

3.1.2. Quartz sand

There are some differences between quartz sand before and after activation such as the shifting of the wave number from 3432 cm^{-1} to 3436 cm^{-1} in the -OH strech group and the wave number from 1628 cm^{-1} to 1635 cm^{-1} in the HOH bend group experiencing increased intensity because impurities have been lost during the activation process so that functional groups become more active. The peak absorption of Si-O-Si asymmetric stretch, Si-O-Si symmetric stretch and O-Si-O bend have decreased intensity which is possible because the pegotor substance on the surface of quartz sand has dissolved by the activation but not to eliminate the absorption of functional groups.

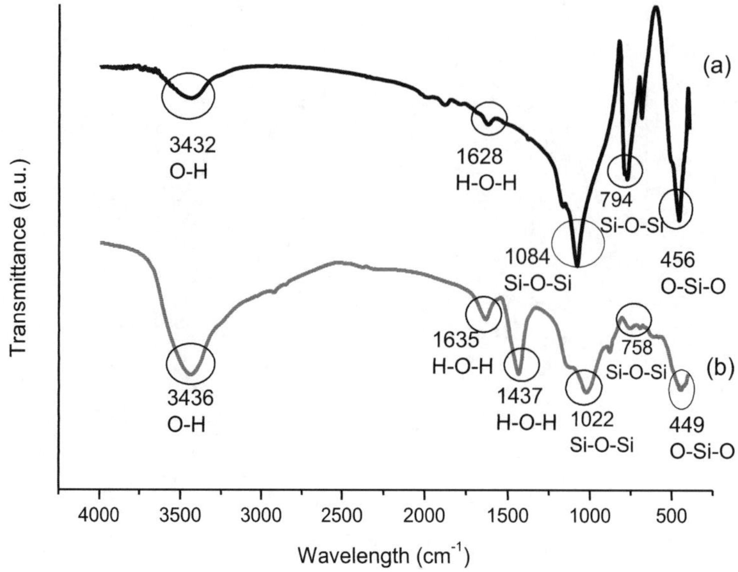

Figure 2. FTIR spectra of quartz sand before (a) and after activation (b)

Table 2. Comparison of FTIR quartz sand adsorption before and after activation

No	Functional Group	Wavelength (cm^{-1})		
		Reference	Before activation	After activation
1	O-H *stretch*	3550-200[15]	3432	3436
2	H-O-H *bend*	1641[16]	1628	1635; 1437
3	Si-O-Si *asymmetric strech*	1076[17]	1084	1022
4	Si-O-Si *symmetric strech*	771[17]	794	758
5	O-Si-O *bend*	461[18]	456	449

3.1.3. Coconut shell charcoal

There are some differences between coconut shell charcoal before and after activation such as shifting the wave number from 3427 cm^{-1} to 3421 cm^{-1} which shows the absorption of the -OH group experiencing peak widening. The absorption peaks of C = O, C = C, and C-O after activation look sharper indicating an increase in absorption intensity. Increased intensity is possible because the impurities in the coconut shell charcoal have dissolved by the activation.

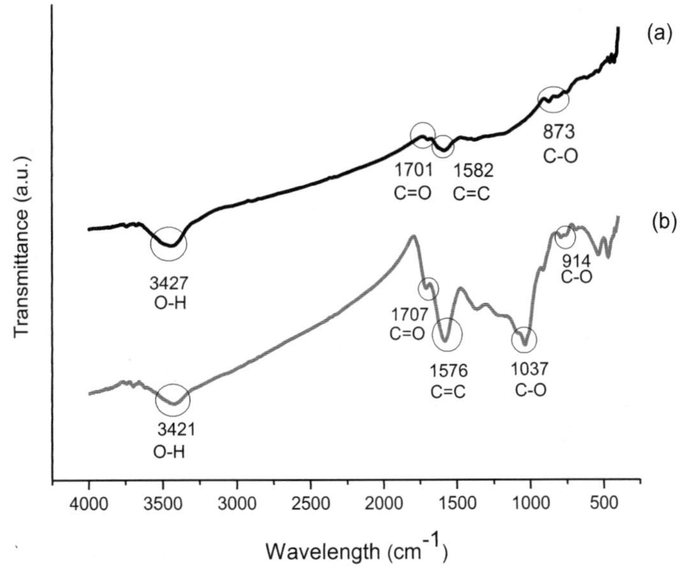

Figure 3. FTIR spectra of coconut shell charcoal before (a) and after activation (b)

Table 3. Comparison of FTIR coconut shell charcoal adsorption before and after activation

No	Functional Group	Wavelength (cm^{-1})		
		Reference	Before activation	After activation
1	O-H	3425[19]	3427	3421
2	C=O	1705[19]	1701	1705
3	C=C	1589[19]	1582	1576
4	C-O	1300-800[15]	873	914; 1037

3.2. XRD Characterization

3.2.1. Claystone

The results of XRD clystone analysis before and after activation showed that the presence of minerals such as kaolinite (K), montmorillonite (M), illit (I) and quartz (Q) as evidenced by the appearance of diffraction peaks at 2θ. Before claystone activation, kaolinite minerals appeared at 2θ = 35.02° and 32.09°, montmorillonite minerals appeared at 2θ = 5.34°; 19.85° and 62.05°, illit minerals at 2θ = 26.70° and quartz minerals at 2θ = 68.29°. On the active claystone the kaolinite mineral appeared at 2θ = 30.20°; 39.65°; 60.11°, montmorillonite minerals appear at 2θ = 5.55°; 20.10°; and 62.05°, illite mineral at 2θ = 26.89° and quartz mineral at 2θ = 68.37°. The diffractogram comparison before and after activation shows a decrease in intensity and the appearance of new peaks. The decrease in the intensity of each mineral shows that the crystal structure becomes amorphous.

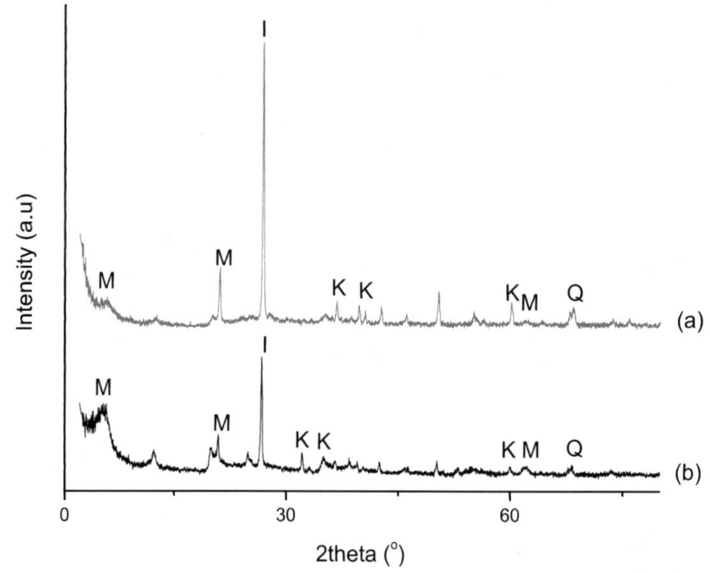

Figure 4. XRD diffractograms of claystone before (a) and after activation (b)

3.2.2. Quartz sand

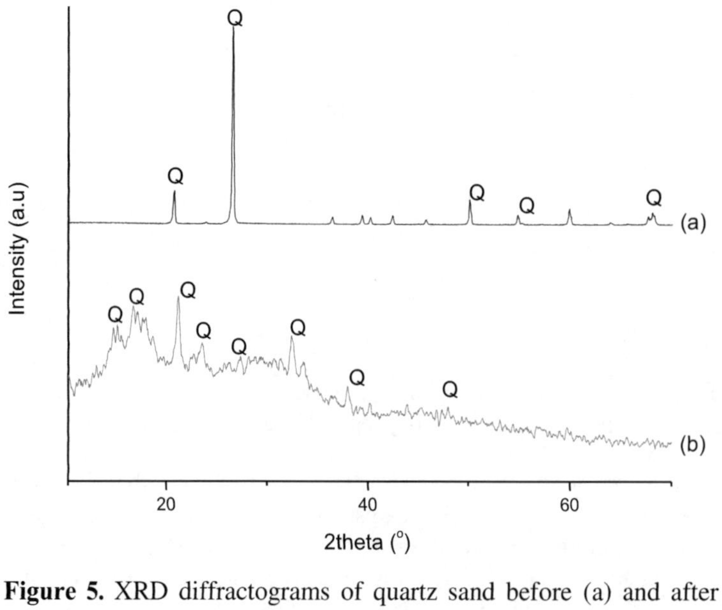

Figure 5. XRD diffractograms of quartz sand before (a) and after activation (b)

The results of XRD analysis of quartz sand before and after activation indicate the presence of quartz mineral content (Q) as evidenced by the appearance of diffraction peaks at 2θ. Figure 5 shows that the quartz mineral peaks appear at 2θ = 20°, 26°, 50°, 54°, and 68°. In active quartz sand quartz minerals appear at 2θ = 14°, 16°, 21°, 23°, 27°, 32° and 38°. The diffractogram comparison before and after activation shows a decrease in intensity and the appearance of new peaks. The decrease in the intensity of each mineral shows that the crystal structure becomes amorphous.

3.2.3. Coconut shell charcoal

The coconut shell charcoal diffractogram was identified in the cristobalite (C), fayalite (F) and manganoan (M) minerals found in the ICSD (Inorganic Crystal Structure Database). The peak of the coconut shell charcoal diffraction before activation is shown at 2θ = 24.22° which is the mineral cristobalite while the peak of the coconut shell charcoal diffraction after activation is shown at 2θ = 21.91° (cristobalite), 26.723° (fayalite), 27.86° (manganoan) and 35.84° (manganoan + fayalite). The diffractogram comparison after activation shows the appearance of a new peak. The appearance of a new peak indicates that the impurity has dissolved due to the activation process. Dissolution of impurities will open the pores on the surface of the coconut shell charcoal and cause an increase in the mineral content.

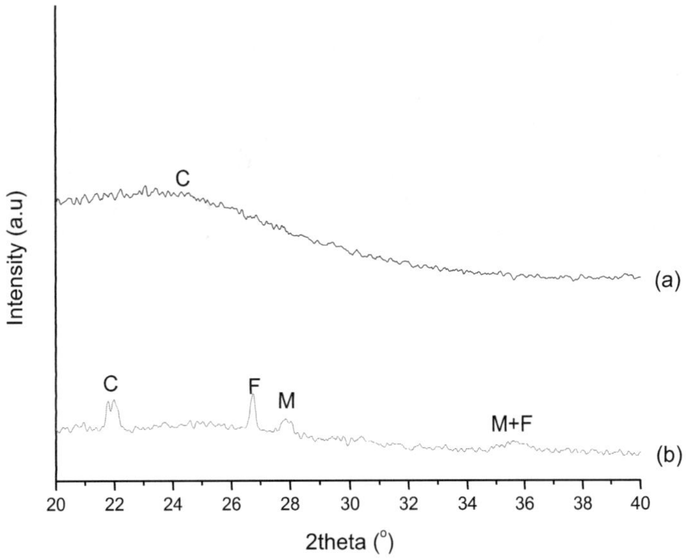

Figure 6. XRD diffractograms of coconut shell charcoal before (a) and after activation (b)

4. Conclusion

The results of FTIR characterization showed thereare active groups in all three materials. The characterizationof claystone showed thereare active groups of cyanol (Si-OH) and aluminol (Al-OH), FTIR characterization of quartz sand showed thereare active groups of silanol (Si-OH) and siloxane (Si-O-Si) and the characterization of coconut shell charcoal showed thereare hydroxyl (OH) and carboxyl (COOH) active groups. XRD claystone characterization showed the presence of montmorillonite, kaolinite, illite and quartz minerals. Characterization of quartz sand shows the presence of quartz minerals, and characterization of coconut shell charcoal shows the mineral content of cristobalite, fayalite and manganoan, so it can be concluded that the three materials have the potential as adsorbents that can be used as absorbers of acid mine water.

Reference

[1] Ranjan R 2019 *JOM FMIPA.* **2(1)** 63-69.
[2] Prianto F A 2016 *Institut Pertanian Bogor* **5**.
[3] Musso T Parolo M, Pettinari G, and Francisca F 2014 *Journal of Environmental Management* **146** 50-58.
[4] Pambayun G S, Yulianto R Y, Rachimoellah M, and Putri E M 2013 *Jurnal Teknik Pomit.* **2** 116-120.
[5] Lesbani A 2015 *Jurnal Penelitian Sains* **14(4)** 41-46.
[6] Rahayu L H, Purnavita S, and Sriyana H 2014 *Jurnal Momentum* **10(1)** 47-53.
[7] Guo X, Yao Y, Yin G, Kang Y, Luo Y, and Zhuo L 2008 *Applied Clay Science* **40** 20–26.
[8] Las T, Florentinus F, and Afit H 2011 *Valensi* **2(2)** 368-378.
[9] Nurhasni, Florentinus F, and Qosim S 2012 *Valensi* **2(4)** 516-525.
[10] Masindi V, Mugera W G, Harganani T, and Marinda D 2017 *Journal of Water Process Engineering* **15** 2-7.
[11] Trabelsi W, and Ali T 2017 *Journal of African Earth Sciences* **129** 647-658.
[12] Sari A, Mustafa T, Demirhan C, and Mustafa S 2007 *Journal of Hazardous Materials* **149** 283-291.
[13] Mahmoudi S, Bennour A, Meguebli A, Srasra E, and Zargouni F 2016 *Applied Clay Science* **127-128** 78–87.
[14] Wogo H E, Febri O N, and Putra J P T 2013 *Jurnal Sains dan Terapan Kimia* **7(1)** 29-41.
[15] Silverstain R M, and Webster F X 2000 *Spectrometric Identification of Organic Compound*, 60th edn. (New York: John Wiley and Sons, Inc).
[16] Dewanto A S, and Munasir 2016 *Jurnal Inovasi Fisika Indonesia (IFI)* **5(2)** 1-4.
[17] Du G H, Liu Z I, Xia X, Chu Q, and Zhang S M 2006 *J Sol-Gel Sci Techn* **39** 285-291.
[18] Ahangaran F, Ali H, and Sirous N 2013 *International Nano Letters* **3(23)** 1-5.
[19] Kurniati F D, Pardoyo, and Suhartana 2011 *Jurnal Kimia Sains dan Aplikasi* **14(3)** 72-76.

Test The Effectiveness and Characterization of Quartz Sand/Coconut Shell Charcoal Composite as Adsorbent of Manganese Heavy Metal

Pranoto[1]*, Tri Martini[1], Fitri Astuti[1], Winda Maharditya[1]
[1]Department of Chemistry, Faculty of Mathematics and Natural Sciences, Sebelas Maret University, Jl. Ir. Sutami 36A Surakarta, 57126 Indonesia

*E-mail: pakpranotomipa@staff.uns.ac.id

Abstract. The research on the effectiveness and characterization of quartz sand/coconut shell charcoal composite as adsorbent of Manganese (Mn) heavy metal has carried out. Quartz sand was previously activated by using Na_2CO_3 while the coconut shell charcoal was activated by using HCl 4 M. The adsorbent characterization was carried out by *Fourier Transform Infrared Analyzer* (FTIR), *X-RayDiffraction*(XRD), *Scanning Electron Microscopy-mapping Energy Dispersive X-Ray* (SEM-EDX), and *Surface Area Analyzer* (SAA). The adsorption experiment of the Mn metal was carried out by varying the composition of quartz sand/ coconut shell charcoal. The variations were in the number of 100:0, 0:100, 25:75, 50:50, 75:25 (b/b) at the degree of pH 1- 6 and contact time 10, 15, 20, 30, 45, 60, 75, and 90 minutes by using *batch* method. The measurement of final concentration at the equilibrium phase was carried out by using AAS. The optimum adsorption condition of the Mn metal occurred in the comparison of quartz sand/coconut shell charcoal 50:50 (b/b), the best condition at the degree of pH 5, contact time during 30 minutes with adsorption capacity of 0.56 mg/gram. The type of adsorption isotherm of composite quartz sand/coconut shell charcoal adsorbent towards the metal of Mn followed the Freundlich isotherm theory.

1. Introduction

Water is generally sourced from ground water and naturally contains inorganic elements, such as ferrous metal (Fe), manganese (Mn), zinc (Zn), lead (Pb) and so on. Mn is an essential heavy metal, where its presence in certain levels is needed by the human body, but in excessive levels can cause toxic effects. In drinking water, the allowable manganese level is 0.4 mg/L [1]. According to Purwonugroho research (2013) [2] the Mn content in ground water reaches 0.8 mg/L. Mn in water can cause turbidity, corrosion and hardness. Mn also causes a yellowish tint on the laundry. Manganese can cause toxic effects in the human body if the concentration is too high. The Mn effect mainly occurs in the respiratory tract and brain. Symptoms of Mn poisoning are hallucinations, forgetfulness and nerve damage. Mn can also cause Parkinson's disease, lung disease and bronchitis. Prolonged exposure to manganese can cause impotence [3]. If Mn metal is absorbed by the body it will be difficult to excrete, as well as if contaminated in the eating environment will be difficult to clean. Therefore, the presence of Mn metal needs to be minimized or even eliminated.

Many methods have been developed to reduce levels of heavy metals from water bodies, one of which is the adsorption method [4]. Adsorption is a relatively simple method, the preparation process is simple, the cost is relatively cheap, efficient, can be done at low concentrations and can use natural adsorbents [5]. Adsorbents that are often used in adsorption include zeolites, natural allophane, chitin, chitosan, biosorbents from algal species, fly ash, activated carbon and cellulose [6]. Coconut shell can be used as an adsorbent because of the availability of abundant coconut shell and is a good adsorbent for purification, removing color, odor, dechlorination, detoxification, filtering, separation and can be used as a catalyst [7]. Coconut shell charcoal also has hydroxyl and carboxyl active groups which influence the adsorption process [8]. Rahmawati and Dony (2016) [9] use activated coconut shell charcoal to adsorb Fe, Mn and Al metals in water up to 20%. Quartz sand has active groups of silanol (Si-OH) and siloxan (Si-O-Si) so that quartz sand can be used as manganese metal adsorbent (Mn). Quartz sand can be used as an adsorbent as previously reported research which shows quite good efficiency. Lindasari et al (2017) [10] conducted a study on the adsorption of chloride ions in water using quartz sand and had an adsorption capacity of 0.2711 mg/g. However, an adsorbent without pre-treatment if applied as an adsorbent gives less than optimal results. Therefore, this research is conducted to increase the activity of adsorbent.

In this research, in addition to activation, physical merging of quartz sand and coconut shell charcoal is also carried out so that quartz sand/coconut shell charcoal will be formed. The merger will add active groups to the adsorbent and increase the surface area of the adsorbent so that it is expected to optimize the performance of the composite of quartz sand/coconut shell charcoal in absorbing the metal of Mn.

2. Materials and Methods

2.1. Materials

Quartz sand, coconut shell, distilled water, Mn 1000 ppm mother solution, HCl p.a (Merck), concentrated HNO_3 p.a (Merck), Na_2CO_3 p.a (Merck), ammonia p.a (Merck).

2.2. Preparation of Adsorbent

Quartz sand is cleaned of impurities, sifted and finely ground. After smooth, proceed with sifting using 150 mesh sieve. Then wash with 0.1 M HNO_3 while stirring with a magnetic stirrer for 1 hour. Washing is done 3 times and dried by aerated [11]. Coconut shell is cut into small pieces, washed 3 times, dried in the sun, purified with 400 °C for 1 hour, grinded and sieved using a 150 mesh size sieve [12]. The quartz sand powder and coconut shell which has dried then are identified.

2.3. Adsorbent Activation

Chemical activation is carried out for quartz sand, 165 grams of quartz sand and 135 grams of sodium bicarbonate are mixed and the mixing process is carried out evenly. Then put in a crucial and then heated to a temperature of 1100 °C for 2 hours. Then the melt is cooled and crushed into powder [13]. Then for the coconut shell, as much as 50 grams of coconut shell soaked with 100 mL of 4 M HCl solution for 24 hours, then filtered and washed with distilled water. The activated carbon produced is then dried in an oven at 110 °C for 3 hours, then cooled in a desiccators [14]. The activated adsorbent is then characterized by FTIR, XRD, SAA and a specific total acidity test.

The next step is to make variations in the composition of the adsorbent between quartz sand and coconut shell, which are 25:75, 50:50, and 75:25 (w/w). Mixing each variation of the composition was carried out using stirring for 1 hour. Then after being filtered the solid phase is washed with distilled water. Furthermore, the resulting composition is heated at a temperature of 105 °C for 4 hours. The dry, solid phase is then crushed with a mortar and sieved with another 150 mesh sieve.

2.4. Adsorbent Effectiveness Test

Optimization of the composition of the adsorbent is done by as much as 0.1 gram of composite quartz sand / coconut shell charcoal with variations in the composition of 100: 0, 75:25, 50:50, 25:75 and 0: 100 (w/w) mixed with 10 mL of solution Mn 6 ppm with pH 5. The solution is stirred for 60 minutes at a constant speed at room temperature. The composite was filtered with Whatman No. filter paper. 42. Filtrate is measured by AAS. The pH optimization of the solution was carried out by as much as 0.1 gram of quartz sand/coconut shell charcoal composite at an optimal composition mixed with 10 mL of 6 ppm Mn solution using a buffer variation of pH 1, 2, 3, 4, 5, and 6. The solution was stirred for 60 minutes at a constant speed at room temperature. The composite was filtered with Whatman No. filter paper. 42. Filtrate is measured by AAS. Then the optimization of the contact time of the solution is done by as much as 0.1 gram of composite quartz sand/coconut shell charcoal at an optimal composition and pH mixed with 10 mL of 6 ppm Mn solution, stirring with contact time variations of 10, 15, 20, 30, 45, 60 , 75 and 90 minutes at constant speed at room temperature. The mixture was filtered with Whatman No. filter paper. 42. Filtrate is measured using AAS.

2.5. Determination of Adsorption Isotherms

There are 0.1 gram of composite quartz sand/coconut shell charcoal at optimum conditions was added with a solution of Mn 2, 4, 6, 8, 10 and 12 ppm as much as 10 mL then stirring with the maximum contact time at room temperature. The adsorption results were filtered with Whatman No. filter paper. 42 and measured by AAS. The results obtained were then analyzed by the Langmuir and Freundlich isotherms.

3. Result and discussion

3.1. Identification of Adsorbent

3.1.1. FTIR Characterization

FTIR characterization is used to determine the functional groups found in raw materials. The results of FTIR analysis of coconut shell charcoal showed the presence of hydroxyl and carboxyl active groups while the results of quartz sand analysis showed the presence of silanol and siloxan active groups. These active groups can be used as adsorbents. FTIR spectra of coconut shell charcoal and quartz sand are shown in Figure 1. Data on functional group analysis is shown in Table 1.

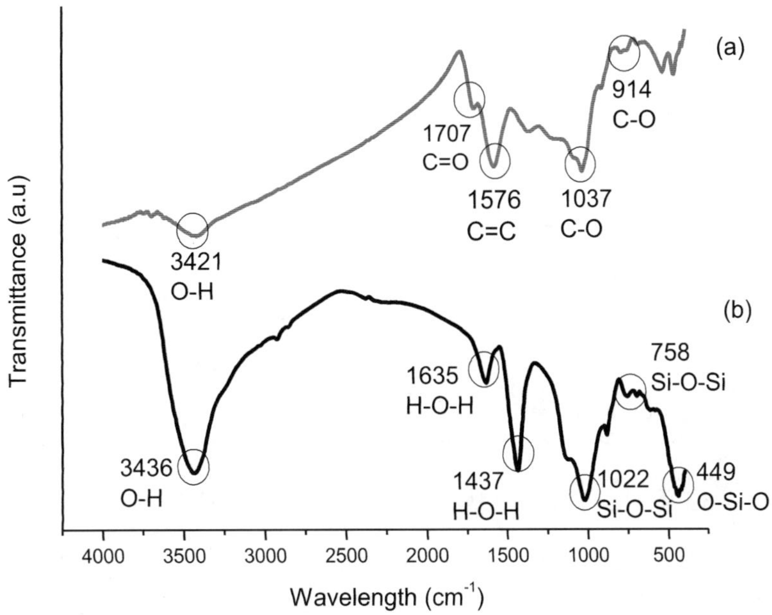

Figure 1. FTIR spectra of coconut shell charcoal (a) and quartz sand (b)

Table 1. Data from the analysis of the functional group coconut shell charcoal and quartz sand

No	Functional Group	Wavelength (cm^{-1})		
		Reference	Coconut shell charcoal	Quartz sand
1	O-H	3550-2000[15]	3421	3436
2	H-O-H	1641[16]	-	1635; 1437
3	C=O	1705[8]	1705	-
4	C=C	1589[8]	1576	-
5	C-O	1300-800[8]	914; 1037	-
6	Si-O-Si *asymmetric strech*	1076[17]	-	1022
7	Si-O-Si *symmetric strech*	771[17]	-	758
8	O-Si-O	461[18]	-	446

3.1.2. XRD Characterization

XRD characterization was used to determine the mineral content of coconut shell charcoal and quartz sand. The results of XRD analysis showed that coconut shell charcoal contains cristobalite (C), fayalite (F) and manganoan (M) minerals while quartz sand contains quartz minerals (Q). XRD diffractogram is shown in Figure 2.

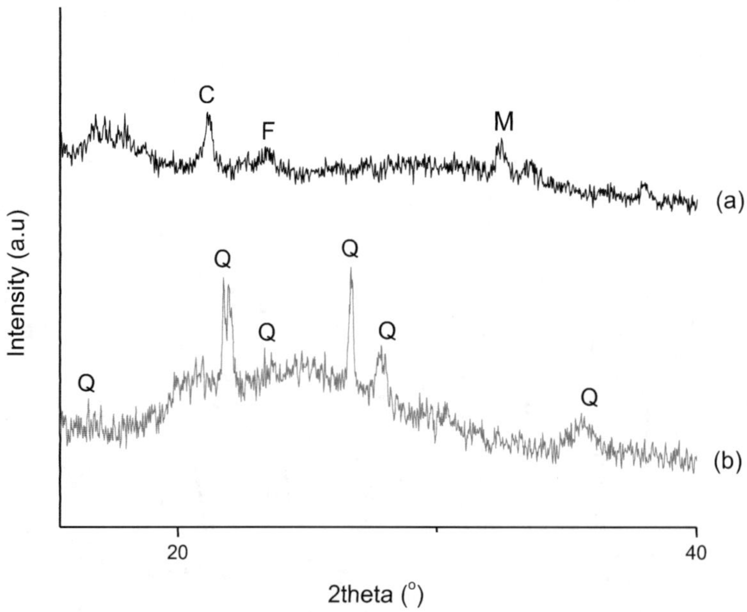

Figure 2. XRD diffractograms of coconut shell charcoal (a) and quartz sand (b)

3.1.3. Surface Area and Acidity Analysis

The surface area will provide an area on the surface of the adsorbent so that the Mn metal can be adsorbed. The specific total acidity analysis aims to show the presence of active sites on the surface of the adsorbent and determine the effect of acidity on the adsorption capacity. More acidic sites indicate that the adsorbent has more active sites so that it will make it easier for the adsorbent to interact with the Mn metal in the adsorption process. The more active sites on the adsorbent, the more likely the adsorption capability will be. The results of surface area and acidity analysis of coconut shell charcoal and quartz sand are shown in Table 2.

Table 2. Surface area and acidity of coconut shell charcoal and quartz sand

Sample	Surface area (m^2/g)	Acidity (mmol/g)
charcoal	51,420	7,647
quartz	17,458	9,882

3.2. Adsorbent Effectiveness Test

3.2.1. Optimization of Adsorbent Composition

Optimization of the adsorbent composition was carried out on a comparison of the composition of quartz sand composites: coconut shell charcoal 100: 0, 0: 100, 25:75, 50:50 and 75:25 (w/w). The optimum composition of quartz sand/coconut shell charcoal adsorbent composites is at a ratio of 50:50 (w/w) with an adsorption capacity of 0.573 mg/g and an absorption percentage of 97.02%. These results indicate that the composite of quartz/coconut shell charcoal with a ratio of 50:50 (w/w) supports each other to form more active sites and increase the surface area so that they are able to absorb Mn metals maximally.

3.2.2. Optimization of pH Conditions

Optimization of pH conditions aims to determine the optimum pH conditions in Mn metal adsorption. In this study the optimum pH of Mn solution was obtained at pH 5 with an adsorption capacity of 0.556 mg/g and an adsorption percentage of 94.59%. At pH 5 the H^+ ions contained in the solution are relatively small so that the active group of adsorbent binds more easily to the metal Mn.

3.2.3. Contact Time Optimization

The optimum contact time of Mn metal adsorption occurred at 30 minutes contact time with adsorption capacity of 0.56 mg/g and the percentage of adsorption of 94.22%. The longer the contact time of the solution with the adsorbent, the greater the adsorption capacity. However, if the contact time is sufficient, the adsorption capacity will be stable.

3.3. Characterization of Adsorbent Composites Before and After Adsorption

3.3.1. FTIR Characterization

FTIR spectra of adsorbents before and after adsorption are shown in Figure 3.

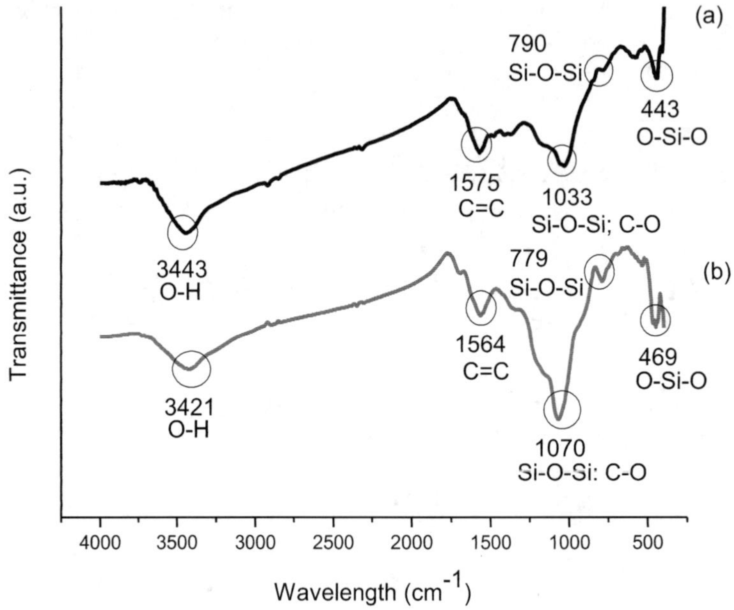

Figure 3. FTIR spectra of composite before and after adsorption

FTIR analysis aims to determine the possibility of interaction between the active group on the adsorbent with Mn metal after the adsorption process. The possibility of this interaction can be seen by the shift of the wave numbers in the FTIR spectra. Active groups such as Si-OH /-COOH will interact with Mn metals to form Si-O-Mn /-COO-Mn. A lower wave number shift indicates that the bond strength decreases while a higher wave number shift indicates an increase in bond strength.

3.3.2. SEM-EDX characterization

The SEM test results are shown in Figure 4 while the EDX test results are shown in Table 3. The appearance of the adsorbent surface before adsorption tends to have many open pores that allow Mn metal to fill the empty space on the surface of the adsorbent. After the adsorption process, the pores on the surface of the adsorbent are more closed which is possible because it is covered by Mn metal. The EDX test results of adsorbent composites after adsorption contained more Mn content than before adsorption that is equal to 1.2% which proves that the adsorbent composites are capable of absorbing Mn.

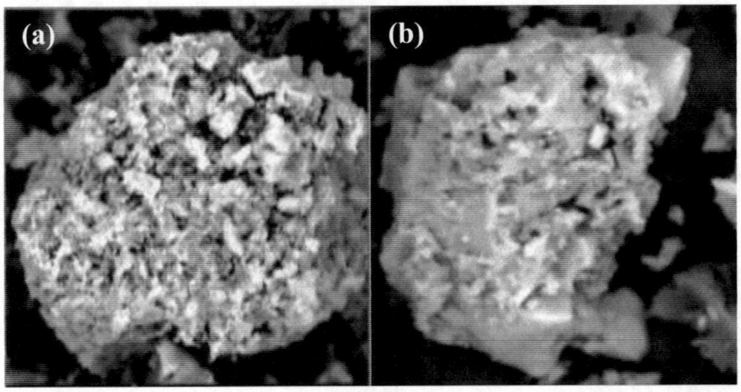

Figure 4. Composite morphology adsorbent quartz sand/ coconut shell charcoal 50:50 (w/w) before adsorption (a) and after adsorption (b)

Table 3. Content of elements in composite adsorbent

No	Adsorbent	Element content (%)			
		Si	C	O	Mn
1	Quartz sand/ coconut shell charcoal composite before adsorption	9,3	38,7	46,9	0,1
2	Quartz sand/ coconut shell charcoal composite after adsorption	12,6	31,6	31,6	1,2

3.4. Determination of Adsorption Isotherms

The adsorption isotherm curve is shown in Figure 5. Based on the isotherm curve, shows that the R^2 value for the Langmuir isotherm is 0.935 with an adsorption capacity of 0.304 mg/g while the R^2 value on the Freundlich isotherm is 0.994 with an adsorption capacity of 0.661 mg/g. The process of adsorption between Mn metal with quartz sand/coconut shell charcoal composites follows Langmuir and Freundlich isotherm models because the R^2 values are equally close to 1. However the Mn metal adsorption capacity is greater in the Freundlich isotherm, so it can be concluded that the adsorption of Mn metal by composite quartz sand/coconut shell charcoal is more dominant in the Freundlich isotherm. The Freundlich isotherm shows that the adsorption process between adsorbents and adsorbates occurs physically through the *van der Waals* force. Mn metal which has a positive charge will interact with O atoms in the negatively charged Si-OH group so that the distance between Mn ions and O atoms is closer due to dipoles. The interaction between adsorbent and adsorbate occurs by the way Mn metal attaches to the pores of the adsorbent with weak binding energy through *van der Waals* forces. The existence of a weak pulling force causes the adsorbate to move from one part of the surface to another part of the surface of the adsorbent so that it can form a multilayer.

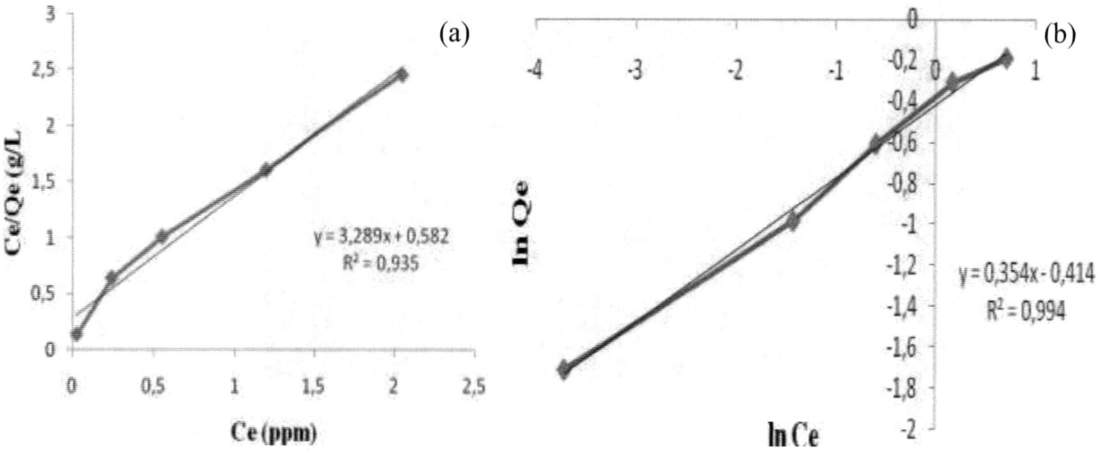

Figure 5. Isotherms curve of Langmuir (a) and Freundlich (b)

4. Conclusion

The results of adsorbent characterization showed the presence of active groups which indicated that the adsorbent was able to absorb Mn metal. The quartz sand/coconut shell charcoal adsorbent composite can adsorb Mn metal with the composition of quartz sand/coconut shell charcoal 50:50 (w/w), the best conditions at pH 5, contact time for 30 minutes with adsorption capacity of 0.56 mg/g and the adsorption percentage of 94.22%. Type of adsorption isotherm composite of quartz sand/coconut shell charcoal adsorbent on Mn metal is more dominant in Freundlich isotherm with adsorption capacity of 0.661 mg/g.

References

[1] Peraturan Menteri Kesehatan Republik Indonesia nomor 492/MENKES/PER/IV/2010 2010.
[2] Purwonugroho N 2013 *Artikel Publikasi Ilmiah.*
[3] Margarita S S, Rahmat G 2017 *Prosiding Seminar Nasional Kimia.* 194-198.
[4] Rochmah V, Agung T P, Triastuti S 2017 *Indonesian Journal of Chemical Science.* **6(2)** 168-172.
[5] Widayatno T, Teti Y, Agung A S 2017 *Jurnal Teknologi Bahan Alam.* **1(1)** 17-23.
[6] Paduraru C, Tofan L 2008 *Enviroment Engineering and Management Journal.* **7(6)** 687-693.
[7] Hidayah N, Muris, Subaer 2015 *Jurnal Sains dan Pendidikan Fisika.* **11(1)** 93-97.
[8] Kurniati F D, Pardoyo, Suhartana 2011 *Jurnal Kimia Sains dan Aplikasi.* **14(3)** 72-76.
[9] Rahmawanti N, Novrian D 2016 *Al Ulum Sains dan Teknologi.* **1(2)** 84-88.
[10] Lindasari S, Rudiyansyah, Kiki P U 2017 *JKK.* **6(1)** 8-16.
[11] Lesbani A 2015 *Jurnal Penelitian Sains.* **14(4)** 41-46.
[12] Rahayu L H, Purnavita S, Sriyana H 2014 *Jurnal Momentum.* **10(1)** 47-53.
[13] Las T, Florentinus F, Afit H 2011 *Valensi.* **2(2)** 368- 378.
[14] Nurhasni, Florentinus F, Qosim S 2012 *Valensi.* **2(4)** 516-525.
[15] Silverstain R M, Webster F X 2000 *Spectrometric Identification of Organic Compound,* 60th edn. New York: John Wiley and Sons, Inc.
[16] Dewanto A S, Munasir 2016 *Jurnal Inovasi Fisika Indonesia (IFI)* **5(2)** 1-4.
[17] Du G H, Liu Z I, Xia X, Chu Q, Zhang S M 2006 *J Sol-Gel Sci Techn.* **39** 285-291.
[18] Ahangaran F, Ali H, Sirous N 2013 *International Nano Letters.* **3(23)** 1-5.

Fabrication of silver-chitosan nanocomposite films and their antibacterial activity

E Susilowati[1], Maryani[2], Ashadi[1], Marwan[1]
[1] Chemistry Department FKIP Universitas Sebelas Maret, Surakarta, Indonesia
[2] Medical Faculty Universitas Sebelas Maret, Surakarta, Indonesia
Jl. Ir Sutami 36 A Surakarta Indonesia 53126

E-mail: endwati@yahoo.co.id

Abstract. The silver-chitosan nanocomposite film fabrication was carried out by the preparation of silver nanoparticles using chemical reduction methods assisted with microwave irradiation with chitosan as a reducing agent as well as a stabilizer. Silver nitrate is used as a precursor and NaOH as an accelerator. Silver-chitosan nanocomposite colloids were converted into films by casting techniques at the room temperature. Based on UV-Visible spectra, colloidal nanocomposites show absorption peak at the range of 409-412 nm. Based on TEM images, the silver nanoparticles are spherical shaped and distributed with a diameter of 6-10 nm. Based on FTIR spectra, interactions between silver nanoparticles and chitosan occur in the hydroxyl group (-OH) and amine group (-NH) of the chitosan polymer chain. Based on SEM images, the silver-chitosan nanocomposite films show a compact but rough surface morphology. The increase in tensile strength and decrease in elongation at break of the film was in line with the increase in the concentration of silver nitrate, while the value of the elasticity of the film fluctuates depending on the tensile strength and elongation at break. The increasing concentration of silver nitrate also causes an increase in the degree of swelling. silver-chitosan nanocomposite films showed better antibacterial activity against Escherichia coli than Staphylococcus aureus.

1. Introduction

Chitosan or β-(1.4)-2-amino-2-dioxy-D-glucose is a type of natural polysaccharide obtained from the chitin deacetylation process. Chitin is the second most biopolymer in nature after cellulose [4]. Chitosan has beneficial properties, including non-toxic, antibacterial, biodegradable, and biocompatible [1, 7]. In addition, chitosan has 3 types of functional groups, namely amine, primary and secondary hydroxyls. The existence of these three functional groups causes chitosan to have a high reactivity [11]. Chitosan has been widely used and applied in various fields, one of which is in the field of making environmentally friendly coating materials or films that have antibacterial activity. To improve the mechanical properties and antibacterial activity of chitosan-based films, we need to combine them with other materials such as nano materials. Nano materials, such as silver nanoparticles, can be selected for dispersion into chitosan biopolymer structures to form silver-chitosan nanocomposites. Chitosan acts as a matrix or binding component while silver nanoparticles act as fillers of components [13].

Dispersion of silver nanoparticles which incidentally has antibacterial properties into chitosan biopolymers will produce a nanocomposite material with excellent antibacterial activity [12, 15]. In addition, the very broad surface of silver nanoparticles will interact with the polymer chains of chitosan, so the mobility of the chitosan polymer chains decreases. This interaction can improve the mechanical properties of the silver-chitosan nanocomposite produced [10].

Chitosan-silver nanocomposite films can be made by first preparing silver nanoparticles. The preparation process can be carried out by chemical reduction methods. Besides being easy and quite effective, the costs required are also relatively affordable. At least, there are two important components that must be involved in the synthesis process of silver nanoparticles using chemical reduction; they are 1) the reducing agent, which will reduce the Ag^+ ions to Ag^0 and 2) the stabilizing agent (stabilizer), which will provide stability to silver nanoparticles that are formed from oxidation, agglomeration and deposition processes [5, 19].

In the synthesis process of silver nanoparticles using a chemical reduction method, materials that are not environmentally friendly are often chosen as reducing agents. The strong reducing agents commonly used such as sodium borohydride and hydrazine may have a negative impact on the environment because they are toxic. Besides, conventional heating is also often used in the synthesis of silver nanoparticles. Besides still involving relatively high energy, the time required is also relatively long. The synthesis of silver nanoparticles can last for 12 hours using a temperature of 60°C [2].

Besides acting as a matrix, chitosan can also act as a stabilizer as well as a reducing agent in the synthesis of silver nanoparticles. This is based on the research by Bozanic [6] and Shah [15], who used chitosan as a stabilizer while Raghavendra [12], in their research used chitosan as a reducing agent. Raghavendra [12], in his research used microwave irradiation for 11 minutes with pauses for 45 seconds every one minute. Thus, the use of microwave irradiation assistance can provide time efficiency in the synthesis process of silver nanoparticles.

Based on the research conducted by Darroudi [9], the use of accelerators can also provide effectiveness and efficiency in the synthesis process of silver nanoparticles. The use of sodium hydroxide as an accelerator can affect the number and size of silver nanoparticles produced. Sodium hydroxide as an accelerator can reduce the size of silver nanoparticles and increase the amount along with the increase in the amount of sodium hydroxide used. The use of sodium hydroxide as an accelerator in the synthesis of silver nanoparticles has also been carried out by Susilowati [18] with the analysis of the results of the same synthesis.

Based on the explanation, this research on the film making from silver-chitosan nanocomposite materials was conducted. The making of silver-chitosan nanocomposite films was carried out by first preparing silver nanoparticles using the reduction method with the help of microwave irradiation. Sodium hydroxide was used as an accelerator. Chitosan was used as a reducing agent as well as a stabilizer. In addition, chitosan was also used as a biopolymer matrix in the synthesis of silver-chitosan nanocomposites. The colloidal silver-chitosan nanocomposite formed was printed into a film with the casting technique at the room temperature.

2. Experimental

2.1. Chemicals

The researcher purchased chitosan with molecular weight (MW) of 1071.579 kDa and degree of deacetylation (DD) of 70.356% from Surindo Cirebon Indonesia Biotech. Silver nitrate ($AgNO_3$), glacial acetic acid (CH_3COOH), and sodium hydroxide (NaOH) were purchased from Merck.

2.2. Preparation of silver-chitosan nanocomposite colloidal

Chitosan solution (1% w/v) in glacial acetic acid solution (1% v/v) was prepared first. The solution was stirred with a stirrer using a medium speed for 2 hours or until completely dissolved. Chitosan solution (1% w/v) that had been produced was then left alone for 24 hours. A total of 12.5 mL of chitosan solution (1% w/v) was poured into a beaker and stirred using a stirrer for 5 minutes. 0.25 mL of $AgNO_3$ solution (0.012 g/mL) was added dropwise using a syringe while stirring with a stirrer for 5 minutes. After that, a 1.75 mL of NaOH (2 M) solution was added dropwise using a syringe while stirring with a stirrer for 5 minutes. The solution was then heated using a microwave with 100 watts of power for 4 minutes. After that, the gel formed was poured into a beaker containing 47.5 mL of chitosan solution (1% w/v), and then stirred using a stirrer until it dissolved to form colloidal silver-chitosan nanocomposites. The same procedure was performed to produce variations in the concentration of $AgNO_3$, which were then named A1 (0.5%), A2 (1.0%), A3 (1.5%), A4 (2.0%), A5 (2, 5%), and A6 (3.0%). In this procedure, concentration of $AgNO_3$ was expressed in percentage or % (w/v, $AgNO_3$/Chitosan).

2.3. Preparation of nanocomposite films
The film was formed from colloidal silver-chitosan nanocomposites using casting techniques at the room temperature. Colloidal nanocomposites were poured into polyethylene molds (10 x 15 cm) and dried at the room temperature until the solvent completely evaporated (approximately 3 days). After drying, the film was neutralized using a solution of NaOH (1 M), and then rinsed using distilled water. After that, the film was dried again on a glass plate and then peeled off and stored in a desiccator. Characterization of film on SEM and FTIR was only carried out on sample films A3 and A6.

2.4. UV-visible spectroscopy analysis
To identify silver nanoparticles, absorption spectra of colloidal silver-chitosan nanocomposites were recorded at a wavelength range of 300 - 600 nm using the Ultraviolet-visible spectrophotometer Shimadzu UV3150. Before measurement, colloidal samples were diluted 10 times using distilled water.

2.5. Transmission Electron Microscopy (TEM) analysis
The shape and size distribution of silver nanoparticles in silver chitosan nanocomposite colloids were identified using TEM (JEM-1400) with a voltage of 80 KV.

2.6. Fourier Transform Infra-Red (FTIR) analysis
To identify the interactions between silver nanoparticles and chitosan in silver-chitosan nanocomposites, the transmittance spectra of nanocomposite films were recorded using FTIR (Shimadzu FT-IR 8201PC) in the wave range of 4000-300 cm^{-1}.

2.7. Scanning Electron Microscope (SEM) analysis
The surface morphology of the silver-chitosan nanocomposite film was observed by using SEM (Zeiss DSM 960) with a resolution of 10 μm, acceleration of 10KV voltage and magnification of 2500 times.

2.8. Tensile measurements
Tensile tests or mechanical properties tests on silver-chitosan nanocomposite films were carried out to determine the value of Tensile Strength, elongation at break, and elasticity (Modulus Young). Measurements were made using Universal Tensile Machine Type Z 0.5 with the stretching speed of 10 mm/min.

2.9. Swelling studies
Investigation of swelling properties or absorptive capacity of silver-chitosan nanocomposite films was determined in a Phosphate Buffer Saline/PBS solution (pH 7.4) at the room temperature. The film was cut first with a size of 2 x 2 cm, and then weighed (W_0). After that, it was soaked in PBS solution for 24 hours, and then dried and weighed. The procedure was repeated 3 times.

2.10. Antibacterial activity

The antibacterial activity test of silver-chitosan nanocomposite films against gram-positive (*Staphylococcus aureus*) and gram-negative (*Escherichia coli*) bacteria was carried out using the disk diffusion method. Before testing, the test bacteria, namely S. Aureus and E. Coli, were first incubated at 37 °C for 24 hours on sterile growth media. The sterile agar solution was poured into a sterilized petri dish (10 cm in diameter), and then left alone to condense. After that, the bacterial colony was placed on a petri dish containing agar evenly and incubated at 37 °C for 24 hours. Antibacterial activity testing was carried out by placing the silver-chitosan nanocomposite film discs on the surface of a filled petri dish so that the test bacteria were overgrown. After that, the incubation was carried out at 37 °C for 24 hours. The antibacterial effectiveness was assessed through measurement of inhibition zones (clear areas) that appear under and around the film discs on the agar media. The inhibition zones that appeared on the media are then named according to the variations in the concentration of $AgNO_3$ used, namely FA0 (chitosan), FA1 (0.5%), FA2 (1.0%), FA3 (1.5%), FA4 (2.0%), FA5 (2.5%), and FA6 (3.0%).

3. Results and Discussion

The silver-chitosan nanocomposite film was made by first preparing silver nanoparticles using the chemical reduction method. Then, the formed silver nanoparticles were dispersed into chitosan biopolymers to form colloidal silver-chitosan nanocomposites. The resulting colloidal color changes to yellow, brownish yellow, to dark brown/reddish, as shown in Figure 1. This indicates that silver nanoparticles have formed in colloidal silver-chitosan nanocomposites [3, 9]. The colloidal nanocomposites formed were then printed into films by casting techniques at the room temperature. The silver-chitosan nanocomposite film produced can be seen in Figure 2.

Figure 1. Colloidal silver-chitosan nanocomposite with varying concentrations of $AgNO_3$

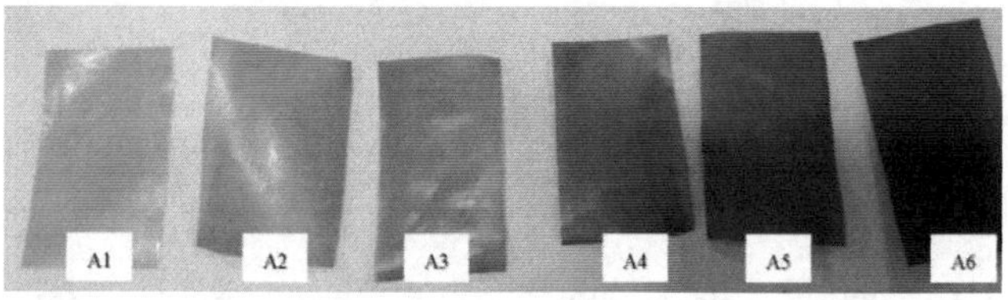

Figure 2. Silver-chitosan nanocomposite films with varying concentrations of $AgNO_3$

3.1. UV-visible spectroscopy analysis

The formation of silver nanoparticles in colloidal silver-chitosan nanocomposites can be identified using UV-Vis spectroscopy. In addition, by using UV-Vis spectroscopy, the size of the formed silver nanoparticles can also be estimated. The UV-Vis spectra of each colloidal silver-chitosan nanocomposite sample can be seen in Figure 3.

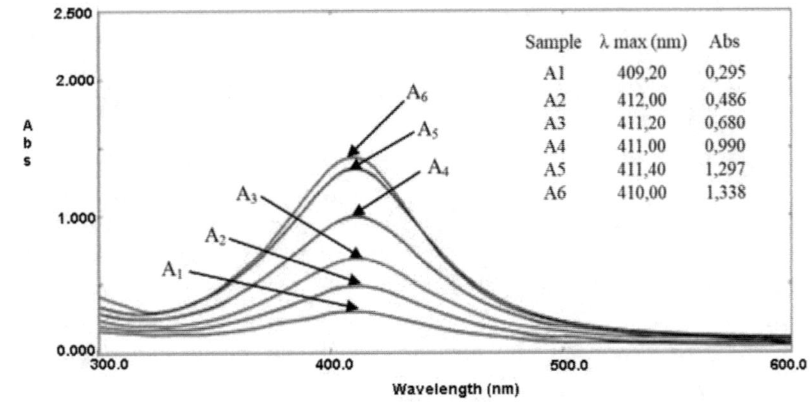

Figure 3. UV-Vis spectra of colloidal silver-chitosan nanocomposites with varying concentrations of $AgNO_3$ (% w/w, $AgNO_3$/Chit): A1 (0.5%), A2 (1.0%), A3 (1.5%), A4 (2.0%), A5 (2, 5%), and A6 (3.0%)

Figure 3 shows that the maximum absorbance of each sample occurs in the range of wavelengths of 409-412 nm. This indicates that silver nanoparticles have been formed. These results are in accordance with the research of Raghavendra [12] and Shah [15], which state that the SPR band silver nanoparticles on UV-Vis spectra can provide characteristic peak absorbance spectra in the wavelength range of 400-450 nm.

Based on UV-Vis spectra (Figure 3), the increase in the concentration of $AgNO_3$ used is generally directly proportional to the increase in absorbance and decrease in the maximum wavelength even though the changes that occur are not so significant for maximum wavelength data. Increased absorbance shows that the concentration of silver nanoparticles formed also increases. The decreasing maximum wavelength indicates that the size of the formed silver nanoparticles also decreases. These results are in accordance with the research conducted by Susilowati [18]. In the sample A1, the sample with the smallest $AgNO_3$ concentration, the maximum wavelength shown is the smallest when compared to other samples. This is possible because the ability of chitosan to reduce the Ag^+ ion which is not too much is still at optimal conditions.

3.2. Transmission Electron Microscopy (TEM) analysis

The shape and size distribution of silver nanoparticles in silver-chitosan nanocomposite colloidal can be determined through TEM images shown in Figure 4 for A3 and A6 samples only to compare on shape and size in high and low content of silver nanoparticles in the colloidal.

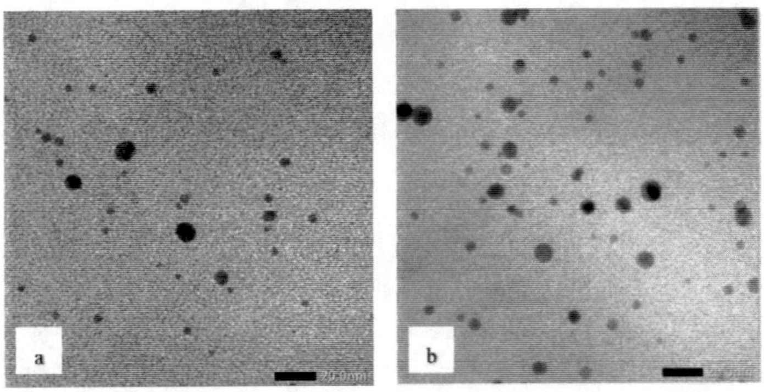

Figure 4. TEM images for (a) sample A3 and (b) sample A6

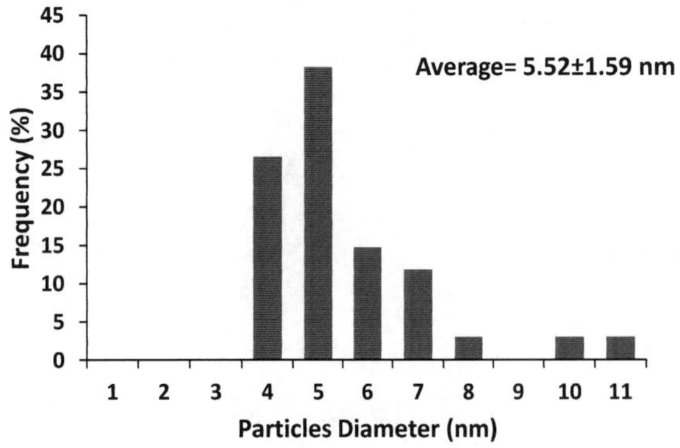

(a)

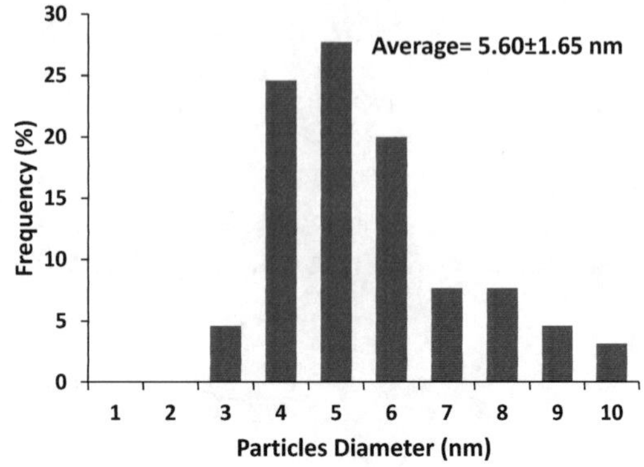

(b)

Figure 5. Distribution of silver nanoparticle size for (a) sample A3 and (b) sample A6

Based on Figure 4, the resulting silver nanoparticles are spherical in shape. Based on Figure 5, silver nanoparticles are distributed in silver-chitosan nanocomposites with varying diameter sizes, which are in the range of 4-22 nm. Based on the frequency, the distribution of silver nanoparticles is dominated by particles with a diameter of 6-10 nm. When sample A3 and sample A6 were compared, it was seen that sample A6 had a higher concentration of silver nanoparticles than sample A3. This provides confirmation of the results of UV-Vis spectra analysis, which states that the increase in the concentration of $AgNO_3$ is directly proportional to the increase in the concentration of silver nanoparticles that are formed. The existence of larger particles is possible because of two things. First, there is overlapping or overlapping between particles [18]. Second, agglomeration of silver nanoparticles has taken place. This agglomerated tendency occurs because silver nanoparticles have a large specific surface area, so chemical bonds between particles form a strong electric dipole [5].

3.3. Fourier Transform Infra-Red (FTIR) analysis

The interactions that occur between silver nanoparticles and chitosan in silver-chitosan nanocomposites can be identified through the identification of functional groups using the FTIR spectrophotometer. FTIR spectra for some samples in this research can be seen in Figure 6.

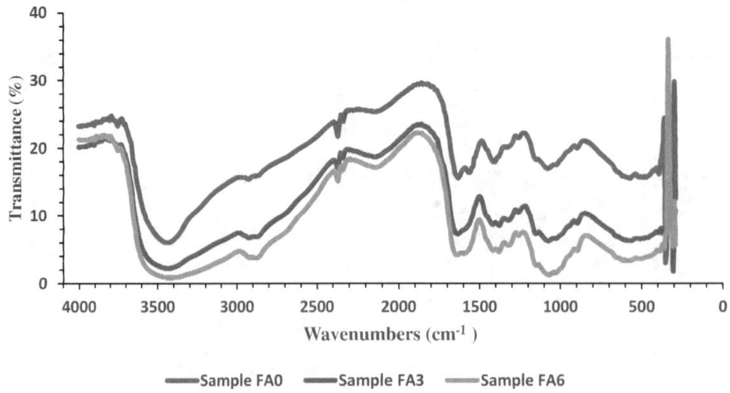

Figure 6. FTIR spectra for samples FA0, FA3 and FA6

Sample FA0 is a sample of chitosan film without silver nanoparticle dispersion. Based on FTIR spectra, sample FA0 shows the absorption peak at the wave number 3448.72 cm^{-1}. This indicates that the sample FA0 has O-H and N-H amine stretching functional groups because the vibration band of the functional group is known to overlap at the absorption peak with a range of wave numbers 3500-

3200 cm^{-1} [8, 14]. When the spectra of the three samples (Sample FA0, FA3 and FA6) were compared, it was seen that there was a shift and widening of the wave number in the peak region of the amine stretching of the O-H and N-H stretching vibrations. The wave number has shifted from 3448.72 cm^{-1} (sample FA0) to 3433.29 cm^{-1} (sample FA3) and 3425.58 cm^{-1} (sample FA6). Thus, the interaction between silver nanoparticles and chitosan occurs in the O-H (hydroxyl) and N-H (amine) functional groups. This is consistent with the results of the research by Raghavendra [12] and Shah [15]. They claim that the interaction between silver nanoparticles and chitosan occurs in hydroxyl and amino functional groups.

3.4. Scanning Electron Microscope (SEM) analysis

The surface morphology of the silver-chitosan nanocomposite film was observed using SEM. SEM images for some samples can be seen in Figure 7. Sample A0 is a pure chitosan sample without silver nanoparticle dispersion.

Based on SEM images from Figure 7, the increasing concentration of $AgNO_3$ causes the film surface to become more uneven. This is possible because the concentration of silver nanoparticles that are formed has also increased, causing the polymer chain from chitosan to expand when interacting or bonding with silver nanoparticles. Based on SEM images, it also appears that in sample FA6, clumping occurs, which indicates that silver nanoparticles have experienced agglomeration. This is due to the high concentration and small size of the formed silver nanoparticles in sample FA6, as explained earlier in the UV-Vis spectroscopic analysis. The increasing concentration and small size of the formed silver nanoparticles will increase the tendency of silver nanoparticles to agglomerate, thus causing the surface of the film to become more uneven [15].

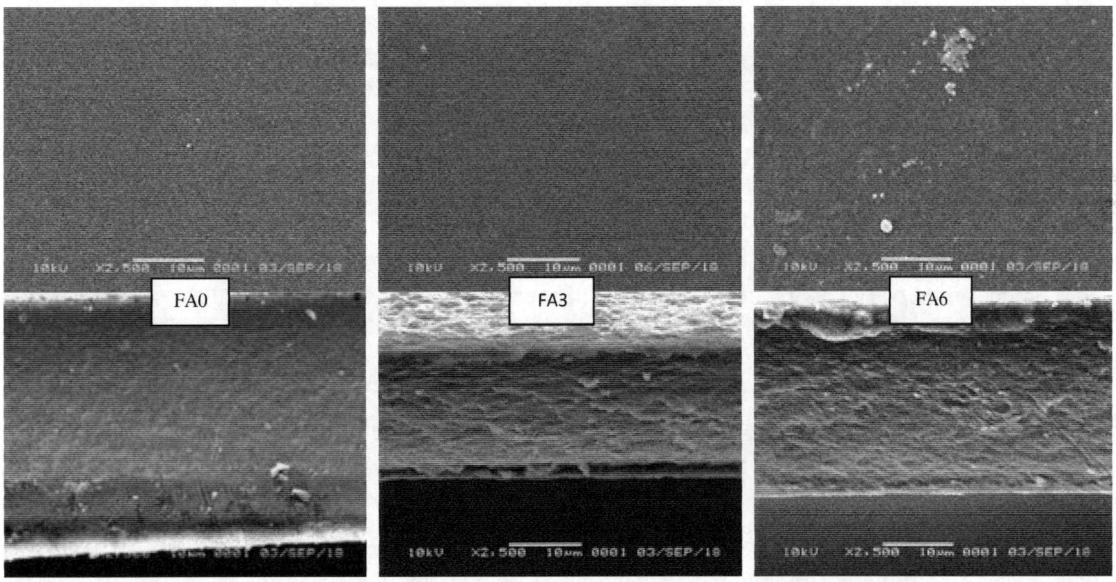

Figure 7. SEM image of surface and cross sections of samples FA0, FA3 and FA6

3.5. Tensile measurements

The tensile test or mechanical properties test of silver-chitosan nanocomposite films was carried out to determine the values of tensile strength, elongation at break, and elasticity (modulus young). The measurement results can be seen in Figure 8. Tensile strength values of silver-chitosan nanocomposite films showed downward trend with increasing $AgNO_3$ concentrations. This is due to the silver nanoparticles cause inhibit the bond between the chitosan biopolymer chain in the film. High surface area of silver nanoparticles can interact and form bonds with the chitosan polymer chain, that decreasing the bonding between the chitosan polymer chains. This causes the strengthening properties of silver-chitosan nanocomposite films to decrease [10, 16, 17].

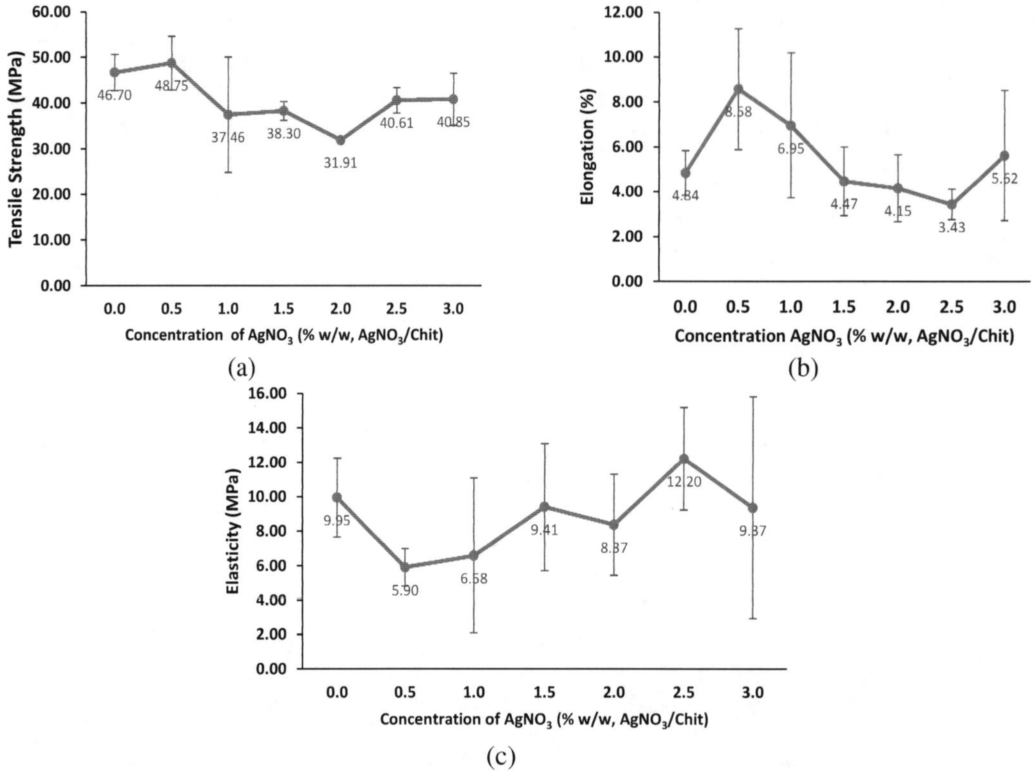

Figure 8. Mechanic properties of silver-chitosan nanocomposite film (a) tensile strength, (b) elongation, and (c) elasticity

In sample FA1 (0.5% w/w), the tensile strength value shown is the highest than other samples. This may be due to the very small size of the silver nanoparticles, even being the smallest of all samples. With its small size, silver nanoparticles can form a strong bond with the chitosan biopolymer chain. In addition, the concentration of silver nanoparticles that are formed is still relatively small, so that the tendency of agglomerated silver nanoparticles becomes smaller. In the FA4 sample (2.0% w/w), the tensile strength value shows the lowest value among all samples. This may be due to the agglomeration or distribution of uneven silver nanoparticles in the chitosan biopolymer. The uneven distribution of silver nanoparticles makes the tensile strength of the film not homogeneous or unequal on each side.

Increasing the concentration of $AgNO_3$ tends to reduce the elongation at break of the film formed. This is caused by interactions that occur between silver nanoparticles and chitosan polymer chains. This interaction causes the mobility of the chitosan biopolymer chain to decrease, so that the strengthening properties of the silver-chitosan nanocomposite film decrease. In samples FA1 to FA5 (0.5% - 2.5% w/w), a reduction in the elongation at break of the silver-chitosan nanocomposite film occurred along with the increased concentration of $AgNO_3$ used. In sample FA6 (3.0% w/w), the elongation at break experienced a greater increase than samples FA3 through FA5. This is possible because of the uneven distribution of nanopaticles in the chitosan biopolymer. Uneven distribution causes the film to have a non-homogeneous tensile strength value, so that when testing, the samples used are not uniform.

The elasticity (Modulus Young) value of the silver-chitosan nanocomposite film is upward trend with the increase in $AgNO_3$ concentration. The elasticity value is influenced by the value of the tensile strength and elongation at break of the film. The higher the tensile strength value is, the higher the elasticity value will be. The higher the elongation at break value is, the lower the elasticity value will be.

3.6. Swelling studies

Swelling test results can be seen in Figure 9. Based on the swelling test results, the increase in the concentration of $AgNO_3$ is directly proportional to the increase in the swelling degree of the silver-chitosan nanocomposite film.

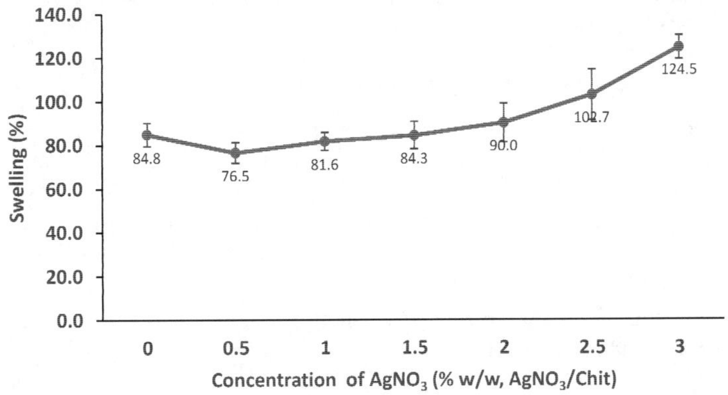

Figure 9. The degree of swelling of silver-chitosan nanocomposite films with varying concentrations of $AgNO_3$

Because the increase in the concentration of $AgNO_3$ is directly proportional to the increase in the concentration of silver nanoparticles and the decrease in size, the degree of swelling has increased with increasing concentration and decreasing the size of silver nanoparticles. This is consistent with the research conducted by Shah [15], which states that the degree of swelling is influenced by the size of the nanoparticles, surface tension, and surface morphology of the nanocomposite film. The growth of silver nanoparticles can cause the chitosan matrix to expand and encourage the formation of an empty space/gap between the chitosan polymer matrices, so it can be easier for liquid to enter and bind to the silver-chitosan nanocomposite film. The contribution of amino and hydroxyl groups from chitosan is also known to cause an increase in the degree of swelling of silver-chitosan nanocomposite films [15].

3.7. Antibacterial activity

The antibacterial activity test of silver-chitosan nanocomposite films against Staphylococcus aureus and Escherichia coli was carried out using the disc diffusion method. Silver-chitosan nanocomposite film discs showed good antibacterial activity against the two bacteria, as evidenced by the appearance of inhibition zones (clear areas) around the film discs on the agar media that had been overgrown with bacteria.

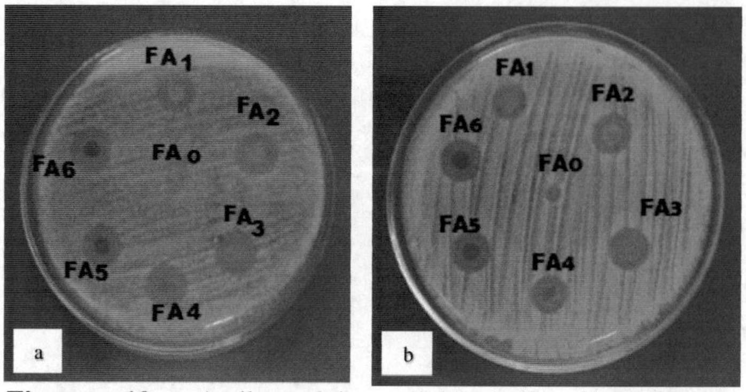

Figure 10. Antibacterial activity of silver-chitosan nanocomposite films with varying concentrations of $AgNO_3$ against bacteria (a) *Escherichia coli* and (b) *Staphylococcus aureus*

The results of testing the antibacterial activity of silver-chitosan nanocomposite films can be seen in Figure 10. The wider clear areas that arise indicate that the inhibition of a compound against bacterial growth becomes stronger. Based on Table 1, the inhibition of silver-chitosan nanocomposite films against Escherichia Coli bacteria is stronger compared to the inhibition against Staphylococcus Aureus bacteria. In addition, the dispersion of silver nanoparticles into chitosan biopolymers is proven to have an influence on the activity of the resulting nanocomposite films.

Table 1. Inhibition zone diameter of chitosan silver-chitosan nanocomposite films against *Escherichia Coli* and *Staphylococcus Aureus* bacteria

Sample	Inhibition Zone Diameter (mm)	
	Escherichia Coli	*Staphylococcus Aureus*
FA0	0	0
FA1	13.13	11.11
FA2	13.75	12.42
FA3	14.38	13.07
FA4	14.38	12.42
FA5	13.75	12.42
FA6	15.00	13.07

The increase in the inhibition zone area for the two bacteria shows a trend that tends to increase along with the increase in the concentration of $AgNO_3$ used, which also means that the increase in the concentration of silver nanoparticles formed is also directly proportional to the increase in the area of the inhibition zone of the silver-chitosan nanocomposite film against the two bacteria. The antibacterial activity of this film nanocomposite solely comes from silver nanoparticles, because without silver nanoparticles (FA0), inhibition zone is zero for two bacteria.

4. Conclusion

Silver-chitosan nanocomposite films can be made using chemical reduction methods with microwave irradiation, where chitosan is used as a reducing agent as well as a stabilizer and NaOH as an accelerator in the preparation of silver nanoparticles. Based on TEM images, the resulting silver nanoparticles are spherical in shape and are distributed with different sizes in the 6-10 nm diameter range. Based on the analysis of functional groups using FTIR, interactions between silver nanoparticles and chitosan in silver-chitosan nanocomposite occur in the hydroxyl group (O-H) and amine group (N-H) of the chitosan polymer. Based on SEM images, the surface morphology of the silver-chitosan nanocomposite film showed an uneven surface in line with the increase in the concentration of $AgNO_3$. The decrease in the tensile strength and elongation at break occur along with an increase in the concentration of $AgNO_3$ used while the elasticity of the film fluctuates depending on the tensile strength and elongation at break. The increasing concentration of $AgNO_3$ used also causes an increase in the degree of swelling of silver-chitosan nanocomposite films. The silver-chitosan nanocomposite film showed high antibacterial activity against Escherichia coli than Staphylococcus aureus. The antibacterial activity of this film nanocomposite solely comes from silver nanoparticles.

Acknowledgements
The author would like to thank DRPM Ristek Dikti who gave financial support through PDUPT Grant 2019 of Universitas Sebelas Maret with contract number: 718/UN27.21/PN/2019.

References
[1] Afonso C R, Hirano R S Gaspar A L Chagas, E G, Carvalho R A, Silva F V, Leonardo G R, Lopes P S, Silva C F, and Yoshinda C M P 2019 *Int. J. Biol. Macromol* **132** 1262-1273
[2] Ahmad M B, Lim J J L, Shameli K, Ibrahim N A, and Tay M Y 2011 *Molecules* **16** 7237-7248
[3] Ahmad M B, Tay M Y, Sahmeli K, Hussein M Z, and Lim J J 2011 *International Journal of Molecular Sciences* **12** 4872-4884
[4] Aranaz, I, Mengibar M, Harris R, Panos I, Miralles B, Acosta N, Galed G, and Heras A 2009 *Current Chemical Biology* **3** 203-230
[5] Ariyanta H A 2014 *Jurnal Media Kesehatan Masyarakat Indonesia* **10** 36-42
[6] Bozanic D K, Trandafilovic L V, Luyt A S and Djokovic V 2010 *React. Funct. Polym* **70** 869-873
[7] Chang A K T, Frias R R, Alvarez, L V, Bigol U G, and Guzman J P M D 2019 *Biocatalysis and Agricultural Biotechnology* **17** 189-195
[8] Dachriyanus 2004 Padang LPTIK Universitas Andalas
[9] Darroudi M , Ahmad M B, Abdullah A H, Ibrahim N A, and Shameli K 2010 *International Journal of Molecular Sciences* **11** 3898-3905
[10] Hadiyawarman A R, Nuryadin B W, Abdullah M, dan Khairurrijal 2008 *Jurnal Nanosains & Nanoteknologi* **1** 14-21
[11] Pratiwi R 2014 *Oseana* **39** 35-43
[12] Raghavendra G M, Jung J, Kim D, and Seo J 2016 *Int J Mol Sci* **84** 281-288
[13] Salari M, Khiabania M S, Mokarrama R R, Ghanbarzadeha B and Kafilc H S 2018 *Food Hydrocolloids* **84** 414-423
[14] Sastrohamidjojo H 2001 Yogyakarta Liberty Yogyakarta
[15] Shah A, Hussain I, and Murtaza G 2018 *Int J Biol Macromol* **116** 520-529
[16] Sriyanti I 2009 *Proceeding of The Third International Seminar on Science Education* 1-6
[17] Sudibyo A, dan Hutajalu T F 2013 *Jurnal Kimia dan Kemasan* **35** 6-19
[18] Susilowati E, Tiyono, Santosa S J, and Kartini I 2015 *Indones J Chem* **15** 29 – 35
[19] Wahyudi T, Sugiyana D, and Helmy Q 2011 *Arena Tekstil* **26** 55-60

IOP Conf. Series: Materials Science and Engineering **858** (2020) 012043 doi:10.1088/1757-899X/858/1/012043

STUDY OF ADSORPTION-DESORPTION Au(III) USING IMPRINTED IONIC MATERIAL (SiO_2 (RHA-BGS) -TMPDT-Au-Imp)

F R Prasetyaningrum[1], S Hastuti[1], S Wahyuningsih[1], and T Martini[1]
[1]Departmenr of Chemistry, Faculty of Mathematics and Natural Sciences, Sebelas Maret University, Surakarta, 57126, Indonesia.

Email : hastuti.uns@gmail.com

Abstract. Research on the adsorption of desorption of Au (III) metal using imprinted ionic $SiO_{2\,(RHA\text{-}BGS)}$ -TMPDT-Au-Imp material has been carried out.
In this research, silica source (SiO_2) from rice husk ash (RHA) and bagasse (BGS) was modified through imprinted ionic technique. Adsorption studies include variations in concentration and selectivity while desorption studies include time variations. Furthermore, the adsorption and desorption processes were characterized using FTIR. The results of the adsorption process followed the Freundlich isotherm model with an adsorption capacity of 981,672 mg/g. The adsorption selectivity shows that the imprinted ionic material ($SiO_{2(RHA\text{-}BGS)}$-TMPDT-Au-Imp) is more selective than the material without imprinted ionic ($SiO_{2\,(RHA\text{-}BGS)}$ –TMPDT). The order of selectivity is Au(III)/Zn(II)> Au(III)/Fe(III)> Au(III)//Cu(II). Study of desorption kinetics Au(III) followed the zero order kinetics. The results of the FTIR show that there are still O-H, C-H and -NH groups. This indicates the imprinted ionic material ($SiO_{2(RHA\text{-}BGS)}$-TMPDT-Au-Imp) is not damaged.

1. Introduction

Gold reserves on earth are decreasing, while market demand continues to increase. This condition is encouraging to move on by finding a second source other than gold ore. In the electronics industry, gold is widely used because it has quite high electrical conductivity and is not easily corroded [1]. At present more than 300 tons of gold are used annually for electronic components such as ICs, contacts and cables. At the end of the use of electronic goods such as computers, mobile phones become waste that has not been used optimally. It is reported that the cellphone contains 300-350 g/ton gold [2]. This encourages efforts to recover gold from electronic waste while helping to provide gold raw materials.
Gold extraction from electronic waste has been carried out, among others, the process of precipitation, ion exchange, solvent extraction and adsorption. Adsorption is one method of extracting gold metal ions which is more effective, simple, and inexpensive. The gold metal recovery process requires high selectivity from the adsorbent material so that it can produce pure gold. According to Sakti et al. (2013) [3] the selectivity of adsorption from silica adsorbents (SiO_2) can be improved by adding imprinted ionic techniques. The imprinted ionic method is judged to have very high selectivity compared to just ordinary modifications. This is because at the time of polymer synthesis there are metal ions which act as templates and monomers that contain functional groups. The release of ions as

IOP Conf. Series: Materials Science and Engineering **858** (2020) 012043 doi:10.1088/1757-899X/858/1/012043

a template from the polymer matrix will cause the formation of imprinted ionic cavities and the arrangement of imprinted ionic materials which will selectively adsorb target ions [4-6].
This research will study about the selectivity of Au(III) adsorption on ionic imprinted materials ($SiO_{2(RHA-BGS)}$-TMPDT-Au-Imp). In addition, Au(III) desorption process was studied to determine the stability of the adsorbent during the desorption process.

2. Experimental

2.1. Materials

The materials used in this study include: sodium hydroxide (NaOH), hydrochloric acid (HCl) (pa), from Merck. The standard solution is Cu(II) 100 ppm, Fe(III) 100 ppm, Zn(II) (pa) 100 ppm from merck. A solution of gold(III) 1000 mgL^{-1} was prepared in our laboratory by dissolving gold in an aquaregia solution.

2.2. Physical measurement

The tools used include glassware, pH meters, AAS (Atomic Absorption Spectrophotometry) Shimadzu AA 6650, FTIR (Fourier Transform Infra-Red Spectrophotometry) Shimadzu IR prestige 21.

2.3. Effect of concentration on the adsorption ability of SiO2 (RHA-BGS) -TMPDT-Au-Imp

$SiO_{2(RHA-BGS)}$-TMPDT-Au-Imp (10 mg) is contacted with 10 mL Au(III) with variations in concentration 0 - 40 (0, 5, 10, 15, 20, 25, 30, 35, and 40) mgL^{-1} at pH 3. Then it is stirred for 60 minutes then filtered. The filtrate was analyzed using SAA. The results are evaluated using the Freundlich (1) and Langmuir (2) equations. From the Freundlich equation obtained KF (maximum capacity) and N. From the Langmuir equation obtained the values of K and Qm. Furthermore, from the KL value adsorption energy can be calculated by calculating Gibbs free energy (3).

$$\text{Log } q_e = \log K_F + 1/n \log C_e \quad (1)$$

$$Ce/q_e = 1/(q_m.K_L) + Ce/q \quad (2)$$

where: Ce dan q_e masing masing adalah konsentrasi dan kapasitas adsorpsi pada saat kestimbangan. K_F adalah kontanta Freundlich yang menunjukkan kapasitas adsorpsinya. The q_m and K_L are the maximum adsorption capacity and constant of the Langmuir isotherm, respectivly. The Free energy (ΔG^o) can be calculated from the Langmuir equilibrium constant (K_L) using equation (4)

$$\Delta G^o = -RT \ln K_L \quad (3)$$

where: R (8,314 J $K^{-1}mol^{-1}$) is a gas constant, T (K) is the adsorption temperature

2.4. Determination of competing adsorption (selectivity) of Au (III) metal ions with other metal ions

$SiO_{2(RHA-BGS)}$-TMPDT-Au-Imp (10 mg) is contacted with 10 mL Au(III) 35 mgL^{-1} with other metal ion pairs such as: Cu(II), Zn(II) and Fe(III) with concentration variations of 5, 10 and 15 mgL^{-1}. The same thing is done for $SiO_{2(RHA-BGS)}$-TMPDT. Then selectivity is calculated using the equation.

$$q = (C_0-C_e) V/M \quad (4)$$

$$D = q/C_e \quad (5)$$

$$\alpha = D_{M_1}/D_{M_2} \quad (6)$$

where q is the adsorption capacity (mg/g), C_0 and C_e are the concentration of the initial metal ion and concentration at equilibrium (mg/L), M is the mass of the adsorbent (g), V is the volume of solution, D_M is the distribution ratio (L/g), and α is the coefficient of selectivity (Sakti et al., 2013).

IOP Conf. Series: Materials Science and Engineering **858** (2020) 012043 doi:10.1088/1757-899X/858/1/012043

2.5. Determination of Optimum Desorption Time

500 mg $SiO_{2(RHA\text{-}BGS)}$-TMPDT-Au-Imp was contacted with 500 mL Au(III) 350 mgL^{-1} at pH 3, then stirred in 60 minutes. Then filtered, then the precipitate was roasted at 60 °C to dry. 10 mg of sludge that has absorbed Au(III) was desorbed using a thiourea-HCl solvent with time variations of 0, 15, 30, 45, 60, 90, 120 and 150 minutes. From the time variation data then evaluated using 0,1 and 2. order kinetics models. The results before desorption and after desorption are then analyzed using infrared spectroscopy (FTIR).

3. Result and Discussion

3.1. $SiO_{2(RHA\text{-}BGS)}$ -TMPDT-Au-Imp Adsorption Performance Test on Au (III) Metal Ion Model Solution.

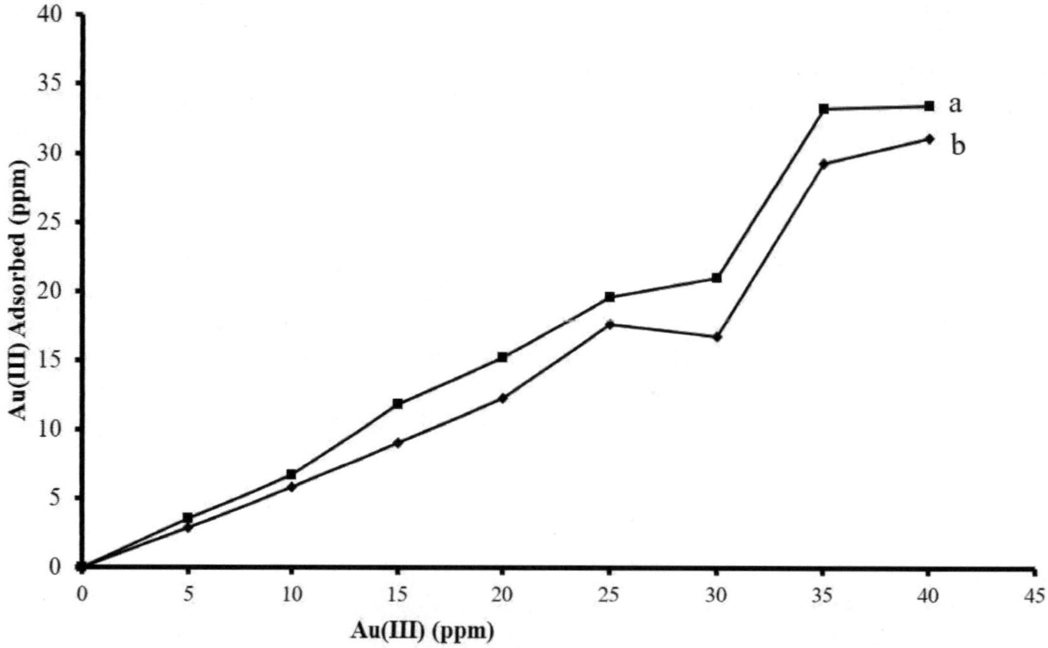

Figure 1. Adsorption Curve in a. $SiO_{2(RHA\text{-}BGS)}$-TMPDT and b. $SiO_{2(RHA\text{-}BGS)}$-TMPDT-Au-Imp on the Concentration of Au (III) Metal Ion Solution.

Figure 1 shows that imprinted ($SiO_{2(RHA\text{-}BGS)}$-TMPDT-Au-Im) material absorbs Au(III) more than without imprinted ($SiO_{2(RHA\text{-}BGS)}$-TMPDT). Furthermore, the time variation data is used to study the adsorption isotherm model. The results are shown in Table 1.

Table 1. Langmuir and Freundlich Isotherm Data.

Adsorption Isotherms	$SiO_{2(RHA\text{-}BGS)}$-TMPDT	$SiO_{2(RHA\text{-}BGS)}$-TMPDT-Au-Imp
Langmuir		
qm (mg/g)	139,341	299,256
R^2	0,0471	0,1266
B	3,10112	2,21277
ΔG	20,0522 kJ/mol	19,9026 kJ/mol
Freundlich		
K_f (mg/g)	473,006	981,672

N	0,95951	1,00857
R^2	0,8173	0,7612

From Table 1 shows that the adsorption of Au(III) on $SiO_{2(RHA\text{-}BGS)}$-TMPDT and $SiO_{2(RHA\text{-}BGS)}$-TMPDT-Au-Imp tend to follow the Freundlich isotherm model. It is shown that the value (R^2) for the Freundlich isotherm model is greater than Lungmuir. The Freundlich model illustrates that the adsorption process occurs multilayer[7]. The adsorption process is not only on the surface of the adsorbent but also enters the adsorbent pore. The interactions that occur are very heterogeneous involving various active sites in the adsorbent. Maximum adsorption capacity (K_f) in $SiO_{2(RHA\text{-}BGS)}$-TMPDT-Au-Imp (981,672 mg/g is greater than $SiO_{2(RHA\text{-}BGS)}$-TMPDT (473,006 mg/g) This is due to the formation of imprinted cavities in the adsorbent so Au(III) is adsorbed more.
From the Langmuir isotherm model the adsorption energy (ΔG) can be estimated. The magnitude ΔG represents the adsorption energy between Au(III) and the active site in the adsorbent. According to the table. 1 ΔG above 20 kj/mol indicates the interaction between Au(III) and the active site of the adsorbent occurs chemically. ΔG on $SiO_{2(RHA\text{-}BGS)}$-TMPDT and $SiO_{2(RHA\text{-}BGS)}$-TMPDT-Au-Imp material is approaching 20 Kj/mol so the interaction is estimated to occur chemically.

3.2. Adsorption Selectivity of Au(III) Metal Ions with Other Metal ions in $SiO_{2(RHA\text{-}BGS)}$-TMPDT-Au-Imp.

Au(III) adsorption Selectivity with other metal Ions in $SiO_{2(RHA\text{-}BGS)}$-TMPDT and $SiO_{2(RHA\text{-}BGS)}$-TMPDT-Au-Imp materials shown in table 2.

Table 2. Selectivity of Au(III) Metals to Cu(II), Zn(II), and Fe(III) Metals.

Variations [Cu(II)]	**Selectivity**	
	$SiO_{2(RHA\text{-}BGS)}$-TMPDT	**$SiO_{2(RHA\text{-}BGS)}$-TMPDT-Au-Imp**
5 ppm	3,20	37,49
10 ppm	153,12	30,46
15 ppm	5,85	114,30
Variations [Zn(II)]	**$SiO_{2(RHA\text{-}BGS)}$-TMPDT**	**$SiO_{2(RHA\text{-}BGS)}$-TMPDT-Au-Imp**
5 ppm	3,36	372,77
10 ppm	4,24	128,96
15 ppm	0,92	1252,12
Variations [Fe(III)]	**$SiO_{2(RHA\text{-}BGS)}$-TMPDT**	**$SiO_{2(RHA\text{-}BGS)}$-TMPDT-Au-Imp**
5 ppm	0	261,93
10 ppm	0	21,19
15 ppm	0	680,71

Au(III) selectivity in $SiO_{2(RHA\text{-}BGS)}$-TMPDT-Au-Imp is higher than $SiO2_{(RHA\text{-}BGS)}$-TMPDT. This is due to the formation of imprinted cavities that match their size with Au(III). At pH 3 the adsorbent active group becomes $-NH_3^+$, so it is expected that the cationic ion Zn(II), Cu(II), Fe(III) is not adsorbed. Part of the metal ion is in fact also distorted. This is due to the adsorption process does not only involve the active site $-NH_3^+$, but also from the pores and s-OH groups of silanol on silica. The order of selectivity in $SiO_{2(RHA\text{-}BGS)}$-TMPDT-Au-Imp is Au/Zn> Au/Fe> Au/Cu.

3.3. Adsorbent Material Desorption Performance Test.

Table 3. Au(III) Metal Desorption on $SiO_{2(RHA\text{-}BGS)}$-TMPDT and $SiO_{2(RHA\text{-}BGS)}$-TMPDT-Au-Imp material.

Time (minutes)	Desorption Au(III)(mg/g)	
	SiO_2-TMPDT	SiO_2-TMPDT-Au-Imp
0	0	0
30	0,78348	0,4279
60	0,64782	0,41824
90	0,64313	0,33258
150	0,52709	0,51107
200	0,77215	0,6898

From the desorption results, it is necessary to know the kinetic model that occurs in the desorption. Determination of the rate of desorption reaction of Au(III) metal ions from each adsorbent is carried out by the kinetic model plot of the desorption with the kinetic model that can be applied to the desorption process is the reaction equation of order 0, order 1, and order 2. From the results of the linear kinetics regression equation the desorption rate will then be determined the coefficient of determination (R^2) of each reaction order equation which is then compared, where the reaction order equation that has the closest R^2 is 1 which best describes the desorption process that occurs.

Table 4. Kinetic Desorption $SiO_{2(RHA\text{-}BGS)}$-TMPDT and $SiO_{2(RHA\text{-}BGS)}$-TMPDT-Au-Imp.

Reaction Kinetics Model	Parameter	$SiO_{2(RHA\text{-}BGS)}$-TMPDT	$SiO_{2(RHA\text{-}BGS)}$-TMPDT-Au-Imp
Orde 0	R^2	0,0176	0,661
	K	0,0204	0,1599
Orde 1	R^2	0,024	0,5989
	K	0,0004	0,0031
Orde 2	R^2	0,0316	0,5127
	K	7×10^{-6}	6×10^{-5}

The results obtained from the linear regression equation show that the kinetic model that is suitable for the desorption reaction of the Au (III) metal ion from the SiO2-TMPDT adsorbent is 2nd order while in the $SiO_{2(RHA\text{-}BGS)}$-TMPDT-Au-Imp adsorbent is zero order.

Functional group analysis on the desorption performance of $SiO_{2(RHA\text{-}BGS)}$-TMPDT and $SiO_{2(RHA\text{-}BGS)}$-TMPDT-Au-Imp material is needed to determine the differences that occur in the two materials before and after desorption using the FTIR (Fourier Transform Infra Red) instrument with the following results :

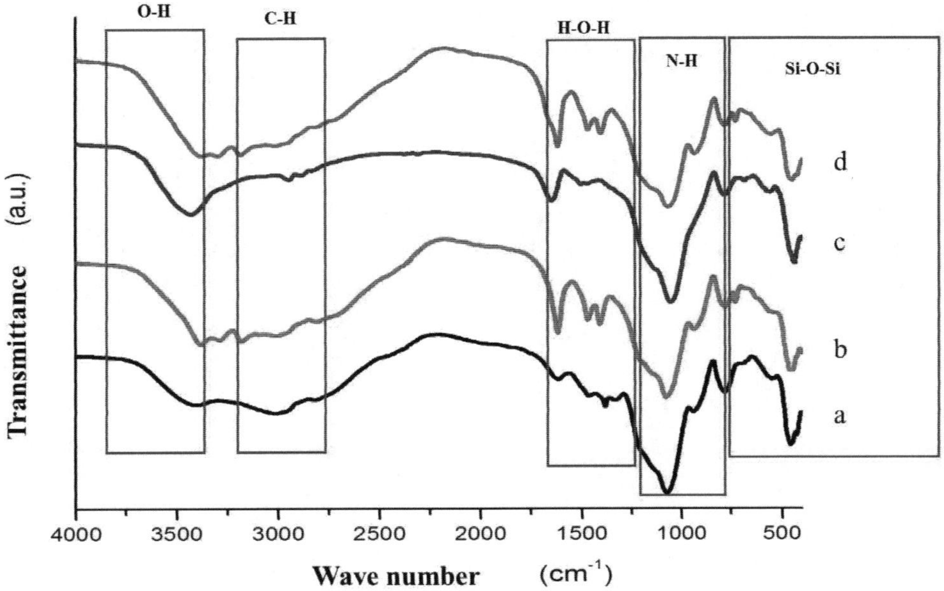

Figure 2. FTIR spectra of a. $SiO_{2(RHA-BGS)}$-TMPDT-Au-Imp before desorption, b. $SiO_{2(RHA-BGS)}$-TMPDT-Imp after Desorption, c. $SiO_{2(RHA-BGS)}$-TMPDT-Au before desorption, and d. $SiO_{2(RHA-BGS)}$-TMPDT after desorption.

Based on Figure 2. above it can be seen that the results of the FTIR spectra of the two adsorbents have peaks that are almost the same as the literature. From the figure it can also be seen that the absorption at wavelengths of absorption –OH, N-H, and C-H is still visible which shows that the $SiO_{2(RHA-BGS)}$-TMPDT and $SiO_{2(RHA-BGS)}$-TMPDT-Au-Imp material has no damage when desorption.

4. Conclusion

The results of the adsorption process followed the Freundlich isotherm model with an adsorption capacity of 981,672 mg/g for $SiO_{2(RHA-BGS)}$-TMPDT-Au-Imp and 473,006 mg/g for $SiO_{2(RHA-BGS)}$-TMPDT. The adsorption selectivity shows that the imprinted ionic $SiO_{2(RHA-BGS)}$-TMPDT-Au-Imp material is more selective than the material without imprinted ionic $SiO_{2(RHA-BGS)}$-TMPDT. The order of selectivity is Au(III)/Zn(II)> Au(III)/Fe(III)> Au(III)//Cu(II). Study of desorption kinetics Au(III) followed the zero order kinetics for $SiO_{2(RHA-BGS)}$-TMPDT-Au-Imp and second order for $SiO_{2(RHA-BGS)}$-TMPDT. The results of the FTIR show that there are still O-H, C-H and -NH groups. This indicates the imprinted ionic material ($SiO_{2(RHA-BGS)}$-TMPDT-Au-Imp) and $SiO_{2(RHA-BGS)}$-TMPDT is not damaged.

Acknowledgement

The author gratefully express acknowledgementto Ministry of Research, Technology and Higher Education (RISTEKDIKTI) of the Republic Indonesia.

References

[1] Roslan, N.A., Suah F.B.M dan Mohamed N. 2017. *Separation and Purification Technology*.**182**.(1–8)
[2] Hagelüken, C., & Corti, C. W. 2010. Gold Bulletin, 43(3), 209-220.
[3] Sakti, S.C.W., Siswanta, D. dan Nuryono. 2013. Pure Aplication Chemistry. 18 (1) : 211-223.

[4] Zhang, N., Hu, B., Huang, C. 2007. Analytica Chimica Acta. 597: 12-18
[5] Li, F., Jiang, H., & Zhang, S. 2007. Talanta, 1487-1493.
[6] Buhani, Nuryono, dan Kunarti, E.S. 2010. *Desalin.* **251**: 83–89.
[7] Andika, I M D., Simpen, I N., Putra, K G D. 2016. **Jurnal Kimia**, **10**(1) : 125-132.

The Degree of Lithium (Li) Stability Compared To Calcium (Ca) and Magnesium (Mg) From Low Lithium Grade Brine Water with Addition of Limestone and Oxalic Acid

F. Firdiyono[1], Latifa Hanum Lalasari[1,a], Ermiziar Tarmizi[2], Eko Sulistiyono[1], Lia Andriyah[1], Tri Arini[1], Nadia C. Natasha[1], Fariza E. Yunita
[1]Research Center of Metallurgy and Materials, Indonesian Institute of Sciences, Kawasan PUSPIPTEK Gedung 470, Tangerang Selatan, 15314, Indonesia
[2]Department of Chemical Engineering, Institut Teknologi Indonesia, Jl Raya Puspiptek Serpong-Tangsel, Indonesia

E-mail: [a]lati003@lipi.go.id; [a]ifa_sari@yahoo.com

Abstract. The purpose of this study was to investigate whether lithium was precipitated along with the removal of Ca and Mg elements from brine water using limestone and oxalic acid. The first stage was Mg removal by adding CaO at room temperature for 3 hours. The second stage was the Ca removal by adding oxalic acid at a ratio of 2; 3; 4; 5; 10; 15; 20; 30; and 40 ml / 100 ml brine water and pH : 0, 1, and 4 at room temperature for 3 hours. The results showed the most optimal addition of CaO was 20 gr /1 l of brine water with the Mg element can be removed above 99 %, while Li was only reduced by about 1 ppm, and the Ca content only increased by around 28 ppm. The use of oxalic acid will also reduce the amount of lithium besides the Ca removal in the form a precipitate of calcium oxalate. The use of oxalic acid at pH 0 and 4 would reduce the Li content more than at pH 1. At the use of oxalic acid up to 40 ml at pH 1, brine water still contained about 78% Li element.

1. Introduction

Bittern water is one of the sources of lithium raw material which is quite potential in addition to hot water and rock specimens. Bittern water is obtained from the evaporation of salt water in the production of salt from sea water or from the evaporation of spring water. In addition to consumption needs, the use of salt which is quite large is for industrial purposes. Indonesia with a large population, the need for salt for consumption will also be large. To meet the needs of salt, Indonesia can obtain it through the process of seawater desalination, because Indonesia is an archipelago with a sea area of two-thirds of its land area. The process of seawater desalination to obtain salt in very large quantities will produce a very abundant bittern water.

Until now the use of bittern water is still not optimal. Meanwhile on the other hand in bittern water contained lithium which has been concentrated. In bittern water there are still many impurities, such as Ca, Mg, K, B and others that must be removed. This bittern water when processed carefully will get lithium concentrate which is very useful in making lithium batteries. Chile, Bolivia, China are the main producers of lithium raw materials. The element of lithium can be found in seawater, spring water and spodumen rocks. In Indonesia supporting data regarding lithium raw material reserves do not yet exist, so an exploration phase is needed to determine the potential of these raw materials, while the exploration work itself requires a very large cost.

Lithium carbonate is the main ingredient in making lithium batteries. Lithium batteries are one source of electricity energy that is widely used for electric cars in the future. The advantages of lithum batteries include:

1. The battery can be recharged quickly
2. The electrical energy generated is quite large, etc.

In this paper we will discuss the process of eliminating the calcium element contained in a brine water solution, as a result of the process of eliminating the element magnesium in a solution carried out using lime. Ca elements that enter the solution will be eliminated again by using oxalic acid. On this occasion the characteristics of the lithium element will be investigated using lime and oxalic acid reagents when the process of removing Mg and Ca from low lithium lithium brine water is carried out.

2. Experimental

2.1. Materials

The raw material used in this experiment was the concentrate of the spring water evaporation process with an initial lithium content of 15-18 ppm. The chemical composition of brine water from the evaporation process was shown in Table 1. The limestone had undergone a roasting process at 900 °C was used as a stage 1 precipitation reagent which has a chemical composition as follows: 99.33% Ca, 0.34% Mg, 0.04% K, 0.24% Si and 0.05% others [1]. Oxalic acid and NaOH are pro-analyzes with Merck specifications used as stage 2 precipitation reagents carried out controlling pH = 0, 1, 4.

Table 1. The chemical composition of brine water samples (ICP-OES analysis)

Al	B	Ca	K	Li	Mg
0.26 ppm	353.32 ppm	7672.48 ppm	1975.56 ppm	79.87 ppm	1569.87 ppm

2.2. Experimental procedure

Brine water used in this experiment was obtained from the evaporation of Tirta Sanita hot water, Ciseeng, Bogor.The removal of Mg element in the solution was done by adding CaO to brine water as previous research by Lalasari et. al (2019)[1]. The variations in the amount of CaO used in this process ware: 0.1; 0.3; 0.5; 0.75; 1; 2; 5; 10; 20; 30; 40; and 50 g/1 1 brine water. The process was carried out at room temperature with stirring time for 3 hours. The next step was the filtration process to separate the solids (residues) and their liquids (filtrates). To find out the shape of the crystals that occurred in the residue, the solids are heated at a temperature of 100 °C for 24 hours and then an XRD and SEM analysis was performed. The content of Mg, Ca, and Li in the filtrates ware then analyzed using ICP-OES. After the content of the Mg, Ca, and Li elements in the filtrate was known, the process continues with the removal of Ca element. The elimination of Ca element was done by adding oxalic acid to the solution. The variations in the amount of oxalic acid used in the process ware: 2; 3; 4; 5; 10; 15; 20; 30; and 40 ml / 100 ml sample water. The pH variations used in this process ware 0, 1, and 4. The process was carried out at room temperature with a stirring time of 3 hours. This experiment will investigate how much of the lithium element settles with magnesium and calcium during the refining process of magnesium and calcium from brine water using reagents of limestone and oxalic acid. The flow diagram of the experiment was shown in Figure 1.

IOP Conf. Series: Materials Science and Engineering **858** (2020) 012044 doi:10.1088/1757-899X/858/1/012044

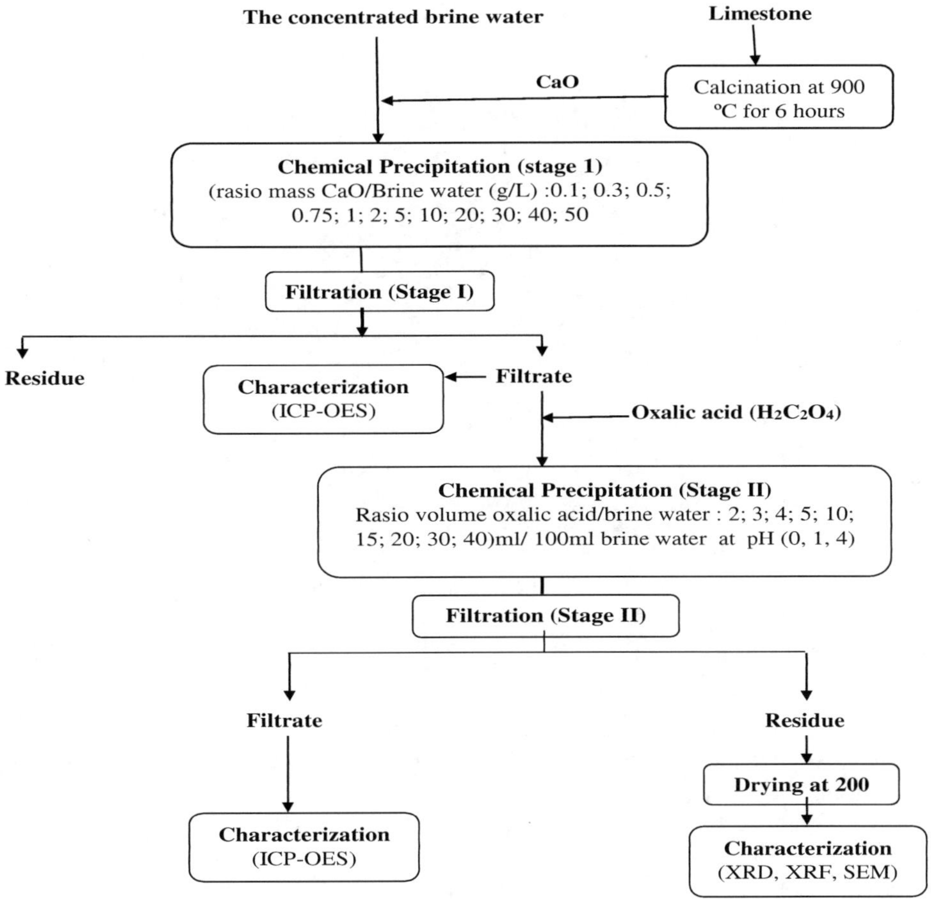

Figure 1. The flow chart of research

3. Result and Discussion

The removal of magnesium element in brine water using limestone as precipitant reagent aims to bind Mg^{2+} elements from brine water so that Mg^{2+} settles. This experimental method is the same as that of Lalasari et al (2019)[1]. However, the chemical composition of the brine water used is different, namely the Mg and Ca content in this experiment are greater as shown in Table 1. Addition of limestone to the solution will occur the deposition of Mg^{2+} in brine water as the following reaction [1] :

$CaO_{(s)} + H2O_{(l)} \rightarrow Ca(OH)_{2(aq)}$ .... (1)

$Ca(OH)_{2(aq)} + Mg^{2+}_{(aq)} \rightarrow Mg(OH)_{2(S)} + Ca^{2+}_{(aq)}$ .... (2)

The above reaction can take place well and spontaneously because the Ksp value of $Mg(OH)_2$ of 1.8×10^{-11} is very much smaller than the price of Ksp Ca $(OH)_2$ which is 6×10^{-6}. The results of these experiments can be seen in Figure 1.

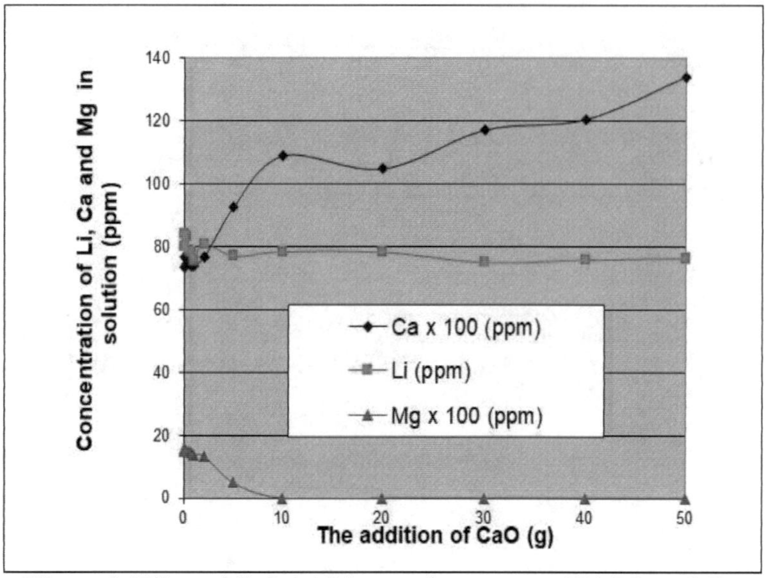

Figure 2. Effect of CaO Addition on the amount of Li, Mg and Ca.

Based on the picture above shows that the addition of CaO into the solution can reduce the Mg content. The use of CaO in the Mg removal process will also result in a reduction in the Li content and vice versa will result in the addition of the amount of Ca in the solution as well. The most optimal addition of CaO is 20 grams because in this addition the magnesium element can be removed as much as 99.45%, whereas Li is only reduced by about 1 ppm, and the Ca content in the solution will only increase by around 2807 ppm, from 7672 ppm to 10479 ppm. According to Lalasari et al (2019), the addition of CaO will cause the Mg^{2+} ions to precipitate to form $Mg(OH)_2$ compounds such as reaction 2 and also precipitate together with Ca to form compounds $(Mg_{0.03}Ca_{0.97})(CO_3)$. Formation of compound $Mg_{0.03}Ca_{0.97})(CO_3)$ will be smaller with an increase in the use of CaO into brine water.

As a result of the precipitation process of $Mg(OH)_2$ by using CaO, the amount of Ca elements in the solution would also increase. The removal of Ca^{2+} from the solution was done by adding oxalic acid to the solution, so that calcium oxalate deposits are formed as follows:

$H_2C_2O_{4(aq)}$ + $Ca^{2+}{}_{(aq)}$ → $CaC_2O_4.H_2O_{(aq)}$ .... (3)

The above reaction can take place well because the Ksp value of CaC_2O_4 of 1.9 x 10^{-9} is slightly smaller than the price of Ksp $Ca(OH)_2$ which is 6 x 10^{-6}. Experiments of precipitation of Ca elements using 1 N oxalic acid were carried out with variations in the amount of oxalate from 2 ml to 40 ml at pH from 0 to 4. According to Dong Jung Kang (2015) [2], the optimal pH of oxalic acid to react with Ca^{2+} ions was at pH 0-1, whereas at pH 4 is expected that oxalic acid will bind Mg^{2+} ions to magnesium oxalate.

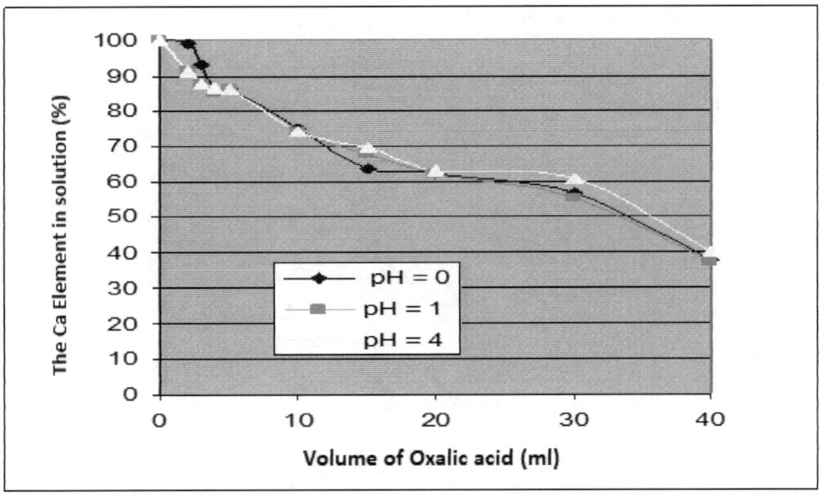

Figure 3. Effect of pH and addition of oxalic acid on Ca content in solution.

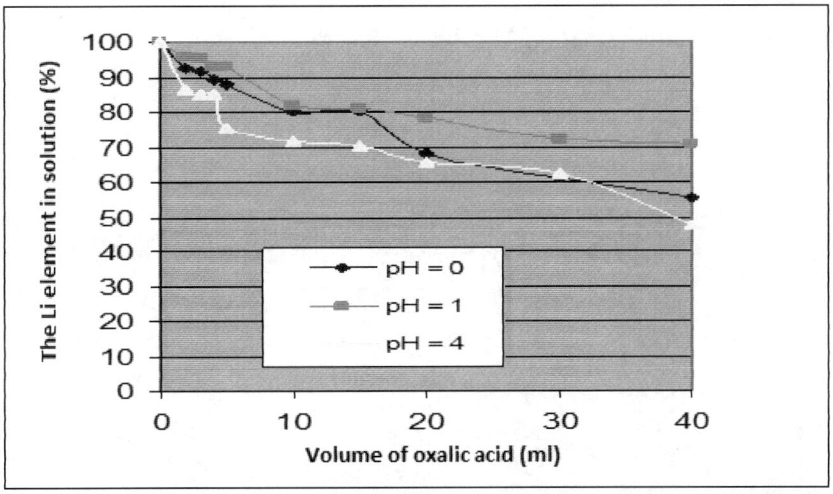

Figure 4. Effect of pH and addition of oxalic acid on Li content in solution.

Figure 3 shows the same tendency in the process of removing Ca elements using oxalic acid. Increased use of the amount of oxalic acid will result in a reduction in the amount of Ca in the solution. The experiment did not show a very significant difference in removal of the Ca element between pH 0, 1 and 4. In the use of oxalic acid up to 40 ml, the dissolution would still contain about 40% Ca element. Figure 4 shows that the use of oxalic acid will also reduce the amount of lithium in solution. The use of oxalic acid at pH 0 and 4 will reduce the amount of element Li in the solution more than at pH 1. At the use of oxalic acid up to 40 ml with pH 1, the dissolution will still contain about 78% Li element. Sulistiyono et al (2018) [3] explained that the chemical precipitation process of brine water using oxalic acid caused the lithium elements to settle together with Ca and Al to form $Al_{1.19}Ca_1Li_{0.81}$ compounds. This causes the Li content in the brine water to decrease.

IOP Conf. Series: Materials Science and Engineering **858** (2020) 012044 doi:10.1088/1757-899X/858/1/012044

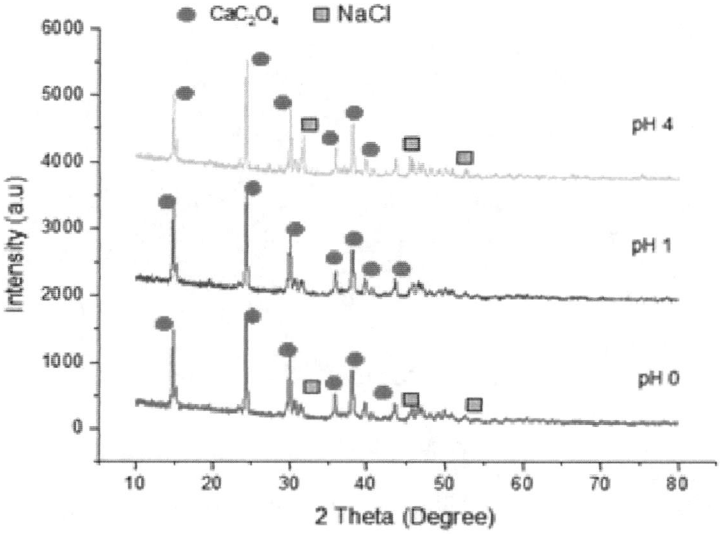

Figure 5. XRD pattern of Ca residues with oxalic acid

The picture above shows the XRD pattern of the residue that is formed when the solution is added 15 ml of 1 N oxalic acid at pH 0, 1, and 4. From the picture is clear that oxalic acid will bind Ca^{2+} ions and form precipitating CaC_2O_4 compounds. Precipitation of Ca element with oxalic acid was apparently not affected by the acidity degree of the solution. CaC_2O_4 compounds that are formed will precipitate at pH 0, 1 and at pH 4.

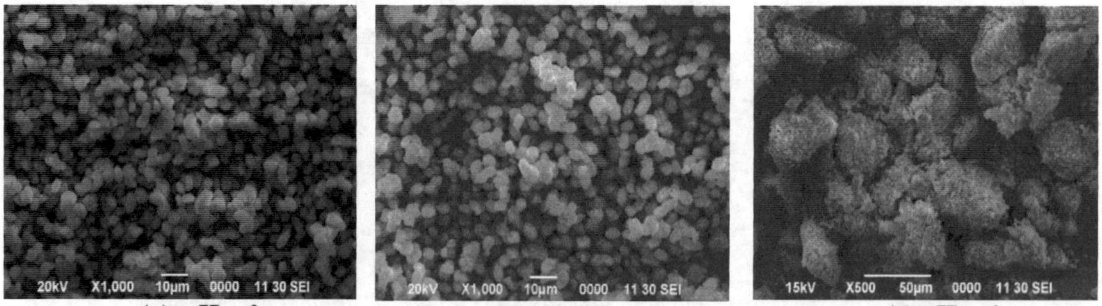

(a) pH = 0 **(b) pH = 1** **(c) pH = 4**

Figure 6. The morphology of residue from the addition of oxalic acid to solution

The picture above shows the results of SEM analysis on the residue resulting from the addition of 15 ml of 1N $H_2C_2O_4$ (pH 0; pH 1; and pH 4) with a magnification of 1000x. That indicates that the residue is dominated by CaC_2O_4 which has a monoclinic and perismatic crystal system. Figure 5a does not provide a significant change in shape with Figure 5b, which means the addition of pH 0 and pH 1 does not provide a significant morphological change in residue. The residual morphology at ph-4 appears to be irregular which is thought to have other elements besides Ca in the residue.

The above experimental results show that oxalic acid can bind the Ca element into Ca oxalate, but the use of oxalate still leaves the Ca element in a large enough solution. In the use of oxalic acid up to 40 ml with pH 1, 2, and 4 then the dissolution will still contain about 40% Ca element

4. Conclusions

CaO reagent is an effective material to eliminate magnesium in the brine water solution, but excessive use of CaO in the Mg removal process will also cause a reduction in Li content and vice versa resulting in an increase in the amount of Ca in the solution. The most optimal addition of CaO is 20 g /1 l of brine water with magnesium element can be removed as much as 99.45%, Li is only reduced by about 1 ppm. Precipitation of Ca with oxalic acid is not affected by the acidity of the solution. CaC2O4 compounds that are formed will precipitate both at pH 0, 1 and at pH 4. The use of aksalat acid will also reduce the amount of lithium in solution. The use of oxalic acid at pH 0 and 4 will reduce the amount of Li element in the solution more than at pH 1. The remaining element Li in the solution is still around 78% on the use of oxalic acid 40 ml / 100 ml brine at pH 1.

5. Acknowledgments

This research was carried out at hydrometallurgy laboratorium and pyrometallurgy laboratorium in Research Center for Metallurgy and Material, Indonesian Institute of Sciences. The authors are grateful for the main financial supports from INSINAS budget 2019.

References

[1] L. H. Lalasari *et al.*, "Magnesium removal from brine water with low lithium grade using limestone , rembang , indonesia," *Int. Conf. Adv. Mater. Better Futur. 2018 ,IOP Conf. Ser. Mater. Sci. Eng.*, 2019.

[2] D. J. Kang, "Patent No. 9169125B2, United States of America," 2015.

[3] E. Sulistiyono, L. H. Lalasari, W. Mayangsari, and A.B. Prasetyo, "Study of Lithium Extraction from Brine Water, Bledug Kuwu, Indonesia by The Precipitation Series of Oxalic Acid and Carbonate Sodium," in *Proceedings of the International Seminar on Metallurgy and Materials (ISMM 2017), AIP Conference Proceedings*, 2018, vol. 010001, no. 1964, pp. 1–6.

IOP Conf. Series: Materials Science and Engineering **858** (2020) 012045 doi:10.1088/1757-899X/858/1/012045

Time and Amplitude Effect on Nano Magnesium Oxide Synthesis from Bittern using Sonochemical Process

F E Yunita[1]*, N C Natasha[1], E Sulistiyono[1], A R Rhamdani[1], A Hadinata[2], E Yustanti[2]

[1]Research Center for Metallurgy and Materials, Indonesian Institute of Sciences, PUSPIPTEK 470, Indonesia

[2]Metallurgical Engineering Department, Faculty of Engineering, Universitas Sultan Ageng Tirtayasa, Indonesia

E-mail: fariza.ekayunita@yahoo.co.id

Abstract. Magnesium in nature can be found in the form of minerals and seawater. Bittern is a sea-salt industry by-product that contains magnesium and potassium salts. Usually, bittern is discharged back to the sea, even though bittern can be further processed to obtain magnesium contained in it. Magnesium oxide (MgO) nanoparticles can be used in a variety of applications because of their good surface recreation properties. In this study, precipitation of Mg2 + ions from bitterns was carried out using sodium hydroxide to produce magnesium hydroxide. Then, it was calcined and went through sonochemical process to produce nano magnesium oxide. Sonication time and amplitude were used as variables. Sample with sonication time of 16 minutes and amplitude of 30% has the smallest particles with an average diameter of 195.7 nanometers.

1. Introduction

Magnesium oxide nanoparticles can be used in wide applications such as sensors and bactericides because of their good surface reactivity. It can be used as a humidity sensor due to its sensitivity to water vapor adsorption [1]. As a bactericide, superoxide ions are produced on the surface of magnesium oxide nanoparticles so that they can react with peptide bonds in bacterial cell walls or spores and destroy it. This bactericidal activity increases with decreasing of magnesium oxide particle size. Compared to titanium oxide nanoparticle which requires light, magnesium oxide nanoparticles can show good bactericidal activity even in the absence of light [2].

Besides being found from magnesium-based minerals, magnesium also can be found in brine water and seawater. With an average magnesium content in seawater of 1300 ppm, the potential of magnesium from seawater is very promising [3]. Bittern is a by-product of the salt processing industry which contains magnesium and potassium salts. From every ton of sea salt produced, there is one cubic meter of bittern as by-product, and every ten cubic meters of bittern contain 1.2 tons of magnesium chloride [4]. Commonly, bittern is discharged back to the sea without further process, whereas bitterns with high salt concentrations will accumulate at the bottom of the sea can endanger ecosystem [5].

In 1943, Robinson et al. patented a precipitation process of magnesium hydroxide from seawater using calcium hydroxide [6]. However, in this process gypsum was formed and become impurities in magnesium oxide product [7], [8]. Another research was conducted by Dong et al. which used ammonia hydroxide to precipitate magnesium hydroxide from desalination plant waste in Singapore which contains 1679 ppm magnesium [9].

Based on research by Jin and Al-Tabbaa, magnesium oxide from bittern has a higher purity and reactivity compared to magnesium oxide from calcined magnesite [10]. Calcination of magnesium hydroxide to produce magnesium oxide has also been studied previously. The higher the calcination temperature, the larger the magnesium oxide size, so that the surface area and its reactivity will decrease [11]–[15].

There are several methods to produce nanoparticles, such as radiation, thermal decomposition, vapor deposition [16]–[18]. However, most of these techniques are relatively expensive and time-consuming. Another simple and cost-effective method is through ultrasonic destruction [19]. The sonochemical process can produce microbubbles in the liquid, followed by the growth and collapse of the bubbles that can raise the surrounding cavity temperatures to more than 5000 K and pressures more than 500 atm. Because bubbles collapse in less than one nanosecond, a cooling rate of more than 1000 K/s can be obtained that can hinder product crystallization. Collapse of the bubbles can also cause shock waves and liquid jets with speeds of more than 400 km/hour. This physical and chemical environment can cause the formation of products with nanoparticle size [20]–[22].

In this research, sodium hydroxide was used to precipitate magnesium ions from bittern. Then, calcination and sonochemical process were performed. We would investigate the effect of sonication time and amplitude as independent variables during sonochemical process to produce magnesium oxide nanoparticles.

2. Materials and Methods

This research was conducted using bittern, sodium hydroxide, and 2-butanol. Bittern was a by-product of salt processing from seawater, while sodium hydroxide and 2-butanol used were PA (pro analysis) materials. Samples characterization was carried out using ICP - OES (Inductively Coupled Plasma - Optical Emission Spectrophotometers), XRD (X-Ray Diffraction), PSA (Particle Size Analyzer), and SEM-EDX (Scanning Electron Microscope-Energy Dispersive X-ray Disorders) Spectroscopy).

The initial stage of this research was to filter 500 ml bittern to remove the solid impurities. Magnesium hydroxide was synthesized by adding sodium hydroxide to the bittern with mol ratio of Mg^{2+} ion : sodium hydroxide = 1: 2 according to its stoichiometric calculation and then be stirred at 300 rpm and temperature of 30°C. The residue that has been separated from the filtrate then was dried in the oven for 24 hours at 100°C. The solid was calcined at 460°C for 2 hours to produce magnesium oxide. Reduction of magnesium oxide particles from micro-size to nano-size was carried out by ultrasonic method using 2-butanol as sonication media with a ratio of 3 grams magnesium oxide / 200 ml 2-butanol. The amplitude was varied between 20%, 30%, and 40% while sonication time was varied between 4, 8, and 16 minutes.

The concentration of magnesium and other impurities in the bittern and filtrate were analyzed using ICP-OES and were calculated based on mass balance so that magnesium concentration in the residue could be obtained. XRD characterization was used to determine phase composition in the residue before and after calcination. After sonochemical process, the samples were analyzed using the free-settling method. Sample that had the lowest height of supernatant was analyzed using PSA and SEM-EDX to determine the size and morphology of the particles obtained.

3. Result and Discussion

In the previous research by Dong et al., magnesium oxide was synthesized from reject brine from desalination plants in Singapore which contain 1,718 ppm magnesium [9]. While in this research, the bittern was a by-product of salt processing from seawater in Lamongan, East Java, Indonesia. Based on the ICP analysis results that can be seen in Table 1, this bittern has greater magnesium content, which is 76,688.9 ppm.

Table 1. Chemical composition of the brine used in this research

Element	B	Ca	K	Li	Mg	Na
Concentration (ppm)	453.34	148.48	2,991.1	29.05	76,688.9	3,126.84

Magnesium hydroxide was synthesized via the reaction between Mg^{2+} ions in the brine and OH^- from sodium hydroxide. The possible reactions were as follows :

$$NaOH_{(aq)} \leftrightarrow Na^+ + OH^- \quad (1)$$

$$Mg^{2+} + 2OH^- \rightarrow Mg(OH)_{2(s)} \quad (2)$$

The filtrate was separated from solid residue and ICP analysis was carried out to determine concentration of the elements in the filtrate which can be seen in Table 2. Magnesium concentration in the residue and percent of magnesium precipitated could be calculated based on mass balance. Based on these data it can be calculated that the percent of magnesium precipitated was 80%.

Table 2. Chemical composition of the filtrate

Element	B	Ca	K	Li	Mg	Na
Concentration (ppm)	12.23	0.87	2,807.1	0.36	15,377.2	8,047.62

The residue that has been separated from the filtrate then was dried in the oven for 24 hours at 100°C. Figure 1 shows XRD diffractogram of the residue obtained from the reaction between the bittern and sodium hydroxide. From this figure, it can be seen that the main crystalline peak positions are magnesium hydroxide. These peaks are followed by several minor peaks of sodium chloride.

Magnesium hydroxide was calcined at 460°C for 2 hours to produce magnesium oxide. XRD diffractogram of the calcined can be seen in Figure 2. The absence of magnesium hydroxide peak indicates that it was fully decomposed at the calcination process resulting in magnesium oxide. These magnesium oxide peaks are also followed by several minor peaks of sodium chloride.

Reduction of magnesium oxide particles from micro-size to nano-size was carried out by ultrasonic method using 2-butanol as sonication media with a ratio of 3 grams magnesium oxide / 200 ml 2-butanol. The amplitude was varied between 20%, 30%, and 40% while sonication time was varied between 4, 8, and 16 minutes.

The equivalent diameter refers to the size of particles that have irregular shapes. One technique for measuring equivalent diameter is Stokes' diameter which is based on sedimentation process. In sedimentation process, as settling rate approaches zero, clear liquid supernatant zone and settle bed zone are formed. This process is illustrated in the schematic diagram in Figure 3. Based on Stokes' law that can be seen in Equation 3, factors that affect sedimentation include particles size and shape, solids weight and volume, fluid viscosity, and specific gravity of solids and liquid. At the same specific gravity, particles that have a larger size will have greater terminal velocity [23].

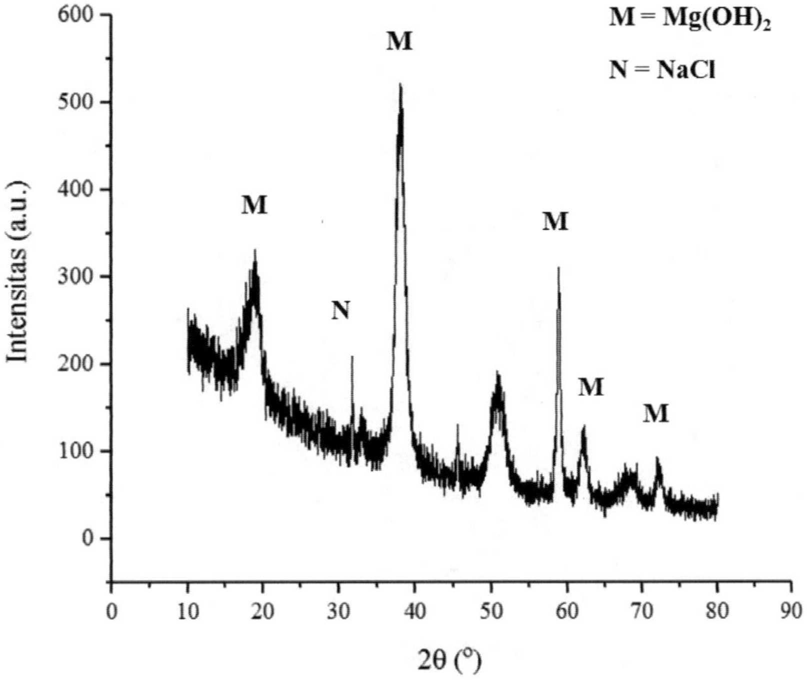

Figure 1. XRD diffractogram of magnesium hydroxide

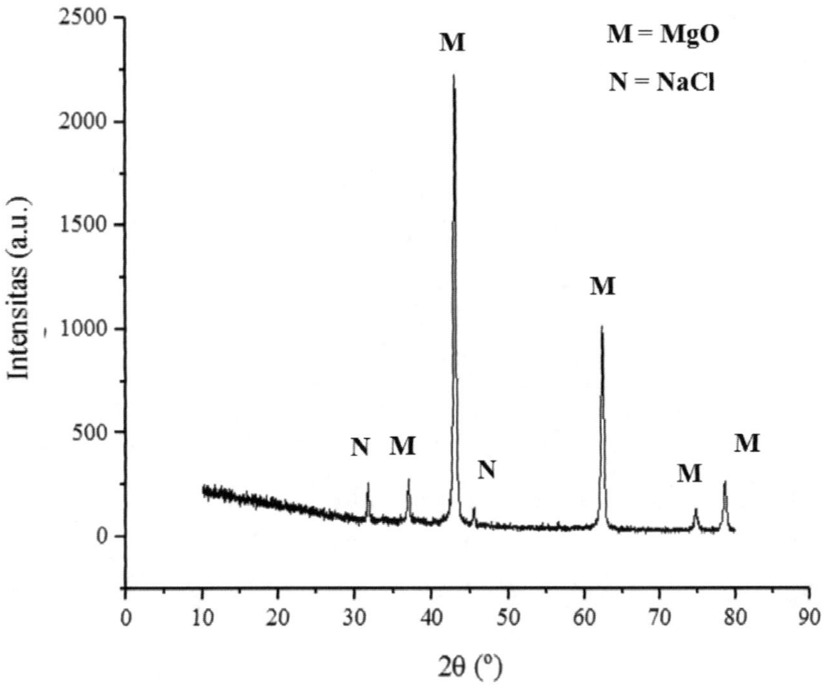

Figure 2. XRD diffractogram of magnesium oxide

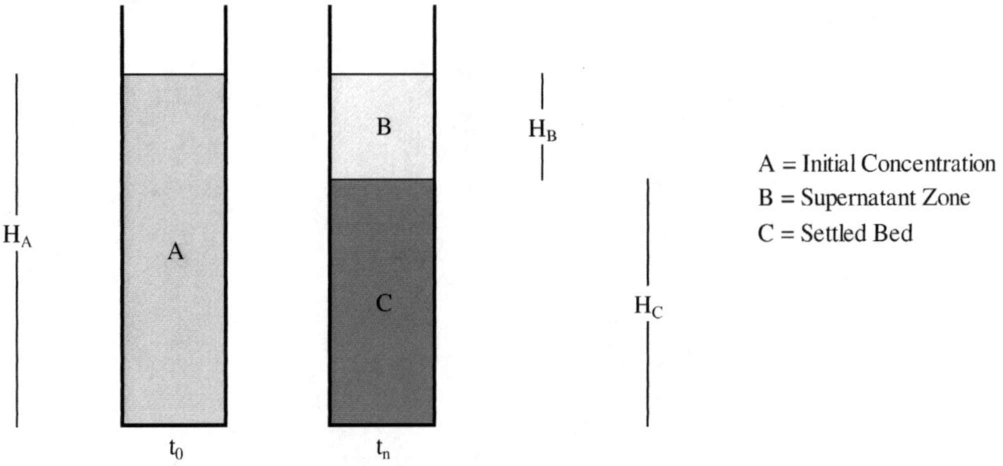

Figure 3. Batch Settling Process

$$v = \frac{d^2 g (\rho_s - \rho_f)}{18\eta} \quad (3)$$

Where :

v – Terminal velocity (m s^{-1})
d = Particle diameter (m)
g = Gravity acceleration (m s^{-2})
ρs = Particle density (kg m^{-3})
ρf = Fluid density (kg m^{-3})
η = Fluid viscosity (Ns m^{-2})

After sonication process, each sample was settled for 7 days. Supernatant zone height can be seen in Figure 4. This figure shows that sample with amplitude variation of 30% and 16 minutes sonication time has the lowest supernatant zone height so that it can be concluded that this sample has the smallest particle diameter. This sample was then analyzed by using PSA.

From Figure 4 we can see that increasing sonication time will decrease the particle size. According to Sáez and Mason, sonication time is one of the parameters that affect particle size during ultrasonic destruction process [24]. Based on Equation 4, dissipated sonic power in liquid is a function of amplitude [25]. However, based on research by Yustanti et al., particles with very small size tend to have agglomeration effect [26]. This causes sample with amplitude variation of 30% is smaller than samples with amplitude variation of 20% and 40%.

From the results of PSA analysis, sample with amplitude variation of 30% and 16 minutes sonication time has an average particle diameter of 195.7 nanometers that shown in Figure 5. In Figure 6, morphology of the sample can be observed through SEM analysis at 5000x magnification. The particles in this figure have spherical shapes, similar to previous study [27].

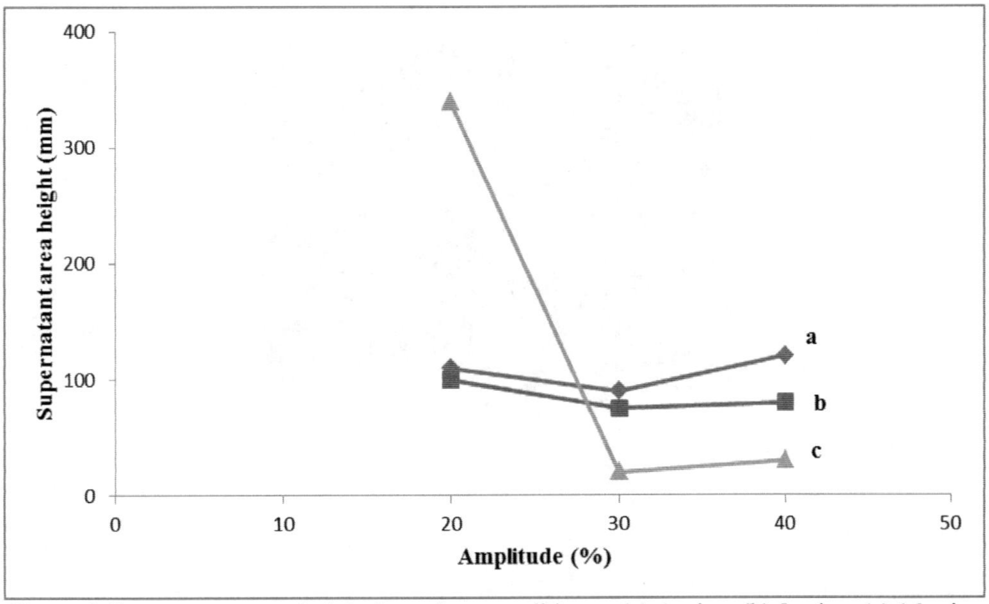

Figure 4. Supernatant zone height in various conditions : (a) 4 mins, (b) 8 mins, (c) 16 mins

$$P_{diss} = \frac{1}{2} A^2 (2f\pi)^2 SZ \qquad (4)$$

Where :

P_{diss} = Dissipated sonic power in liquid (W)
A = Amplitude (m)
f = Frequency (Hz)
S = Emitter surface (m^2)
Z = Acoustic impedance (kg m^{-2} s^{-1})

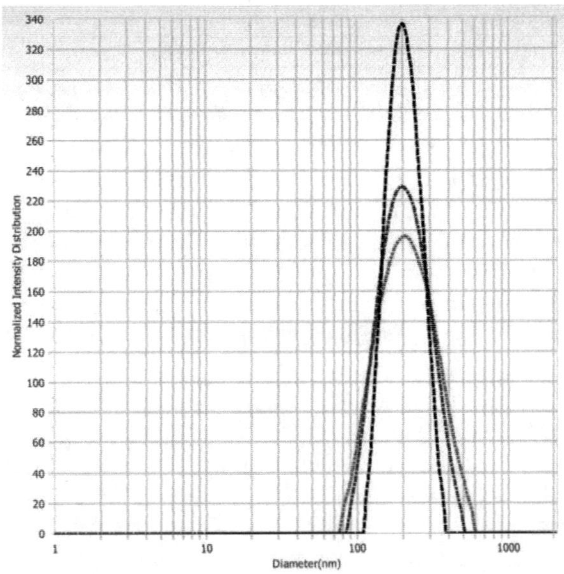

Figure 5. Magnesium oxide particle size distribution

Figure 6. SEM images of magnesium oxide

4. Conclution

The results obtained in this present research show that amplitude and sonication time affect the size of the particles produced in the ultrasonic destruction process. Sample with amplitude variation of 30% and 16 minutes sonication time has the smallest particle size, with an average particle diameter of 195.7 nanometers. Increasing sonication time will decrease the particle size. However, particles with very small sizes tend to have an agglomeration effect.

5. Acknowledgment

The authors wish to thank Tematik grant from Indonesian Institute of Sciences for funding this research.

6. References

[1] S. K. Shukla *et al.*, "Nano-like magnesium oxide films and its significance in optical fiber humidity sensor," vol. 98, pp. 5–11, 2004.
[2] L. Huang, D. Q. Li, Y. J. Lin, M. Wei, D. G. Evans, and X. Duan, "Controllable preparation of Nano-MgO and investigation of its bactericidal properties," *J. Inorg. Biochem.*, vol. 99, no. 5, pp. 986–993, 2005.
[3] J. F. Anthoni, "The Chemical Composition of Seawater," 2006. [Online]. Available: http://www.seafriends.org.nz/oceano/seawater.htm#composition.
[4] A. A. Hussein, K. Zohdy, and M. Abdelkreem, "Seawater Bittern a Precursor for Magnesium Chloride Separation: Discussion and Assessment of Case Studies," *Int. J. Waste Resour.*, vol. 07, no. 01, pp. 1–6, 2017.
[5] A. M. O. Mohamed, M. Maraqa, and J. Al Handhaly, "Impact of land disposal of reject brine from desalination plants on soil and groundwater," *Desalination*, vol. 182, no. 1–3, pp. 411–433, 2005.
[6] H. A. Robinson, R. E. Friedrich, and R. S. Spencer, "MAGNESUM HYDROXDE FROM SEAWATER," US 2405055, 1943.
[7] J. W. An, D. J. Kang, K. T. Tran, M. J. Kim, T. Lim, and T. Tran, "Recovery of lithium from Uyuni salar brine," *Hydrometallurgy*, vol. 117–118, pp. 64–70, 2012.
[8] T. Karidakis, S. Agatzini-Leonardou, and P. Neou-Syngouna, "Removal of magnesium from nickel laterite leach liquors by chemical precipitation using calcium hydroxide and the potential use of the precipitate as a filler material," *Hydrometallurgy*, vol. 76, no. 1–2, pp. 105–114, 2005.

[9] H. Dong, C. Unluer, E. H. Yang, and A. Al-Tabbaa, "Synthesis of reactive MgO from reject brine via the addition of NH4OH," *Hydrometallurgy*, vol. 169, pp. 165–172, 2017.
[10] F. Jin and A. Al-Tabbaa, "Characterisation of different commercial reactive magnesia," *Adv. Cem. Res.*, vol. 26, no. 2, pp. 101–113, 2014.
[11] L. Mo, M. Deng, and M. Tang, "Effects of calcination condition on expansion property of MgO-type expansive agent used in cement-based materials," *Cem. Concr. Res.*, vol. 40, no. 3, pp. 437–446, 2010.
[12] E. Alvarado, L. M. Torres-Martinez, A. F. Fuentes, and P. Quintana, "Preparation and characterization of MgO powders obtained from different magnesium salts and the mineral dolomite," *Polyhedron*, vol. 19, no. 22–23, pp. 2345–2351, 2000.
[13] J. K. Bartley, C. Xu, R. Lloyd, D. I. Enache, D. W. Knight, and G. J. Hutchings, "Simple method to synthesize high surface area magnesium oxide and its use as a heterogeneous base catalyst," *Appl. Catal. B Environ.*, vol. 128, pp. 31–38, 2012.
[14] W. R. Eubank, "Calcination Studies of Magnesium Oxides," *J. Am. Ceram. Soc.*, vol. 34, no. 8, pp. 225–229, 1951.
[15] K. Itatani, K. Koizumi, F. S. Howell, A. Kishioka, and M. Kinoshita, "Agglomeration of magnesium oxide particles formed by the decomposition of magnesium hydroxide - Part 2 Agglomeration at fixed temperatures," *J. Mater. Sci.*, vol. 24, no. 7, pp. 2603–2609, 1989.
[16] Y. H. Kim, D. K. Lee, B. G. Jo, J. H. Jeong, and Y. S. Kang, "Synthesis of oleate capped Cu nanoparticles by thermal decomposition," *Colloids Surfaces A Physicochem. Eng. Asp.*, vol. 284–285, pp. 364–368, 2006.
[17] A. A. Ponce and K. J. Klabunde, "Chemical and catalytic activity of copper nanoparticles prepared via metal vapor synthesis," *J. Mol. Catal. A Chem.*, vol. 225, no. 1, pp. 1–6, 2005.
[18] C. Yeh, "Formation and Characteristics of Cu Colloids from CuO Powder by Laser Irradiation in," pp. 6851–6857, 2000.
[19] V. Zin, B. G. Pollet, and M. Dabalà, "Sonoelectrochemical (20 kHz) production of platinum nanoparticles from aqueous solutions," *Electrochim. Acta*, vol. 54, no. 28, pp. 7201–7206, 2009.
[20] S. J. Doicrycz and K. S. Suslick, "Interparticle Collisions Driven by Ultrasound," vol. 2, no. 67, pp. 1987–1989, 1987.
[21] K. S. Suslick, "The Chemical Effects of Ultrasound," *Sci. Am.*, pp. 80–86.
[22] A. Gedanken, "Using sonochemistry for the fabrication of nanomaterials," vol. 11, pp. 47–55, 2004.
[23] B. A. Wills, J. A. Finch, B. A. Wills, and J. A. Finch, "Chapter 4 – Particle Size Analysis," *Wills' Miner. Process. Technol.*, pp. 91–107, 2016.
[24] V. Sáez and T. J. Mason, "Sonoelectrochemical synthesis of nanoparticles," *Molecules*, vol. 14, no. 10, pp. 4284–4299, 2009.
[25] Ratoarinoro, F. Contamine, A. M. Wilhelm, J. Berlan, and H. Delmas, "Power measurement in sonochemistry," *Ultrason. Sonochem.*, vol. 2, no. 1, pp. 43–47, 1995.
[26] E. Yustanti, M. A. E. Hafizah, and A. Manaf, "Synthesis of strontium substituted barium titanate nanoparticles by mechanical alloying and high power ultrasonication destruction," *AIP Conf. Proc.*, vol. 1725, no. 2016, pp. 0–6, 2016.
[27] M. A. Alavi and A. Morsali, "Syntheses and characterization of Mg(OH)2 and MgO nanostructures by ultrasonic method," *Ultrason. Sonochem.*, vol. 17, no. 2, pp. 441–446, 2010.

IOP Conf. Series: Materials Science and Engineering **858** (2020) 012046 doi:10.1088/1757-899X/858/1/012046

Anti fouling coating characterization with purified Zirconia from West Kalimantan Zircon Sand for Boiler application

Yulinda Lestari[1], Anne Zulfia[2, a], Radhian Krisnaputra[3], Nareswari Novita Satiti[3], Eko Sulistiyono[4] and Efendi Mabruri[4]

[1]PhD Candidate at Metallurgy and Material Engineering Department, University of Indonesia
[2]Metallurgy and Material Engineering Department, University of Indonesia
[3]Vocational school of mechanical engineering department, Gajahmada University
[4]Research Center for Metallurgy and Material, Indonesia Institute of Sciences

[a] E-mail: anne@metal.ui.ac.id

Abstract. Zircon sand is a mineral material for zirconia purification which can be found in Indonesia. Zirconia has various functions, one of the function applied at Boiler high temperatures. The low coefficient of zircon thermal expansion indicates high thermal shock and spalling resistance. In this study, zirconia comes from Kalimantan zircon sand was applied to an anti-fouling coating. The purified zirconia is compared with zircon sand and synthetic zirconia. The purified zirconia yields a 82.48% Zr content, compared to the zircon sand has 70.85% Zr content and 89.43% in synthetic zirconia. The substrate used for coating is ferritic steel. The coating method uses a brush coating then sintered at a temperature of 400°C and 800°C. Coatings that contained purified zirconia have a rougher surface compared to coatings that contained zircon sand and synthetic zirconia. It is because the zirconia powder granules produced have a larger size. The microstructure also shows the surface of the coating with zirconia from the purification has a tenuous structure with large granules. It shows further treatment is needed in the purification of zirconia so that the grain size is homogeneous when applied to the coating.

1. Introduction

The choice of anti fouling coating material is related to thermal conductivity which is inversely proportional to fouling resistance. If the fouling resistance is high then the thermal conductivity of the material is not too high but enough to conduct heat. Based on this, zirconia (ZrO_2) is selected as matrix and graphite lubricant is selected as filler in ceramic based composite coating material [1, 2]. Graphite has good thermal stability, the price is relatively cheap and easily combined with various matrices. In addition to the superiority of the characteristics, another reason for the use of zirconia (ZrO_2) in this study is to increase the added value of local mineral resources, zircon sand ($ZrSiO_4$) from West Kalimantan [3]. Zircon sand contains ZrO_2 and other metal oxides such as SiO_2, TiO_2 and Al_2O_3 which are commonly used as coating materials [4-6]. The zirconia is a mineral in Indonesia that has gone through a refining process. The low coefficient of zircon thermal expansion indicates high thermal shock resistance and resistance to spalling. Spalling is the deformation of a material due to heat shock [7]. A review of several anti-fouling coatings has been carried out by Banerjee, et al [8] but most are for applications in marine and the material is dominated by polymer coatings. While anti-fouling coatings on boilers and heat exchangers that operate at high temperatures are summarized by Gomes da Cruz, et al [9].

2. Experimental

2.1 Materials

The methodology stages in this study consisted of purification of West Kalimantan zircon sand, zirconia powder characterization resulting from purification and commercial zirconia powder, substrate coating with ceramic slurry and characterization of coated substrates.

2.2 Method

2.2.1. Purification of West Kalimantan Zircon Sand

ZrO_2 purification was carried out by zircon decomposition using an alkaline fusion method [3, 10]. The stages of the ZrO_2 purification process are zircon sand sifting, zircon decomposition, water leaching, acid leaching, precipitation, and calcination.

2.2.2 Characterization of Zirconia Powder

The zirconia powder characterization was done by XRF testing. This test is carried out to determine the percentage of elements (% wt) contained in ZrO_2 powder. Calculation of conversion of elemental percentages to percentage of oxide can be done to find out what levels of ZrO_2 and other oxides.

2.2.3. Substrate was coated with ceramic slurry

Ferritic steel as a substrate has the dimensions of 20x15x7 mm. Before the coating process, sand blasting is carried out on the surface of the substrate. Sand blasting aims to strengthen the bond between the steel surface and the coating [2]. Ceramic slurry is made with the composition listed in Table 1. The following composition refers to research conducted by Wang, et al. 2017. The mixing process was carried out with a magnetic stirrer for 30 minutes until evenly mixed, then ultrasonic milling was carried out on the mixture for 3 minutes as a deagglomeration process [11]. The mixture is processed further with an ultrasonic bath for 1 hour [12]. The ultrasonic bath process is followed by a mixture filtration process to minimize water content. Filtration is done to get a thicker mixture called slurry. Brush painting is the simplest method and is very possible to do when working conditions are needed where not much coating material is wasted [13]. Sintering of coated steel is carried out at 400°C and 800°C. The temperature selection follows the optimum temperature in a preliminary study conducted by Xiao et al. 2018 [14].

Table 1. Composition of Ceramic Slurry Coating [2]

Materials	%w
Zirconia Powder	12
Graphite	2
Potash Waterglass	18
SLS Powder	5
Aquades	Balanced

2.2.4. Characterization of Coated Substrates

SEM-EDAX testing was carried out to observe surface morphology coatings on steel surfaces. From EDAX results it can be seen what elements are present at a test point. The SEM-EDAX test results were used to analyze the effect of ZrO_2 powder types used as a mixture of ceramic slurry on the surface morphology coating that was formed, by comparing the test samples with and without sintering.

3. Results and Discussion

3.1. ZrO_2 Content

The XRF testing was carried out in 3 powder samples such as Kalimantan zircon sand, purified zirconia and commercial zirconia.

Table 2. XRF Test Result

Oxide	% Weight		
	Kalimantan Zircon Sand	**Purified Zirconia**	**Commercial Zirconia**
ZrO_2	59.212	67.986	76.887
SiO_2	29.755	7.923	1.661
HfO_2	1.251	1.387	1.584
TiO_2	1.143	0.5	0.058
Al_2O_3	1.403	2.813	0.253
Fe_2O_3	0.143	0.489	0.024
P_2O_5	5.906	16.583	18.758
MgO	0.3	0.473	0.397
Y_2O_3	0.377	0.310	0

Table 2 shows zircon oxide content and other impurities. Zirconia powder contains other metal oxides which are also often used as coating materials, such as SiO_2, TiO_2 and Al_2O_3. HfO_2 is an oxide material which has high oxidation resistance that is resistant to ablation and spallation.

3.2. Visual Image of Coated Substrates

The substrate has been coated with ceramic slurry using the brush painting method, then it dried in an oven at 70°C for 30 minutes. Figure 1 shows the visual images of coated substrates.

(a) (b)

Figure 1. Visual Image of Coated Substrates. (a) Purified Zirconia, (b) Commercial Zirconia

From Figure 1, it shows visual differences in surface coating. Figure 1(b) has a smoother surface than Figure 1(a). In figure 1(a), the coating surface has powder particles which are aglomerate in several areas.

3.3. Comparison of Microstructures between Purified Zirconia and Commercial Zirconia Ceramic Coating

The microstructure of the surface coating was seen by SEM using a 500x magnification secondary electron imaging. The surface topography of the coating can be seen in the following image:

IOP Conf. Series: Materials Science and Engineering **858** (2020) 012046 doi:10.1088/1757-899X/858/1/012046

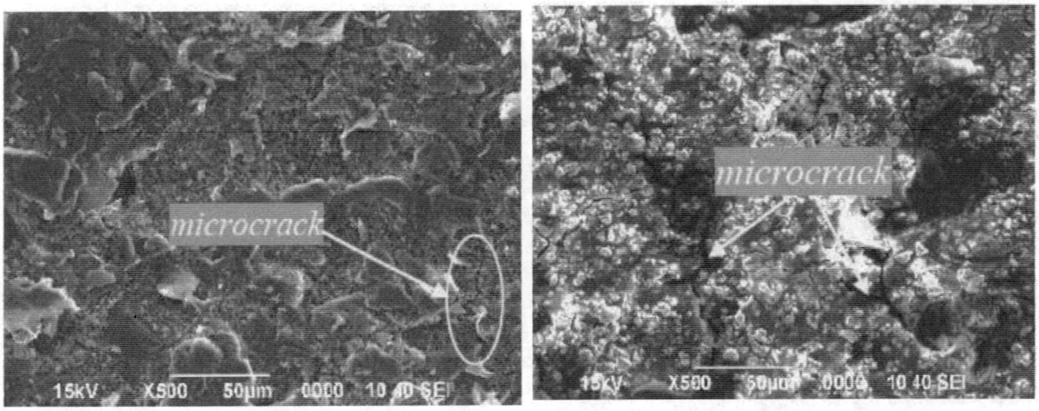

(a) (b)

Figure 2. Microstructure of Coating Surface. (a) Purified Zirconia, (b) Commercial Zirconia

Figure 2 shows difference morphology surface of coating samples using commercial zirconia and purified zirconia. The surface of the commercial zirconia sample coating looks smoother and even distribution of coating particles. There are no graphite sheets which accumulate on the surface of the coating. The presence of porosity and microcrack is seen on the coating surface in Figures 2 (a) and 2 (b). ZrO_2 on the coating surface of figure 2 (a) is seen with light colors in small amounts and not spread evenly. This is supported by EDAX data of the elements scattered on the following surface coatings:

Table 3. EDAX Test Result Not Sintered Coating

Element	Purified Zirconia Coating (% Mass)	Commercial Zirconia Coating (% Mass)
C	62.16	20.46
O	19.39	29.28
Na	1.12	1
Al	0.88	-
Si	12.51	17.85
K	0.75	0.75
Fe	2.55	-
Zr	0.64	30.66

From table 3, it can be seen that the purified zirconia coating is dominated by the element carbon (C) where carbon is the main constituent of graphite used as a lubricant in the ceramic slurry mixture. The oxygen element (O) represents the presence of various oxide bonds in zirconia powder samples used as a mixture of ceramic slurry. The level of Zr detected is so small that no distribution is seen on the surface of the coating. Whereas in commercial zirconia coating, zirconium element (Zr) became the most dominant constituent element with the largest mass percentage of 30.66%, followed by oxygen (O) with a mass percentage of 29.28%, carbon element (C) with mass percentage 20.46%. This supports the surface of the microstructure, that shows the very even distribution of ZrO_2 on the commercial zirconia coating when compared to the purified zirconia coating.

3.4. Effect of Sintering on the Microstructure Surface Coating

Sintering aims to eliminate volatiles in the ceramic layer at high temperatures. As porosity increases, there will be an increase in the contact area of the particles and their mechanical strength increases.

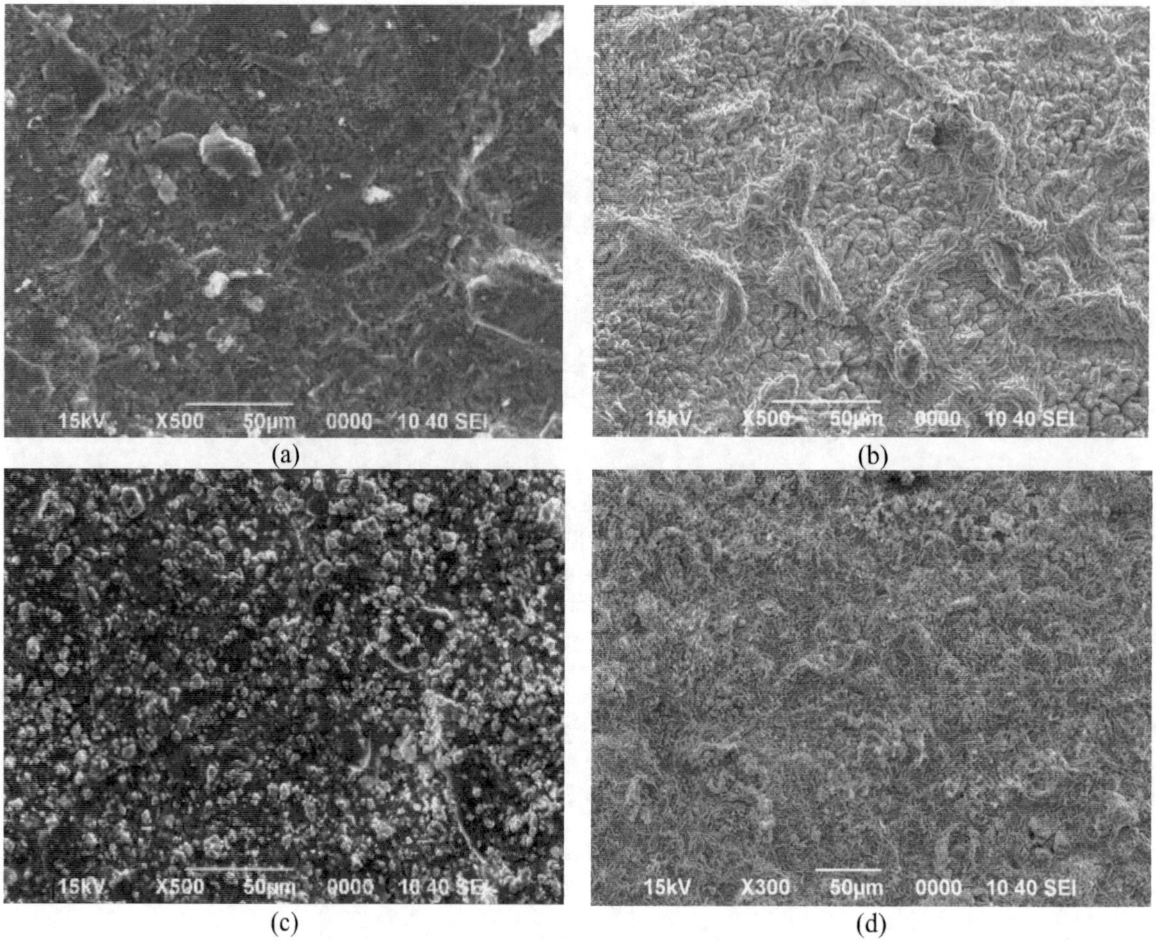

Figure 3. Microstructures of Sintered Coating Surfaces. (a) Purified Zirconia sintered at 400°C, (b) Purified Zirconia sintered at 800°C, (c) Commercial Zirconia sintered at 400°C, and (d) Commercial Zirconia sintered at 800°C.

Figure 3a, 3b, 3c, 3d shows microtrustures of sintered purified zirconia and sintered commercial zirconia surface coating. The microstructure show that the coating particles formed are more fused and compacted so that the sample surface becomes smoother and more even than zirconia coating without sintering. The sintering coating surface has an even distribution of coating particles, a denser surface, and invisible graphite sheets. The coating surface of commercial zirconia (Figure 3c and 3d) after sintering is denser and more even when compared to samples without sintering which can be seen from the size of the microcrack that becomes narrower. This is supported by research which states that sintering causes the combining of particles which is indicated by the expansion of the particle boundary area so that a continuous area is formed [15]. But at sintering temperature of 800°C (figure 3b and 3d), visible light-colored particles begin to fade and even disappear. This can be related to the EDAX results in table 4, that the elements of Carbon and Zircon are much reduced. The dominant element on the surface of the substrate coating is Fe, this shows that the coating is peeling off and damaged. When the coating layer is too thin and sintered at very high temperatures, cracking and volume shrinkage are possible during crystallization and phase transformation [16].

Table 4. EDAX Test Result Sintered Coating

Element	Purified Zirconia Coating Sintered at 400°C (% Mass)	Purified Zirconia Coating Sintered at 800°C (% Mass)	Commercial Zirconia Coating Sintered at 400°C (% Mass)	Commercial Zirconia Coating Sintered at 800°C (% Mass)
C	44.71	3.07	12.27	5.34
O	28.25	19.29	31.43	21.26
Na	1.56	8.51	2.64	11.36
Al	1.06	-	-	-
Si	20.2	1.12	16.42	0.85
K	1.23	-	0.98	-
Mn	-	-	-	2.37
Fe	2.14	68.01	2.22	57.77
Zr	0.86	Very low	34.04	1.04

4. Conclusion

The result of coating with purified zirconia has a morphology rougher than commercial zirconia. Commercial zirconia has a more even coating paticles and coating-forming elements. Commercial zirconia has a higher and more even zircon element when applied as a coating. This is because zirconia from purification has granular particles which are still coarse. So that purified Zirconia is produced from the alkaline fusion method still requires further treatment if it is to be applied as a coating. Whereas Sintering temperature of 400°C produces a finer micro-coating structure and closes the micro-crack formed on the coating that is not sintered. But Sintering temperature of 800°C damages the coating on the substrate so that a lot of peeling is indicated by the forming elements of the coating being slightly on the substrate.

Acknowledgement
The Authors would like to thank to PDD Funding of Kemenristekdikti who have funded this research.

References

[1] Bahaaideen F B, Ripin Z M and Ahmad Z A 2010 *Electroless Ni-P-Cg(graphite)-SiC Composite Coating and its Application onto Piston Rings of a Small Two Stroke Utility Engine* (Journal of Scientific & Industrial Research) 69, p 830-834

[2] Wang J, Yuan Y, Chi Z, and Zhang G 2017 *Development and application of anti-fouling ceramic coating for high-sodium coal-fired boilers* (Journal of the Energy Institute)

[3] Artisantri L 2018 *ZrO_2 Formation through NaOH Alkali Decomposition and HNO_3 Leaching* (Thesis, Sultan Ageng Tirtayasa University)

[4] Nguyen M D, Bang J W, Kim Y H, Bin A S, Hwang K H, Pham V H, and Kwon W T 2018 *Slurry Spray Coating of Carbon Steel for Use in Oxidizing and Humid Environments* (Ceramics International) 44 (7), p 8306-8313

[5] Oldani V, Del Negro R., Bianchi C L, Suriano R, Turri S, Pirola C, and Sacchi B 2015 *Surface properties and anti-fouling assessment of coatings obtained from perfluoropolyethers and ceramic oxides nanopowders deposited on stainless steel* (Journal of Fluorine Chemistry) 180, p 7-14

[6] Srinivasulu K and Sa MV 2016 *Advanced Ceramic Coatings on Stainless Steel: A review of Research, Methods, Materials, Applications and Opportunities* (International Journal of Advanced Engineering Technology) 7, p 141

[7] Ayala L 2015 *Technical handbook on zirconium and zirconium compounds* (Zircon Industry Association)

[8] Banerjee L, Pangule R C, and Kane R S 2011 *Antifouling coatings: recent developments in the design of surfaces that prevent fouling by proteins, bacteria, and marine organisms* (Advanced Materials) 23, p 690-718

[9] da Cruz L G, Ishiyama E, Boxler C, Augustin W, Scholl S, and Wilson D 2015 *Value pricing of surface coatings for mitigating heat exchanger fouling* (Food and Bioproducts Processing) 93, p 343-363

[10] Liu J, Song J, Qi T, Zhang C and Qu J 2016 *Controlling the formation of Na2ZrSiO5 in alkali fusion process for zirconium oxychloride production* (Advanced Powder Technology) 27(1), p 1-8

[11] Bittmann B, Haupert F, and Schlarb A K 2009 *Ultrasonic dispersion of inorganic nanoparticles in epoxy resin* (Ultrasonics Sonochemistry) 16, p 622-628

[12] Billotte C, Fotsing E R, and Ruiz E 2017 *Optimization of Alumina Slurry for Oxide-Oxide Ceramic Composites Manufactured by Injection Molding* (Advances in Material Science and Engineering, Hindawi Publishing Corporation)

[13] Ulaeto S B, Rajan R, Pancrecious J K, Rajan T P D and Pai B C 2017 *Developments in smart anticorrosive coatings with multifunctional characteristics* (Progress in Organic Coatings) 111, p 294-314

[14] Xiao K, Xue W, Li Z, Wang J, Xueming Li, Dong C, Wu J, Xiaogang Li, and Wei D 2018 Effect of Sintering Temperature on the Microstructure and Performance of a Ceramic Coating Obtained by the Slurry Method (Ceramics International) 44(10), p 11180-11186

[15] Nguyen P, Harding S, and Ho S Y 2007 Experimental studies on slurry based thermal barrier coatings (Eng. Aust) 1, p 545-550

[16] Dai Y, Yin Y, Xu X, Jin S, Li Y, and Harmuth H 2018 *Effect of the phase transformation on fracture behaviour of fused silica refractories* (Journal of the European Ceramic Society) 38(16), p 5601-5609

IOP Conf. Series: Materials Science and Engineering **858** (2020) 012047 doi:10.1088/1757-899X/858/1/012047

Dissolution of quartz sand in sodium hydroxide solution for producing amorphous precipitated silica

Eni Febriana[1], Uli Artha B Manurung[2], Agus Budi Prasetyo[1], Murni Handayani[1], Ersan Y Muslih[3], Firdaus Nugroho[3], Eko Sulistiyono[1], and F Firdiyono[1]

[1]Research Center for Metallurgy and Material – Indonesian Institute of Sciences, Serpong, Tangerang Selatan, Indonesia 16314
[2]Metallurgical Engineering Department, Faculty of Engineering, Sultan Ageng Tirtayasa University, Cilegon 42435, Indonesia
[3]PT Powertech Nano Industri, Jakarta, Indonesia

E-mail: enif001@lipi.go.id

Abstract. This research studied the dissolution of quartz sand in sodium hydroxide solution as an effort to form silica precipitate. The dissolution process was carried out using the leaching method directly at atmospheric pressure. The process variables in this study included the concentration of sodium hydroxide solution, temperature, dissolution time, and stirring rate. The four process variables had a positive influence to enhance silica precipitate formation, The result showed that the extraction rate was still low even though the SiO_2 concentration in the starting material was very high. Temperature provided the most significant influence among other process variables. The result exhibited that silica precipitate content was 13. 6% obtained in the optimum condition at a temperature of 90 °C with a stirring speed of 800 rpm using a 7.5 M of sodium hydroxide solution.

1. Introduction

Silica is an important oxide which is abundant, about 60% of the earth's crust layer contains silica both as silica itself and in the form of silicates with metal oxides. This material is commonly found as beach sand or river banks, desert sand, silica rock, and other minerals [1]. Silica is widely used as a raw material for the glass, ceramic, cement, sandblasting industries, as well as supporting materials for the metal casting, oil and mining industries, and refractory bricks [2]. Along with the development of technology, silica, especially in the form of amorphous and furthermore the size of the nanocrystal, can be applied to the polymer composite amplifier industry, electronic devices, bitumen, insulators, catalysts, enzyme encapsulation, batteries, etc[3].

In general, silica processing can be carried out using an alkaline fusion method at a temperature over 1000 °C with the addition of sodium carbonate through the roasting stage to form sodium silicate, dissolving sodium silicate into the water, silica deposition using acid[3,4]. Other sources of silica can be obtained from biomass materials such as rice husks or baggage. The processing method for this biomass material, in general, is direct dissolution using base [5] or acid solutions [6].

The processing of silica from non-biomass material using a base dilution method has been carried out by several researchers before. Riyas et al. reported that 80% of pure precipitated silica can be obtained by reacting desert sand and NaOH at 150 °C for 45 minutes in an open system [4]. Conversion obtained is still quite low considering the NaOH used as many as 2 stoichiometry based on the weight of quartz sand. Munasir et al. comparing the results of SiO_2 nanopowder obtained using the 5-7 M NaOH dissolution method directly at 90 °C for 2 hours, with the base roasting method using NaOH at a temperature of 500 °C [2]. The study did not explain the percentage of extraction from each method used.

This research will study the extraction of silica from quartz sand using the dissolution method with sodium hydroxide. Factors that influence the formation of sodium silicate will be analyzed based on the amount of quartz sand residue after dissolution.

2. Experimental

The raw material used as a source of silica for this research was quartz sand from Sukabumi Indonesia. Other ingredients were NaOH and HCl 37% from Merck, and deionized water. Figure 1 shows a flow diagram of the experimental process carried out. Quartz sand was first dried in an oven at a temperature of 110 °C for 8 hours, crushed with a disk mill, then sieved until a particle size of -200 mesh is obtained. (<74 μm). Quartz sand was then analyzed using XRF and XRD to determine the chemical composition and crystal structure respectively. The dissolution process was carried out using a sodium hydroxide solution in a glass beaker equipped with a magnetic stirrer, a thermometer to determine the temperature of the process, and given a plastic wrapping cover to prevent the solvent evaporation. The heat source during the dissolution process was obtained from the hot plate.

A certain amount of sodium hydroxide was dissolved in aquademin. The solution was then heated to a certain temperature. After the process temperature had been reached, a certain amount of quartz sand was put into the hot sodium hydroxide solution and the process temperature was kept stable at a process condition determined by ±5 °C intolerance.

After the dissolution process was carried out over a while, the solution was filtered to separate the insoluble material. The residue obtained was washed several times then dried in an oven at 110 °C for 4 hours and weighed to determine the extraction of silica from quartz sand. The filtrate was stored in a volume flask with a certain volume and then precipitated as precipitated silica by titrating sodium silicate solution using concentrated HCl dropwise while stirring with a magnetic stirrer to reach pH = 3. The precipitate formed was filtered, washed repeatedly until the pH of the washing water was near the neutral condition, then dried in an oven at 110 °C for 24 hours and weighed. The precipitated silica products formed were analyzed using XRF and XRD as well as the quartz sand raw material. XRD analysis (X-ray diffractometer) using Shimadzu XRD 7000 Maxima-X with Cu-kα radiation in the range of 2Θ 10-80 °, while the elemental composition was analyzed using XRF (X-ray fluorescence) S2 Puma Bruker.

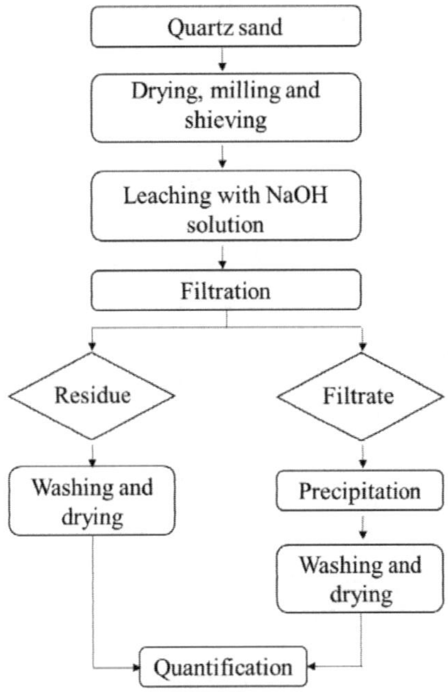

Figure 1. Flowchart of the experiment

3. Results and Discussions

3.1. Characterization of quartz sand

The chemical composition based on the results of the XRF analysis is shown in Table 1. SiO_2 is the dominant oxide component in quartz sand samples. Most impurities in the form of Al_2O_3 are only 4.69%, and other impurities such as MgO, K_2O, Fe_2O_3 Cr_2O_3, CaO, P_2O_5, and TiO_2 are in small amounts, each under 1% with a total of about 1.6%. The XRD analysis results of the quartz sand samples used in this study are shown in graph Figure 2. The diffraction pattern in Figure 2 indicates that the quartz sand samples are dominated by the crystalline quartz phase with the highest peaks at 2ϴ of 26.6, 20.8, and 59.9° (3.34, 4.26, and 1.54Å) based on PDF2 data 46-1045, according to previous research by Trabelsi et al.[7] dan Sdiri et al.[8].

Table 1. XRF analysis of quartz sand

Component	Cl	MgO	Al_2O_3	SiO_2	P_2O_5	SO_3	K_2O	TiO_2	MnO	Fe_2O_3	Cr_2O_3	CaO
Raw Quart Sand	0.025	0.27	4.69	93.7	0.011	0.042	0.41	0.091	0.011	0.64	0.067	0.031

IOP Conf. Series: Materials Science and Engineering **858** (2020) 012047 doi:10.1088/1757-899X/858/1/012047

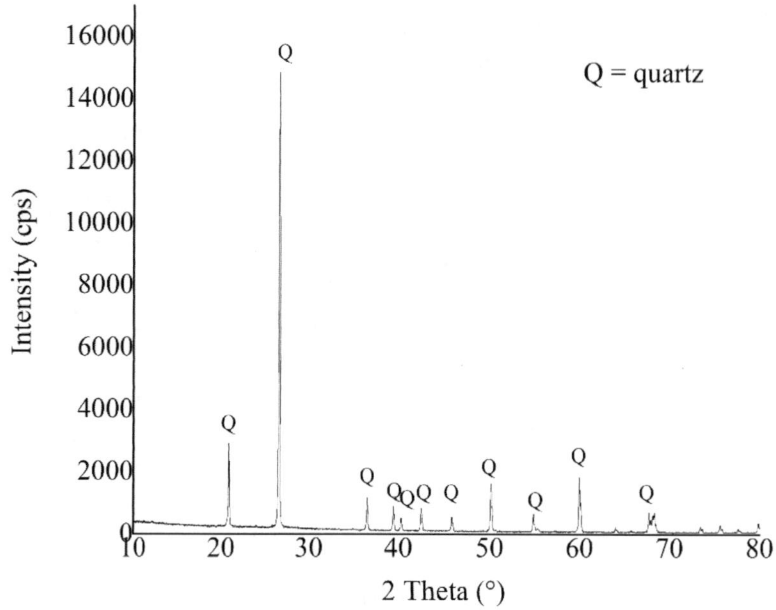

Figure 2. X-ray diffraction pattern of quartz sand

3.2. Effect of sodium hydroxide concentration

Figure 3 shows the effect of different concentrations of sodium hydroxide solution (6.25 - 7.5 M) on the dissolution of quartz sand to sodium silicate. The dissolution process was carried out at a temperature of 90 °C for 3 hours with a stirring speed of 300 rpm and a solid : liquid ratio of 2 : 10 (wt/v). From the experimental results, it can be seen that the concentration of sodium hydroxide solution has a positive effect on silica solubility. The higher the concentration of sodium hydroxide solution, the higher the silica dissolved as sodium silicate [9]. A sodium hydroxide concentration of 6.25 M is equivalent to 1 stoichiometry for the calculation of reactions on a SiO_2 basis in the quartz sand sample used. But this amount has not been able to dissolve the silica contained in quartz sand even when the concentration of sodium hydroxide solution used exceeds the stoichiometric requirements. The process of dissolving silica is strongly influenced by the bond configuration between Si and O. Therefore it is very different from amorphous silica. Quartz, the most stable silica phase, is very difficult to dissolve in sodium hydroxide solution [9]. The stability of quartz sand can be shown in diffractogram Figure 2 where quartz has a very high intensity with a sharp peak [10]. Peak width due to crystallite size varies inversely with crystallite size, as the crystallite size gets smaller, the peak gets broader. The highest extraction rate that can be obtained from this dissolution process is 10.9%.

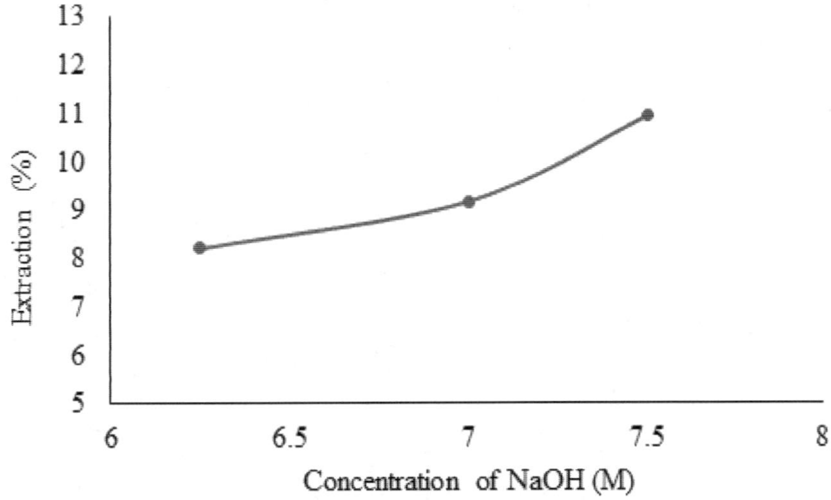

Figure 3. Effect of sodium hydroxide concentration

3.3. Effect of dissolution temperature

In general, the temperature is very influential in a chemical reaction. To find out the level of dissolving silica in a sodium hydroxide solution, the reaction temperature is varied from 70 to 90 °C. The dissolution process was carried out for 3 hours with a stirring speed of 300 rpm and a solid: liquid ratio of 2:10 wt/v. The results are as seen in Figure 4. The increase in temperature gives a significant influence on the level of silica solubility. The highest extraction rate of 11% was obtained in the dissolution process with a temperature of 90 °C. The heating of the reaction aims to increase the occurrence of collisions between molecules in the solution so that it can accelerate the dissolution process. The higher the temperature, the viscosity of sodium hydroxide solution decreases. This facilitates mass transfer between quartz sand and its solvent material thereby accelerating the silica dissolution reaction [11].

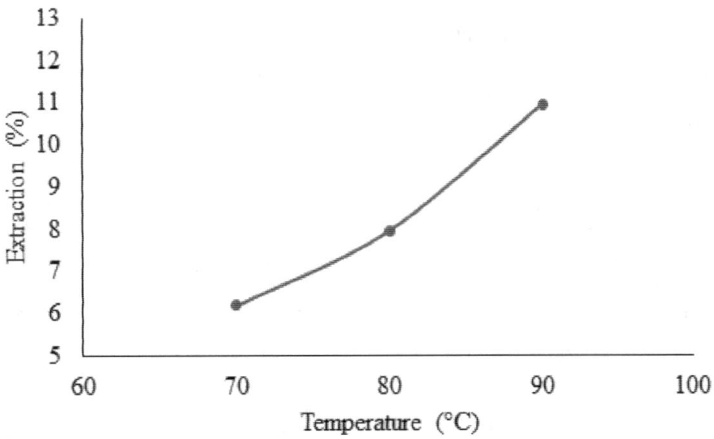

Figure 4. Effect of dissolution temperature

3.4. Effect of dissolution time

The effect of quartz sand dissolution time on an atmospheric dissolution system was carried out using a 7.5 M sodium hydroxide solution. The dissolution time varies from 1 to 5 hours. The dissolution temperature is kept constant at 90 °C, the stirring speed is 300 rpm, the solid : liquid ratio is 2 : 10 wt/v. From the experimental results shown in Figure 5, it is known that the dissolution time also has a positive effect on silica solubility [9]. The longer the dissolution time, the more silica is dissolved into sodium silicate. Dissolving time up to 5 hours can dissolve as much silica as 12.2%.

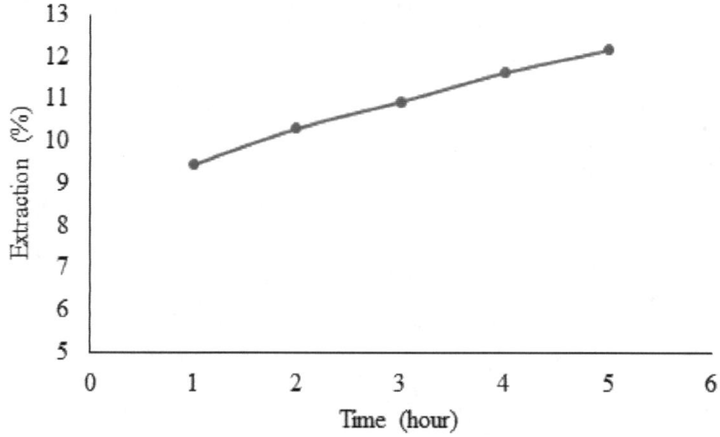

Figure 5. Effect of dissolution time

3.5. *Effect of stirring speed*

Stirring speed in a chemical reaction process, in general, will be able to increase the reaction rate. In addition to accelerating contact between the reacting substances and reducing deposition, this stirring can also help distribute temperature during the extraction process.

In this reaction of dissolving quartz sand with a 7.5 M sodium hydroxide solution, the stirring speed is varied from 300 to 800 rpm. Other operating conditions are temperature 90 °C, solid:liquid ratio 2:10 wt/v. As we can see from Figure 6, the stirring speed can increase the percentage of silica extraction even though increasing the extraction rate is not so significant with higher stirring speed. The possibility that occurred in the dissolution reaction of quartz sand with sodium hydroxide solution in this study is that the reaction is controlled by chemical control so that the stirring speed does not significantly influence the percent of silica extraction [12,13]. However, stirring speeds up to 800 rpm is capable of providing a percentage of silica extraction of 14%. This is quite different from the results of the study of Mufakhir et al in the process of leaching silica from ferronickel slag material which states the opposite result, that the rate of silica dissolution decreases with increasing stirring speed due to reaction controlled by diffusion through the product layer [14].

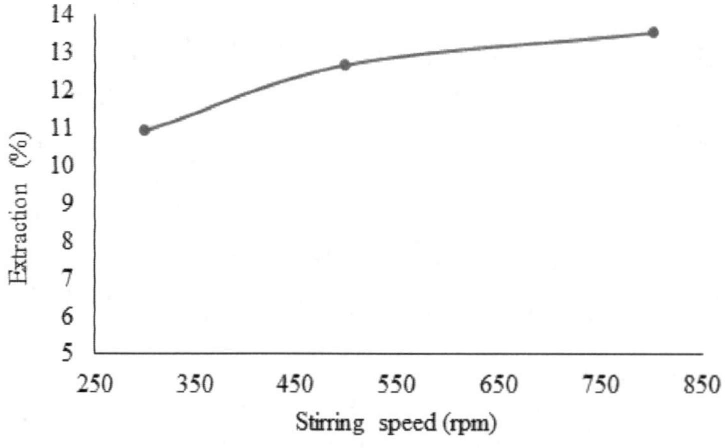

Figure 6. Effect of stirring speed

3.6. Characterization of silica precipitate

XRF analysis of precipitate silica products obtained from the deposition process in the form of white powder with SiO_2 content of 96.4% with Al_2O_3 main impurities of 2.56%. As shown in Figure 7, XRD diffractogram of a precipitated silica sample, the broadening of the peak centered at a 2Θ angle between 20-30° indicates that formed precipitated silica has an amorphous phase [15][16][17], and no crystalline phase was detected indicating that the washing process was effective in releasing precipitated silica from sodium chloride [16].

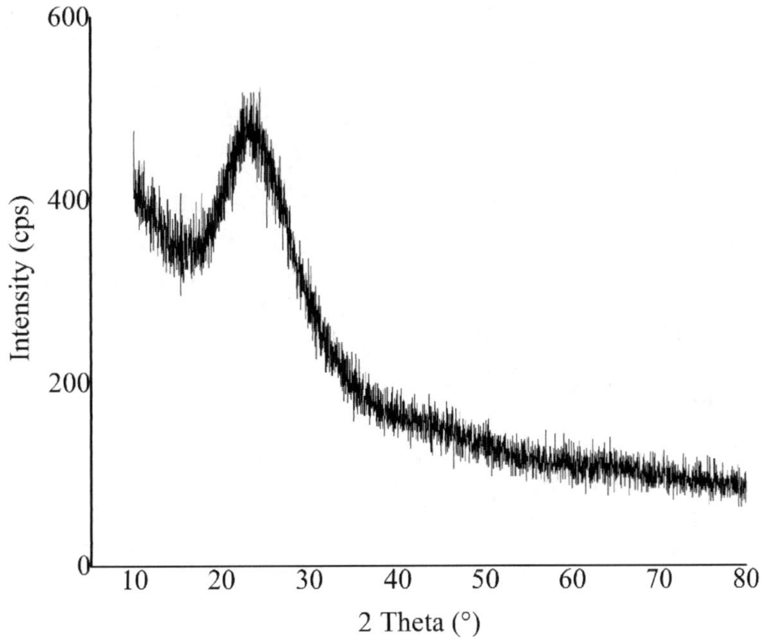

Figure 7. Diffractogram of silica precipitate

4. Conclusion

From the overall operating condition factors studied in the process of dissolving quartz sand in sodium hydroxide solution, it can be concluded that temperature, time, concentration of sodium hydroxide, and stirring speed have a positive influence on the level of silica solubility of quartz sand. The process of dissolving silica from quartz sand with sodium hydroxide solution is quite difficult to do under atmospheric conditions due to the stability of the SiO_2 phase in the quartz sand. Extra time may be needed, but the process will be ineffective. The highest extraction rate was obtained under the operating conditions of dissolving quartz sand with a 7.5M sodium hydroxide solution at 90 °C for 3 hours and a stirring speed of 800 rpm. From the dissolution process can be obtained amorphous precipitate silica with purity ±96%.

Acknowledgment
The authors would like to thank the Ministry of Research, Technology and Higher Education of the Republic of Indonesia through Program Pengembangan Teknologi Industri (PPTI) scheme for funding the research Pengembangan Nano Silika Presipitat Sebagai Bahan Baku Industri Konstruksi.

References
[1] Pradyot P 2002 *Handbook of inorganic chemicals*
[2] Munasir, Triwikantoro, Zainuri M and Darminto 2015 *Mater. Sci. Pol.* **33** 47–55

[3] Arunmetha S, Karthik A, Srither R, Vinoth M, Suriyaprabha R, Manivasakan P and Rajendran V 2015 *RSC-Advances* **5** 47390–7
[4] Sharafudeen R, Al-Hashim J M, Al-Harbi M O, Al-Ajwad A I and Al-Waheed A A 2017 *Silicon* **9** 917–22
[5] Huang G, Liu J, Wang L and Song Z 2016 *Int. J. Miner. Process.* **148** 116–23
[6] Liou T H and Yang C C 2011 S *Mater. Sci. Eng. B Solid-State Mater. Adv. Technol.* **176** 521–9
[7] Trabelsi W, Benzina M and Bouaziz S 2009 *Phys. Procedia* **2** 1461–7
[8] Sdiri A, Higashi T, Bouaziz S and Benzina M 2014 *Arab. J. Chem.* **7** 486–93
[9] Srivastava K, Shringi N, Devra V and Rani A 2013 *Int. J. Innov. Res. Sci. Eng. Technol.* **2** 2936–42
[10] Sheng L, Zhang Y, Tang F and Liu S 2018 *Microporous Mesoporous Mater.* **257** 9–18
[11] Chen G, Wang J, Wang X, Zheng S L, Du H and Zhang Y 2013 *Hydrometallurgy* **139** 46–53
[12] Habashi F 1970 *Extractive Metallurgy Vol 1: General Principles*
[13] Oediyani S, Ariyanto U and Febriana E 2019 *IOP Conference Series: Materials Science and Engineering* vol 478
[14] Mufakhir F R, Mubarok M Z and Ichlas Z T 2018 *IOP Conference Series: Materials Science and Engineering* vol 285 (Institute of Physics Publishing)
[15] Mourhly A, Khachani M, El Hamidi A, Kacimi M, Halim M and Arsalane S 2015 *Nanomater. Nanotechnol.* **5** 1–7
[16] Ayegba C O, Makinde T, Obigwa P and Orijajogun J 2015 *Int. J. Mater. Chem.* **5** 123–6
[17] Sarikaya M, Depci T, Aydogmus R, Yucel A and Kizilkaya N 2016 *IOP Conf. Ser. Earth Environ. Sci.* **44**

IOP Conf. Series: Materials Science and Engineering **858** (2020) 012048 doi:10.1088/1757-899X/858/1/012048

Identification of Quartz Sand From the Hills of Gunung Walat at Sukabumi Regency for Raw Materials of Nano Silica Precipitate

E Sulistiyono[1], M Handayani[1], A B Prasetyo[1] , Y Irawan[1], E Febriana[1] , S N Sembiring[2] , E Yustanti[2]

[1] Research Center of Metallurgy and Materials, Indonesian Institute of Sciences

[2] Department of Metalurgy and Material, Faculty of Engineering, Sultan Ageng Tirtayasa University, Cilegon

E-mail: eko221068@gmail.com

Abstract. Gunung Walat Hills Formation, Sukabumi Regency is an area that has abundant natural quartz sand resource potential. Quartz sand from the Gunung Walat hills is partly in the form of Quartz Wacke which is composed of monochrystaline and polychristaline quartz minerals associated with clay minerals so it has high Al_2O_3 and Fe_2O_3 impurities. Until now the use of quartz sand from the hills of Gunung walat has only been used as building material so that it has a low sale value. In this paper, we will examine the potential of quartz sand from Gunung walat hills for the raw material of nano silica precipitates so that it can increase the added value of natural resources of quartz sand. In this paper the results of identification are described by taking samples from unprocessed quartz sand, the results of washing by mining companies and dissolving with sulfuric acid. From the results of the X-Ray Fluorescent analysis it is seen that quartz sand from mining has a silica composition of 85.869% wt, water leaching results by the mining company 93.702% and sulfuric acid leaching results obtained levels of 95.672% wt. The results of SEM-EDX analysis show that impurities Al_2O_3 and Fe_2O_3 are in the mineral sludge, not in one complex compound bond on the same grain so it is relatively easier to separate. From the identification results it can be seen that quartz sand from the Gunung Walat hills can be used as a raw material for making nano silica precipitates.

1. Introduction

One of the main producers of quartz sand from Sukabumi comes from the Gunung Walat hill formation in the western area of Sukabumi City. Its location adjacent to the highway and business center makes quartz sand from the hills of Mount Walat much mined. In general, the mining of quartz sand from the hills of Gunung Walat is used as a raw material for the building construction industry so that the added value of quartz sand is very low. Based on the results of a report from the Department of Mining and the Environment Sukabumi Regency in 2003 to 2014, it is recorded that the production of quartz sand reached 100,000 tons to 250,000 tons [1]. Quartz sand reserves in Sukabumi Regency reach 2.990 billion tons, is the third largest mineral reserves after mountain rocks and limestone [1].

The silica content of quartz sand from Sukabumi Regency is quite low below 90%, for example the silica sand content in Cibadak is around 89%, Ciemas is around 49.8% and Cisaat is around 93.33% [1]. Quartz sand that is in the mountains of walat is mostly of Quartz Sand Quartz Wacke which is composed of monochrystaline and polychristaline quartz minerals associated with clay minerals so it has high Al_2O_3 and Fe_2O_3 impurities [2]. Alumina impurity is derived from hallosite minerals and brasal iron oxide impurities from geotite minerals [2]. Quartz sand from the hills of Gunung Walat is composed of alternations between sandstone and clay stone in the form of brownish white solids, clastic texture, lamination and normal gradation [2]. Quartz sand that is in the mining area is in the form of lumps like rock but is easily crushed into coarse sand grains, this can be seen from the picture below [2]:

Figure 1. Anita Yulianti et al survey results of the Gunung Walat quartz sandstone outcrop both natural and artificial outcrops [2].

Walat Formation in the southern area of Sukabumi Regency is generally a distinction between sandstone, conglomerate and clay stone containing a small amount of coal [3]. The Walat Formation is composed of paleogene-aged or late Eocene Early Oligocene sedimentary rock, equivalent to the Bayah formation of Oligocene age [3].

Based on the above data, in order to increase the added value of quartz sand from Gunung Walat, a washing process is carried out to separate it from the impurities [2]. Removal of impurities such as iron in quartz sand can increase the value added by tens of times. This is due to an increase in the function of quartz sand for example quartz sand with Fe_2O_3 content of 0.2% for refractory and glass materials, Fe_2O_3 content of 0.02% for special glass materials, Fe_2O_3 content below 100 ppm for premium ceramics, Fe_2O_3 content below 10 ppm for optical glasses and Fe_2O_3 levels below 1 ppm for optical fibers [4].

To improve the quality of quartz sand through the removal of impurities, many efforts have been made, among others through the process of washing with water which is assisted with a magnetic separator, flotation and ultrasonic cleaner [5-7]. Washing process with water is the cheapest, simplest and most environmentally friendly, because it does not use chemicals. However the washing process with water cannot increase high purity, in general it is only able to eliminate alumina and potassium compound compounds [7]. Washing process with water though less effective but able to increase the level of quartz sand from 96% to 97.75% in quartz sand obtained from mining results [8]. The process of washing water with the help of ultrasonic waves is able to take heavy metal impurities by adding sodium carbonate solution [9]. The process of washing with water does not produce a higher level of purity so it needs to be followed by an acid leaching process, such as sulfuric acid and hydrochloric acid. This acid leaching process is usually for the removal of impurities from the element groups of aluminum, iron, calcium, magnesium and other elements [10]. The results of the acid washing process cannot be directly used because of the new bond between the impurities and acidic materials such as sulfur and chloride, therefore the acid washing results must be washed immediately with water bersih [10]. The acid leaching process has evolved using oxalic acid which is able to take iron impurities by forming iron oxalate bonds so that iron impurities dissolve easily [8, 11, 12, 13]. The effectiveness of the use of oxalic acid is far better than the use of sulfuric acid and hydrochloric acid, where with oxalic acid obtained quartz sand with a high level of purity. The results of the washing process of oxalic acid in quartz sand from Mojosari with silica content of 95% can be increased to 97.79% [13]. Then with the help of ultrasonic waves, the washing process with oxalic acid can increase silica levels from 97.75% to 99.37% [8].

2. Experimental

2.1. Row Materials

The object of the study was quartz sand from the hills of Gunung Walat, Sukabumi Regency. In this activity, samples of quartz sand were taken from a quartz sand mining company that is under the hills of Gunung Walat. The samples come from two different parts, namely the samples in the reservoir that have not been washed with water and the samples that have been washed one time with river water and then rinsed again with clean water. The quartz sand mining company that we visited processes quartz sand mixed with clay with water to obtain clean quartz sand that is ready for sale for construction materials such as lightweight concrete and brick.

Figure 2. The location of sampling is the quartz sand mining company, which is the quartz sand pile from the mining and the location of the quartz sand washing process.

2.2. Characterizations

Quartz sand samples taken from the quartz sand mining company are then carried out the characterization process and the washing process with sulfuric acid and washing again with water to remove the acid in the quartz sand. The quartz sand obtained was divided into three samples, namely samples from mining products, samples from the first washing and the results of the second washing. Then the results of the second washing were washed with sulfuric acid with a concentration of 10 N and 5 hours with a ratio of dissolving 50 g per 100 mL solution. After the four samples were obtained the characterization process was carried out using X-Ray-Fluorescent (XRF), X-Ray-Diffraction and SEM-EDX.

3. Result and Discussion

3.1. Analysis of Silica Content with X-Ray Fluorescent (XRF)

Silica content analysis of the quartz sand from Gunung Walat hills was carried out using the X-Ray Fluorescent method. The results of the analysis conducted at the LIPI Metallurgical and Material Research Center are as follows:

Table 1. Silica content analysis of the quartz sand from Gunung Walat hills

Component	Natural Quartz Sand (% wt)	Water Dissolution Process I (% wt)	Water Dissolution Process II (% wt)	Sulfuric Acid Dissolution 10 N (% wt)
SiO_2	85,869	89,753	93,702	95,672
Al_2O_3	11,706	7,769	4,691	2,877
K_2O	1,068	0,625	0,414	0,369
Fe_2O_3	0,505	1,098	0,641	0,166
MgO	0,468	0,376	0,273	0,192
TiO_2	0,189	0,138	0,091	0,252
LOI	0,049	0,036	0,091	0,472

From the XRF test results, it appears that quartz sand from the hills of Gunung Walat, Sukabumi Regency has a low quality. It is indicated by silica levels below 90% and there are many impurities. Low levels of quartz sand due to the large number of impurities makes the selling value of quartz sand from the hills of Gunung Walat very low. By business actors, the quartz sand produced by mining is washed with water from the river in the first washing step and then with clean water in the second washing step. From the results of washing with water, it appears that an increase in silica levels is quite high from 85.869% by weight to 93.702% by weight in the second stage of washing. The impurity element that has decreased during the water washing process is Al_2O_3, while the Fe_2O_3 content has not decreased. This shows that most of the alumina is in the clay mineral group that is easily suspended in water. While Fe_2O_3 is a goethite mineral with has the same density as silica. In the dissolution process with 10 N of sulfuric acid, silica levels increases to 95.672%, a decrease in alumina content of 50% to 2.877% wt and a decrease in Fe_2O_3 levels by 80% to 0.166%. After washing with sulfuric acid, it is seen that quartz sand from the hills of Gunung Walat is suitable as a silica precipitate with a slightly reduced Al_2O_3 content to below 2% wt and Fe_2O_3 levels below 0.1% wt [14]. Therefore it is necessary to wash using other types of acids such as hydrochloric acid and oxalic acid or by using magnetic separator, flotation and ultrasonic waves.

3.2. Analysis result with XRD

From the results of tests with X-Ray Diffraction on quartz sand from the hills of Mount Walat shows that the peak formed is only the peak of the Quartz compound (SiO_2) in all the peaks obtained and other peaks do not exist. From the results of this analysis showed that there is no complex compound between SiO_2 and Al_2O_3 even though Al_2O_3 content are quite high at around 11% wt. This gives an indication that the impurities in quartz sand can be destroyed by physical processes such as washing, dissolving, flotation, magnetization and ultrasonic processes.

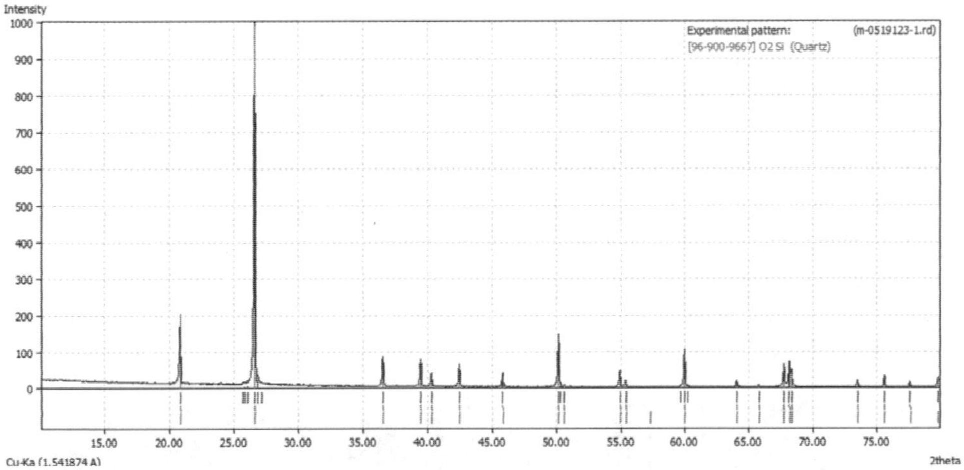

Figure 3. XRD analysis results on raw materials of quartz Sand from Gunung Walat Hills, Sukabumi Regency.

3.3. Analysis result with SEM-EDX

SEM analysis was performed to see the appearance of the granules and the location of the main elements of silica and the concentration of impurities present in the granules. Then the mapping analysis of the elements is taken on three grains that show different appearance. Images of grain appearance can be seen in the image below.

Figure 4. Results of SEM Mapping on Quartz Sand samples from the mining results of Gunung Walat hills, where items 1, 2 and 3 were tested by EDX

By looking at the shape of the grains, it is seen that quartz sand grains are a mixture of various types of rock even though the grains are dominated by quartz. In SEM images, there are three different forms of buitran, after an EDX analysis it is seen that each granule has a different elemental content. Granule (1) is dominated by zircon elements with 38% content, granule (2) is dominated by silica with 26% content and granule (3) is dominated by iron with 30% content. By looking at this result, it show that impurities do not bind fully with silica so that it is more easily separated physically.

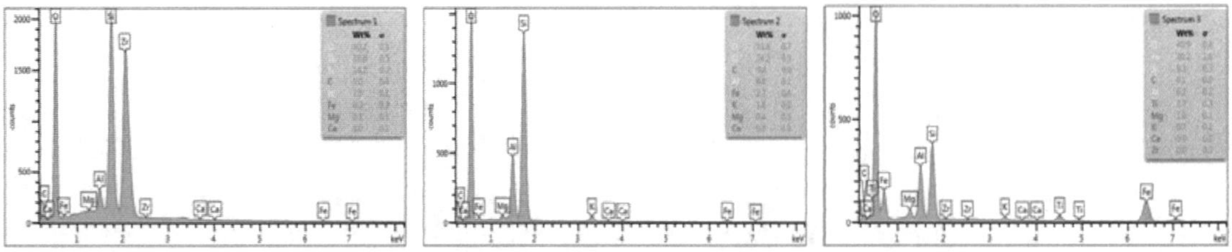

Figure 5. The results of EDX analysis of natural quartz sand granules

The results of SEM analysis of quartz sand that has been washed with water twice shows the same thing in granules as before washing process. Results of grain appearance can be seen in the picture below:

Figure 6. The results of the analysis of the appearance of grains on quartz sand that has been washed 2 times with water.

In grain (1) shows a mixture of silica and alumina with levels of 21.9% and 10.8%. Then in grain (2) shows a mixture of 30% iron and 18% silica. In item (3) it shows that silica is 34% and Carbon is about 6%. From these results it appears that impurities are mostly on smaller grains and silica is on grains with relatively larger sizes.

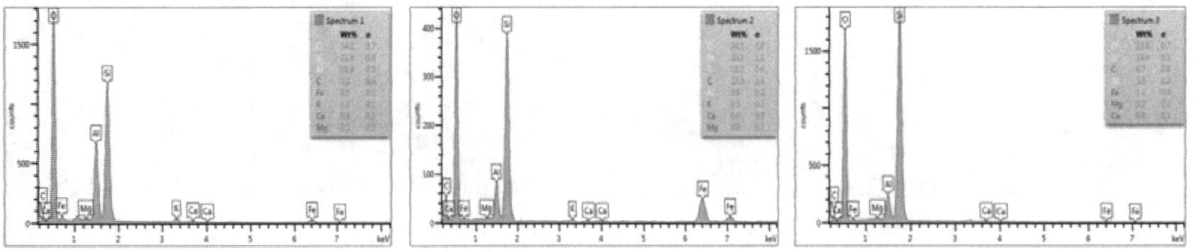

Figure 7. EDX analysis results per grain from the SEM image of quartz sand from the water washing process twice.

4. Conclusion

The results of the study indicate that quartz sand from the area has the potential as a raw material for the manufacture of nano silica precipitates. With a number of treatments, namely washing water twice and followed by washing using sulfuric acid, quartz sand has not been obtained that meets the requirements for raw materials of nano silica precipitate. However, quartz sand from the hills of Mount Walat has the potential to produce better quartz sand because there is no bond between impurities and silica that forms complex bonds. The requirement for raw material of nano silica precipitate is to be able to be processed into sodium silicate which has requirement of Al_2O_3 content under 1.5% and Fe_2O_3 content below 0.1%. Therefore it is necessary to do a better experiment using oxalic acid and using the help of physical processes such as flotation, magnetization and ultrasonic..

References

[1]. Anonim 2014 *Laporan Akhir Penyusunan Neraca Sumberdaya Mineral Kabupaten Sukabumi Neraca Sumberdaya Minera*l (Sukabumi : Dinas Pertambangan dan Lingkungan Hidup)
[2]. Yulianti A, Linjewas L, Erlangga B D and Herawan W 2016 *Prosiding Geotek Expo Puslit Geoteknologi LIPI*
[3]. Vos K, Vandenberghe N and Elsen J 2014 *Journal Earth Science Reviews* **128**(2014) 93-104
[4]. Cammas E, Panias D and Taxiarchou M 2001 *Proceeding Conference IX, Balkan Mineral Processing Congress 2001*, Volume IX, 289-295.
[5] Wahyuningsih S, Ramelan A H, Suharty N S, Handayani M, Firdiyono F, Sulistiyono E, Munawaroh H, Sari P P and Kristiawan Y R 2018 *IOP Conf. Ser: Mater. Sci. Eng.***333** 012050
[6] Wahyuningsih S, Suharty N S, Pramono E, Ramelan A H, Sasongko B, Dewi A O T, Hidayat R, Sulistyono E, Handayani M, and Firdiyono F 2018 *AIP conf. Proceedings* **1964**, 020021
[7]. Haghi H, Noaparast M, Tonkaboni S Z S and Mirmohammadi M 2016 *Journal Minerals* **6** 121
[8]. Sumarno, Novarita P, Januarty M and Yuniarti Y 2015 *Prosiding Teknik Kima Kejuangan* **G12** 1-8
[9]. Farmer A D, Collings A F and Jameson G J 2000 Journal Ultrasonics Sonochemistry **7**(2000) 243-247
[10]. Boussa S A, Kheloufi A, Zourar N B and Bouachma S 2016 *Journal Acta Physica Polonica* **132**(2017) 3-11
[11]. Veglio F, Passariello B and Abbruzzese C 1999 *Journal Industrial and Engineering Chemistry Research* **38**(11) 4443-4448
[12]. Taxiarchou M, Panias D, Douni I, Paspaliaris I and Kontopoulos A 1997 *Journal Hydrometalurgy* **46**(1997) 215-227.
[13]. Suratman 2015 Indonesian Mining journal **18**(3) 133-141
[14]. Aini S, Nizar U K, Amelia N S T and Efendi J 2018 *IOP Conf, Series: Materials Science and Engineering* **335**(2018) 012025

IOP Conf. Series: Materials Science and Engineering **858** (2020) 012049 doi:10.1088/1757-899X/858/1/012049

Pitting Corrosion Resistance of CA6NM and 410 Martensitic Stainless Steels in Various Environments

E Mabruri[1], H M Sigit[2], M S Anwar[1], M A Prasetyo[1], A Nikitasari[1], A De Fretes[2]
[1]Research Center for Metallurgy and Materials, Indonesian Institute of Sciences (LIPI), Tangerang Selatan, Indonesia
[2]Department of Mechanical Engineering, Atmajaya University, Tangerang Selatan, Indonesia

Email : effe004@lipi.go.id

Abstract. Pitting corrosion is an importance phenomena in power generation steam turbine, as it is one of the main causes for blade failure, particularly in the last row of low pressure blading. In this work, the pitting corrosion resistance of the widely used blading alloy of wrought 410 stainless steel and that of the candidate alloy of cast CA6NM stainless steel were investigated. Cyclic polarization measurements for all samples were conducted in simulated solution of thermal and geothermal power plants. The experimental results show that the pitting potentials of the cast CA6NM steel were higher compared to those of the wrought 410 steel in all simulated solution. Both steels exhibited higher pitting potentials in CO2 containing simulated geothermal water than those in the other solutions. These results were correlated with microstructure of the steels observed by Scanning Eectron Microscope (SEM).

1. Introduction

Pitting is one of the localized corrosion attacks on metals surface due to the passive film breaking in corrosive environments. It is an importance phenomena in power generation steam turbine, as it is one of the main causes for blade failure, particularly in the last row of low pressure blading [1-2]. Therefore, the pitting resistance of the steam turbine blade materials is expected to be high for ensuring longer life of the blade. The improvement of pitting resistance as well as mechanical properties of the blade materials has been continuously investigated by chemical composition modification or by optimizing heat treatment processes of the martensitic stainless steel type 403/410 as the standard alloys for steam turbine blades. The Ni and Mo were incorporated and optimized into the standard 403/410 stainless steel and the improved properties were exhibited by the modified alloys [3-6].

The CA6NM alloys is cast martensitic stainless steel originally containing Ni and Mo which is widely used as the blade materials for hydro turbine due to its high cavitation erosion resistance. There were no intensive application of this CA6NM steel for steam turbine blade, although with the high content of Ni (4%) and reasonable Mo (1%). This might be caused by the inferior mechanical properties combination of the cast product (CA6NM) compared to the wrought product (410). However, for the application of low pressure blading, particularly in geothermal steam turbine with lower temperature of steam compared to coal-fired steam turbine, the cast product with better pitting corrosion resistance is appropriate and relatively more economical choice. This work aimed to assess the pitting resistance of the cast CA6NM in the geothermal environments and in the chloride containing media and compared it to that of the 410 in similar environments. The pitting parameters were determined from the cyclic polarization experiments by using Gamry G750 Corrosion Measurement System (CMS).

2. Experimental

2.1. Materials

The steels for the experiments were cast CA6NM stainless steel prepared by induction melting furnace in the laboratory and the 410 stainless steel rod purchased from the market. The composition of the two steels are shown in the Table 1. Both steels were experienced by similar heat treatment process consisting of quenching and tempering before the cyclic polarization measurement. The steels were austenitized at temperature of 1050°C for 2h and finished by quenching in oil. Subsequent tempering was conducted at temperature of 650°C for 3h and finished by air cooling. The samples from heat treated steels were prepared for the hardness testing (Rockwell C), microstructural observation by optical microscope and cyclic polarization. The samples for microstructural observation were prepared by standard metallographic preparation consisting of mounting, grinding and polishing until 0,3 micron of alumina particle. The polished samples were etched with Kalling's reagent for revealing the microstructures. The samples for cyclic polarization were machined to the sizes of 1 x 1 cm^2 and attached with copper cable by soldering before they were mounted with resin. The surface of the mounted samples was then ground with SiC abrasive paper until size 1200.

Table 1. Chemical composition of the investigated steels (wt.%).

Steel	C	S	P	Mn	Si	Cr	Mo	Ni	Fe
CA6NM	0.04	0.008	0.02	0.40	0.35	12.83	1.02	4.17	Bal.
410	0.07	0.005	0.02	0.30	0.44	10.93	0.01	0.12	Bal.

2.2. Cyclic Polarization Test

The cyclic polarization test was conducted by Gamry G750 Corrosion Measurement System at room temperature in three kinds of solution separately. These solutions were the simulated geothermal solution without CO2, the simulated geothermal solution with CO2 and the NaCl solution. The composition of the simulated geothermal solution is displayed in the Table 2. Three kinds of electrode were used for the measurement, i.e. the working electrode of sample, the auxiliary electrode of graphite rod and the Saturated Calomel Electrode (SCE) as a reference electrode. The cyclic polarizations were carried out for a potential range of -600 to 1200 mV with forward scan rate of 5 mV/s and reverse scan rate of 2.5 mV/s. After the measurement, the samples were clean with water and dried. The pitting occurred in the samples were observed by Scanning Electron Microscope (SEM) imaging.

Table 2. Chemical composition the simulated geothermal solution.

Chemical Species	g/l
NaCl	224.60
NaSO4.10H2O	12.07
MgCl.6H2O	10.62
CaCl2.2H2O	6.45
KCl	0.39
KOH	0.32

NaHCO3	0.05

3. Results and Discussion

The mirostucture of CA6NM and 410 stainless steels in quenched-tempered condition are shown in Figure 1. Both steels exhibits dominantly martensite phase as a typical microstructure of the martensitic steels. The difference for the two martensite structures are in the size and fraction, where smaller size and larger fraction are observed in the CA6NM steel. The metal carbides exist in both steels with no distinc appearance in their sizes and distribution due to lack of resolution and magnification of optical microscope imaging. In addition, delta ferrite phase are absence in these two steels due to the absence of Mo (which acting as ferrite stabilizer) in 410 steels and due to the presence of high Ni (4%) in CA6NM steels (with 1% Mo) which suppressed the formation of delta ferrite.

Figure 2 show the hardness of the two investigated steels in non quenched-tempered (Non-QT) and in quenched-tempered conditions. The Non-QT condition for CA6NM steel refers to as cast condition, while for 410 steel refers to annealed condition. Therefore, for Non-QT condition the hardness of CA6NM steel is higher than that of 410 steel. This is due to the martensite phase exists in the typical as cast martensitic steels, while for annealed condition the microstructure consists of soft ferrite phase and globular carbides. In QT condition, the CA6NM steel consistently exhibits higher hardness compared to that of 410 steel. The microstructure of the steels in Figure 1 confirms the higher hardness of CA6NM steel due to larger fraction of martensite phase.

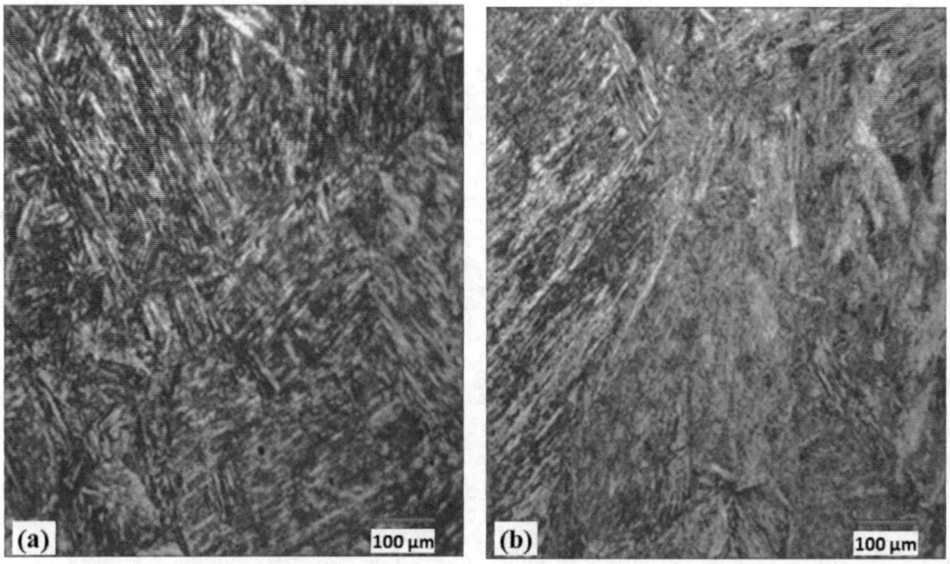

Figure 1. Optical microscope images of the microstructure of the (a) CA6NM (b) 410 stainless steels in quenched-tempered condition.

The pitting corrosion behaviour of the two martensitic stainless steels can be assessed from the cyclic polarisation measurements. The cyclic polarisation curves of the steels are shown in Figure 3. Both steels display hysteresis curves resulted from forward and reversed scans which describe the passive region of the steels in the respective solution. The pitting behaviour of the steels can be evaluated by determining the pitting potential, repassivation potential and how large the difference between them. The steels with higher pitting and repassivation potentials exhibit higher pitting resistance. Whereas, the steels with smaller potential differences (small hysteresis) will have lower susceptibility to pitting corosion. At the pitting potential the passive film breaks forming pit and the current density abruptly increases through the forward scan. In the reverse scan the current density decreases again and the repassivation of pit occurs when crossing the forward line. As shown in Figure 3 most of the curves do not exhibit this crossing potential indicating that the pits may continue to grow.

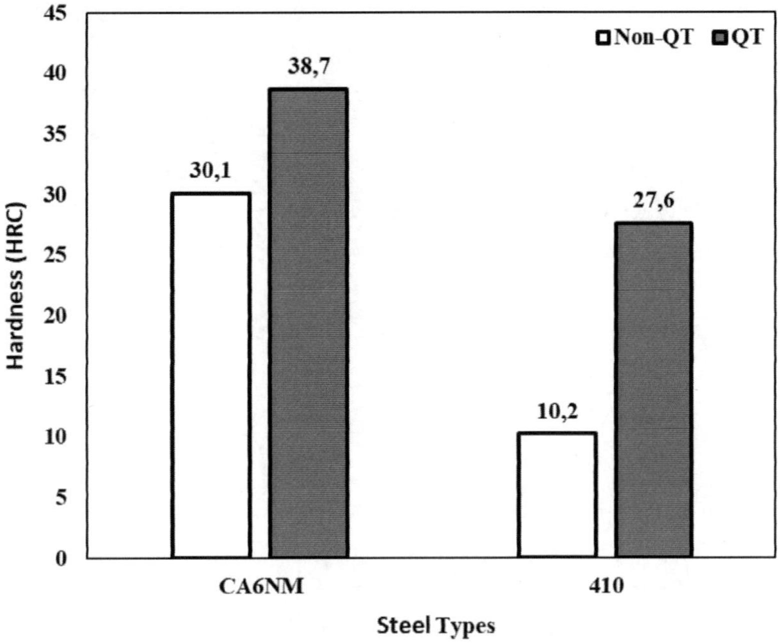

Figure 2. Hardness of the CA6NM and 410 stainless steels in non quenched-tempered (Non-QT) and in quenched-tempered (QT) conditions.

The pitting potentials of the steels in difference solution determined from the curves in Figure 3 are shown in Figure 4. The CA6NM steel has pitting potential higher than the 410 steel has in all solutions. This may be due to the higher content of Cr in the CA6NM steel compared to that in 410 steel, where Cr is the element responsible for passive film formation. On the other hand, the pitting potentials of both steels in CO_2 containing solution is higher than those in other solutions. The CO_2 bubblling into the solution might promote the formation of Na/Mg/Ca carbonates salts in simulated geothermal solution which inhibited the chloride ions attacks. S.A.M Rafae *et al* reported that the addition sodium carbonates increased the pitting resistance of the stainless steel by the inhibition effect of (Fe,Cr)CO3 film [7]. The lowest pitting potential is exhibited by the 410 steel in the simulated geothermal solution with Epit is -475.3 mV.

Figure 5 and Figure 6 show the SEM images of the surface of the respective stainless steels CA6NM and 410, after cyclic polarisation in the three simulated solutions. The number of pits in CA6NM is larger than that in 410, although the pitting potential of the CA6NM steel is higher. This can be explained by considering other factors such as the carbides content and the quality of the steels. The CA6NM steel with higher content of Mo may have larger content of carbides due to accceleration of carbides precipitation caused by Mo[8]. These carbides are potentially as an initial location for pitting corrosion [9]. With respect to the steels quality, the cast product tends to have inhomogeneity or chemical segregation higher than the wrought product have. In addition, it can be observed also that the pits in 3.5% NaCl are larger and deeper in sizes, particularly in 410 steel. This indicate that higher content of NaCl (in this case 3.5% NaCl is higher than 224.60 g/l in simulated geothermal solution) increases the corrosion severity of the solution. Furthermore, general corrosion are observed in sumulated geothermal solution, particularly in CA6NM steel. This may be due to higher hardness of the CA6NM steel compared to the 410 steel, considering that the hardness value has correlation with the corrosion resistance as reported by S. Weng *et al.*[10].

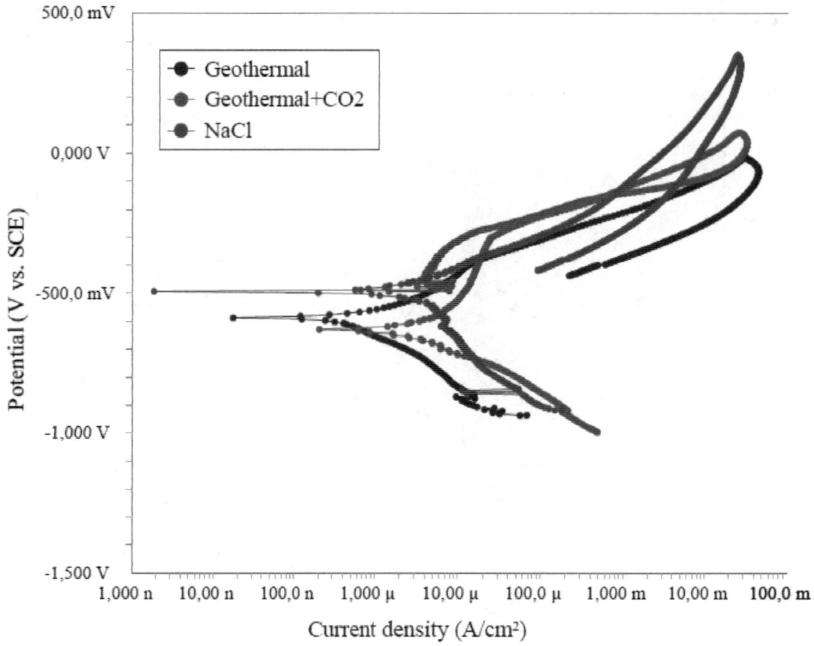

(a)

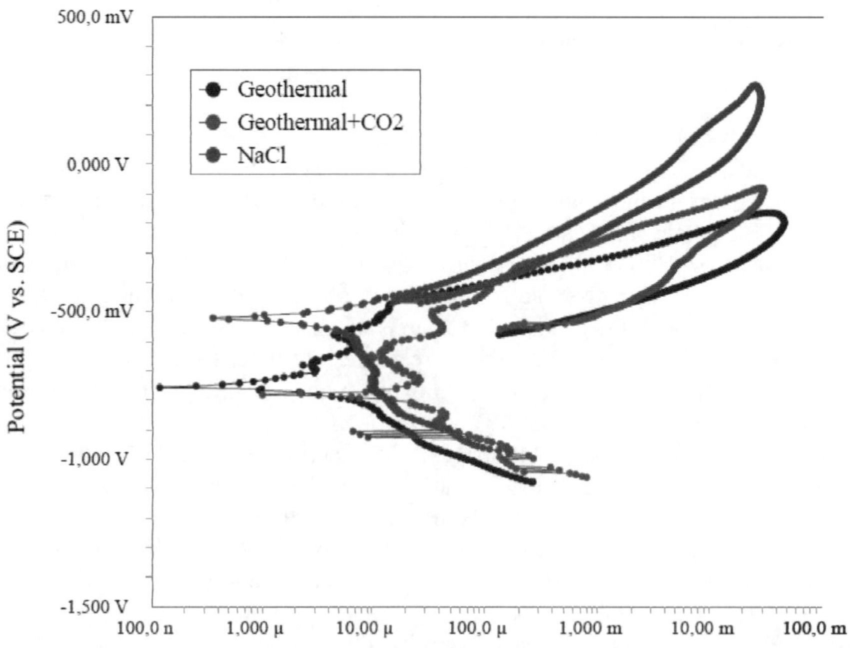

(b)

Figure 3. Cyclic polarization curves of the (a) CA6NM (b) 410 stainless steels in various simulated solutions.

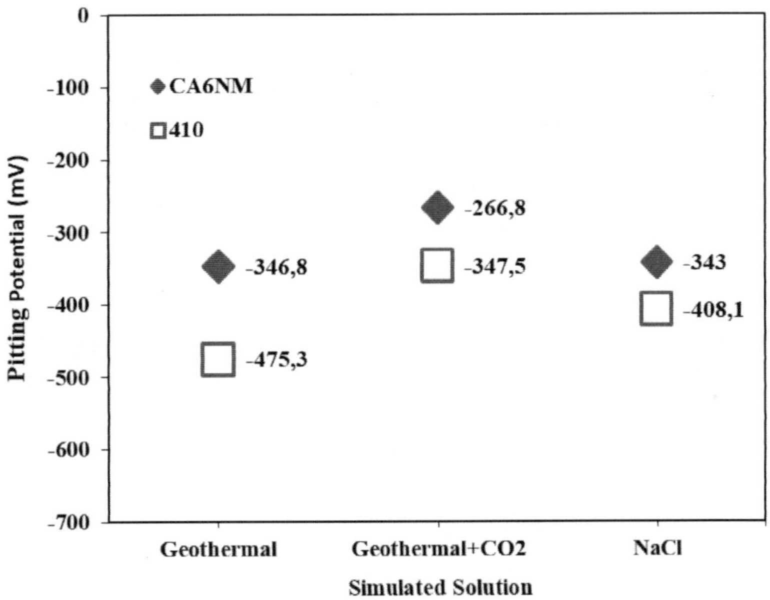

Figure 4. Pitting potential of the CA6NM and 410 stainless steels in various simulated solutions.

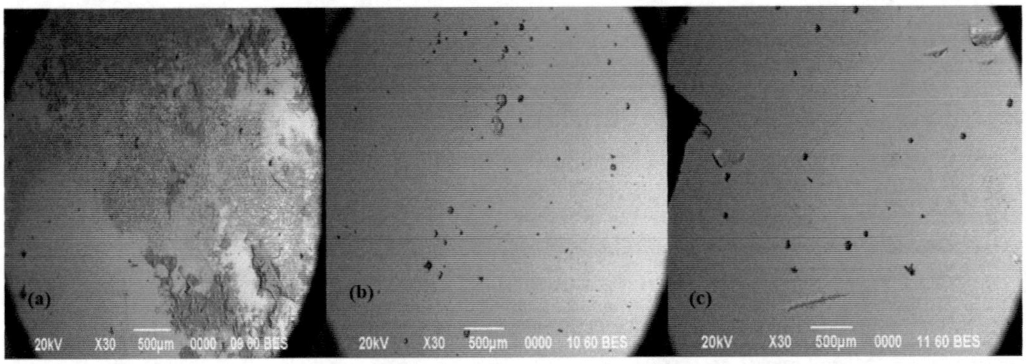

Figure 5. SEM images the surface of the CA6NM stainless steel after cyclic polarisationin in (a) geothermal (b) geothermal+CO2 and (c) NaCl simulated solutions.

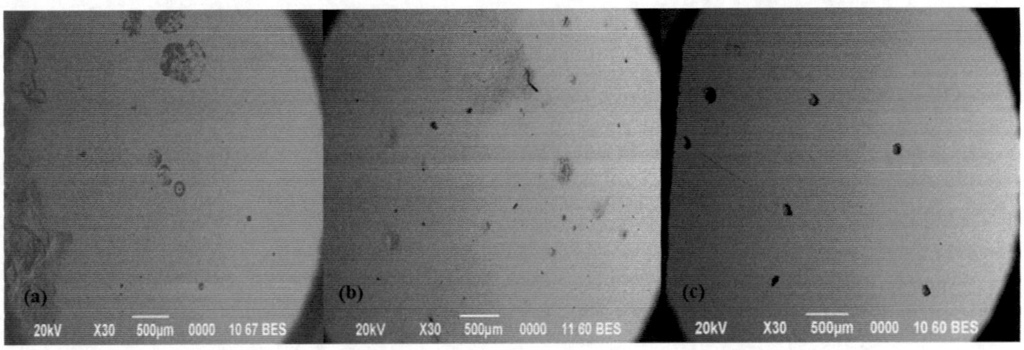

Figure 6. SEM images of the surface of the 410 stainless steel after cyclic polarisationin in (a) geothermal (b) geothermal+CO2 and (c) NaCl simulated solutions.

4. Conclusion

The pitting corrosion resistance of the cast CA6NM stainless steel in the geothermal environments and in the chloride containing media has been assessed and compared it to that of the 410 stainless steel in similar environments. Pitting potential of CA6NM steel is higher than that of 410 steel in simulated geothermal and in 3.5% NaCl solutions, however, the number of pits in CA6NM steel are larger than that in 410 steel according to SEM images observation. The CO_2 increases pitting potentials of the two steels in simulated geothermal solution. Furthermore, the pits of two steels in 3.5%NaCl are larger and deeper in sizes than those in sumulated geothermal solution.

Acknowledgement

This work is a part of the research on the development of high temperature materials for power plants. The authors thank Research Center for Metallurgy and Materials, Indonesian Institute of Sciences (LIPI) for financial supporting of this research.

References

[1] Harison M C A, Swamy M, Pavan A H V, Jayaraman G 2016 *Trans. Indian Inst. Met.* **69** 2 659–663
[2] Kasl J, Matějová M and Mrštík J 2018 *MATEC Web. Conf.* **188** 04020
[3] Mabruri E, Anwar M S, Prifiharni S, Romijarso T B and Adjiantoro B 2016 *AIP Conf. Proc.* **1725** 020039
[4] Mabruri E, Sujianto, Anwar M S, Romijarso T B and Adnyana D N 2019 *Mater. Today Proc.* **13** 1 121-126
[5] Anwar M S, Romijarso T B and Mabruri E 2018 *Inter. J. Electrochem. Sci.* **13** 15-1526
[6] Mabruri E, Anwar M S, Prifiharni S, Romijarso T B and Adjiantoro B 2015 *Metalurgi* **30** 30
[7] Refaey S A M, Taha F, Abd El-Malak A M 2005 *Appl. Surf. Sci.* **242** 1–2 114-120.
[8] Wang Z, Zhang H, Guo C, Liu W, Yang Z, Sun X, Zhang Z and Jiang F 2016 *J Mater Sci* **51** 4996–5007
[9] Jiang Z, Feng H, Li H, Zhu H, Zhang S, Zhang B, Han Y, Zhang T and Xu D 2017 *Mater.* **10** 861
[10] Weng S, Huang Y H, Xuan F Z and Luo L H 2015 *Procedia Engineering* **130** 1761 – 1769

IOP Conf. Series: Materials Science and Engineering **858** (2020) 012050 doi:10.1088/1757-899X/858/1/012050

Modal Characterization of Planar Waveguide Formed by Ag^+/Na^+ Ion Exchange in Soda Lime Glass

Ahmad Marzuki, Seran Daton Gregorius
Department of Physics, Sebelas Maret University, Surakarta 57126, Indonesia

*E-mail: amarzuki@mipa.uns.ac.id

Abstract. Modal characterization carried out reported in this paper was aimed determining the refractive index and thickness profile of Ag^+/Na^+ Ion Exchanged planar waveguides formed in Soda Lime Glass. Modal characterization was carried out using prism coupling technique and applied to all waveguides fabricated at temperature of 240 ^{0}C, 260 ^{0}C and 280 ^{0}C and processed for different duration. It was shown that the number of modes tends to increase with the increase of both temperature and duration of ion exchange process.

Keywords: Waveguide, Planar waveguide, Ion exchange planar waveguide, refractive index profile, prism coupling characterization.

1. Introduction

An optical planar waveguide is designed structurally to guide light through a particular path within small region just below the surface. Optical planar waveguides are developed for many applications, such as optical sensing [1, 2], optical amplifiers [3, 4], optical splitter [5] and integrated optics [6, 7].

Various techniques for planar waveguide fabrication have been developed [8]. Among them, ion exchanged planar waveguide is of interest since this method provides cheap and easy mass production. In order to design these divices to meet the technological requirement, better understanding on controlling parameter process fabrication is required. For this purpose, several research on ion exchange planar waveguide have been reported [9, 10, 11].

There are two type of planar waveguide: step index planar waveguide and graded index planar waveguide. Step index planar waveguide has abrupt change in index of refraction, i.e., the abrupt change from high refractive index corresponding to the guiding layer to a lower refractive index which corresponds to its surrounding transparent materials. For a given step index planar waveguide, its refractive index profile can be simply determined using optical microscope or scanning electron microscopy. The boundary between them can be clearly seen. In graded index planar waveguide, however, distinct boundary does not exist. Refractive index changes gradually from a point with the highest refractive index to that of the lowest refractive index corresponding to the refractive index of substrate. Consequently, simple refractive index profiling by applying optical microscopy cannot be done. Considering that formation of waveguiding layer by ion exchange is a complex restructuration process involving not only two main atomic ions, such as Ag^+ and Na^+ ions in Ag^+/Na^+ ions exchange, refractive index profile cannot also be resumed from individual atomic profiling. Other change in concentration profile involving other components of glass matrix do occur, although is not as significant as the main exchanged ions [12, 13].

In this work, we present planar waveguides formed by Ag^+/Na^+ ions exchange. Modal characterizations were performed to derive the refractive index profiles of the samples as function of temperature and duration of the ion exchange process. For this purpose, prism coupling method was applied.

2. Experiment

Commercial soda lime glass from Sail Brand was used as substrate. Glass was cut for the size of 10 mm x 15 mm. Ion exchange process was carried out by immersing the glass into a molten salt of 100 mol% of $AgNO_3$ at three different constant temperatures: 240^0C, 260^0C and 280^0C; each processed for 15, 60, 135, 240, 375, 540 and 735 minutes.

Refractive index of the samples, before and after ion exchange, were measured using ABBE Refractometer using light source generated from red LED. Modal characterization was carried out by applying prism coupling method. Figure 1 is the experimental setup. Light from He-Ne laser was launched to the base of the prism sitting at the centre of spectrometer. the spectrometer table was then rotated such that the incident angle formed by laser light at the base of prism is gradually change. Whenever strong coupling occurs, spot light with dark line at its centre is formed. The number of modes for a given planar waveguide can be obtained from the number of this occurrence.

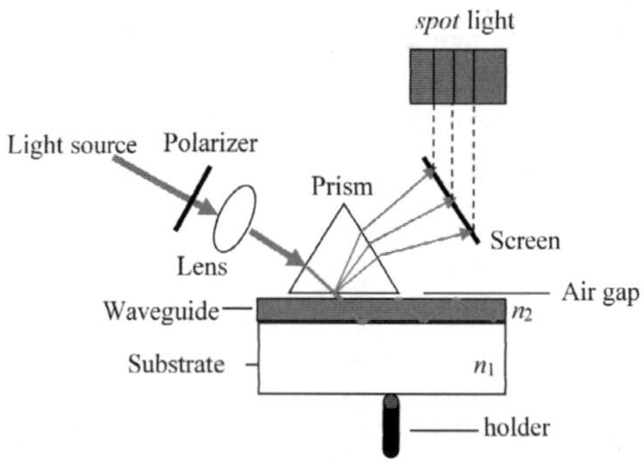

Figure 1. Experimental setup used for modal characterization

3. Results and Discussion

Table 1 shows the change in refractive index and numerical aperture of all ion exchanged glasses treated at different processing time and temperature. Refractive index of all glass surfaces after ion exchange process are increased indicating that Ag^+ ions diffuse into the glass to substitute the light ions already exist in the glass. Since the main components of the glass are SiO_2 (69 – 74 %), CaO (5 -14 %) and Na_2O (10 -16%); the most probable ion exchanged with Ag^+ ions is Na^+ ions. Na^+ ion is lighter and therefore is more mobile than the heavier components Si^{4+} and Ca^{2+}. In addition, Na^+ ion is weakly bonded to the glass structure and therefore less energy required to break its binding to other ions. The refractive index increase appearing after Ag^+/Na^+ ion exchange process is because Ag^+ ion possesses higher ionic polarizability (2.4 $Å^3$) compared to that of Na^+ (0.41 $Å^3$) [10, 14, 15, 16]. For all processing temperature, the refractive index change increases for processing time up to 240 minutes and fluctuates at around 0.06 for processing time longer than 240 minutes.

IOP Conf. Series: Materials Science and Engineering **858** (2020) 012050 doi:10.1088/1757-899X/858/1/012050

Table 1. Refractive index change Ag^+/Na^+ ion exchanged planar waveguide processed at different time and temperature.

Temperature (^0C)	Time (minute)	Refractive index Before (n_1)	Refractive index After (n_2)	Refractive index change ($\Delta n = n_2 - n_1$)	Numerical Aperture (NA)
	15	1.528	1.532	0.004	0.11
	60	1.528	1.533	0.005	0.12
	135	1.530	1.536	0.006	0.14
	240	1.525	1.533	0.008	0.16
240	375	1.528	1.534	0.006	0.14
	540	1.528	1.535	0.007	0.15
	735	1.530	1.536	0.006	0.14
	15	1.525	1.529	0.004	0.11
	60	1.525	1.530	0.005	0.12
	135	1.525	1.5315	0.0065	0.14
	240	1.525	1.532	0.007	0.15
260	375	1.525	1.530	0.005	0.12
	540	1.525	1.532	0.007	0.15
	735	1.525	1.531	0.006	0.14
	15	1.522	1.527	0.005	0.12
	60	1.522	1.528	0.006	0.14
	135	1.522	1.529	0.007	0.15
	240	1.522	1.5275	0.0055	0.13
280	375	1.522	1.528	0.006	0.14
	540	1.522	1.528	0.006	0.14
	735	1.522	1.527	0.005	0.12

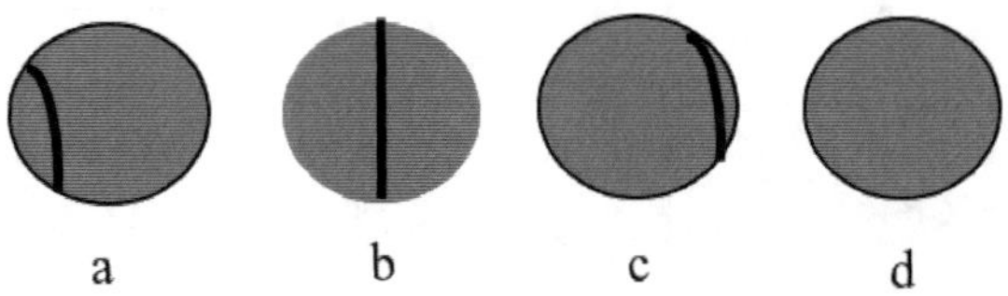

Figure 2. Typical change of the spot light appearance (from a to d) as the incident angle striking the base of the prism is changed and strong coupling occurs.

Figure 2 shows the typical change of the spot light appearance observed on the screen in prism coupling experiment (figure 1). Strong coupling of light into the waveguide occur only at a particular synchronism angles (θ_{m-i}). In Figure 2, spot light pattern indicating the occurring of strong coupling is shown by the pattern (b). If the incident angle (θ) is gradually changed, there might be several θ_{m-I} with *i*, equals 1, 2, 3, 4, etc., is the number of mode present for each waveguide. If the light intensity corresponding to the appearance of θ_{m-I} is represented by a point at "0" and spot with no line is represented by a point at "1" [17], a series of "V" shape pattern can be drawn (intensity Vs incident angle) as shown in Figure 3. Number of modes for each waveguide is shown by the number of valley (lower dot for each "V" shape) and the results are tabulated in Table 2. As seen, the number of modes increases with the increase of the processing time. Since the number of modes (M) at a given wavelength λ_0 is related to the numerical aperture (NA) and effective diffusion depth (d) as given by:

$$M \doteq 2\frac{d}{\lambda_0} NA \quad (1)$$

Where

$$NA = \sqrt{{n_{co}}^2 - {n_{cl}}^2} \quad (2)$$

Here, n_{co} and n_{cl} are refractive index of core (the surface of the waveguide) and substrate, respectively.

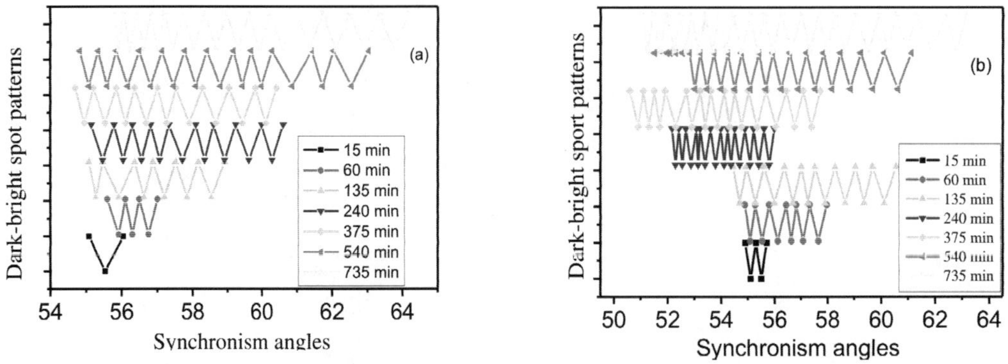

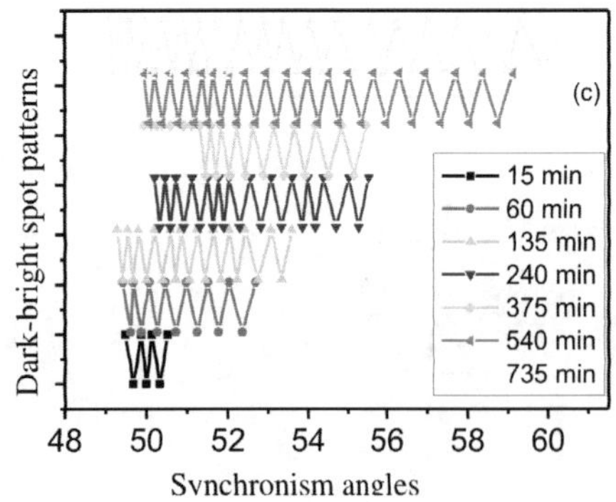

Figure 3. Mode patterns of Ag^+/Na^+ ion exchanged planar waveguide as function of the processing times taken from simple prism coupling experiment as described above for the processing temperature; (a) 240^{0}C, (b) 260^{0}C, (c) 280^{0}C.

Figure 4 is the effective diffusion depth calculated based on Tables (1) and (2). For a given processing temperature, it can be seen that the effective depth increases with the increase of the square root of the processing time, which is easily understood since $d = 2\sqrt{Dt}$ with D is diffusion coefficient whose value depends on temperature as expressed by:

$$D = D_0 exp\left(-\frac{\Delta H}{kT}\right) \quad (3)$$

where ΔH is activation energy and k is Boltzmann constant.

IOP Conf. Series: Materials Science and Engineering **858** (2020) 012050 doi:10.1088/1757-899X/858/1/012050

Table 2. Number of mode of Ag^+/Na^+ ion exchanged planar waveguide possessed at different time and temperature.

Processing time (minutes)	Number of modes for processing temperature		
	$T = 240^0C$	$T = 260^0C$	$T = 280^0C$
15	2	3	4
60	4	7	8
135	7	10	12
240	10	13	14
375	10	14	17
540	13	19	19
735	15	21	20

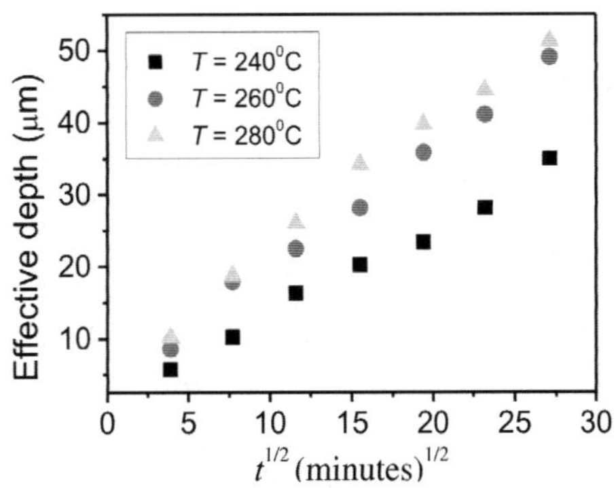

Figure 4. Effective diffusion depth at processing temperature 240°C, 260°C and 280°C as function of $t^{1/2}$ (square root of processing time)

Using refractive index data (Table 1) and equation (1) and (2), refractive index profiles of the samples can then be calculated from equation:

$$n(x) = \Delta n \text{ erfc } (x/d) + n_s \quad (4)$$

where n_s is refractive index of the substrate and the results are shown in Figure 5. For all samples treated at a constant temperature (Figure 5-a, 5-b, 5-c), the waveguiding depth increases with the increase of the processing time. In addition, the gradient concentration decreases with increasing the processing time, which is a direct consequence of the third law of thermodynamics relates movement of particles towards a new equilibrium concentration.

IOP Conf. Series: Materials Science and Engineering **858** (2020) 012050 doi:10.1088/1757-899X/858/1/012050

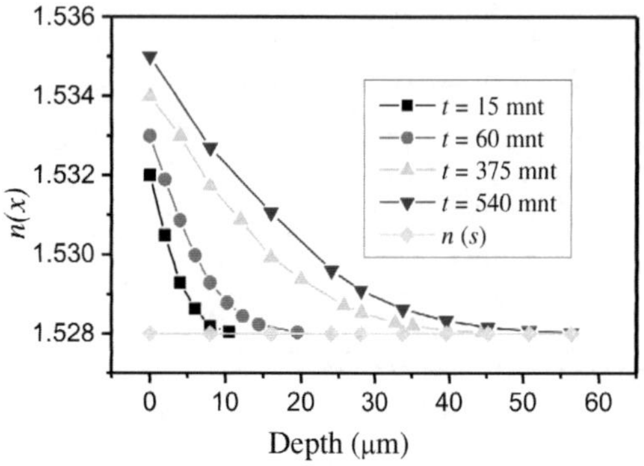

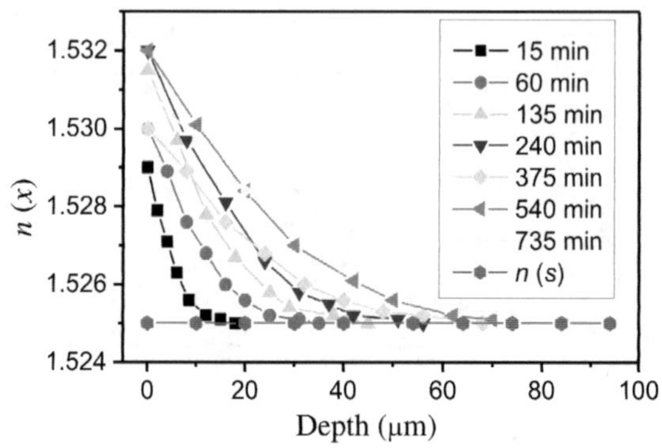

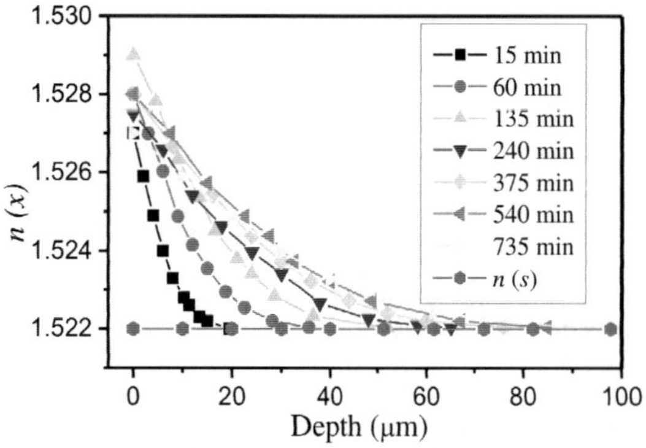

Figure 5. Refractive index profiles of Ag^{+}/Na^{+} ion exchanged planar waveguide as function of the processing times for the processing temperature; (a) 240^{0}C, (b) 260^{0}C, (c) 280^{0}C.

4. Conclusions

Ag^{+}/Na^{+} ion exchanged planar waveguides have been fabricated and characterized utilizing prism coupling technique. It has been shown that the number of modes increases with the increase of both processing time and temperature. Combining both the data of refractive index change and the number of modes, the refractive index profile of each waveguide has also been drawn.

References

[1] Rezem M, Günther A, Rahlves M, Roth B and Reithmeier E 2016 *Proc. Technol.* **26** 517-523. https://doi.org/10.1016/j.protcy.2016.08.064.

[2] Lavers CR, Itoh K, Wu SC, Murabayashi M, Mauchline I, Stewart G and Stout T 2000 *Sens. Actuators B Chem.* 69 (1–2) 85-95. https://doi.org/10.1016/S0925-4005(00)00412-3.

[3] Zhang J, Guo W-T, Tang C-Y, Yan S, Li W-N and Liu C-X 2019 *Opt. Commun.* **453** 124344, https://doi.org/10.1016/j.optcom.2019.124344.

[4] Atar G, Eger D, Bruner A, Sfez B and Ruschin S 2016 *Opt. Mater.* **55** p. 49-54. https://doi.org/10.1016/j.optmat.2016.03.017.

[5] Chen Q, Milanese D, Chen Q, Ferraris M and, Righini GC 2008 *J. Non-Cryst. Solids* **354** (12–13) 1230-1234. https://doi.org/10.1016/j.jnoncrysol.2006.11.045.

[6] da Silva J C and Rodriguez-Esquerre V F 2020 *Optik* **212** 164756.

https://doi.org/10.1016/j.ijleo.2020.164756.
[7] Bader M A and Marowsky G 2001 *Synth. Met.* **124** (1) 141-143. https://doi.org/10.1016/S0379-6779(01)00452-0.
[8] Righini G C and Chiappini A 2014 *Opt. Eng.* **53** (7) 071819. https://doi.or/10.1117/1.OE.53.7.071819.
[9] Yulianto M, Marzuki A and Suryanti V 2017 *JPCS* **909** 012022. http://doi.org/10.1088/1742-6596/909/1/012022.
[10] Martin M, Videau J J, Canioni L, Adamietz F, Sarger L and Flem G L 2000 *Appl. Opt.* **39** (3) p. 435 -440. https://doi.org/10.1364/AO.39.000435.
[11] Yulianto M, Marzuki A and Suryanti V 2017 *AIP Conf. Proc.* **1868** 060006. https://doi.org/10.1063/1.4995170.
[12] Marquez H, Salazar D, Villalobos A, Paez G and Rincon J M 1995 *Appl. Phys.* **34** (25) 5817 – 5822. https://doi.org/10.1364/AO.34.005817.
[13] Gonella F, Caccavale F, Bogomolova L D and Quaranta A 1998 *J. Appl. Phys.* **83** (3) 1200-1206. http://dx.doi.org/10.1063/1.366816
[14] Tessman J R, Kahn A H and Shockley W 1953 *Phys. Rev.* **92** (4) 890-895. https://doi.org/10.1103/PhysRev.92.890.
[15] Rehouma F and Bentouila O 2020 *J. King Saud Univ. Sci.* **32** (1) 433-435. https://doi.org/10.1016/j.jksus.2018.07.001.
[16] Dimitrov V and Komatsu T 2010 *J. Univ. Chem. Technol. Metallurgy* **45** (3) 219-250. https://dl.uctm.edu/journal/node/j2010-3/1_Veselin_Dimitrov_219-250.pdf
[17] Marzuki A, Gregorius S D, Widhianingsih I, Lestari S and Suryawan J 2015 *AIP Conf. Proc.* **1699** 040004-1–040004-4. http://doi.org/10.1063/1.4938319.

Decision Support System for Detection of Skin Diseases in Smart Health development planning

Hartatik[1], Abdul Aziz[1], Rudi Hartono[1], Rizal Abdilah[1]
[1]Informatics Engineering Departement, Faculty of Mathematics and Natural Sciences, Sebelas Maret University, Surakarta, Indonesia

E-mail: hartatik.uns@gmail.com

Abstract. Skin disease is a disease that is quite dangerous and is often found in tropical countries like Indonesia. Lack of knowledge about the types of skin diseases and do not know how to prevent them can cause a person suffering from acute skin disease. Detection of skin diseases is usually done by a dermatologist but actually the detection of skin diseases can be done alone or with the help of others by observing the diagnoses that arise. The expert system is a solution for using computers to help in the early detection of possible skin diseases. Dipen-ku is a expert system which designed to detect naive skin diseases with Bayes. System output is a type of skin disease which is a diagnosis of this type of disease. This expert system is a new innovation in innovation in drug materials that combines treatment technology and bigdata technology to support the design of smart city development

1. Introduction

The skin is one of the five human senses and the first part that can receive external stimuli. Therefore skin health is something are very important to be maintained, so that the skin does not experience various diseases. Various skin diseases can be caused by several factors such as a less clean environment, viruses, bacteria, immune system, and others. Determination of skin diseases must be done by skin experts or skin specialists, because skin diseases can be something serious if an error occurs. From the writer's observations, public awareness of the dangers of skin disease is currently considered lacking. Machine learning plays an important role in the medical field for the automation of many processes[1] in a previous study developed detection of skin diseases [2].

From the problems that the authors found, the creation of an expert system of skin disease detection using the Naive Bayes method was felt to be very necessary to help all people who want to know about skin diseases that are being experienced or need information about skin diseases. Algorithm The Naive Bayes method is one of the algorithms can be use for found in classification problem[3]. The Naive Bayes algorithm is a technique for classification with use probability concept and statistical methods. The Naive Bayes algorithm is considered a result to be suitable for the manufacture of a skin disease expert system that will be created because of the workings of the Naive Bayes Algorithm which can classify large amounts of data so that the results of the diagnosis are

IOP Conf. Series: Materials Science and Engineering **858** (2020) 012051 doi:10.1088/1757-899X/858/1/012051

correct[4,5]. With the existence of this information system for skin diseases, it is expected to help patients with skin diseases in general who need it.
For this application is called dipen-ku , a patient can be ceck and detect about skin disease and alternative drug as a one of solution about their disease. So dipenku can be called as a drug technology in healty especially skin disease.

2. Experimental

Skin Disease Detection Expert System Using the Web-Based Naive Bayes Method is an expert system that can be used by the community to detect diseases based on perceived indications[6]. The user makes a diagnosis by accessing the system and conducting consultations illness by choosing a feeling that is felt without using the system login[7]. The methos of the research can be seen in Figure 1 .

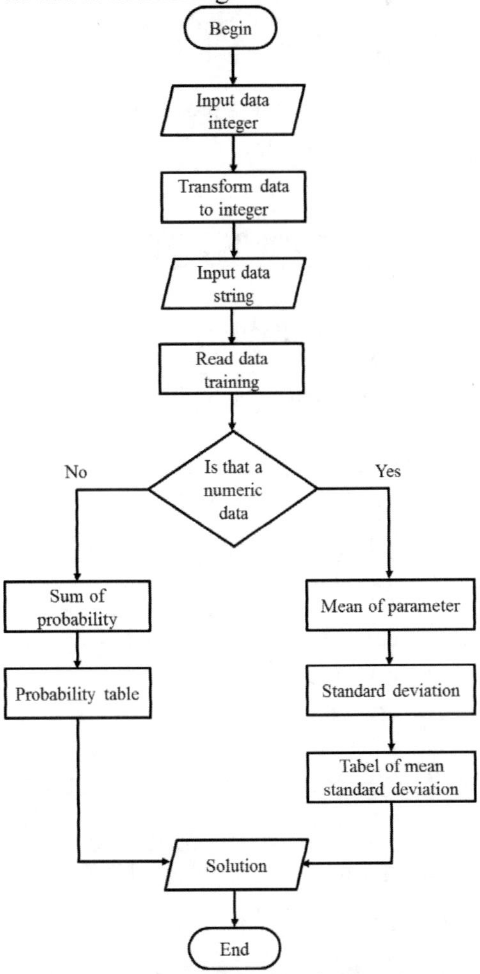

Figure 1. Flowchart of business process

3. Result and Discussion

The development of a decision support system for skin diseases in this study was based on an expert (expert) one of the skin doctors in Solo. Based on the results of interviews with Narasuber, a health information system was developed, a decision support system for skin disease detection. Business process of dicesion support system can be seen in Figure 2. The system are related to the process functionality of the system carried out by actors who have access to the system process.

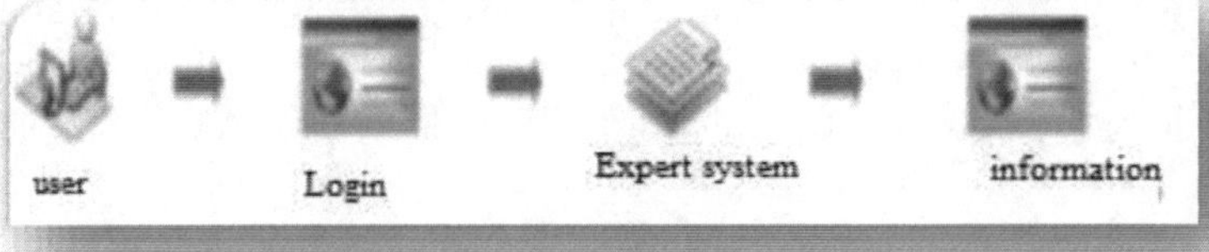

Figure 2. business proses in Dipen-ku

System functional requirements for decision support system that describe the database of system can be seen in Table 1.

Tabel 1. Functional requirements.

Code	Description	Actor
FS-01	The system can display disease data from Admin disease table	Admin
FS-02	The system can manage data (add, change, and delete) diseases in the disease table	Admin
FS-03	The system can display all symptom data	Admin
FS-04	from the Admin symptom table	Admin
FS-06	The system can manage data (add, change, and delete) all indications in the Admin symptom table	Admin
FS-05	The system can manage Admin patient data	Admin, user
FS-06	The system can display Admin expert profiles	Admin, user
FS-07	The system can display the results of processing data entered by the Admin, user,	Admin, user

Non-functional needs are requirements related to user interaction with the system created. Non-functional system needs can be seen in Table 2.

Table 2. Non-functional requirements.

Code	Description
NF-01	Use a web browser to access the system
NF-02	To manage data in the system using login
NF-03	Password on the system using MD5 technology

The making of an expert system to diagnose skin diseases begins with system design including the design of knowledge base (Knowledge Base), Use Case, Activity Diagram, Sequence Diagram, Class Diagram and ERD .
At the prosses of making expert systems, facts and knowledge are the important factor which related with Indications of skin diseases. This indications will be used in making a conclusion. These facts and knowledge are obtained from interviews with experts or other sources such as books, journals, internet pages and others. The facts and knowledge that have been obtained will be translated by the system maker into a knowledge base stored in the expert system that will be created [9,10] , can be seen in figure 3.

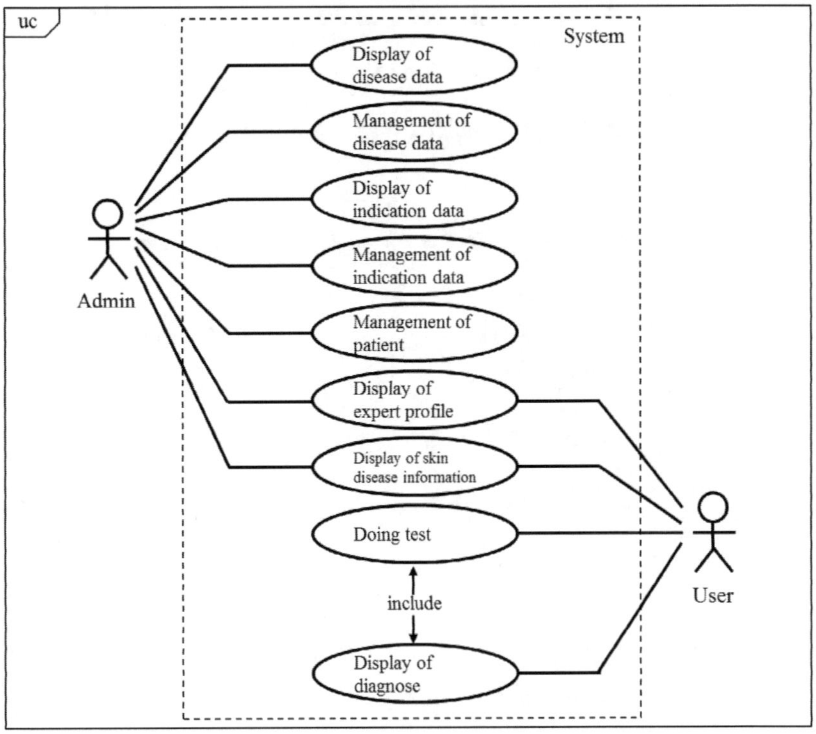

Figure 3. Use case diagram.

The home page is the initial page that will appear after the admin has successfully logged into the system. On the left there are menus available on the system. The picture of the admin home page can be seen in Figure 4 and Figure 5. The process of sequencing the system rules using Naive Bayes for classify the indications of disease in the system and then decide the results of the system classification in figure 6.

If there is an addition of disease data by the admin, the disease data entered can affect the results of the tracing if the admin has entered the indications data. Sometimes not all indications of a disease are met, so if 60% of the indications entered by the user are in accordance with the indications of one disease, then the patient can be categorized into skin diseases according to the indications entered. If the user inputs symptom data randomly or regardless of the child's condition, the system is very possible to make errors in the tracing or bring up the default value because there is no disease data that matches the indications entered by the user.

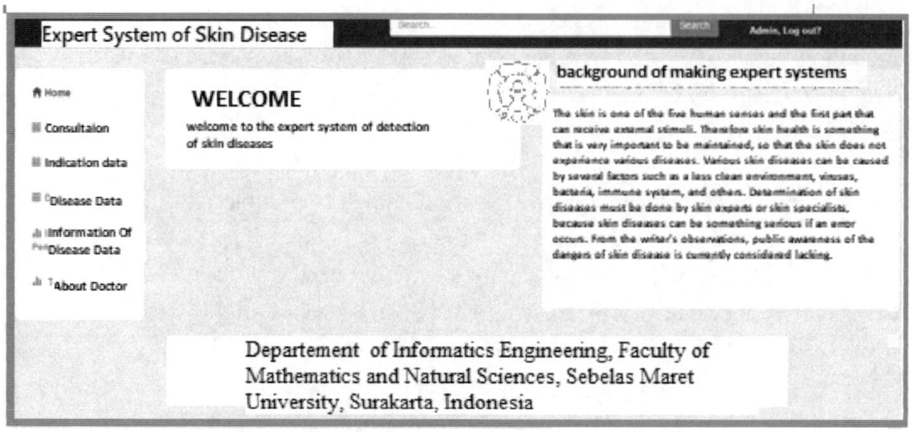

Figure 4. Implementation of home page Dipenku.

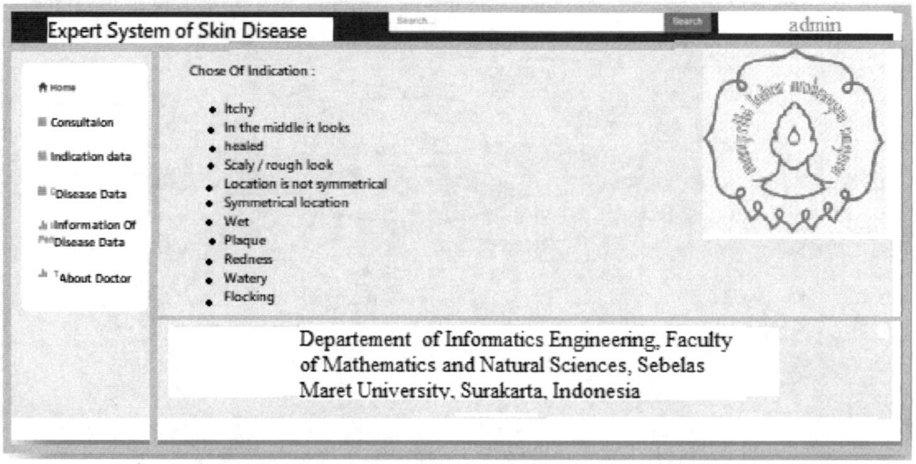

Figure 5. Implementation of admin page Dipenku.

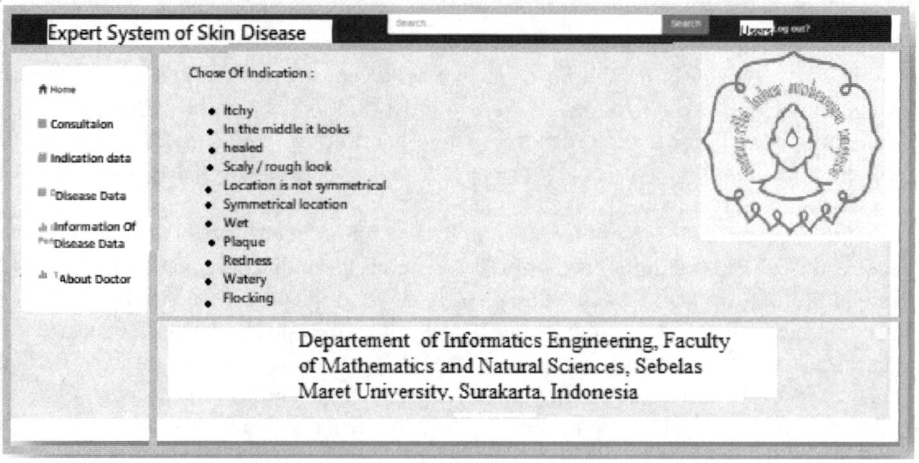

Figure 6. Implementation of consultation Dipenku.

The script of dipen-ku is a script for admin login which the admin can later manage data after login which resulting interface in figure 5 [11].

Expert system that is made is tested by conducting interviews with respondents. The respondent tried the expert system that was made and then asked questions. From survey result that 65% percent stated that this expert system was very helpful, this can be seen on Figure 7 .

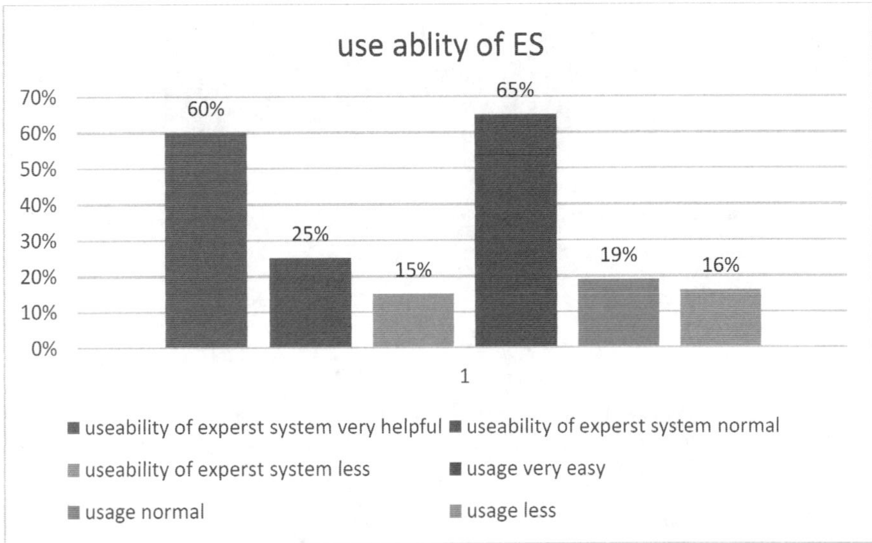

Figure 7. Survey result for Dipen-ku

The application is tested in general to 101 respondents and most of the respondents stated that it is very helpful for students to learn to write which is 60 % said that the application is very helpful in figure 8[10,11].This is can be information that dipenku is one of riil action for smart health development planning.

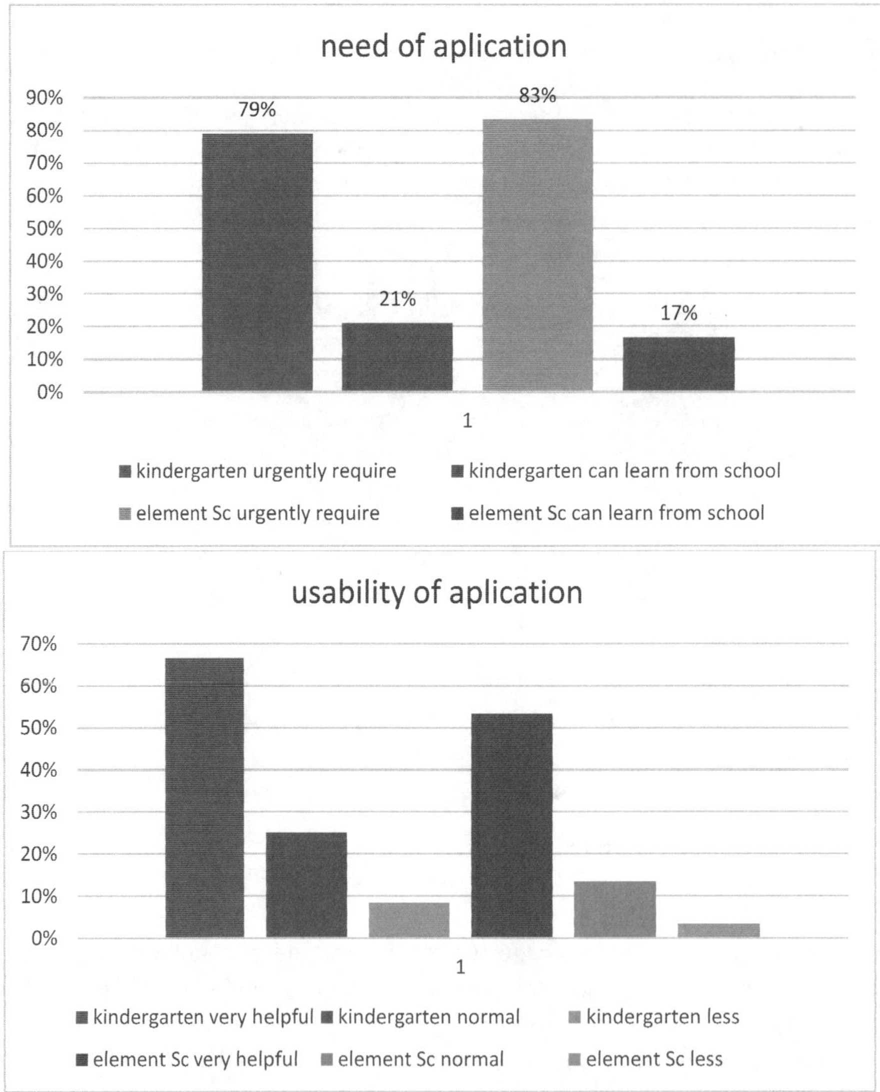

Figure 8. public test result

4. Conclusion

Based on the results of the above study it can be concluded that dipenku is a expert System for Detecting Skin Diseases and a drug technology in smart health can be successfully created which uses the PHP programming language with the CodeIgniter framework and the MySQL management system database with the Naive Bayes method .

Acknowledgments

The authors acknowledge their thanks to Sebelas Maret University, especially thanks to Departement of Informatics Engineering for this research opportunity.

References

[1] S. R, M. Suhil, and D. S. Guru 2015 Segmentation and Classifications of Skin Lesions for Disease Diagnosis International Conference on Advanced Computing Technologies and Applications (ICACTA-2015) Mysore.

[2] A. K. Mittra and R. Parekh 201 Automated Detection of Skin Diseases Using Texture Features *International Journal of Engginering Science and Technology (IJEST)* vol. 3, pp. 4801- 4808 .

[3] Hartatik 2017 Naïve Bayes Approach For Design Of Expert System For Identification Of Children Leather Based On Android *IOP Conference Series: Materials Science and Engineering* Vol. 333.

[4] Zainudin, M., Erna Zuny Astuti 2017 Application of the Naive Bayes Algorithm For Classifying the Feasibility of Prospective Customers *Pt.Bni Semarang Insurance. Article Thesis Dian Nuswantoro University.*

[5] A. Madsen 2010 Bayesian Networks for Disease Diagnosis *B. Thomas Golisano College of* Computing *and Information Sciences.*

[6] M Z Asghar , M J Asghar , S M Saqib , B Ahmad , S Ahmad and H Ahmad 2011 Diagnosis of Skin Diseases using Online Expert System *(IJCSIS) International Journal of Computer Science and Information Security, Vol. 9, No. 6.*

[7] Kusrini 2007 Concept and Application of Decision Support Systems Andi publisher Yogyakarta.

[8] Hustinawaty and R Aprianggi 2014 The Development of Web Based Expert System for Diagnosing Children Diseases Using PHP and MySQL *International Journal of Computer Trends and Technology (IJCTT)* – **Vol** 10 number 4 – Apr 2014.

[9] N. Arbaiy and S. T. Chong 2012 Android mobile application for medical diagnosis expert system: a knowledge dissemination tool *Proc. 1 st International Conference on Mobile Learning, Applications, and Services* pp. 31-35

[10] N Fitri, Yoga, H. A and Endah, N. Y **2016** Expert Application for Diagnosis Skin Disease Using the Forward Chaining Method *Al Arif Skin Care of Ciamis Regency.*

[11] Durkin. J. 1994.*Expert System Design and Development*, Prentice Hall International Inc, New Jersey. United States: McGraw Hill Professional

IOP Conf. Series: Materials Science and Engineering **858** (2020) 012052 doi:10.1088/1757-899X/858/1/012052

Chemistry of Natural Dyes of Batik Crafts Colouring Process

S Wahyuningsih[1*], Rahmawati[2], S R Handayani[3], Setyaningsih[4], Ponimah[5]
[1,4,5] Faculty of Mathematics and Natural Sciences, Sebelas Maret University, Surakarta, Indonesia
[2] Economic and Business Faculty, Sebelas Maret University, Surakarta, Indonesia
[3] Faculty of Visual Arts and Designs (FSRD), Sebelas Maret University, Surakarta, Indonesia

E-mail: sayekti@mipa.uns.ac.id, rahmaw2005@yahoo.com,

Abstract. By Fixation technique, we found the increasing stability of natural color. The aim of the study was to evaluate the performance of dyes anthocyanin, mangiferin, curcuma longa, Indigi fera and tannin (tingi) in batik dyeing. The medium was optimized by conditioning of natural dyes from amount quantity of extracted dyes in several condition: acid condition, base condition and also complexation during fixation process usually done in Batik dyeing. It was concluded that the color values were found to be influenced by the addition of namely mordants as complexation agent, consequently different fashion hues were obtained from the same dye extract using different mordants. The chemistry aspect such as an absorption in UV Visible region of each dyes and Infra Red absorption have been studied, and also a typical of fixation model should be noticed as a chemistry of dyes.

1. Introduction

Batik products with natural colours have potential as export commodities. The quality of batik products continues to be developed not only in business management, networks or marketing channels, increasing skills of craftsmen but also production facilities, production process technology, motive design innovations, and colouring materials innovations. Raw material for batik production including fabric (prime cotton, mori cloth, primis fabric, super silk fabric, chemical colorants, and wax fragments, batik mixture (caustic soda, sodium carbonate, HCl, TRO, Al_2O_3, and starch). By understanding of chemical processing occurring along processing, quality and colour behaviour will be maintenance well.

Natural textile dyes are generally obtained from extracts of various parts of plants such as roots, wood, leaves, seeds or flowers. Batik craftsmen have known many plants that can color textiles, such as: Indigo tree leaves (Indigofera), in of soga tingi trees (Ceriops candolleana arn), tegeran wood (Cudraina javanensis), turmeric (Curcuma), tea (Tea), noni roots (Morinda citrifelia), soga jambal skin (Pelthophorum ferruginum), kesumba (Bixa orelana), guava leaves (Psidium guajava). However, until now there has never been reported standardization of coloring preparations using natural dyes, as well as testing the color quality resulting from natural coloring. Standardization of natural coloring preparations from various source materials will greatly determine the homogeneity of coloring.

Natural dyes are comprised of dyes and pigments that are obtained from various parts of plants including roots, bark, leaves, flowers, and fruit [1]. The major applications are coloring of food, leather, wood, and natural fibers like wool, silk, cotton, and flax. Natural dyes may have a wide range of shades. Dyeing with natural dyes, however, normally requires the use of mordants, which are metallic salts of aluminum, iron, chromium, and copper, among others, to ensure a reasonable color fastness to sunlight and washing [2]. The metal ions of these mordants can act as electron acceptors for electron donors to form coordination bonds with the dye molecules, making them insoluble in water [3]. The use of natural dyes cuts down significantly on the amount of toxic effluent resulting from synthetic dye processes. The use of natural dyes in textile applications is growing rapidly, reflecting the strict environmental standards being established in many countries, and the concern about the health hazards associated with synthetic dyes, for example, the recent ban on the use of azo dyes by the European Union, has also increased the scope for the use of natural dyes [4].

2. Experimental

The natural dyeing techniques and materials that have been conducted and characteristic colour and their spectrum have been studied. The material and their processing including:

a. Natural plant source for natural colouring include: anthocyanin dyes from flowers, tannin from tingi wood and also secang wood, chlorophyll from green leaf, indigo from indigo fera fermentation
b. Extraction of natural dyes was carried out by about 1000 grams of natural coloring sources were put into 10 liters of water and boil so that the remaining volume of water is 5 liters. Then filtered to get a solution of natural dyes
c. Dyeing processing technique: dyeing processing technique for physical and chemical interaction between dyes and cotton materials. In mordaning process, the cloth was boiled in water with a ratio of alum:water about (100 g alum :17 liters of water) to tie the Fe^{2+} ions to the fabric (wool, cotton, silk) for 1 hour. Leave it for overnight, them dried out. For colouring, natural dyes used as a comparison of cloth: dye = 1:30 (w/w). The coloring process was done by repetitive dipping technique (10 times or more)
d. Chemical reaction of fixation: alum ($Al_2(SO_4)_3$), tunjung ($FeSO_4.7H_2O$), or lime (CaO) solution for binding process were prepared for fixation through complexation (also copigmentation) mechanism.
e. Influence of pH condition during colouring processing techniques from natural plant was carried out by the addition of alum, tunjung, acetic acid or lime solution in natural dye source in an exact amount, the pH will be decrease or increase to compensatied the H^+ and OH balance along colour expression of dyes.

3. Result and discussion

3.1 Chemistry behaviour of natural plant source for batik colouring process

3.1.1. Anthocyanin

There are many natural colours for colouring batik processing including anthocyanin group, tannin group and also chlorophyll group. Anthocyanins are responsible for the red, purple, and blue colours found in a variety of fruits, vegetables, cereal grains, and flowers. Unfortunately, anthocyanin pigments are highly unstable and are totally unsuitable for dyeing textiles. It should be related to the structure and the interaction between dye and cotton materials. The mechanism behind the destruction of anthocyanins in acid condition has been the subject of some debate, but in acid condition anthocyanin give a intensive red color [5]. Anthocyanin in various pH have chemical structure illustrated in Figure 1.

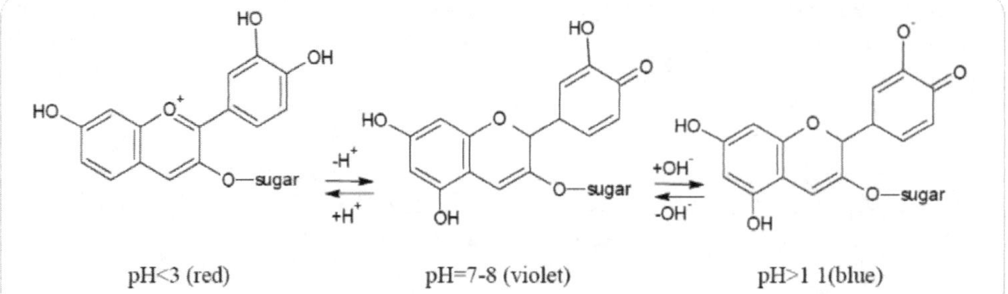

(a) (b) (c)

Figure 1. Structure of Anthocyanin transformation as a function of pH in natural color source in a variety of fruits, vegetables, cereal grains, and flowers

At a lower pH, anthocyanidins such cyanidin is highly soluble in water due the formation of flavylium cation which appear as red (Figure 1(a)). But in high pH (Figure 1(a)), formation of anion form is dominan, because of the deprotonation OH^- group. Anthocyanins degraded to each other by direct interaction of anthocyanin with acid, but it will be probably, by free radical formation pathway, cleave the flavylium core structure of anthocyanins. This condition triggers instability of anthocyanin and results in color destabilization dye. Stabilization of anthocyanins can be attempted by copigmentation using metal ions. Metal complexation is the clear pathway for increase stability. This idea is as in the fixation in Batik colouring process. Ascorbic acid also acts as a co-pigment. Thus, improved colors are obtained with the intrinsic protection of ascorbic acid from oxidation, and also in fixation process, usually using Alum ($xAl(SO_4)_2 \cdot 12\ H_2O$) X= K^+, NH_4^+, tunjung ($FeSO_4$) or lime (CaO). Temperature also can help to retain the color intensity of anthocyanin, witch the increase in temperature the greater degree in anthocyanin destruction.We suggest that the speedy destruction of anthocyanin in higher temperatures could be due to hydrolyzation of 3-Glycoside structure, which has a protective effect in unstable anthocyanin [6].

Anthocyanins are stable at low pH. It becomes less stable when exposed to heat, causing a loss of color and browning. In solution, anthocyanins molecules are present in an 1equilibrium between the colored cationic form and the colorless pseudo base. pH is very important for the color of anthocyanins, some anthocyanins are red in acid solutions, violet or purple in neutral solutions, and blue in alkaline pH. The reason for this is right here this structure of the anthocyanins (red circle) is called flavylium cation the reason, at low pH the cyanidin molecule is protonated and forms a positive ion or cation, as the pH increases the molecules become deprotonated, at high pH the molecule forms a negative ion or anion [7].This is the reason that most colorants containing anthocyanins can only be used at pH values below pH 4. Figure 2 show the UV Vis Spectras of anthocyanin before and after isolated from rose flower. Fourier Transform Infra Red absorption (FTIR) of anthocyanin was shown in Figure 3.

IOP Conf. Series: Materials Science and Engineering **858** (2020) 012052 doi:10.1088/1757-899X/858/1/012052

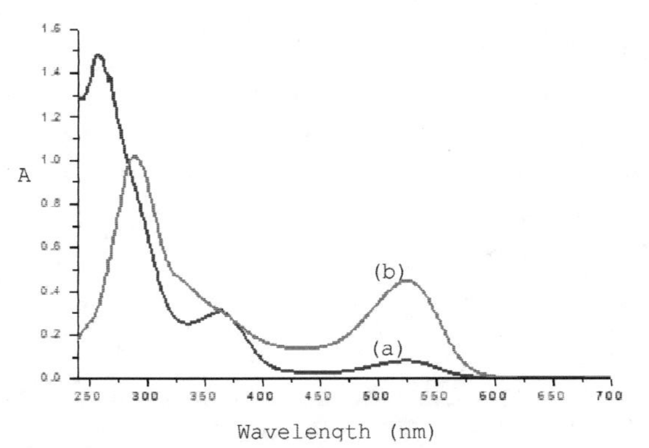

Figure 2. UV Visible Absorption spectra of (a) extracted anthocyanin dye (b) isolated anthocyanin dye. Anthocyanin dye extracted and isolated from rose flower. UV-Vis Absorption spectra was measured in acid condition of 0.1 M HCl in methanol.

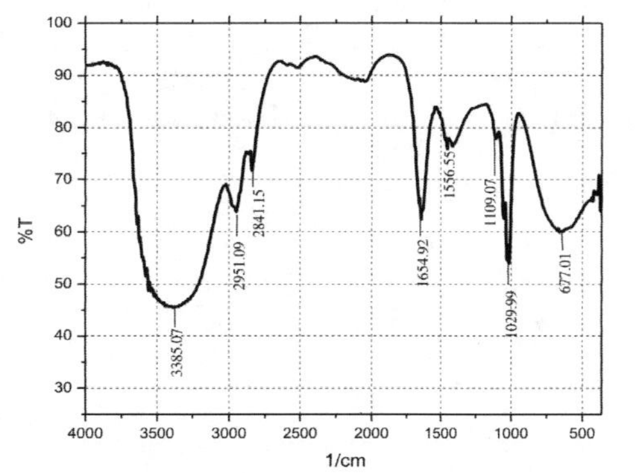

Figure 3. FTIR Absorption of Antocyanin dye isolated from rose flower

The basic colours anthocyanin as blue, purple, red and orange colours has correlation with the number of hydroxyl groups and with the number of methoxyl groups in anthocyanin. The number of methoxyl groups in anthocyanin in different source is different. By fixation using acidification, colour expression will be different that depend on the amount of acid have been used. But the heating step in colouring process will decrease stability of anthocyanin structure. Another strategy for improving colours of anthocyanin based natural dyes is purification step in initial step before colouring processing. By this way, it will reduce sugar content and also many compounds that influence colour stability. Table 1 give an illustration about a shifted of absorbance of extracted and isolated anthocyanin that influence of colour and colour intensity.

Table 1. List of maximum absorbance UV Visible of extracted and isolated anthocyanin dye related to Figure 2.

Wavelength(nm)	Anthocyanin	
	Extracted	Isolated
λ_1	256	288
λ_2	329	-
λ_3	526	526

3.1.2. Mangiferin dye

Mango (Mangifera indica L.) is a medicinal plant that belongs to the Anacardiaceae family. Mangiferin is one of the dominant polyphenols in mango leaves that has a function as a natural dye for batik craft and has several pharmacological properties such as antioxidants, anti-bacterial, and anti-fungi [8-9]. Thus, the use of material in natural textile coloring is expected to provide 2 (two) benefits at once, namely coloring and providing protection from attacks by free radicals and microorganisms. Figure 4 show the UV Vis Spectras of mangiferin fron mangi leaves and FTIR of mangiferin was shown in Figure 5.

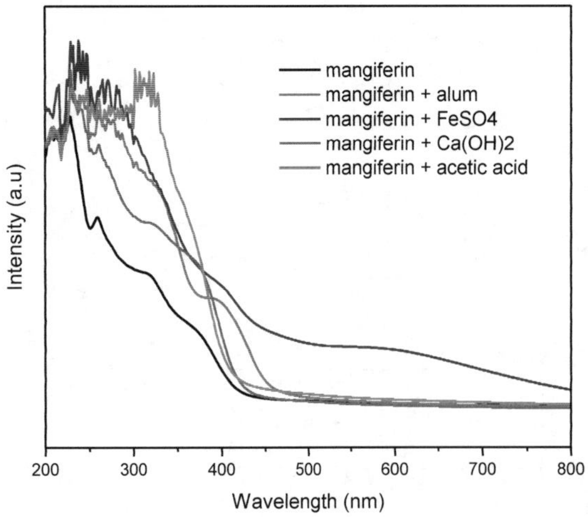

Figure 4. UV Visible Absorption spectra of extracted mangiferin dye. Mangiferin dye extracted from mango leaves and measured with addition alum ($Al_2(SO_4)_3$), tunjung ($FeSO_4$), lime ($Ca(OH)_2$ and acetic acid in water solution.

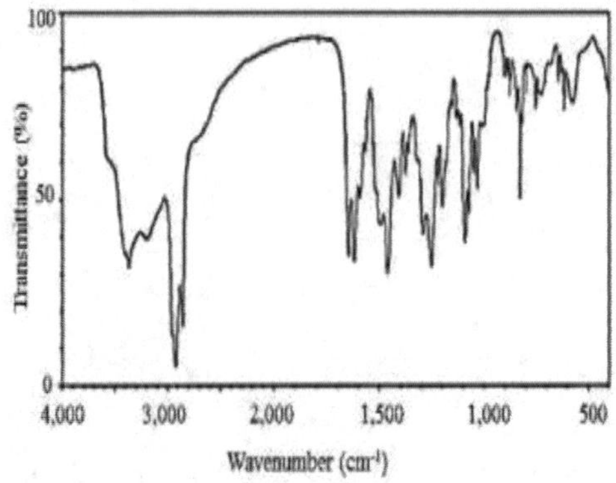

Figure 5. Fourier Tranform Infra Red Absorption of mangiferin dye extracted from mango leaves

The structure of mangiferin from mango leaves was shown in Figure 6.

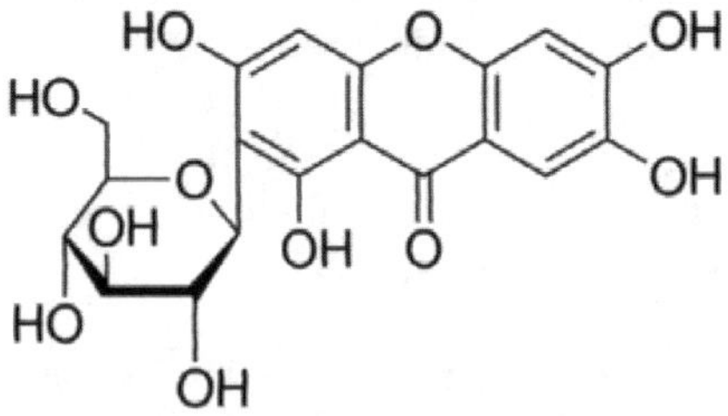

Figure 6. Structure of magiferin from mango leaves

3.1.3. Javanese cumin

Javanese cumin (Curcuma longa L)) have curcuminoids structure that are one of the producers of natural textile dyes, yellow, resulting from the extraction of turmeric root (Curcuma longa L). As well as mangiferin, a large number of previous studies have also been successful in proving the functionality of curcumin, which is not only able to color silk fibers inbrilliant yellow but is also capable of providing other functional properties, such as antibacterial activity [10] and antioxidants [11].

We have observed of the colours of curcuma dye in various pH in Figure 7. Curcuma extracts were further treated by ball milling process due to decreasing the particle size and also improved solubility curcuma in water. Differences of colour and intensity were occurred related to structural change in pH value.

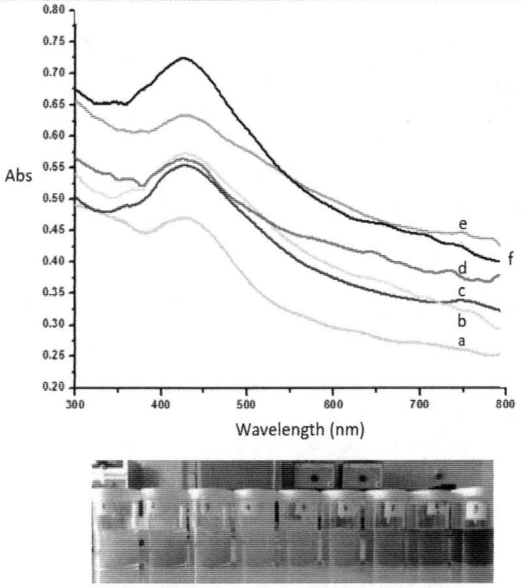

Figure 7. UV Visible Absorption spectra of extracted curcuma dye. Curcuma prepared in nanosize using ball milling methoda and measuared in pH (a) pH 9, (b) pH 7 (c) pH 5 (d). pH 3 (e). pH 1 (f). without milling

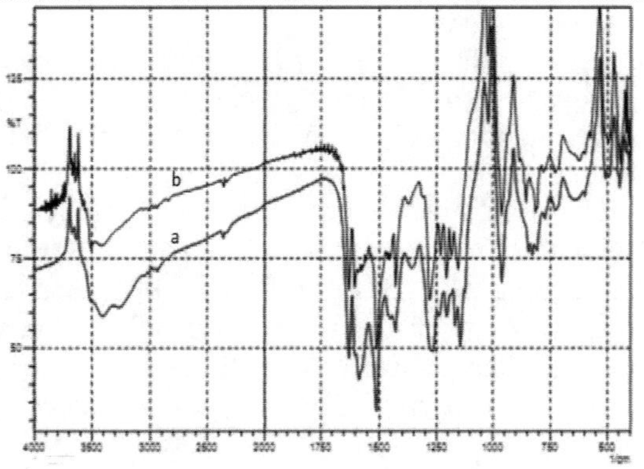

Functional groups	Wavenumber (cm^{-1})	
	Curcumin	Curcumin-Al
O-H Streching	3500,8	3406,28
Vibrasi C=O	1627 1602	1625 1585
Vibrasi C-O	1282	1267

Figure 8. FTIR Absorption of curcuma (a) After complextion with alum (b) before complexation with alum

Protonated form was occurred in low pH, isoelectric form was occurred in medium pH, while deprotonated form was occurred in high pH (Figure 9). The FTIR spectra of curcuma before complexation and after complexation with alum were shown in Figure 8. In fixation process for dying the Al-curcuma was formed as in Figure 10.

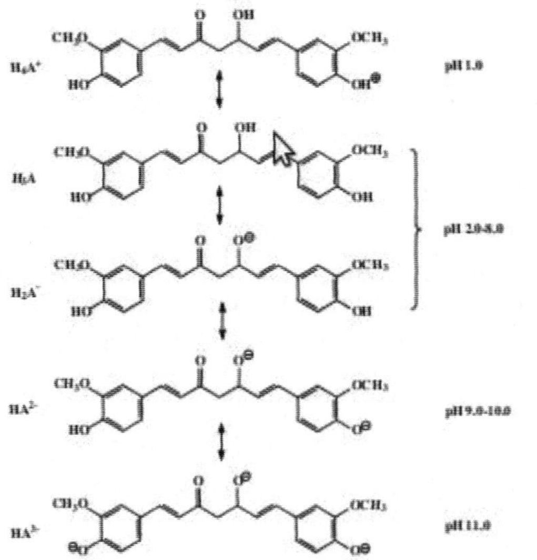

Figure 9. Protonated form was occurred in low pH, isoelectric form was occurred in medium pH, while deprotonated form was occurred in high pH.

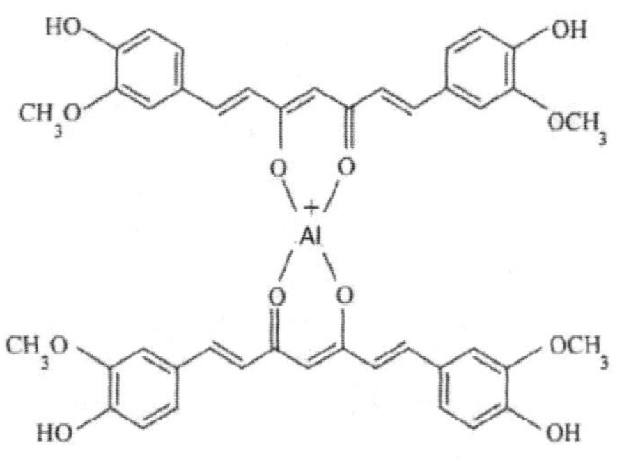

Figure 10. Complex formation of Al-curcuma in fixation process of batik dying.

IOP Conf. Series: Materials Science and Engineering **858** (2020) 012052 doi:10.1088/1757-899X/858/1/012052

3.1.4. Indigo fera

Surprisingly the only satisfatory long-lasting natural blue dye is indigo. However, indigo is not a naturally occurring plant pigment, and must be manufactured by fermenting plants that contain the appropriate precursor chemicals. In principle the production of indigo is a simple process; however in practice it is very difficult to produce a deep blue dye.

Indigofera arrecta, commonly known as Bengal Indigo, Indigo, Natal Indigo, or Java Indigo, is a tropical small shrub of herb growing up to 4 m in height with partly woody stems and slightly hairy leaves. It has pink or brown flowers. It is a major source of the blue dye 'indigo'. Indigofera arrecta, originated in tropical Africa but is now distributed widely throughout the tropics. Widely distributed throughout the continent from Senegal to Somalia, south to S. Africa. Also extending into Arabia. According to Heyne, 1927, the Dutch introduced it to Java in about 1860. It tolerates altitudes up to 2700 m [12].

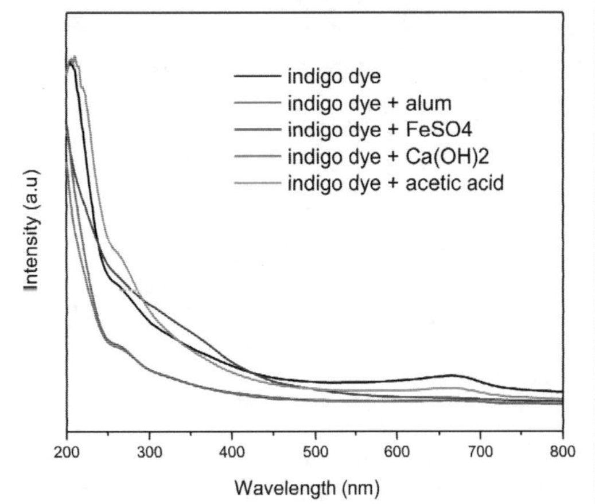

Figure 11. UV Visible Absorption spectras of indigo dye and indogo dye after addition fixation compounds of Alum($xAl(SO_4)_2\cdot12H_2O$); x = K^+, NH_4^+), Tunjung ($FeSO_4$), lime ($Ca(OH)_2$ and acetic acid.

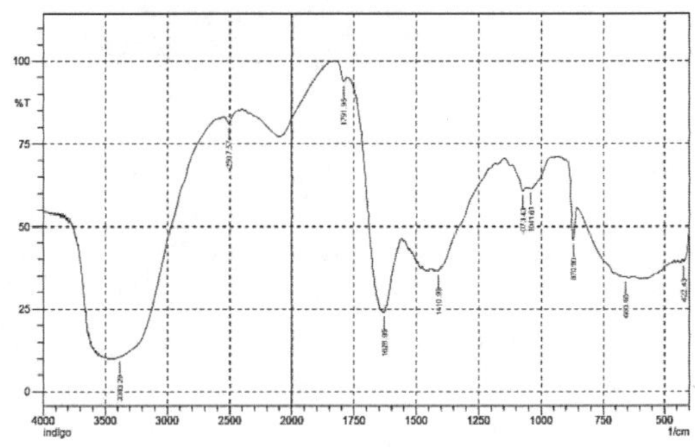

Figure 12. FTIR absorption spectras of indigo dye

Functional group	Wavenumber (cm^{-1})
O-H, $-NH_2$, -NH stretching	3500 – 3100
C=C aromatic	1600
C-N amine	1350 – 1000
C-H aromatic	900 – 690

Based on indigo structural in acid and in basic condition (Figure 13), the structure were difference in acid and base conditions, due to protonated and deprotonated form. Alum, tunjung and acetic acid give acid condition and accelerate protonated form indogo structure that give a spesific color intensity.

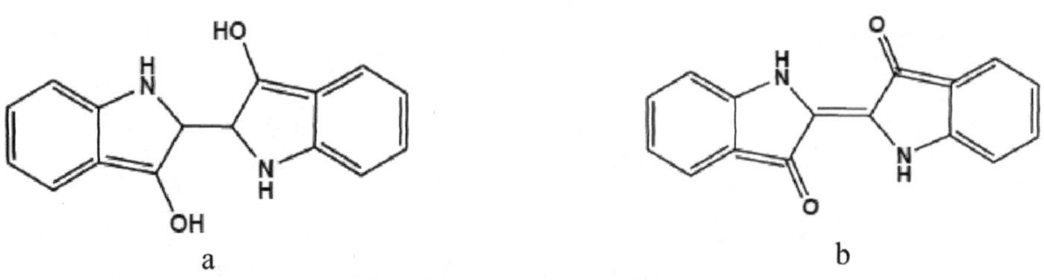

a b

Figure 13. Basic structure of Indigo (a) acid condition (b) basic condition

The UV Visible absorption of indigo dye show that the maximum absorption peak is at a wavelength in the range 550 – 700 nm. The strong absorption in almost visible range give deep blue color of indigo. The permitted electronic transition of $\pi \rightarrow \pi^*$ and also $n \rightarrow \pi^*$ of chromophore group of indigo dye give high intense of blue absorption [13]. Color intensity related to the fixation condition using alum ($xAl(SO_4)_2 \cdot 12H_2O$); x = K^+, NH_4^+), tunjung ($FeSO_4$), and acetic acid.

3.1.5. Tannin from tingi

Tannic acid (tannin) is a common mordant used in the dyeing process for cellulose fibres such as cotton, often combined with alum and/or iron (tunjung). Quercitannic acid is one of the two forms of tannic acid found in oak bark and leaves. The other form is called gallotannic acid and is found in oak galls. Tannic acid is a specific commercial form of tannin.

The bark of soga tingi which contains tanin as natural coloring subtances can be used as a substitute for synthetic dyes for Batiks. Tingi (Ceriops tagal) bark was highly potential as tanning materials due to high tannin concentration. The results showed that the tingi bark contains tannin level 70.91% and non-tannin levels 10.63%. Tannins from tingi bark were classified as proanthocyanidins (condensed tannins) type and the absorption values obtained maximum value at wavelength 490 nm. The FTIR spectral data indicated that tingi tannins contained hydroxyl group (v O-H; v N-H) in the area (3467.418-3057.025) cm^{-1}, aromatic group (v C-H) in the area of 2875,733 cm^{-1}, v C=O (ester group in tanned material) in the area (1747.442-1612.422) cm^{-1}, v-OH; R-COO- in the area of 1444.626 cm^{-1} and $v(SO_4)_2$-; $R\text{-}SO_3$; $R\text{-}SO_3H$ in the area (1112.823-1062.729) cm^{-1}. The coloring substances are subsequently analyzed by UV-Vis spectrophotometry. The coloring substances made of soga tingi bark is applied by adding fixation compound such as alum, lime, and tunjung. Fixation soga tingi dye with tunjung produce black color, lime produces a brown color, and alum produces a reddish-brown color. The color of fixation dyes were occured from chemical interaction produce complexes compound between them.

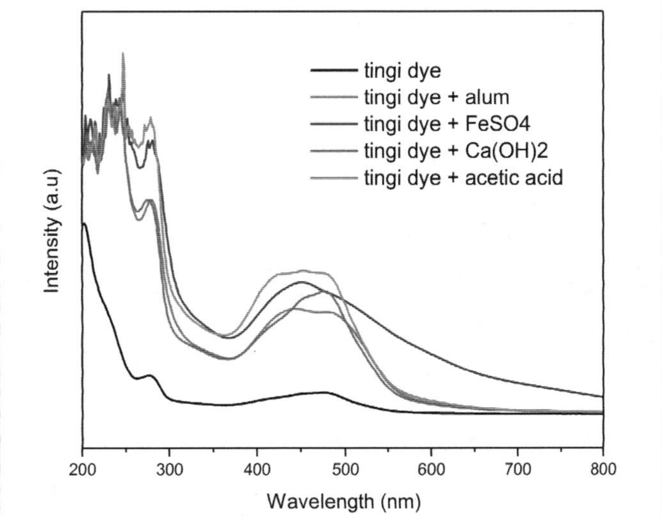

Figure 14. UV Visible Absorption spectras of tingi dye before and after addition fixation compounds of Alum ($xAl(SO_4)_2 \cdot 12H_2O$); x = K^+, NH_4+), Tunjung ($FeSO_4$), lime ($Ca(OH)_2$ and acetic acid.

IOP Conf. Series: Materials Science and Engineering **858** (2020) 012052 doi:10.1088/1757-899X/858/1/012052

As a resume, anthocyanin (rose flower), mango leaves (Mangifera indica), javanese curcumin (Curcuma zanthorrhiza), indigo fera, soga tingi bark (tannin), and javanese soga bark (secang) gives the final color as listed in Table 2.

Table 2. Natural dye applied and final color applied

Natural dyes	Color
Anthocyanin (rose flower)	Red
Mango leaves (Mangifera indica)	Light brown
Javanese curcumin (Curcuma zanthorrhiza)	Yellow
Indigo fera	Blur
Soga tingi bark(tannin)	Brownish red
Javanese soga bark (secang)	Red

4. Conclusion

In conclusion, natural dye has different chemical characteristics. The treatment process in the fixation in the batik coloring process has the respective influence on the color and color intensity. So the traditional coloring treatment process still requires chemical knowledge related to color expression.

Reference

[1] Allen R L M 1971 *Colour Chemistry* (London UK: Nelson)
[2] Robertson S M 1973 *Dyes From Plants* (New York USA: Van Nostrand Reinhold)
[3] Mongkholrattanasit R, Kryštůfek J, Wiener J, and Studničková J 2011 *Fibres and Textiles in Eastern Europe* **19**(2) 90–95
[4] Sivakumar V, Vijaeeswarri J, and Anna J L 2011 *Industrial Crops and Products* **33**(1) 116–122
[5] Wahyuningsih S, Ramelan A H, Wardani D K, Aini F N, Sari P L, Tamtama B P N, Kristiawan Y R 2017 *IOP Conf. Ser.: Mater. Sci. Eng* **193** 012047
[6] Laleh G H, Frydoonfar H, Heidary R, Jameei R and Zare S 2006 *Pakistan Journal of Nutrition* **5**(1) 90-92
[7] Wahyuningsih Wulandari S L, Wartono M W, Munawaroh H, and Ramelan A H 2017 *IOP Conf. Ser.: Mater. Sci. Eng.* **193** 012047
[8] Masibo H and He Q 2008 *Compr. Rev. Foof Sci. Food Saf* **7** 309-319
[9] Ribeiro S M R and Schieber A, *Bioactive compound in mango (Mangivera indica L.)* R R Waston, V R Preedy ed London Academic Press
[10] Ghoreishian S M, Maleknia L, Mirzapour H and Norouzi M 2013 *Fiber Polym* **14** 201-207
[11] Zhou Y, Zhang J, Tang R, and Zhang J 2015 *Ind Corp Prod* **64** 224-232
[12] Plan for a future. Indigofera hochst ex a rick. https://pfaf.org/user/Plant.aspx?LatinName=Indigofera
[13] Wahyuningsih S 2017 *IOP Conf. Ser.: Mater. Sci. eng* **193** 0120048

IOP Conf. Series: Materials Science and Engineering **858** (2020) 012053 doi:10.1088/1757-899X/858/1/012053

A novel method to study the twist dynamics in chiral nematic liquid crystals by texture analysis

R Suryantari[1], **Flaviana**[2]
[1,2]Department of Physics, Faculty of Information Technology and Science, Parahyangan Catholic University, Jl Ciumbuleuit 94, Bandung , Indonesia 40141

E-mail: ristisuryantari@unpar.ac.id

Abstract. We have experimentally studied the twist dynamics as a function of temperatures in chiral nematic liquid crystals by applying texture analysis to the image data observed by polarized light microscopy. These microscopic textures are analyzed using MATLAB software. The irregularity of texture at different temperatures observed by statistical parameter such as entropy indicate the changes of director orientation. Moreover, the changes in textural as a function of temperatures are useful to investigate the order parameter values of chiral nematic liquid crystals also to predict the behavior of twist dynamics. The order parameter values of the samples at different temperatures are obtained based on images intensity along twist-line using simple calculation then the behavior of twist dynamics are visualized. The study of dynamics of twist mode chiral nematic liquid crystals is important to observe the electro-optic effect in display technology or other applications.

1. Introduction

Liquid crystals are compounds having both characteristics like liquids and long range order like crystals. Two main classes of liquid crystals namely thermotropic and lyotropic liquid crystals. Typical of liquid crystal thermotropic phases are nematic, smectic, and cholesteric phases [1][2]. The cholesteric phase usually involves cholesteryl compounds shown in figure 1 [3]. The cholesteric liquid crystals have chirality of the molecules or twisted molecular arrangements, resulting in helical structures with certain periodicity. Another type cholesteric liquid crystal with non- cholesteryl compounds is chiral nematic liquid crystal. The phase may be categorized into nematic phases and called a chiral nematic phase [1][2][3].

On chiral or twisted molecular structure, it consists of quasi-nematic layers whose individual directors are turned by a fixed angle from one layer to the next layer. Each layer has different molecular direction. The distance of two layers with same orientation angle are equivalent to 2π namely pitch (p) as shown in figure 1. The helical pitch is thermally responsive to changing reflective colors [3].

Cholesteric liquid crystal or chiral nematic liquid crystal possesses a number of properties of unique optical effects referred that it is birefringent because of anisotropic characteristic [1] [3]. For any anisotropic materials, different orientations respect to the molecular alignment may cause the incident light be transmitted with different velocities. When unpolarized light enters to chiral nematic, incident light selectively polarized, commonly become two directions namely ordinary polarized ray and extraordinary ray polarized [2][4][5]. In the case of chiral nematic liquid crystal sample with the homogeneous planar alignment placed between two crossed polarizers, the texture appears completely black when the director is either parallel or perpendicular to the transmission axis of the polarizer or analyzer [2].

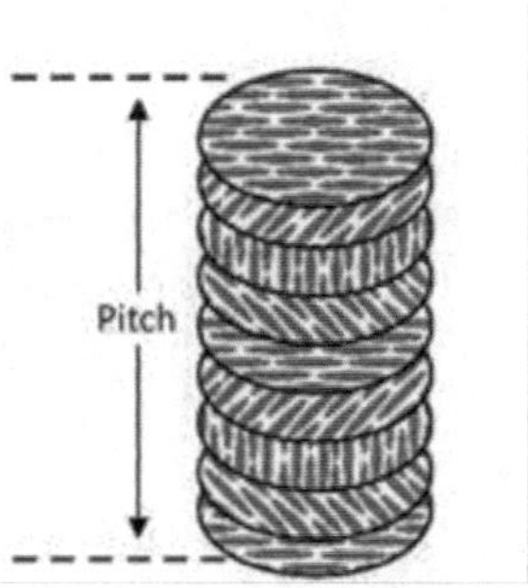

Figure 1. Molecules alignment of cholesteric or chiral nematic structure

The study of dynamics of cholesterics is important in the context of applications that make use of the electrooptic effect [6][12]. A technique which combines image analysis and polarising optical microscope (POM) is a useful tool for the physical investigation of liquid crystals. Textural features of the samples provided the information approximately suitable to determine the optical properties of the samples such as optical transmittance, absorption coefficient, birefringence, and order parameter [6][12]. A new image processing method is developed in order to predict the behaviour of twist dynamics of chiral nematic liquid crystals based on entropy parameter using feature extraction and the order parameter value using simple calculation taken from statistical data.

2. Experimental

The chiral nematic liquid crystals with 50 μm spacing are obtained from sfcx Inc without any treatment on orientational order. Temperature dependent textural investigations are carried out by polarizing optical microscope attached to hot stage. The color image detected by the camera which represents the true color pixel tone that ranges from 0 to 255 in RGB colors under crossed polarizer. To compute thermo optical properties, textures in RGB images converted to monochromatic images. We choose HSV image (hue, saturation, value) and only focus on value parameter to simplify the value of pixel tone that ranges from 0 to 1. Value (V) parameter represent the brightness of the color or on the other hand it is a representation of intensity. The higher value show the greater in intensity [13].

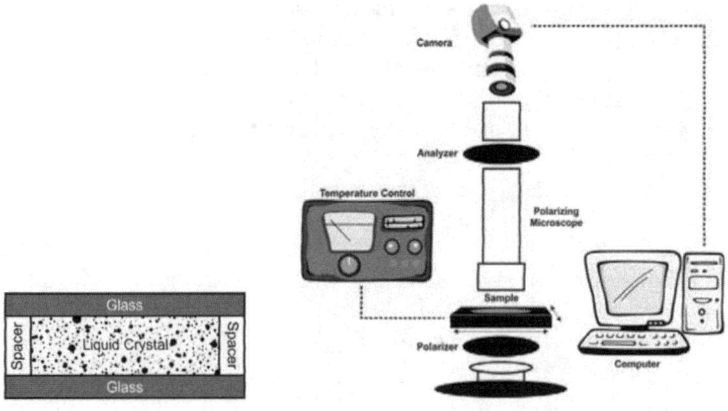

(a) (b)

Figure 2. (a) Side view of sample. (b) Instrumentation set-up

IOP Conf. Series: Materials Science and Engineering **858** (2020) 012053 doi:10.1088/1757-899X/858/1/012053

3. Results and Discussions

Figure 3 shows chiral nematic liquid crystal monochromatic textures at different temperatures. We observe no visual change until the temperature T = 31.5 °C. The textures change significantly at T = 31.5 °C to T = 32°C. As increasing on temperature, we observe the regularity of patterns. The crossed patterns appear here caused by birefringence effect of chiral nematic liquid crystals and the consequence of polarizing light by crossed polarizer [1][3][15]. For further literature review, this crossed patterns namely double twist [16-19]. To verify this observation, we take statistical data based on images intensity as shown in figure 3, and applied first order feature extraction by entropy parameter using Matlab [13][14]. The graph of entropy value shown in figure 4.

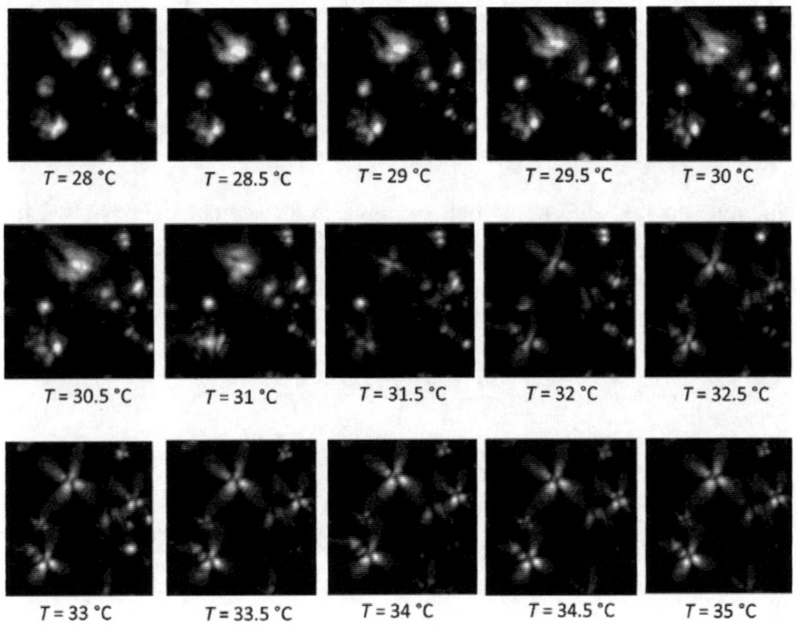

Figure 3. Experimental textures of an chiral nematic liquid sample at different temperatures.

Figure 4 shows the Entropy value as a function of temperatures. Entropy is a measure of the randomness of image intensity. The greater of entropy, the more irregular intensity of each pixel [13][14]. Near critical temperature (T_c), at T_c = 31.5, indicated by dash line at figure 5, the entropy extremely decrease which represent the regularity of pattern. We note that phase transition as shown in figure 3 is properly with the quantity result of entropy measures. In accordance with the visual observation based on figure 3, the maximum decreasing in entropy occurs until double twist patterns achieved experimentally at T_c = 31.5 °C.

It is obviously seen the change of texture through the increasing of temperature by figure 3. It might be possible to predict the dynamics of twist pattern by statistical data of intensity along twist-line. By simple method, we plot the various value of intensity along twist-line at different temperature and compare the results. It gives information about the intensity that transmitted by sample, the different intensity value means the different of director orientation. Hence, we can determine the director orientation by θ angle to the initial director ($\vec{n}$) [1][2][15].

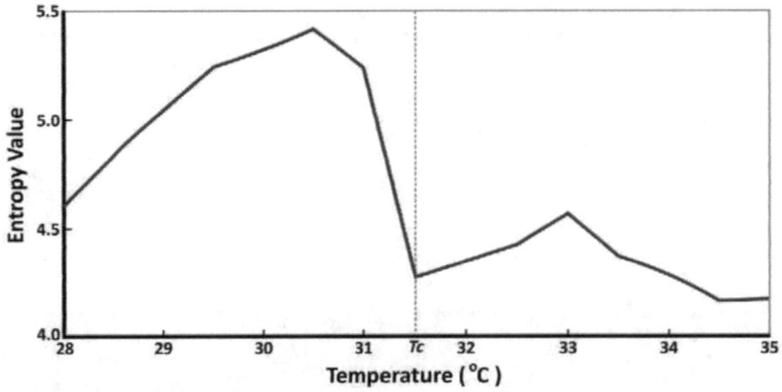

Figure 4. Entropy value as a function of temperatures.

Schematic diagram showing the orientation of single rod-like molecules shown in figure 5. The orientation of a rod-like molecule can be represented by a unit vector ($\vec{a}$) a which is attached to the molecule and parallel to the long molecular axis [15].

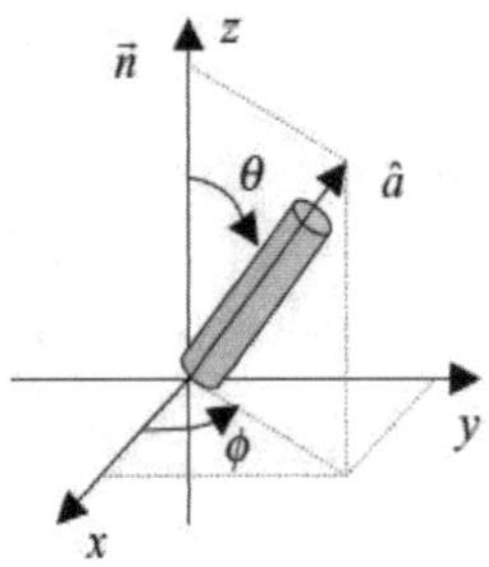

Figure 5. Schematic diagram showing the orientation of single rod-like molecules [15]

Chiral nematic liquid crystal consists of quasi-nematic layers whose individual directors are turned by a fixed angle from one layer to the next layer. Each layer has different molecular direction [3]. A simple approach for observing the twist dynamics in chiral nematic liquid crystal consists by taking the order parameter equation in nematic liquid crytal by argument that the different intensity of images as a result of different orientation of director and the different orientation of director could be represented by order parameter (*S*). The order parameter is usually defined in such a way that it is zero in the high-temperature unordered phase and non-zero in the low-temperature ordered phase. The average value of the second-order Legendre polynomial for the order parameter for nematic liquid crystal shown in equation 1 [15].

$$S = \left\langle \frac{1}{2}\left(3\cos^2\theta - 1\right) \right\rangle \quad (1)$$

Based on equation 1 we note that S =1 is a condition for the orientation of molecules parallel with the initial director ($\vec{n}$) and correspondent to the maximum intensity value of images. Otherwise, S = -0.5 is for isotropic states and correspondent to the minimum intensity value of images [15].

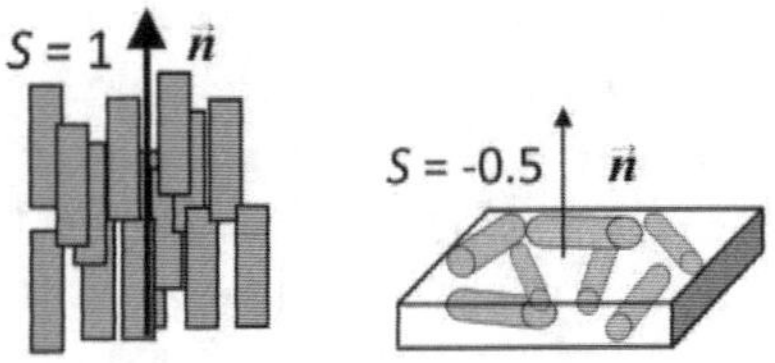

Figure 6. Schematic diagram showing the states with different order parameters for $S = 1$ and for $S = -0.5$ [15]

Applying the normalization of intensity value along line segment of twist-line we get θ value variation between 0 – 90° and then using equation 1 to calculate the S value. The example of taking a line segmentation along double twist pattern shown by figure 7.

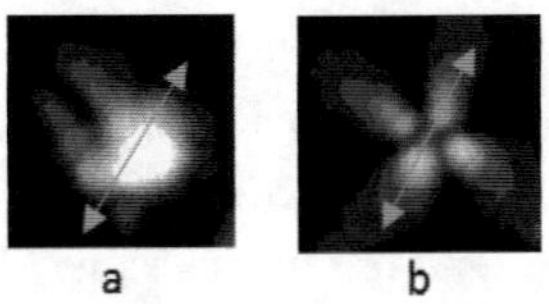

Figure 7. Line segmentation along double twist pattern for (a) T = 28 °C and (b) T= 35 °C shown by arrow symbols

Figure 8 illustrate the various value of order parameter between -0.5 to 1. For a line segmentation in figure 7, we note that $S = 1$ defines the bright images correspondent to θ = 0° and S = -0.5 for dark image correspondent θ = 90 ° as illustrated in figure 6.

To simplify the analysis we only choose the result for T = 28 °C and T = 35 °C which initial and final temperature in the range of experimentally set up. Consider only the region between dash lines vertical in figure 7, the different of order parameter value between dash lines vertical show the dynamic of twist-line at higher temperature. For T = 28 °C the similar value of S represent the similarity of intensity image along line segmentation of twist pattern, it means that the all the director orientation parallel to $\vec{n}$, based on figure 7a it obviously seen as bright image or maximum intensity along line segmentation.

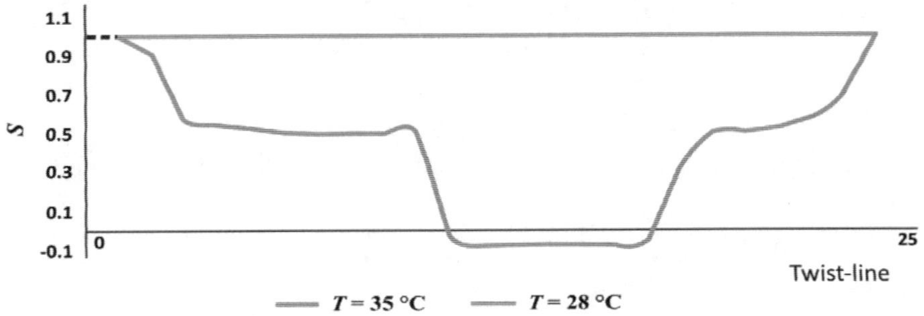

Figure 8. Comparison of order parameter along twist-line for T = 28 °C and T = 35 °C. The different on value between dash lines vertical show the dynamic of twist-line at higher temperature. The decreasing of order parameter represent to the different orientation of director along twist-line.

Reinitzer in 1888 observed under a polarizing microscope a mesogenic compound made of chiral molecules and noticed blue colour phenomenon, which appears and then quickly disappears but still liquid. It was exhibited in the narrow temperature range between the chiral nematic and the isotropic phases. Thus blue phases appear as the three-dimensional counterpart of chiral nematics. The blue phases are generated by double-twisted cylinders separated by defect lines. De Gennes introduce the schematic of director orientation in double twist pattern as shown in figure 8 [2][16][17][18][19].

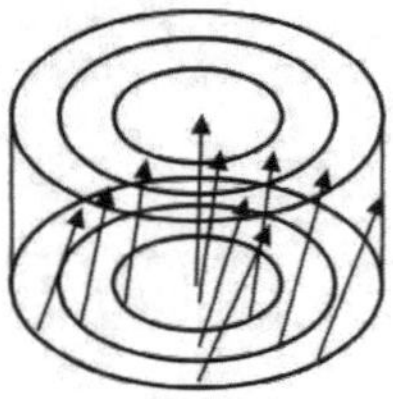

Figure 9. Schematic representation of a double twist.[2]

As temperature increased or decreased, the texture change depending on alignment of molecules in sample. At the lowest temperatures the texture remains same like solid phase (shown in figure 7a) with S = 1, and at the highest temperatures the texture remains tend to isotropic phase by appearance of dark color (shown in figure 7b) with S << 1. In isotropic phase, the destruction of molecular alignment from their respective phases causes the disappearance of birefringence property and it leads to the decrement of Birefringence value. On the other hand, the various value of S illustrated in figure 7 at T_{max} = 35 °C bring to the decrement of Birefringence value. The minimum of S indicate the absence of Birefringence and it cause the appearance of dark color.

For further analysis we can predict the behavior of molecular alignment based on the graph of figure 7. There is no change in molecular alignment at T = 28 °C, all directors have similar orientation. If we choose vertical axis for director orientation at T = 28 °C then directors orientation for T = 35 °C will various in θ , which θ is the angle director orientation in vertical axis to the final directors orientation as shown in figure 5. Figure 9 is an illustration of molecular alignment at T = 28 °C and T= 35 °C.

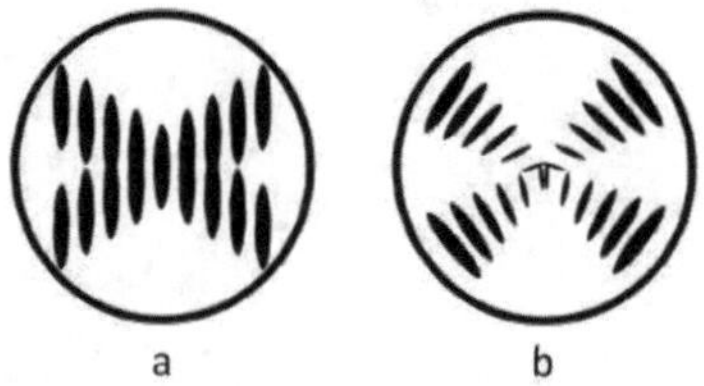

Figure 10. Schematic of molecular alignment at (a)T = 28 °C and (b) T = 35 °C

The schematic shown in figure 9 is an approach for one double twist pattern, it is clear that there are many patterns that appears on microscopic image in chiral nematic liquid crystal sample. Theoretically, the value of order parameter S =1 means all of directors align at similar orientation [15]. However we only draw molecular orientation in figure 9a in accordance with the molecular orientation in figure 9b so that we can distinguish the patterns easier. The illustration in figure 9b is not exactly similar to the order parameter value in figure 7, in fact it might be different. However, the difference of schematic as shown in figure 9a and 9b illustrate qualitatively the twist dynamics between range of temperatures at T = 28 °C until T = 35 °C. It approximately suitable to the theoretical as derived as double twist schematic in figure 8 [2].

4. Conclusions

Textural of the samples microscopic images as function of temperatures provide the information approximately suitable to determine the optical properties of the chiral nematic liquid crystal sample based on the difference of intensity of texture. As increasing on temperature, the change of patterns apprear caused by birefringence effect of chiral nematic liquid crystals and the consequence of polarizing light by crossed polarizer. Entropy parameter useful to illustrate the irregularity of intensity value in monochromatic images. It derive to the argument that there are difference in director orientation of molecules alignment on sample. The graph of entropy as temperature function gives us information about critical temperature from chiral nematic to isotropic phase, which a double twist pattern appear. Moreover, Approach to nematic liquid crystal to calculate the order value parameter. The various of order parameter value derive the prediction of twist dynamics in range of temperatures set up. Based on the behavior of order parameter value, we can illustrate the schematic of molecular alignment to observe clearly the twist dynamics at lowest and highest temperature.

Acknowledgment

The authors would like to thank to Research Department of Parahyangan Catholic University for the financial support and Material Laboratory of Gadjah Mada University for the facilities of research.

References

[1] S. Chandrasekhar. *Liquid Crystal*. Cambridge University Press. New York. 1992
[2] de Gennes PG, Prost J. *The Physics of Liquid Crystals,* 2nd ed. Oxford University Press. Oxford.1993
[3] Hallcrest. Handbook of Thermochromic Liquid Crystal. Glenview, IL. Hallcrest. 1991
[4] Redouane Borsali, et al. 1998 *Phys. Rev. E* **58**
[5] Whitfield, C.A, et al. 2017 *The European Physical J. E* **40** 50
[6] B. Montrucchio et al 1998 *Liquid Crystals* **24** 6
[7] Sparavigna A. Mello and B. Montrucchio 2006 *Phase Transitions* **79** 293
[8] Ch. Venkayamma and S. Sreehari Sastry. 2017 *Int. J. Eng. Research and Tech* **6** 03
[9] S. Sreehari Sastry, et al 2013 *Liquid Crystals* **40 (3)** 384
[10] S. Sreehari Sastry et al 2012 *Liquid Crystals* **39** 12
[11] H. Y. D. Sigaki, R. F. de Souza, R. T. de Souza, R. S. Zola, and H. V. Ribeiro. 2019 *Phys. Rev.E* **99**
[12] Alexander Ryabchun, et al 2019 *Appl. Material and Interfaces* **11** 10895
[13] Gonzales, R.C., Woods, R.E. *Digital Image Processing*, 2nd ed. Prentice Hall. 2002
[14] Information on https://www.mathworks.com/help
[15] Yang, Deng-Ke & Wu, Shin-Tson. *Fundamentals of Liquid Crystal Devices*. John Wiley&Son Ltd. 2006
[16] Tanaka S, et al. Double-twist cylinders in liquid crystalline cholesteric blue phases observed by transmission electron microscopy. *Scientific Report* 2015
[17] Kikuchi H. (2008). Liquid crystalline blue phases. Liquid Crystalline Functional Assemblies and Their Supramolecular Structures. Vol 128 ed T Kato (Berlin: Springer) pp 99–117
[18] MD Asiqur Rahman et al 2015 *Sci. Technol. Adv. Mater.*
[19] Sala-Tefelskaa, MM et al. (2018). Influence of cylindrical geometry and alignment layers on growth process and selective reflection of blue phase domain._

IOP Conf. Series: Materials Science and Engineering **858** (2020) 012054 doi:10.1088/1757-899X/858/1/012054

Effect of DVB on Adsorption of Methylene Blue using Coply(Eugenol-Strearyl acrylate-DVB)

D S Handayani[1], A Masykur[1], N Nareswari [1]
Department of Chemistry, Faculty of Mathematics and Natural Sciences, Sebelas Maret University

E-mail: desisuci@staff.uns.ac.id

Abstract. This research aims to study the effect of divinylbenzene (DVB) on copoly(Eugenol-Stearyl Acrylate-Dvinylbenzene)/Copoly (Eg-SA-DVB) on the adsorption of methylene blue dyes. Variations of DVB ware added as much as 2%, 4% and 6% to the weight of eugenol. Adsorption was carried out by batch method in variations of acidity (pH), contact time and initial concentration of methylene blue. The results showed that the optimum pH of methylene blue dye adsorption ware pH 5.0 for copoly(Eg-DVB-SA) 2% and 6%; while for 4% was pH 3. The optimum adsorption contact time was 60 minutes for copoly(Eg-DVB-SA) 2% and 6%, while for 4% at 50 minutes. Study of adsorption kinetic presented the adsorption of methelene blue dyes using 2%, 4% and 6% copoly (Eg-DVB-SA) adsorbents followed the Largregen kinetic model, pseudo-first order. Study of isotherm revealed the adsorption of methylene blue dyes tends to follow the Langmuir isotherm. The maximum capacity of methylene dyes adsorption using copoly(Eg-DVB-SA) 2%, 4% and 6% were 31,2; 36,2 and 14,7 mg/g, with the adsorption energy 31.1; 33.6 and 31.3 KJ/mol respectively.

1. Introduction

Since the 19th century, synthetic dyes began to take over the role of natural dyes because their production is easier, the price is cheaper, the range of color variations is more high and the ability and stability of coloring is higher [1] . Dyes are organic compounds. The structure of the compound is designed to have resistance to time, exposure to sunlight, water, soap, microbial invasion and oxidizing agents so that it cannot be treated with conventional wastewater treatment processes [2]. Dyes can be classified into several types based on their use namely, acidic (anionic), basic (cationic), dispersive, direct, reactive, solvent, sulfur and vat [3].

Metylen blue (MB) is a synthetic cationic dyestuff with an aromatic amine structure that is often used in the textile industry for coloring leather, mori, cotton and wool fabrics [4]. Prolonged direct contact with MB can cause vomiting, nausea, anemia and hypertension [5]. The increasing level of contamination of dyes in water bodies requires the development of exploration methods of wastewater treatment that can reduce and eliminate levels of contaminants either physically, biologically or chemically [6].

The study of the method of handling waste containing cationic methylene blue (MB) in Indonesia that has been developed is by Susilawati [4] with the adsorption method using active bentonite clay. Dwiasi et al. [7] carried out MB photodegradation in batik waste using the Fe_2O_3-UV-H_2O_2 system. Is Fatimah et.al. [8] adsorbs MB using activated carbon from the melinjo shell. Badriyah and Putri [9] performed MB adsorption with eggshells. Rizqi and Purnomo [10] tested the biodegradation activity of Daedalea dickinsii fungi against MB dyes. Saadon et al. [11] used palm ash as an MB dye adsorbent. Latupeirissa et al. [12] carried out MB adsorption using activated carbon from the pecan skin.

Adsorption method is a method that has been widely explored to reduce levels of dyes from water bodies. The study of adsorption in Indonesia is prioritized in the search for adsorbents that are economical, effective, efficient and can be made using natural resources or modified waste [13]. One of these adsorbents can be made from eugenol modifications. Eugenol is a major component of clove oil which is one of the leading products with large commodities in Indonesia [14]. Eugenol has a vinyl functional group that allows it to be modified with other compounds through a polymerization reaction. Previous studies that have been carried out regarding the use of eugenol-based polymers as adsorbents were carried out by Silvianti et al. [13] the event of adsorption of dyes with polieugenol-derived adsorbents can occur because there is the formation of hydrogen bonds between $[Fe(OH)]^{2+}$ with the —OH group of eugenol.

In this study copoly(Eugenol-Stearil Acrylate-Divinylbenzene) or copoly (Eg-SA-DVB) as adsorbent for methylene blue dye. Eugenol compounds are modified with stearyl acrylate (SA) and divinylbenzene crosslinking agents (DVB), where the addition of DVB crosslinking agents directs the formation of 3-dimensional structured polymers such as nets allowing particles of a size corresponding to the size of the net to be trapped [15]. In addition, the anion carbonyl as active group from the SA monomer and the —OH group from the eugenol monomer allow copoly (Eg-SA-DVB) to be used as an adsorbent for methylene blue cationic dyes through the formation of hydrogen bonds and electrostatic interactions.

This research studies the adsorption parameters of coply (Eg-SA-DVB) are the influence of pH, contact time and adsorbate concentration; besides also studying the kinetics and isotherms adsorption.

2. Experimental

2.1. Ingredient

All chemical used in research were procured fram E Merck, expect special ones, Copoly(Eg-SA-DVB) 2 %, 4 % and 6 % (synthesis product), NaOH, HCl

2.2. Instrument

Spectrophotometer UV *Visible Shimadzu,* Spectrometer *Fourier Transform Infra Red* (FTIR) *Shimadzu Type* FTIR-8201 PC KBr, *Scanning Electron Analysis* (SEM) (JEOL JSM 6360 EDX (JED-2200))

2.3. Prosedure

2.3.1. Adsorbent Stability Test

A total of 10 mL Copoly (Eg-SA-DVB) was put into the solution with a pH variation at 2, 4, 6, 8 and 10. The mixture was stirred for 2 hours, then filtered. The final mass of the adsorbent is weighed and compared to the initial mass.

2.3.2. Effect of pH on adsorption of methylene blue

Methylene blue with a concentration of 10 ppm of 10 mL each was adjusted to pH 1, 2, 3, 4, 5 and 6 using HCl and NaOH. Then copoly (Eg-SA-DVB) of 5 mg was added to each solution and stirred for 30 minutes. Sample concentrations before and after adsorption were analyzed by UV Vis spectrophotometer at maximum wavelength. The analysis results are used to determine the optimum pH.

3. Results And Discussion

The concentration of Methylene blue (MB) adsorbed was determined by UV Vis spectrophotometer at a wavelength of 664 nm. Adsorption of MB dye using copoly(Eg-SA-DVB) at 2%, 4% and 6% DVB variations was assessed based on several adsorption parameters including pH, contact time and initial concentration of the dye.

IOP Conf. Series: Materials Science and Engineering **858** (2020) 012054 doi:10.1088/1757-899X/858/1/012054

Effect of pH on Adsorption

The acidity (pH) of the solution is one of the factors that influence the adsorption process of MB cationic dyes using copoly(Eg-SA-DVB) in the variation of DVB. The effect of the pH of the MB solution on the adsorption capacity was shown in Figure 1.

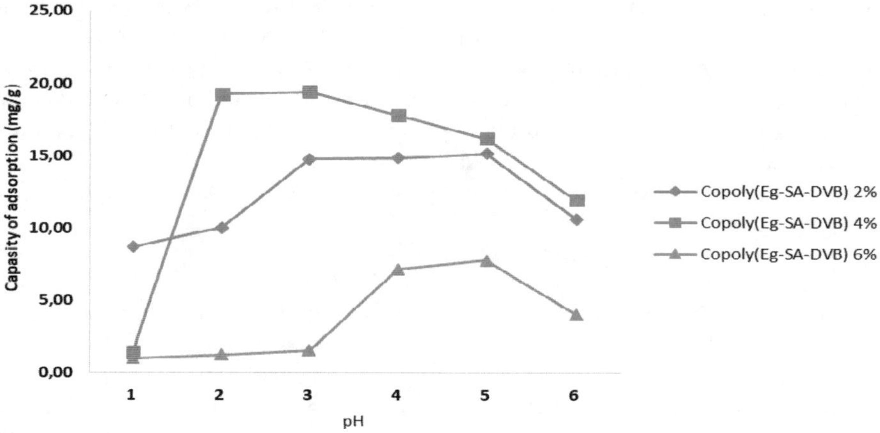

Figure 1. Effect of pH on adsorption of MB dyes

Copoly (Eg-SA-DVB) at variation of DVB 2% and 6% had an optimum pH of 5 with adsorption capacities at 15.18 and 8.21 mg/g, whereas MB Adsorption with copoly(Eg-SA-DVB) the variation of DVB 4% had an optimum pH of 3 with an adsorption capacity of 19.45 mg/g. The negatively charged active group on the surface of copoly(Eg-SA-DVB) can be protonated at low pH so that it can cause electrostatic repulsion between MB cation and the adsorbent surface which are both positively charged. This was demonstrated in MB adsorption by copoly(Eg-SA-DVB) at 6% DVB variation, where the adsorption capacity was higher with increasing pH. In addition, it is possible that MB can also be protonated at low pH to colorless leuco-methylene blue (LMB).

The copoly adsorption capacity (Eg-SA-DVB) at variations of DVB 2% and 4% was higher than 6%, this indicates that the higher DVB composition of the adsorbent leads to the formation of copolymer structures with the number of negatively charged active groups on the surface, so the MB dyed adsorption decreases.

Effect of Contact Time on Adsorption

The effect of contact time on adsorption of methylene blue (MB) dyes is shown in Figure 2. The adsorption rate is higher at the initial adsorption time because the surface area of the adsorbent available for MB adsorption is still large. After reaching a certain time, the adsorption rate decreases as the surface active side of the adsorbent decreases.

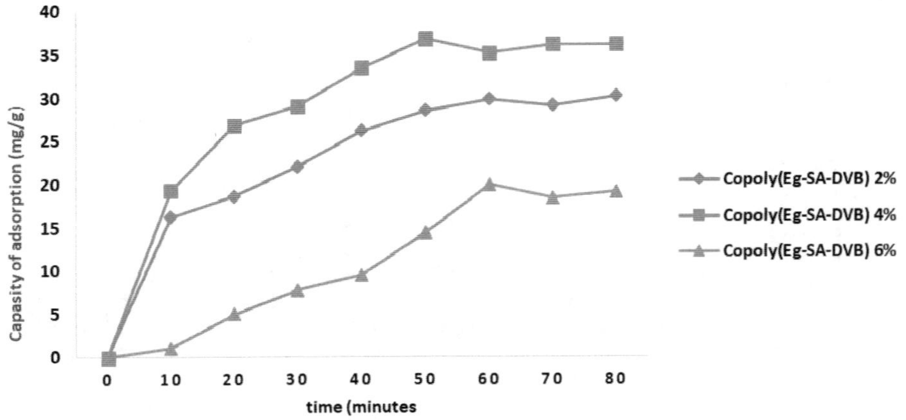

Adsorption of MB using copoly(Eg-SA-DVB) at a variation of 4% DVB had an optimum contact time at 50 minutes with an adsorption capacity of 36.84 mg/g. As for copoly(Eg-SA-DVB) at variations of 2% and 6% DVB showed an optimum contact time of 60 minutes with adsorption capacities of 14.92 and 19.95 mg/g. The adsorption rate decreased at the end of the experiment because the adsorbent was saturated after equilibrium was reached.

The adsorption kinetics can be assessed from experimental data on the influence of contact time. The kinetics test was performed on the pseudo first order (lagergren) and pseudo second order (Ho) models. The results of the kinetics model of the adsorption of the three copoly(Eg-SA-DVB) against MB dyes are presented in Table 1.

Table 1. Kinetic of adsorpsi MB with coply(Eg-SA-**DVB**) at variation of DVB

Adsorben	**Lagergren**		**Ho**	
coply(Eg-SA-DVB)	K (g/mg.min)	R^2	K (g/mg.min)	R^2
DVB 2 %	5.78×10^{-2}	0.95	6.5×10^{-3}	0.68
DVB 4%	5.64×10^{-2}	0.98	3.1×10^{-3}	0.84
DVB 6%	2.40×10^{-2}	0.90	11×10^{-3}	0.77

The adsorption kinetics model was determined based on the higher coefficient of determination (R^2) between the Lagergren and Ho kinetics models. The three copoly(Eg-SA-DVB) at variations of DVB 2%, 4% and 6% have higher R^2 values in the Lagergren pseudo first order kinetics model. This shows that the adsorption process using the three copoly(Eg-SA-DVB) against MB dyes is more likely to follow the Lagergren kinetics model (first-order pseudo).

Effect of Initial Concentration on Adsorption

The effect of initial concentration of methylene blue (MB) dyes on adsorption with copoly(Eg-SA-DVB) on the variation of DVB 2%, 4% and 6% is shown in Figure 3.

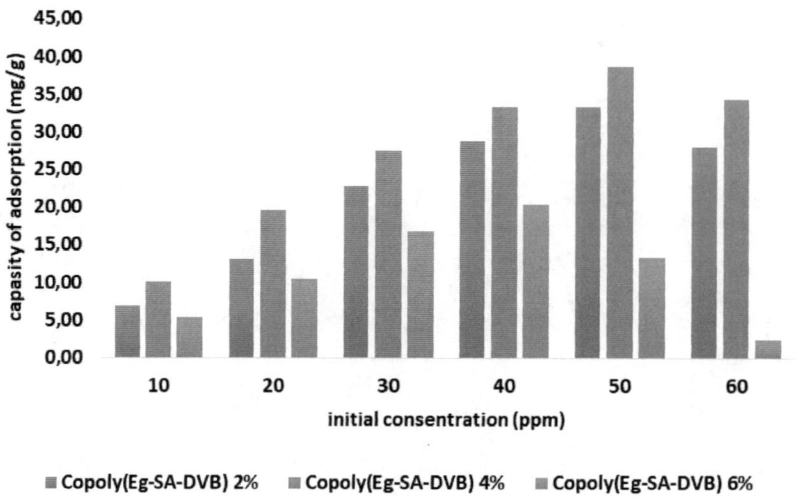

Figure 3. Effect of initial concentration on adsorption of MB

Figure 3 shows that the adsorption capacity increases with the initial MB concentration increase. The adsorption capacity continues to increase until the optimum adsorption capacity was obtained, when the adsorption process has reached equilibrium. The decrease in MB adsorption at higher concentrations of concentration is likely due to MB cation monomers forming dimeric compounds or other aggregates. The results of determining the effect of the initial concentration of the dye were used to study the MB adsorption isotherm with copoly (Eg-SA-DVB). The results of the adsorption isotherm study are shown in Table 2.

Tabel 2. Isoterm adsorption of MB with copoly(Eg-SA-**DVB**)

Isoterm Adsorpsion	Copoly(Eg-SA-**DVB**)		
	DVB 2%	**DVB 4%**	**DVB 6%**
Langmuir :			
R^2	0.97	0.99	0,92
X_m (mg/g)	31.20	36.22	14,66
K_L (L/mol)	24.11×10^4	64.94×10^4	25.30×10^4
E_{ads} (kJ/mol)	31.12	33.60	31.28
Freundlich :			
R^2	0.50	0.61	0.35
n	3.11	3.80	2.93

Isotherms adsorptionof MB using all three copoly(Eg-SA-DVB) tend to follow the Langmuir isotherm (higher R2 value). The Langmuir model assumes that the monolayer layer is formed when adsorption takes place without interaction between the absorbed MB molecules. Adsorption energy can be used to estimate the type of adsorption, if the value is below 8 kJ / mol shows a tendency towards physical adsorption, while for 8-16 kJ / mol indicates adsorption can be explained through ion exchange reactions, and if> 16 kJ / mol shows more adsorption tends to occur more strongly by chemical adsorption than ion exchange reactions. It can be concluded that the three adsorbents show that the adsorption process occurs chemically.

4. Conclusion

Adsorption of MB using copoly(Eg-SA-DVB) in the variation of DVB 2%, 4% and 6% tends to follow the kinetics of Lagergren pseudo first-order and Langmuir isotherm.

Acknowledgment
The financial support from Sebelas Maret Univervesity through the research project of Fundamental Reasearch 2019 is greatly appreciated.

References

[1] Shabbir, M., Islam, S. U., Bukhari, M. N., Rather, L. J., Khan, M. A., and Mohammad, F. 2016 *Textiles and Clothing Sustainability* **2(1)** : 1-9.
[2] Khodaie, M., Ghasemi, N., Moradi, B., dan Rahimi, M. 2013 *Journal of Chemistry*, 1-6.
[3] Berradi, M., R Hsissou, M Khudhair, M Assouag, O Cherkaoui, A El Bachiri, A El Harf, 2019 Heliyon **5**, 1-11
[4] Susilowati 2014 *Jurnal Riset Kimia*, **7(2)** : 169-173
[5] Shahryari, Z., Goharrizi, A.S., dan Azadi, M., 2010. *International Journal of Water Resources and Environmental Engineering*, **2(2)** : 16-28
[6] Lachheb, H., Puzenat, E., Houas, A., Ksibi, M., Elaloui, E., Guillard, C., dan Herrmann, J.M., 2002 *Applied Catalysis B : Environmental*, **39(1)** : 75-90.
[7] Dwiasi, D.W., Setyaningtyas, T., dan Riyani, K., 2018. *Jurnal Rekayasa Kimia dan Lingkungan*, **13(1)** : 78-86
[8] Is Fatimah , A. Yahya , R A T Sasti, 2017 AIP Conference Proceedings 1823, 020125 : 1-6
[9] Badriyah, L., dan Putri, M.P., 2017. *Alchemy*, **5(3)** : 85-91.
[10] Rizqi, H.D., dan Purnomo, A.S. 2014 *Jurnal Seni dan Sains*, **2(1)** : 1-6.
[11] N Saadon, N Razali, M Mat Yashim and N A Yusof 2016 *ARPN Journal of Engineering and Applied Sciences* 11(5) : 6191-6194
[12] Latupeirissa, J., Tanasale, M., dan Musa, S., 2018 *Indonesian Journal of Chemical Researrch*, **6** : 524-533.
[13] Silvianti, F., Siswanta, D., Aprilita, N., dan Kiswandono, A.A., 2017 *Jurnal Natural* **17(2)** : 108-117
[14] Wijaya, C., Jayuska, A., dan Alimuddin, A.H., 2015 *JKK* **4(4)** : 15-20.
[15] DS Handayani 2004 Biopharmacy Journal of Pharmacological and Biological Sciences 2(20 : 54-57

IOP Conf. Series: Materials Science and Engineering **858** (2020) 012055 doi:10.1088/1757-899X/858/1/012055

Power efficiency analysis in various types of coil design

N Sari[1,2], M Khoiro[1], AY Rohedi[1], G Yudoyono[1], and YH Pramono[1]
[1] Physics Department, Faculty of Scientics, Institut Teknologi Sepuluh Nopember Surabaya, Indonesia
[2] Physics Study Program, Faculty of Mathematics and Natural Science, University of Lambung Mangkurat, Indonesia

E-mail: nurmasari01@ulm.ac.id

Abstract. The efficiency of power produced by magnetic flux in the coil has been analysed using various forms of coil section. Magnetic flux was produced by magnets which were moved radially with a certain rotational speed. Meanwhile, a coil placed close to the magnets will produce an electromotive force (emf) according to Faraday's law. Besides affecting the magnitude of emf, the shape of coils also affected the power loss due to partial of magnetic field radially dispersed or didn't go through a coil cross-section. It caused only a portion of the magnetic field would be converted. Therefore, this study aims to reduce the power loss by various type of coil design i.e. round-shaped coil, square-shaped coil, and cone-shaped coil. The coil was formed by the same wire with a length of 60 m to produce the same resistance. The results showed that at a speed of 1100 rpm, the voltage which obtained using round-shaped coil, square-shaped coil, and cone-shaped coil respectively of 1.26 V and 1.01 V and 1.5 V and 1.88 V.

1. Introduction

Eddy currents are electric currents that circulate in conductor induced by changing the magnetic field, flow in closed loops and perpendicular to the magnetic field. It can be produced either when the conductor moved in a fixed magnetic field or when changed magnetic field induced the conductor. It is corresponding to law of Lens. The currents can cause repulsion, traction, impulses, and a warming effect. Eddy currents are also known as the main cause of energy loss in transformers and electric motors [1]. It due to Eddy's tendency that against the cause.

Many researchers have been worked to reduce the loss of energy. The effect of eddy currents had been reduced both in simulation and experiment [2-4] by designing the place of permanent magnets in the rotor, substituting the material of magnets and varying thickness of the rotor [3]. Varying shape of the coil also have potential to reduce the eddy current effect. Coil shape had been analysed in many previous studies [5-9]. The analysis is carried out using coils in many shapes such as circles, rectangular and triangular. However, conical shaped coils have not ever been analysed. In this paper, varied coil's shape which is circular shaped coil, a rectangular shaped coil and conical coil shape was analysed to reducing Eddy current effect and improving power efficiency. The efficiency can observe from the value of voltage obtained in each shaped of coil. The shape of coil that obtain highest voltage and longest retardation time potentially used in permanent axial magnetic flux generator applications.

IOP Conf. Series: Materials Science and Engineering **858** (2020) 012055 doi:10.1088/1757-899X/858/1/012055

2. Experiment

2.1 Materials and design

Experimental of this study was made using some equipment such as power supply, speed control devices, motors, rotors, coils, and multimeters. The rotor in this study is similar to the rotor on a disc-shaped axial flux permanent magnet synchronous generator. The rotor was contained ten neodymium (NdFeB) permanent magnets implanted on the disc. The magnets are coin-shaped with dimensions of 20 x 10 mm. The magnets are arranged symmetrically with the poles in the opposite direction to provide *N-S* magnetization as shown in Figure 1. The disk of rotor was made of acrylic material with a diameter of 24 cm.

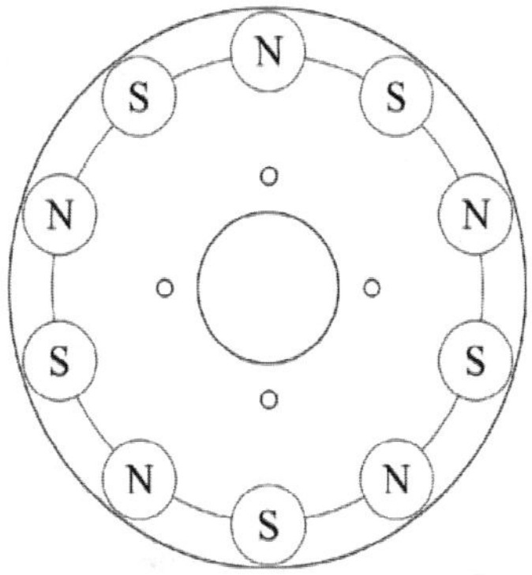

Figure 1. Magnet arrangement in disk of rotor

The coils were made from copper wire (Polyurethane Email wire - UEW type) with a diameter of 0.5 mm and length of 60 m. Email type of wire is chosen due to more flexible, resistant to repeated loops, and the protective layer which easier to peel and clean. The emf was analyzed in varied shaped coils i.e. circular coil, square coil and conical coil as shown in figure 2. The rectangular coil has a size of 5 x 5 cm. Diameter and depth of the circular coil are 5 cm and 5 cm respectively. The cone-shaped coil has two face sections which are front face and back face with diameter of 5 cm and 2.5 cm respectively. It has effective depth of 5 cm.

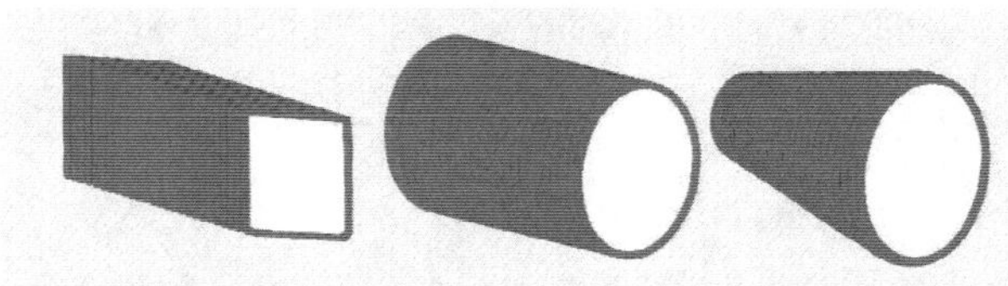

Figure 2. Circular coil (left), square coil (center) and conical coil (right)

2.2 Metode

While the rotor rotated, the coil was placed close to the outer side of the rotor and parallel with the section of magnet. Experiment had been done using the equipment scheme as shown in Figure 3. The wire at the end of coil is connected to multimeter to measure the generated voltage. In this study, measurement of voltage was obtained in the range of rotor's speed from 200 rpm to 1100 rpm. The deceleration data was recorded after the voltage had stabilized and the power supply was turned off. the deceleration data is the required time when the power supply was turned off until the rotor stops.

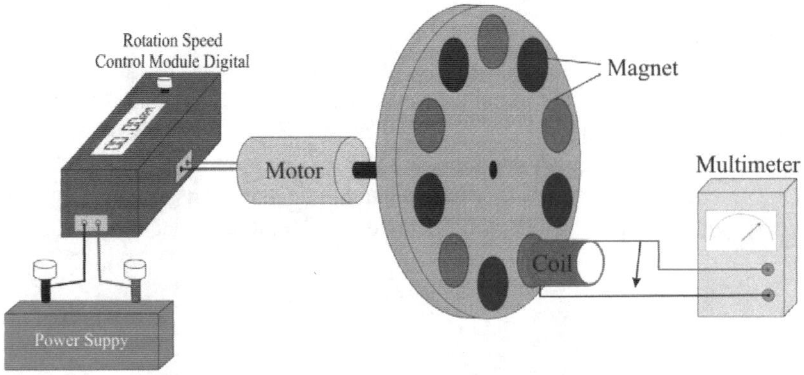

Figure 3. Arrangement of measurement equipment

3. Results and Discussion

Coils with round-shaped coil, square-shaped coil, and cone-shaped coil made of wire of the same diameter and length have the same resistance value in each coil. In general, the voltage increases when the rotation speed value on the rotor is getting bigger. At the same rotation velocity value, the voltage value from the smallest to the largest is obtained sequentially provided by square-shaped coil, round-shaped coil, a wider section coil and a smaller section cone coil (Figure 4). The results showed that at a speed of 1100 rpm, the voltage which obtained using round-shaped coil, square-shaped coil, a wider section coil and a smaller section cone coil respectively of 1.26 V and 1.01 V and 1.5 V and 1.88 V. The voltage represents changed flux corresponded to change rotation velocity in each shaped of coils. Higher voltage indicated higher efficiency.

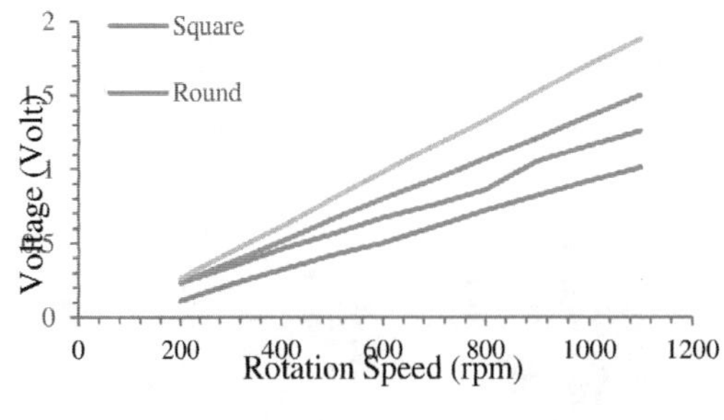

Figure 4. Voltage as measurement results corresponding to rotation velocity

The retardation data obtained from the experiment shows that with the same voltage value the small face cone coil has a greater retardation value. (Figure 5). This is due to the magnetic flux produced by magnets can generate current eddy. Meanwhile, the coil placed near the magnet will produce an electromotive force (emf) according to Faraday's law. In addition to influencing the magnitude of the emf, the shape of the coil also affects power loss because some magnetic fields are radially dispersed or not through the cross section of the coil. That causes only a portion of the magnetic field to be converted. In the smaller section of cone-shaped, magnetic flux entering the coil is more focus. The statistical and dynamic nature of magnetic topology systems can provide a tight relationship with geometric structures [10] and field strength depends on the length, dimensions, and placement of the source coil [11]

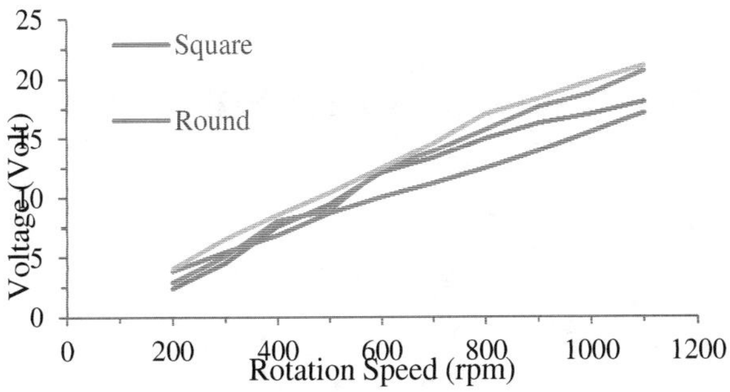

Figure 5. Retardation time corresponding to rotation velocity using short circuit

5. Conclusion

Magnetic flux generating electric motion force can be dependent by shape of coils. Cone-shaped coil can obtain higher electric motion force. However, if the smaller section placed near magnet, electric motion force would be higher due to the magnetic flux passed coils is more focus. When short circuit had connected to coils, the magnetic flux can generate current eddy that can also affected retardation of rotation. Therefore, the slowest rotation had been also obtained using cone-shaped coil with smaller-section placed near magnet.

Acknowledgements

The authors are thankful to scholarship of BUDI-DN (Beasiswa Unggulan Dosen Indonesia Dalam Negeri) by Kemenkeu Indonesia.

Reference

[1] Aoukili A and Khamlichi A 2018 Procedia Manufacturing **22** 209–214.
[2] Azaman K N Sophian A.and Nafiah, F 2017 IOP Conf. Ser.: Mater. Sci. Eng. **260** 012001.
[3] Krishna GL Anantha, and KM Sathish Kumar 2018 *Mater.Today: Proceedings* **5** 2575.
[4] Wu T Bowler J R. and Theodoulidis T P 2017 IEEE Transactions on Magnetics **53**,1-9.
[5] Tytko G and Dziczkowski, L 2015 IEEE Transactions on Magnetics **51** 1-4..
[6] Wakiwaka H., Ito D, Tashiro K, Yajima, H., Manta Y, Kanazawa T and Fujiwara N. 2014. In Materials Science Forum **792**. 215-220.
[7] Fava J O and Ruch M C 2006 NDT & e International, **39** 414-424.
[8] Song-Ling H K S H and Shen Z W W 2011. Acta Physica Sinica **7** 118
[9] Burke S K, Ditchburn R J and Theodoulidis T P 2014. NDT & E International 64, 1-6.
[10] Teixeira A W, Castillo-Sepúlveda S, Vojkovic S, Fonseca J M, Altbir D, Núñez Á S and Carvalho-Santos V L 2019 Journal of Magnetism and Magnetic Materials **478**, 253-259.
[11] Nataraj C, Khan S and Habaebi M H 2018 Sādhanā **43** 67.

IOP Conf. Series: Materials Science and Engineering **858** (2020) 012056 doi:10.1088/1757-899X/858/1/012056

Application To Measuring Food Business Opportunities For The Development Of Smart Cities

Hartatik and Putri Ayu W
Informatics Engineering Departement, Faculty of Mathematics and Natural Sciences, Sebelas Maret University, Surakarta, Indonesia

E-mail: hartatik.uns@gmail.com

Abstract. In the development of technology, more people are using products from technology development to facilitate their daily activities especially in food industry. However, to measuring and determine opportunities when opening a food business covering food material and type of culinary. People usually choose what food ingredients are of public interest and estimates materials and cost to determine a business opportunity. In this situation a Decision Support System (DSS) can be developed, namely apus-kul is a decision makers by using certain data and models to solve some unstructured problems using Simple Additive Weighting (SAW) algorithm. This system is expected to be a means to assist the community when making decisions to choose materials and food businesses for the development of smart city governance systems.

1. Introduction

Citizen have a variety of jobs, one of which is a culinary business. To determine the opportunity when opening a culinary business, people usually choose with their respective estimates without using a system that is able to help the community to determine a business opportunity. This is also happened in Purwantoro District, Wonogiri Regency, Central Java Province. There is no Decision Support System yet that is able to help communities in Purwantoro to choose and determine culinary business opportunities and culinary business locations. By utilizing the advanced of technology can be one way to create a system that can help the community to determine culinary businesses that are in accordance with conditions and criteria. With this system, it is hoped that the community will be easier to choose the right culinary business in micro businesses. This research is an idea for developing smart cities, namely Smart Living. Smart City as an effective integration between physical infrastructure, digital systems and HR skills to build an environment that provides hope for a sustainable, prosperous and inclusive future [1][2].

In previous research, about Smart Governance Planning said that smart government is the spearhead of Smart planning City [3][4]. This provides inspiration for how to build smart government related to the management of culinary businesses in the region. So, in this research are developed the application based a web. This is a Decision Support System using the CodeIgniter (CI) framework. The system can be managed by an administrator. The system will be created using the Simple Additive Weighting (SAW) method in its calculations. The existence of DSS in companies or organizations is not to replace the tasks taken, but is a facility that helps them in making a good decisions about materials of food business and location [5]. The criteria used for business ranking are business funds, profits, number of competitors, and number of buyers. The criteria used for business place ranking recommendations are the distance between the village and the market, the distance between the village and the main road, and the population of the village.

2. Experimental

The research methodology that be used in this study is the Waterfall model. Waterfall method can be seen in Figure 1. Below the phase of waterfall method:

a. Analysis
b. Design
c. Coding
d. Testing
e. Support and Maintenance

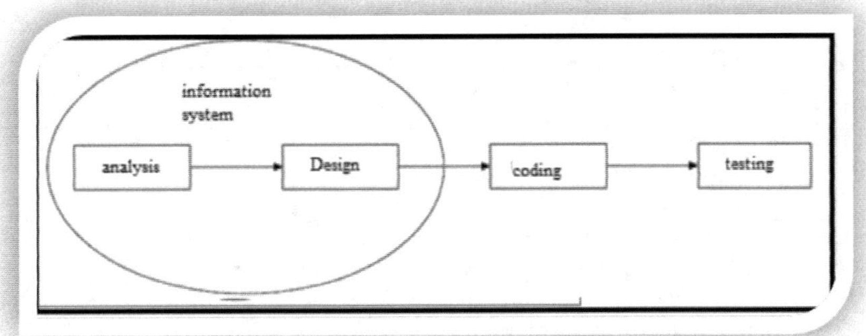

Figure 1. Business Proses In Apus-kul

3. Result and Discussion

Functional requirements are used to determine the functions performed by actors who have access to the system. Functional Needs in the Application of Determining Culinary Business Opportunities can be seen in Table 1.

Table 1. Functional requirements

Code	Description	Actor	*Dependency*
FR01	Login	Admin	
FR02	Admin can change alternative culinary business data	Admin	FR03
FR03	The system can display a list of alternative culinary business data	System	
FR04	Admin can change the alternative data for culinary business locations	Admin	FR05
FR05	The system can display a list of alternative culinary business location data	System	
FR06	The system can display a list of culinary business criteria data	System	
FR07	Admin can change the criteria of culinary business criteria	Admin	FR06
FR08	The system can display a list of culinary business location criteria data	System	
FR09	Admin can change the criteria for	Admin	FR08

	culinary business location		
FR10	The system can display a list of business fund attribute data	System	
FR11	Admin can change the business fund attribute data	Admin	FR10
FR12	The system can display a list of business profit attribute data	System	
FR13	Admin can change the business profit attribute data	Admin	FR12
FR14	The system can display a list of attribute data for the number of buyer	System	
FR15	Admin can change the attribute data of the number of buyer	Admin	FR14
FR16	The system can display a list of attribute data for the number of business competitors	System	
FR17	Admin can change the attribute data of the number of business competitors	Admin	FR16
FR18	The system can display a list of data attribute business distance to the market	System	
FR19	Admin can change the data attribute of the distance of the business location to the market	Admin	FR18
FR20	The system can display a list of attribute data of the distance of the business location to the highway	System	
FR21	Admin can change the data attribute distance of business location to the highway	Admin	FR20
FR22	The system can display a list of attribute data for the number of residents in the business location	System	
FR23	Admin can change the attribute data on the number of residents in the business location	Admin	FR22
FR24	The system can process calculations using the SAW method	System	FR26
FR25	Admin can see the results of the calculation of the culinary business data SAW method	Admin	FR24
FR26	Admin can see the results of the calculation of the culinary business location data method	Admin	FR24
FR27	User accesses the user dashboard page	User	
FR28	User can accesses the list of alternative business sequences according to the results of calculations and filters used	User	FR27
FR29	User can accesses the list of alternative business locations according to the results of calculations and filters used	User	FR27

IOP Conf. Series: Materials Science and Engineering **858** (2020) 012056 doi:10.1088/1757-899X/858/1/012056

The Code field content is unique for each system's functional requirements. The Description field contains a description of the functional requirements of the system. In the Actor field contains actors in the form of people or systems that perform or relate to a function. Then the Dependency field contains the Code from other functions, where a function has a dependency on the function whose unique code is listed in the Dependency field.

Non-functional requirements are used to analyze application interactions outside of functional requirements. Non-functional needs of the Culinary Business Opportunity Determination Application can be seen in Table 2.

Table 2. Non-functional requirements

Code	Descriptiom
NFR01	The system can be run by web browser software
NFR02	The system can provide information when input data does not comply with the rules

The non-FR Code column contains unique codes for each non-functional system requirement. The FR Description column contains a description of the non-functional requirements of the system.

In this study there are 2 actors involved in the system, namely admin and user. Admin and user have different access rights and features in the system. One of the differences between admin and user access is that the admin must log in if he wants to access the admin page and the user does not need to log in to access the login page. Admin and user can see the results of the order of culinary business recommendations and culinary business location based on calculations using the Simple Additive Weighting method.

Entity Relationship Diagram (ERD) or E-R Model or is a model used to describe data in the form of entities, attributes, and relationships between entities[6] . ERD in this study can be seen in Figure 2.

IOP Conf. Series: Materials Science and Engineering **858** (2020) 012056 doi:10.1088/1757-899X/858/1/012056

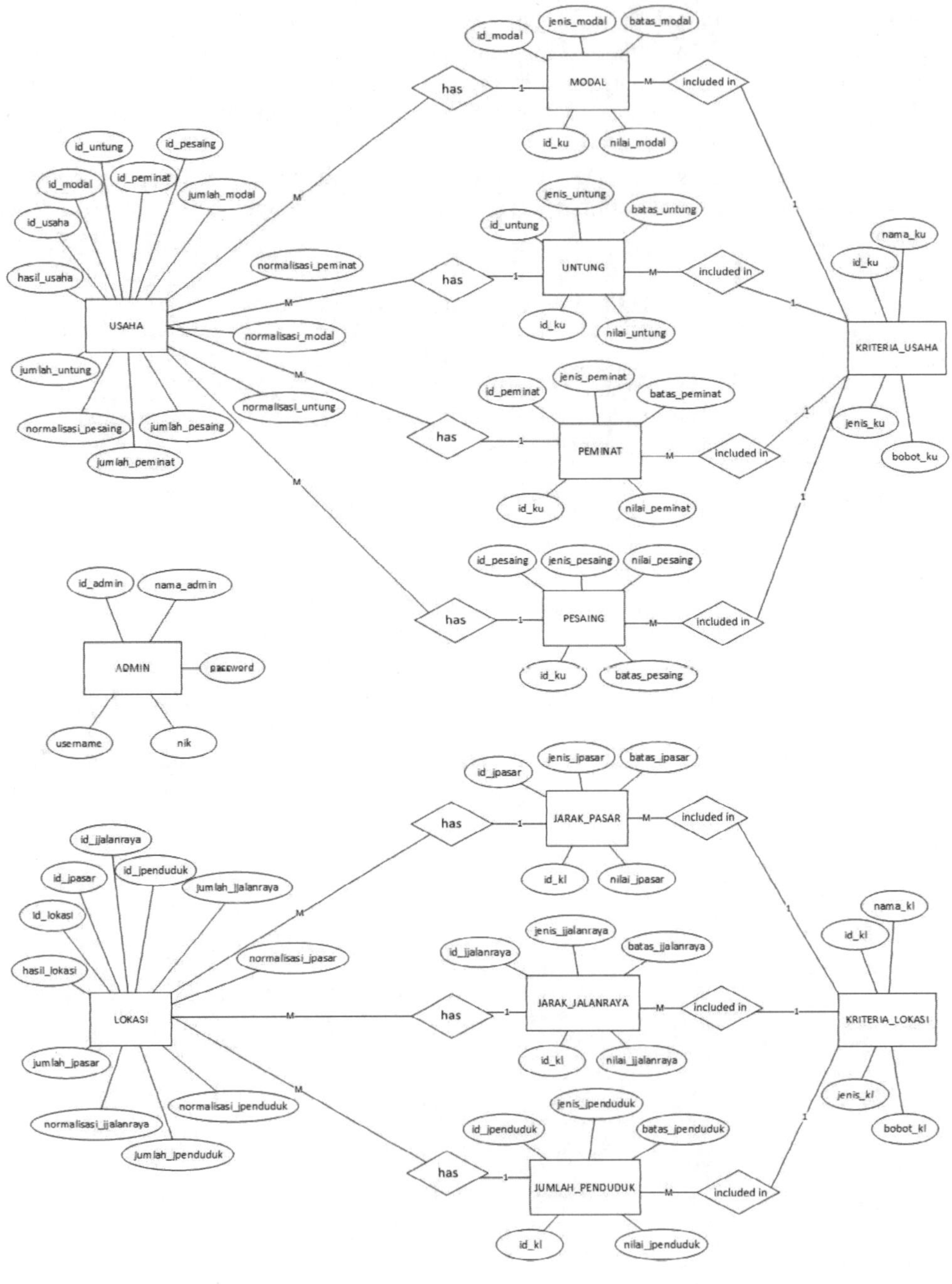

Figure 2. Entity Relationship Diagram

Implementation of the design for the admin page can be seen in Figure 3 and Figure 4. And the result of Simple Additive Weighting (SAW) is a weighted sum method. The Saw method is used in supply chain vendor evaluation [7] and also for the selection of personnel within an organization [8]. The basic concept of Simple Additive Weighting (SAW) is to find the weighted sum of the performance ratings for each alternative on a criterion. The Simple Additive Weighting (SAW) method requires the process of normalizing the decision matrix (X) to a scale that can be compared with all existing alternative branches. The calculation step for the SAW method is[9] :

1. Determine the alternative (Ai)

2. Determine the criteria (Ci) that will be used as a reference in decision making
3. Determine the weight of preference or level of importance (W) for each criterion. W = [W1 W2 W3 ... W4]
4. Make a match rating table for each alternative on each criterion
5. Making the decision X matrix formed from the match rating table of each alternative (Ai) for each predetermined criterion (Cj) where, I = 1,2, ..., m and j = 1,2, ..., n.
6. Normalize the decision X matrix by calculating the normalized performance rating (rij) value of alternative (Ai) on performance (Cj)
7. The results of the normalized performance rating (Rij) form a normalized matrix (R).

Implementation of the design for the admin page can be seen in Figure 5. Application development uses codeegniter(CI). CI source code that is equipped with comments in it so that it clarifies the function of a program code and the resulting CI is very clean and search engine friendly (SEF)[10][11].

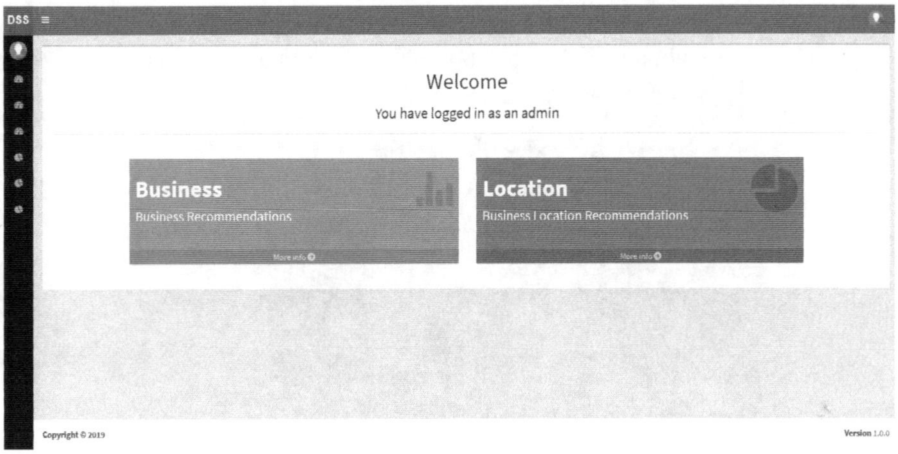

Figure 3. Dashboard admin page

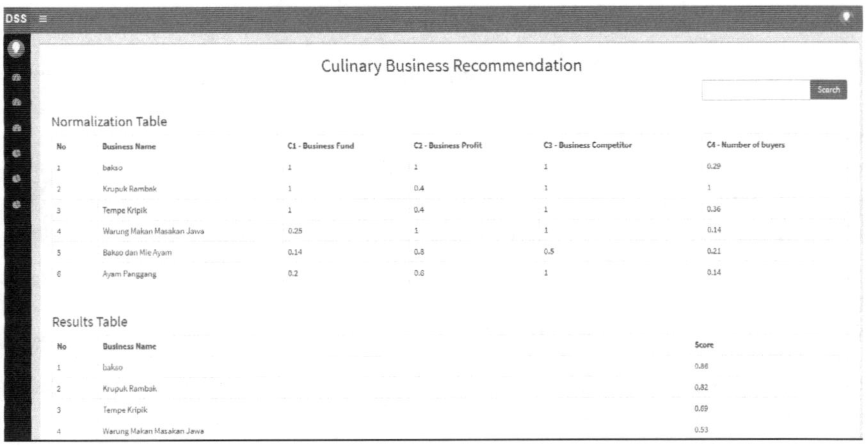

Figure 4. Alternative data menu page

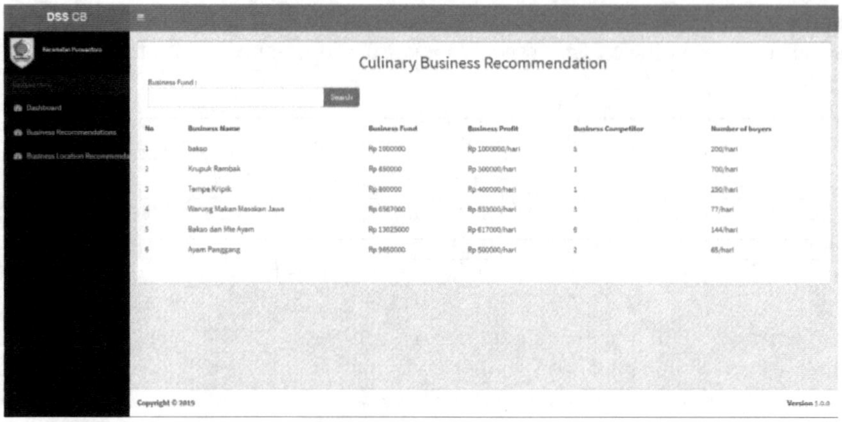

Figure 5. Culinary business recommendation page on the user page

The test results of the Culinary Business Opportunity Determination Application based on the opinion of respondents during the system public test can be seen in Figure 6 .

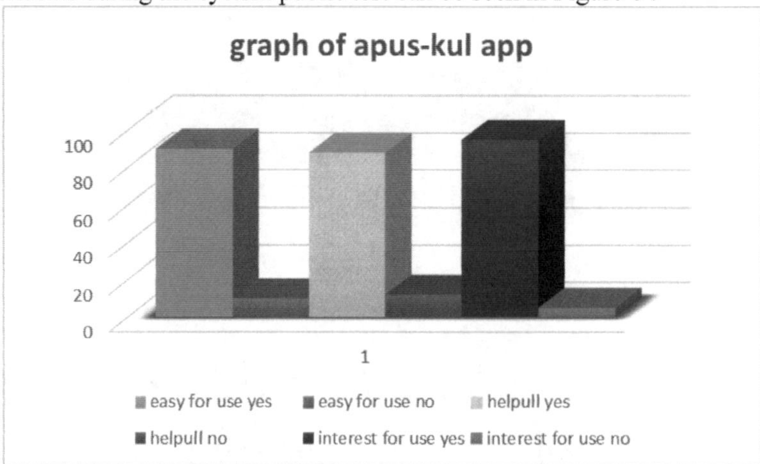

Figure 6. result public test about apus-kul

4. Conclusion

The research that was carried out produced conclusions that had been made developing a system for a development smart cities with the title Application of Culinary Business Opportunities Determination apus-kul) to measuring and determining food material and type of business .

References

[1] Y Arafah1 2 3, H Winarso1 2017 IOP Conf. Series: Earth and Environmental Science 70 (2017) 012065

[2] Muhammad Sani Roychansyah1 and Sushardjanti Felasari2 2018 IOP Conf. Series: Earth and Environmental Science 126 (2018)

[3] R. P. Dameri and C. Rosenthal-Sabroux, Eds 2014 "Smart City and Value Creation," in Smart City (Cham: Springer International Publishing) pp. 1–12.

[4] M. Angelidou 2014 Smart city policies: A spatial approach , Cities, vol. 41, pp. S3–S11.

[5] P Pangaribuan and A Beniyanto 2018 IOP Conf. Series: Materials Science and Engineering 407 (2018) 012045

[6] A. Kadir 2009 Dasar Perancangan & Implementasi Database Relasional (Yogyakarta: ANDI).

IOP Conf. Series: Materials Science and Engineering **858** (2020) 012056 doi:10.1088/1757-899X/858/1/012056

[7] Gupta, A., Gupta A. 2012 Interscience Management Review (IMR) 2(3) pp 10-16
[8] Afshari A., Mojahed M., Yusuff, R. M. 2010 International Journal of Innovation, Management and Technology 1(5) pp 511-515
[9] Harsiti and H. Aprianti 2017 Journal of Information Systems vol. 4.
[10] Al-Fedaghi S. 2011 International Journal of Software Engineering and Its Applications 5 57-68
[11] M. Destiningrum and Q. J. Adrian, 2017 TEKNOINFO vol. 11.

AUTHOR INDEX